Toxic
Waste
Sites
An Encyclopedia of
Endangered America

Toxic Waste Sites

An Encyclopedia of Endangered America

Mark Crawford

ABC-CLIO

Santa Barbara, California
Denver, Colorado
Oxford, England

Maps by William Nelson

Library of Congress Cataloging-in-Publication Data
Toxic waste sites : an encyclopedia of endangered America / Mark Crawford.
p. c.m.
Includes bibliographical references and index.
1. Hazardous waste sites—United States—States. I. Title.
TD1040.C73 1997
363.738'4'02573—dc21
97-17303
CIP

ISBN 0-87436-934-7

02 01 00 99 98 10 9 8 7 6 5 4 3 2

ABC-CLIO, Inc.
130 Cremona Drive, P.O. Box 1911
Santa Barbara, California 93116-1911

This book is printed on acid-free paper.
Manufactured in the United States of America

CONTENTS

INTRODUCTION

Superfund sites are the worst areas of pollution in the United States. Collectively they form the National Priorities List, or NPL. This book summarizes each of the more than 1,300 sites, briefly describing each site's location, history, the type of contamination that affects the soil and groundwater, the number of people who are threatened, the clean-up work that has been conducted or that is planned, the parties responsible for the contamination, the legal actions that have been taken, and sources for more detailed information. It describes sites that have been monitored for years, including a few sites that have been deleted from the NPL but are still under long-term monitoring. It also introduces sites that have most recently come to the attention of the EPA, including 13 sites that were proposed to the NPL in the summer of 1996. All current sites have been included in Appendix A, which categorizes sites by state and type.

The Environmental Protection Agency (EPA) has divided the United States into ten regions, each with a regional office that is responsible for a different group of states (see Appendix D). Each regional office publishes NPL summaries, progress reports, and data sheets that vary in style and content. The primary sources for this volume are the annual reports for each state, entitled *Progress at NPL Sites* (PNPLS); some entries also include information from the local project managers. The PNPLS citation is generally followed by the address of the local EPA repository, usually the nearest public library or town hall, which has a complete set of records for the site. Some sites do not have local repositories, and therefore an address does not follow the PNPLS citation.

The level of detail in the PNPLS series varies from state to state. The annual PNPLS updates may be up to a year behind schedule for some states; therefore the regional EPA office or project manager should be contacted about the most recent developments for particular sites.

The purpose of the book is to provide an overview of the Superfund program and basic information on the NPL sites. It describes the kinds of activities that created these contaminated areas, the complicated, lengthy, and costly evaluations and clean-up techniques, and the extent to which these sites threaten the health of large population groups and the natural and physical environment.

To learn more about a particular Superfund site, consult the listed source or the appropriate regional office listed in Appendix D. Updates and site summaries for all the NPL sites are also available from the EPA on the Internet at www.epa.gov/superfund/oerr/impm/products/nplsites/usmap.htm.

The History of Superfund

The post–World War II years witnessed unprecedented economic confidence and business growth in the United States. With this new development came unforeseen impacts on the environment; huge volumes of solid, liquid, and hazardous wastes were generated by a growing population and evolving industry. These chemical wastes were simply buried in unlined landfills, gravel pits, quarries, strip mines, or low-lying areas and wetlands, often under only a few feet of earth.

Many gasification plants that had operated since the 1880s went out of business in the 1940s and 1950s, leaving behind large expanses of toxic coal tars. Smelters damaged the surrounding land with sulfur- and metal-bearing smoke that contaminated yards and homes and killed thousands of acres of forest. Acid rain polluted lakes and wetlands. Millions of gallons of industrial wastewater were pumped daily into waterways, dry wells, sewer systems, and holding ponds. Radioactive compounds, explosives, chemical warfare agents, and other dangerous wastes were buried by the U.S. military on sprawling bases across the United States.

This unregulated disposal of waste was a procedure accepted by government and industry because the negative, long-term environmental and health impacts had not been seriously considered. Poisonous chemicals had been moving into waterways, drinking water supplies, wetlands, and wildlife habitats for decades before the federal government passed the National Environmental Policy Act (NEPA) in 1969. The Environmental Protection Agency (EPA) was created the following year. The National Environmental Policy Act (PL 91-190) was the first attempt by the federal government to establish a national environmental policy with long-term goals. It also required the completion of environmental impact statements before any major federally funded construction projects could begin.

As our environmental awareness increased and the negative impacts of unregulated disposal became more apparent, more new laws were adopted to protect the country's water supplies. The Water Pollution Control Act of 1972 (PL 92-500, commonly called the Clean Water Act) and the Clean Water Act Amendments of 1977 (PL 95-217) were approved to "restore and maintain the chemical, physical and biological integrity of the nation's water." Goals included making waters "fishable and swimmable" by 1983 and regulating all industrial discharges by 1985. The Safe Drinking Water Act of 1974 (PL 93-523) and amendments to it were passed to protect important drinking water aquifers from contaminants being injected below the water table. Under the Resource Conservation and Recovery Act (RCRA) of 1976 (PL 94-580) the federal government began to monitor and regulate the processing, storage, and disposal of solid waste by waste generators, transporters, and managers.

The effects of uncontrolled dumping became more and more visible by the late 1970s. Shocking stories of neighborhoods and communities poisoned by polluted water, air, and soil were suddenly in the news. For example, hundreds of people were evacuated from their homes in the Love Canal area near Niagara Falls in New York, where children received chemical burns from contaminated soil and health problems soared, including cancer and birth defects. Times Beach, Missouri, became a ghost town encircled by fencing, barbed wire, and 24-hour security patrols when widespread dioxin contamination was discovered. Homes and businesses across the United States had been built on or near old landfills and industrial sites—who would be next? Would the drinking water cause cancer? Was the air safe to breathe? Was the food here safe to eat?

To deal with this growing problem, the U.S. Congress passed the Comprehensive Environmental Response, Compensation, and Liability Act (CERCLA) (PL 96-510) in 1980. Also known as "Superfund," it launched the evaluation of over 35,000 hazardous waste sites in the United States. The program is administered by the EPA. The first Superfund budget was $1.6 billion; by 1986, as more and more sites were identified, it had jumped to $8.5 billion.

CERCLA allows the federal government to take immediate action when the environment is threatened by an actual, or even a potential, release of hazardous materials. It also requires industries and handlers of hazardous waste to identify dump and spill areas. Failure to comply with CERCLA is punishable by severe financial penalties.

The Present

Contaminated sites are brought to the EPA's attention by private citizens, state agencies, community and environmental groups, and corporations. The EPA uses existing databases to identify potential sites; for example, it examines aerial photographs to locate waste pits and lagoons. About half of the NPL sites have been identified by state testing programs. The EPA uses a Hazardous Ranking System (HRS) to "score" a site according to the threat that the potential release of hazardous chemicals would pose to human health and the local environment. The most dangerous are the Superfund sites, all of which contain excessive levels of various industrial chemicals in the groundwater, soils, nearby streams, and occasionally air. Chemicals from most of these sites have already been detected in neighboring properties, private and public wells, soils, stream water and sediments, wildlife, and aquatic life.

Once identified, a site is "screened" by the EPA. The first step is a preliminary assessment that includes checking background and records, data searches, interviews, and the identification of potentially responsible parties (PRPs). An on-site inspection that includes geochemical sampling and the documention of impacts from the contamination follows. A high level of contamination will require more detailed assessments, which for the most dangerous properties usually lead to an NPL listing.

The EPA has identified over 1,400 Superfund sites since the program began in 1980. The pollution is most often the result of unregulated industrial waste discharge or waste disposal that was compliant with federal laws 20 to 30 years ago but is now recognized to have been dangerously underregulated. Of course, there are always some illegal or deliberately negligent operations, but for the most part, the companies were operating within the laws that existed at the time. About 50 to 100 new sites are added each year, and it is expected that there will be over 2,100 Superfund sites by the year 2,000. To date, 108 sites have been removed from the NPL. Presently there are 1,310 active NPL or NPL-nominated sites.

Superfund's mission is ambitious: to clean up the worst pollution sites in the United States permanently. Any immediate threats to human health and the environment are addressed first through emergency response actions, such as removing exposed contaminated soils and waste, building surface runoff controls, capping the area of contamination, supplying clean water to affected residents, removing leaking tanks or drums, erecting fencing, and other security measures. This work often takes 1 to 4 years to complete.

After the site is secured and the immediate threats removed, the longer-term studies on the site begin. After a site is placed on the NPL, it takes an average of 8 years before the clean-up work can actually begin. This time period includes the emergency response phase, 4 to 6 years of investigation, and 1 to 2 years of engineering and design. The more detailed site investigations include the installation of groundwater monitoring wells, comprehensive sampling, and groundwater flow modeling. Groundwater moves slowly, from inches to 100 feet per year, and it takes years of monitoring and sampling to construct reliable flow models. Once the site dynamics and contaminants are identified, the EPA selects a remediation plan best suited for the site from an array of modern clean-up techniques.

After the plans are engineered and approved, environmental contractors undertake the necessary construction. The total average cost to clean up a Superfund site is about $35 million. Some are much more expensive: The Love Canal site cost about $250 million. Clean-up costs for the Rocky Mountain Arsenal near Denver are projected to exceed $1 billion. Depending on the extent of the contamination, a groundwater extraction and treatment system can operate for 25 years or longer.

Construction of the final clean-up remedy is under way at about 500 of the current Superfund sites. Once completed, the sites will be monitored for years afterward to assure clean-up standards have been met. If a site is primarily contaminated by petroleum, the responsibility for it can be transferred to the federal Leaking Underground Storage Tanks (LUST) Trust Fund. The fund was built by a 0.1 cent tax on each gallon of motor fuel sold in the United States between 1986 and 1995; it presently contains about $1.64 billion. The money is used by the EPA to undertake or oversee corrective actions at leaking tank sites.

Who Pays?

The EPA believes that "the polluter should pay." CERCLA and the Superfund Amendments and Reauthorization Act of 1986 (SARA) (PL 99-499) assign liability to owners, operators, and past and present site owners regardless of whether the owner had knowledge of past operations. A generator of hazardous waste can be held responsible even if the company was not aware of how the transporter had disposed of its waste. According to CERCLA's cost recovery provisions, any local, state, or federal government agency or private citizen who spends money cleaning up hazardous substances can sue the party believed to be responsible for the pollution.

The Superfund program also allows the federal government to order the potentially responsible parties (PRPs) to take action if there is an "imminent hazard" as determined by the EPA. Companies who refuse to comply can be fined up to $25,000 per day. If the federal government is forced to conduct the clean-up actions because of a company's noncompliance, the government can sue for treble damages—three times the amount of money that it required to clean up the site.

This enactment of "retroactive liability" is especially chilling to corporations, which believe it is a violation of the Constitution's prohibition of laws that make conduct unlawful "after the fact." Many PRPs were obeying the laws that existed at the time they disposed of the hazardous waste. The costly fines compel companies to fight expensive court battles with the federal government. Proponents of retroactive liability maintain that the policy will encourage polluters to stay informed about current laws and maintain compliance.

A few PRPs—generally those who purchased contaminated property in the 1940s and 1950s—have won lawsuits against the EPA by using the "innocent landowner liability" defense. The argument here is that the owner followed the standard real estate practices at the time the property was purchased, which did not include environmental testing. Because of the well-known liability burdens of aquiring contaminated property today, many real estate transactions are contingent upon an environmental audit that checks the prospective property for signs of contamination. This hopefully prevents buyers and lenders from acquiring land that could be an expensive liability in the future. However, if the audit is in error, and contamination is found at a later date after purchase, the owner is still liable, not the environmental consultant.

The EPA identifies PRPs by reviewing records, collecting data, and interviewing former employees, suppliers, and neighbors. An evidence checklist is used to define the elements of the PRP's liability. The PRP then receives a "notice of potential liability," which begins what is generally a long process of negotiations and legal actions. The majority of agreements are negotiated voluntarily between the EPA and the PRPs. Consent Orders outline a course of action that the PRP has agreed to carry out, with stipulated penalties if the goals are not met. Unilateral Administrative Orders do not require the consent of the PRP, and if the PRP fails to comply with the order, it is subject to paying treble damages. After the site investigations have been completed, a judge issues Consent Decrees that define the extent of the remedial actions, usually to be carried out by the PRPs.

Bankruptcies occasionally result from the litigation process. The EPA has focused on "viable" PRPs who are profitable enough to be able to afford clean-up costs. The bigger companies regard this as unfair and have often reacted by suing the smaller PRPs, who in turn sue more distant and questionable PRPs such as suppliers. Ultimately profitable companies still in existence often pay most of the cost for remediating a site that many dozens of companies may have legally used 20 years ago.

One reason it takes so long to implement a clean-up plan is that companies find it more cost-effective to force lawsuits with the EPA and Department of Justice, and this delays the entire procedure for years. Smaller polluters frequently negotiate a minimal settlement. The Hazardous Substance Response Trust Fund managed by CERCLA uses federal funds to pay for the clean-up costs that are not recovered from the responsible parties.

Sites on land owned by the U.S. government are evaluated and remediated by other federal agencies, usually the U.S. Army Corps of Engineers, and occasionally under EPA oversight.

What Kinds of Property Are Superfund Sites?

Superfund sites occur in all states and most U.S. territories (See Appendix A). Nevada has the fewest sites (1), closely followed by North Dakota (2), Guam (2), and the Virgin Islands (2). As might be expected, those states with the greatest number of Superfund sites are some of the most heavily populated and industrialized: New Jersey (109), Pennsylvania (104), and California (95).

About 28 percent of the sites (370) are landfills or waste dumps. Most of these operated in the 1950s and 1960s, were unlined, and received all kinds of municipal and industrial wastes, including waste that is now recognized as "hazardous."

About 57 percent of the sites (747) are related to various industries and businesses. These include wood treatment facilities, chemical plants, dry cleaning operations, coal gasification plants, petroleum-related businesses, recycling centers, and metal-processing, electroplating, and smelting operations.

About 11 percent (147) of the sites are related to government activities—usually the military or the Department of Energy, including nuclear weapons and energy research and development. These bases and research complexes contain hundreds of waste disposal sites.

About 4 percent (46) of the sites are related to mining activities that often date back to the mid-1800s. Piles of waste rock and tailings were left beside the mine openings, and acidic runoff entered nearby drainages, polluting the water. Smelter emissions carried heavy metals that settled out, contaminating large expanses of the countryside.

How Dangerous Is Dangerous?

Hundreds of chemical compounds at Superfund sites are a threat to human health and the environment. (Common contaminants are categorized in Appendix B.) Clearly no one would want drinking water or air to contain any of these chemicals. How poisonous a compound is depends on its chemical composition, its concentration in the soil, water, or air, "exposure time," and how sensitive one is to its effects. Levels of acute toxicity (sudden death) and chronic toxicity (illness) are known for most compounds. Bioaccumulation rates, or how quickly a chemical builds up in the tissues and organs of the human body, are also a factor.

Exposure to some compounds will bring death in a few seconds—fortunately these chemicals are rare. Even though these are deadly, then, the overall risk to human health is very low because the chemicals are not "loose" in the environment.

The greatest risks are from those compounds that occur in even the lowest levels but are so common that we are exposed to them on an almost daily basis, such as petroleum products. Petroleum and gasoline and their additives (most notably benzene) and derivatives are highly carcinogenic. We are even more at risk from these petrochemicals because they are volatile organic compounds (VOCs), which means they vaporize easily and can be breathed in with air.

Some of the more common compounds at Superfund sites, and how they affect our health, are listed below:

Pesticides include compounds such as endrin, aldrin, dieldrin, chlordane, DDT, DDE, and DDD. Exposure usually results from eating or drinking contaminated food, milk, or water. Brief, high-level exposure can lead to convulsions and death; longer-term low-level exposure will damage the liver and the nervous, digestive, and immune systems.

Benzene is a component of plastics, detergents, pesticides, tobacco smoke, and gasoline. Because it is so volatile, it is easy to inhale. Brief, high-level exposures can result in death; longer-term low-level exposures have been linked to leukemia, anemia, internal bleeding, and reproductive problems.

Heavy metals such as arsenic, chromium, lead, zinc, copper, and mercury can lead to a variety of conditions. Arsenic poisoning, for example, can cause skin damage, cancer, and death. Other health problems include skin irritations, liver and kidney problems, birth defects, mental and physical impairments in young children, heart disease, memory loss, and depressed immune systems.

Cyanide is commonly found in wastewater, other industrial discharges, and vehicle exhaust. Short, high-level exposure can impair the central nervous system, respiratory system, and the heart; longer-term low-level exposures can result in deafness, vision problems, thyroid impairment, and loss of muscular coordination.

Phthalates are derived from plastics and can enter food from plastic wrapping. They cause cancer in laboratory animals and are thought to be linked to liver damage, fertility problems, and birth defects in humans.

Polycyclic Aromatic Hydrocarbons, or PAHs, result from the incomplete combustion of coal, oil, gas, and other organic materials. They are common in tobacco, wood, and barbecue smoke. PAHs directly affect the spleen, kidney, and liver and are suspected carcinogens. Laboratory tests have shown that PAHs cause tumors, reproductive problems, and immune system dysfunction in mice.

Pentachlorophenol, or PCP, a pesticide and wood preservative, is common at wood treatment sites. High-level exposure can damage all the internal organs and result in death. Testing has shown that it affects the kidneys and the immune and nervous systems.

Phenol is a common ingredient in consumable medicines. Consumption of large amounts will result in paralysis or death. Phenols are a suspected link to birth defects, cancers, and tumors.

Polychlorinated Biphenyls, or PCBs, though banned since 1977, are still a problem and cause skin rashes, liver cancer, and reproductive problems.

Tetrachloroethene, or TCE, one of the most common VOCs, is a popular degreaser and dry cleaning solvent. It is thought to be related to nervous disorders, liver and kidney impairments, birth defects, and leukemia.

Future Directions

When Superfund clean up was initiated, the EPA's goal was "zero detection"—to carry out the steps necessary to eliminate the contaminants from the environment completely. Costs quickly became staggering. Many violators realized it was more cost-effective to fight the EPA in court and stay in business than to pay the huge cost of the remediation work. Great sums have been spent in legal fees.

Increasingly, criticism of the Superfund program questioned the high cost of clean ups and the value of reaching the clean-up levels the EPA aimed to achieve. Is zero detection possible? Is it really necessary? Does the EPA need to be involved in all the steps? In response, the EPA adjusted its goals in the early 1990s to reflect risk-based assessments and responses, rather than zero-detection objectives. Those sites that pose serious health threats will still receive maximum remediation efforts. But other sites that are less dangerous will be evaluated from a more cost-effective perspective. A good example is a contaminated groundwater plume that is not used for domestic purposes and is located in an isolated area. Instead of spending $30 million to build a sophisticated treatment system, a cost-effective alternative would be simply to impose deed restrictions that would prohibit use of the groundwater in the area. The contamination will dissipate to acceptable levels over time through natural attenuation. This takes longer, but the time factor is not as important since the risk factor is so low.

After ten years of evaluating, testing, and remediating Superfund sites, the EPA has also identified the remediation techniques that work most effectively under given conditions and has learned what clean-up limits are actually attainable in the field. With the learning curve now so abbreviated, the EPA is anxious to streamline the remediation process, keep costs down, and clean up sites more quickly.

This new line of thinking led to the Superfund Administrative Reform Overview in 1995, which calls for "smarter clean up choices that protect public health at less cost." The agency is striving to establish cost-effective thresholds for remediation that are attainable (including choosing the most cost-effective technologies) and is allowing PRPs more latitude in performing risk assessments. The EPA also wants to reform NPL listing and deletion policies, reduce litigation with PRPs by better negotiations with principal PRPs, improve settling strategies with smaller PRPs, and allow greater input by state and community-based groups.

In order to appease the larger PRPs, the EPA is focusing on distributing Unilateral Administrative Orders to all PRPs, regardless of their viability. Smaller PRPs who are cash-poor may be able to contribute to the clean-up cost by supplying personnel, material, equipment, or other services. More time will be spent on "litigation risk analysis"—estimating the cost of a lawsuit against a PRP relative to the amount of dollars a PRP could contribute. Other factors, such as the type and size of business, sometimes result in a decision not to pursue a PRP at all.

Also as a result of the Superfund overhaul in 1995, individual states have much greater control over the listing of new Superfund sites. The EPA cannot add any new sites to the NPL without the approval of the governor of the state, and this has slowed down the listing process. Some states, such as New York, New Jersey, and Texas are eager to list NPL sites. Others, such as the New England states, are resisting listing new sites. They don't like the stigma of having another Superfund site and would prefer to clean it up using their own resources.

Beware!

There are few areas in the United States, if any, that are free from the threat of contamination. The EPA has logged over 35,000 sites to date. Although most are not Superfund sites, they can still be harmful to those who live or work nearby. And these are the large ones. There seem to be a countless number of smaller sites, and tens of thousands more will be discovered in the future.

This book describes the 1,310 worst sites in America. If you live close to a Superfund, industrial, or landfill site, you are likely familiar with some of the negative impacts it creates. There are no guarantees to the safety of your water, soil, and air, *wherever* you are. Do you know the history of your community? Do you know where the old buried landfill is? Do you know if the groundwater flows from that landfill toward your home? Have you had your water tested? Do you need a water filter? Do you know where former industrial sites are located?

Our environmental situation has vastly improved since Superfund was enacted. Industry is better regulated and more conscientious, recycling is more prevalent, advances in technology make clean ups more effective, and previous damage is being reversed. It is still possible, however, to be poisoned by contaminants from a point source as small as a leaking underground storage tank. Although a site may look inviting and unspoiled, the underlying soil and groundwater may contain unwanted contaminants from a long-closed operation on the site. Research the historical land use in your area before buying property and consider an environmental assessment report by a certified environmental consultant. By becoming familiar with the industrial and municipal history of your community, you can better protect yourself against the contamination that could be in your own backyard.

ALABAMA

Alabama Ammunition Plant
Talladega County
This U.S. Army facility near Childersburg operated from 1941 to 1991. Activities included manufacturing explosives from 1941 to 1945. Related wastes, including recycled acids, were buried or discharged on the 5,170-acre property. Testing has shown that the groundwater and soil contain nitrates, trinitrotoluene (TNT), asbestos, lead, and other heavy metals. The contamination threatens Talladega Creek and the local drinking water supplies for about 3,000 residents.

Response actions included a series of site investigations by the U.S. Army. Recommendations included soil excavation and incineration and soil stabilization and reburial (completed 1993–1995). Groundwater studies are nearing completion.

Sources: PNPLS-AL; Rainwater Memorial Library, 112 Ninth Avenue Southwest, Childersburg, AL, 35044.

Anniston Army Depot (Southeast Industrial Area)
Calhoun County
U.S. Army vehicles and military equipment are stored, maintained, repaired, modified, or decommissioned on this 600-acre parcel in Anniston. Operations began in 1948. Various production wastes, including chemical cleaning solutions, paints, and electroplating sludges, were buried or discharged on the base. Testing in 1979 showed that the groundwater and soil contain heavy metals, chlorinated solvents, volatile organic compounds (VOCs), and phenols. The contamination threatens Dry Creek, Choccolocco Creek, Coldwater Spring, the Coosa River, and the drinking water supplies for about 75,000 residents.

Response actions included installing an on-base air-stripper (1987) and excavating contaminated soil (1993). A groundwater extraction and treatment system became operational in 1992. Other studies are nearing completion.

Sources: PNPLS-AL; Anniston Public Library, 108 East Tenth Street, Anniston, AL, 36202.

Ciba-Geigy Corporation, McIntosh Plant
Washington County
Pesticides and organic chemicals have been manufactured on this site near McIntosh since the 1950s. Various wastes were buried in landfills on the 1,500-acre property. Other wastewaters were discharged into the Tombigbee River prior to 1965. The soil and groundwater contain chromium, mercury, VOCs, phenols, DDT, lindane, and other pesticides. Lindane has been detected in residential wells. The contamination threatens wetlands, the Tombigbee River, and water wells that supply about 3,500 nearby residents.

Response actions included the installation of a groundwater extraction and treatment system. In 1991 the EPA recommended excavation of contaminated sludge and soil, thermal treatment of soil, solidification and reburial of soil, and in-situ soil flushing. Soil remediation will begin in 1997. Other studies are in progress.

Ciba-Geigy Corporation, the potentially responsible party (PRP), has undertaken the site studies and clean-up work.

Sources: PNPLS-AL; McIntosh Town Hall, Commerce Street, McIntosh, AL, 36553.

Interstate Lead Company
Jefferson County
Lead was recovered from used batteries and smelted on this site near Leeds. Various production wastes including slags were buried on the property or used as construction fill. Testing in the 1980s revealed that the soil and groundwater contain lead, cadmium, chromium, nickel, and arsenic. The metals were also found in Dry Creek and other surface waters, and lead was detected in air samples. The contamination threatens water wells that supply about 30,000 residents in the area.

Response actions included removing lead-bearing wastes from nearby properties in 1984. Contaminated soil, slag, and buildings were removed from the site in 1992. Other plans included additional soil and debris excavations, capping, and long-term groundwater monitoring to assure that the contamination plume is naturally attenuating. Other studies are in progress.

The PRP signed a Consent Order that required the company to conduct the site investigations. The company went bankrupt in 1992, and the remediation work is being carried out with federal funds.

Sources: PNPLS-AL; Leeds Public Library, 802 Parkway Drive Southeast, Leeds, AL, 35094.

Monarch Manufacturing
Lauderdale County
Ceramic tile and glaze have been manufactured on this site in Florence since 1954. Heavy metals were used as colorants in the production process. Contaminated wastes were discharged into the city sewer system or buried in the city landfill and on-site trenches. Sampling in the 1980s revealed that the groundwater and soil contain lead, zinc, cadmium, and barium. The heavy metals were also detected in Cox Creek and Sweetwater Creek. The contamination threatens Pickwick Lake, the Tennessee River, and public and private water wells that supply about 100,000 residents.

ALABAMA

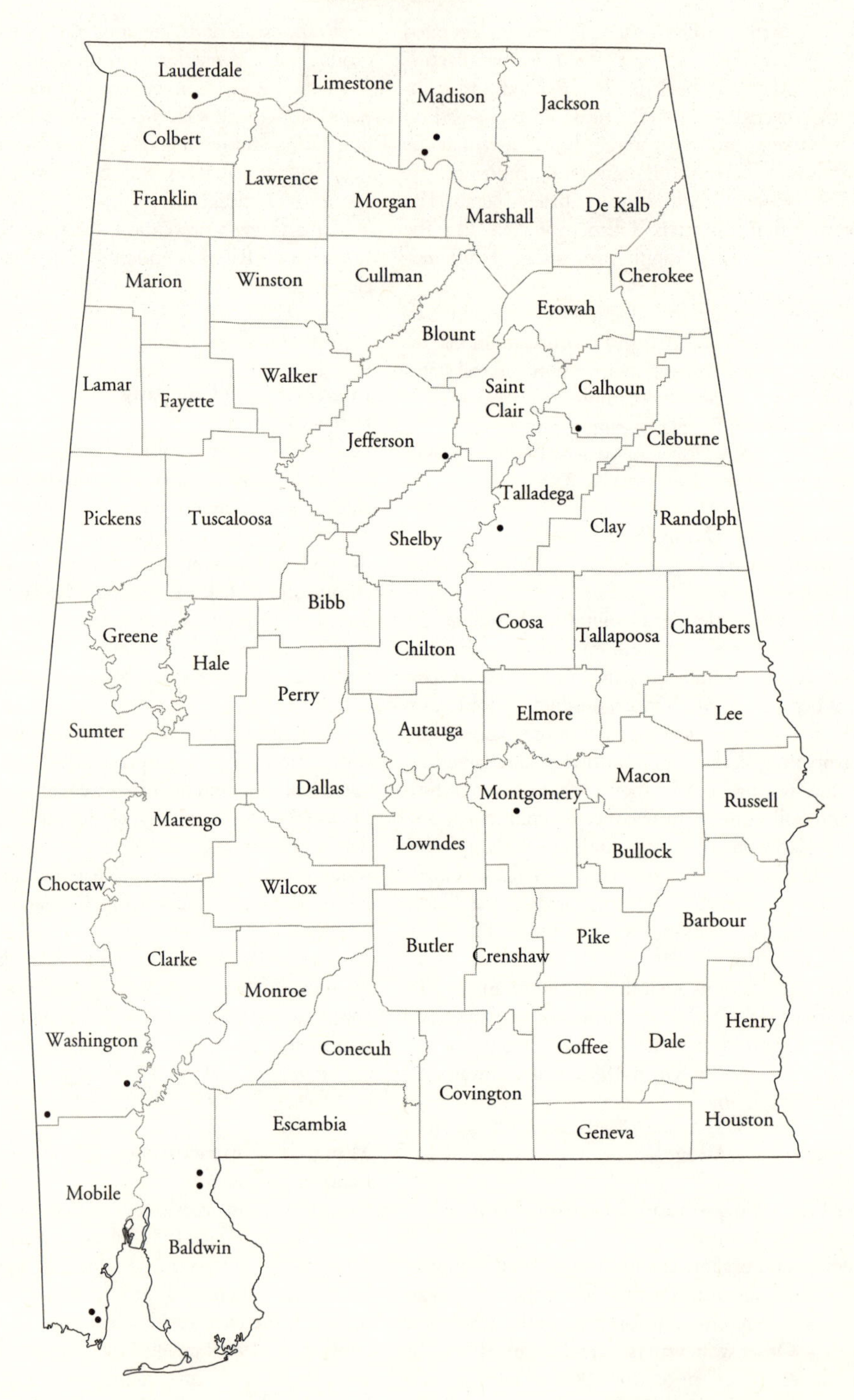
Lauderdale
Limestone
Madison
Jackson
Colbert
Lawrence
Franklin
Morgan
Marshall
De Kalb
Marion
Winston
Cullman
Cherokee
Etowah
Blount
Lamar
Fayette
Walker
Saint Clair
Calhoun
Cleburne
Jefferson
Talladega
Pickens
Tuscaloosa
Shelby
Clay
Randolph
Bibb
Coosa
Tallapoosa
Chambers
Greene
Hale
Chilton
Perry
Elmore
Lee
Sumter
Autauga
Macon
Dallas
Montgomery
Russell
Marengo
Lowndes
Bullock
Choctaw
Wilcox
Barbour
Pike
Clarke
Butler
Crenshaw
Monroe
Henry
Washington
Conecuh
Coffee
Dale
Covington
Escambia
Geneva
Houston
Mobile
Baldwin

Site investigations will be initiated.

Sources: PNPLS-AL; Florence-Lauderdale Public Library, 218 North Wood Street, Florence, AL, 35630.

Olin Corporation, McIntosh Plant

Washington County

Chlorine, pesticides, organic chemicals, and caustic soda have been manufactured on this site in McIntosh since the 1950s. Processing chemicals included mercury and chlorinated aromatic compounds. Testing has shown that the groundwater and soil contain mercury, chlorinated aromatic compounds, and VOCs. The contamination threatens the Tombigbee River, associated wetlands, and public and private wells that serve local residents.

Response actions included capping (1984), fencing, and excavating contaminated soil (1990). A site investigation completed in 1994 recommended a groundwater pump-and-treat system, cap upgrading, and groundwater monitoring. Wetland studies are in progress.

The PRPs have participated in the site studies and remediation work.

Sources: PNPLS-AL; McIntosh Town Hall, Commerce Street, McIntosh, AL, 36553.

Perdido Groundwater Contamination

Baldwin County

A railroad train derailment near Perdido in 1965 contaminated this area with benzene. Local residents complained about the quality of well water in 1981. Testing revealed that the groundwater contains benzene. The contamination threatens agricultural wells and the local drinking water supplies for about 500 residents.

Affected residents were connected to clean water supplies in 1983. In 1988 the EPA recommended a groundwater extraction and air-stripping/carbon adsorption system that became operational in 1992.

CSX Transportation, the PRP responsible for the derailment, constructed the groundwater treatment system.

Sources: PNPLS-AL; Perdido Water Board, Route 1, Perdido, AL, 36562; Bay Minette Public Library, 119 West Second Street, Bay Minette, AL, 36507.

Redstone Arsenal

Madison County

This U.S. Army base near Huntsville has been in operation since 1941. Activities have included producing chemical weapons (including mustard gas and lewisite), pesticides and industrial chemicals, chlorine, and rocket motors and solid rocket propellants for NASA. Related liquid and solid wastes were leaked, spilled, discharged, or buried on the 38,300-acre site. Testing has shown that the soil and groundwater contain VOCs, sulfur monochloride, ethylene, brine, caustic soda, thionyl chloride, DDT, and arsenic. The contamination threatens the Tennessee River, Indian Creek, the Wheeler National Wildlife Refuge, and private and public wells that serve about 50,000 residents.

Interim response actions have included capping and the installation of two groundwater pump-and-treat systems. Site investigations are planned.

Sources: PNPLS-AL; Huntsville-Madison County Library, 915 Monroe Street Southwest, Huntsville, AL, 35805.

Redwing Carriers

Mobile County

This chemical transporting business near Saraland operated from 1961 to 1971. Variously contaminated rinsewaters from cleaning the tanker trucks were discharged on the site. An apartment complex was later developed on the property. In the mid-1980s residents observed a tarlike leachate on the premises. Testing revealed that the soil and groundwater contain various chemicals, including VOCs. The contamination threatens the drinking water supplies for the 20,000 residents of Saraland.

Response actions included removing contaminated soil and sludge. A site investigation was completed in 1992. The EPA recommended excavating and treating contaminated soil and sludge, demolishing selected buildings, and treating groundwater. Local residents need to be relocated before clean up can begin.

Various agreements with PRPs require the companies to contribute to the site studies and remediation work.

Sources: PNPLS-AL; Saraland Public Library, 111 Saraland Loop Road, Saraland, AL, 36571.

Stauffer Chemical Company, Cold Creek Plant

Mobile County

Pesticides called thiocarbamates are produced on this site near Saraland, about 20 miles north of Mobile. Various leaks, spills, and discharges have contaminated the groundwater, soil, and surrounding wetlands with these chemicals. Cold Creek Swamp is also polluted with mercury. The contamination threatens private water wells supplying water to residents within 3 miles of the site.

In 1989 the EPA recommended groundwater extraction and treatment, groundwater monitoring, and closure of contaminated wells. The groundwater treatment system has been operational since 1993. Reconstruction of the wetlands is also planned. Other studies are in progress.

The PRPs, Zeneca Corporation and Akzo Chemicals, are participating in the site studies and remediation work.

Sources: PNPLS-AL; Saraland Public Library, 111 Saraland Loop, Saraland, AL, 36571.

Stauffer Chemical Company, Le Moyne Plant
Mobile County
Various chemicals have been produced on this site near Saraland since the 1950s, including carbon disulfide, carbon tetrachloride, chlorine, caustic soda, and industrial inorganic compounds. Various solid and liquid waste products—solvents, heavy metals, acids, and organic compounds—were discharged or buried in landfills on the 950-acre site. Testing has shown that the groundwater and soil contain VOCs and heavy metals. Sediments in Cold Creek Swamp contain mercury. Thiocyanates were detected in Halby Pond. The contamination threatens nearby streams and wetlands and the local groundwater that supplies drinking water to about 5,000 residents.

Response actions included installing a groundwater treatment system in 1980. A groundwater study completed in 1989 recommended expanding the groundwater extraction system, closing contaminated wells, and long-term monitoring. In 1993 a wetlands study recommended excavating contaminated sediments, capping, and revegetation. Other studies are in progress.

An Administrative Order on Consent (AOC) (1986) and Consent Decree (CD) (1990) require the PRPs to undertake the site studies and clean-up work.

Sources: PNPLS-AL; Saraland Public Library, 111 Saraland Loop, Saraland, AL, 36571.

T. H. Agriculture & Nutrition Company, Montgomery Plant
Montgomery County
Pesticides, insecticides, and herbicides were distributed on this site near Montgomery. Various production or storage wastes were buried in pits and trenches on the property. Testing has shown that the groundwater and soil contain pesticides such as DDT, lindane, and toxaphene. The contamination threatens public and private wells that serve about 255,000 residents within 3 miles of the site.

Response actions included removing contaminated soil in 1981. A recently completed site investigation (1991–1995) will recommend a final remediation plan. An interim groundwater pump-and-treat system should be operational by 1997.

An AOC (1991) with the PRP requires the company to conduct the site investigations. The PRP also voluntarily conducted the removal actions in 1981.

Sources: PNPLS-AL; Air University Library, Maxwell Air Force Base, Montgomery, AL, 36112.

Triana/Tennessee River
Madison County
This 1,400-acre site near Triana is located along a 20-mile stretch of the Tennessee River and its tributaries. Chemical manufacturing upstream between 1947 and 1970 contaminated this stretch of the Tennessee River drainage basin with DDT. Fish and aquatic life also contain DDT and are consumed by local residents. The river was used as a source of drinking water until 1967.

Response actions included diverting the river, burying the most contaminated river channels, and monitoring the health of fish and wildlife. Construction activities were completed in 1987, and monitoring will continue until 1998. So far, DDT concentrations in fish have decreased by about 75 percent.

The Olin Company, the PRP, has participated in the clean-up work since 1983.

Sources: PNPLS-AL; Town Hall of Triana, 640 Sixth Street, Madison, AL, 35758.

Alaskan Battery Enterprises
Fairbanks North Star Borough
Battery manufacturing and recycling have been conducted on this site on the outskirts of Fairbanks since 1961. Battery casings were dumped in wetlands, and liquid waste was discharged into an on-site septic system and drain field. Tests in 1986 found lead and acid in soils on-site and in the surrounding area. Schools and wetlands are located nearby.

Response actions included the removal of contaminated soil (1988–1989). An EPA site investigation in 1992 found that the groundwater was not contaminated. Remaining contaminated soil was soil-washed in 1993.

Settlements and lawsuits against 36 PRPs (1992–1996) will recover a portion of the EPA's clean-up costs.

Sources: PNPLS-AK; Alaska Department of Environmental Conservation, 1001 Noble Street, Suite 350, Fairbanks, AK, 99701.

Arctic Surplus
Fairbanks North Star Borough
Located 6 miles southeast of Fairbanks, this facility began operations as a salvage yard in 1944. Military equipment, asbestos, oil, batteries, and transformers were processed. A site inspection in 1988 discovered that the soil and groundwater contain high levels of lead, VOCs, and polychorinated biphenyls (PCBs). About 1,000 residents within 3 miles of the site use the local groundwater.

Response actions included removing on-site bulk wastes and fencing the site (1989). Contaminated soils were excavated in 1990. Additional contaminated soil and debris were removed between 1991 and 1994. An EPA site investigation was completed in 1995 and will recommend a final remediation plan.

Sources: PNPLS-AK; Defense Reutilization & Marketing Office, Building 5001, 1/4 Mile Badger Road, Fairbanks, AK, 99703.

Eielson Air Force Base
Fairbanks North Star Borough
This U.S. Air Force base has been in operation since 1944 and covers nearly 20,000 acres. Waste products have been disposed of in a number of unlined landfills, trenches, and lagoons. The groundwater and soil are contaminated with VOCs, petroleum, and heavy metals from these burial sites and from on-site leaks and spills. The contamination threatens the recreational Tanana River and the water supplies of about 6,000 people living within 3 miles of the site.

Joint EPA-USAF studies from 1991 to 1996 have led to the selection of a remediation plan that includes bioventing and soil vapor extraction to clean fuel-contaminated soil.

Source: PNPLS-AK; Noel Wien Library, Fairbanks, AK, 99709.

Elmendorf Air Force Base
Anchorage Borough
This U.S. Air Force facility covers about 13,000 acres near Anchorage. Wastes such as spent solvents, lead, batteries, paints, transformer oils, and pesticides have been discarded in a number of landfills on the base. The soil and groundwater contain free product petroleum, VOCs, polycyclic aromatic hydocarbons (PAHs), PCBs, pesticides, and heavy metals (lead). The contamination threatens the water supply of over 100,000 residents within 3 miles of the site as well as sensitive wildlife habitats, including moose and salmon.

Response actions included removing storage tanks and contaminated soil (1990–1991), installing a leachate collection system (1993), installing underground bioventing for petroleum-soaked soil (1993–1994), and constructing a wetland to collect petroleum-rich seepage (1995–1996). Final clean-up recommendations for other parts of the base are still pending.

Sources: PNPLS-AK; Bureau of Land Management, Alaska Resource Library, 222 West 7th Avenue, No. 36, Federal Building, Anchorage, AK, 99513.

Fort Richardson
Anchorage Borough
Built in 1940 by the U.S. Army, the fort is bordered by the city of Anchorage, Cook Inlet, and Chugash State Park. It is the main base for military forces in Alaska. Three sites (including a landfill) on the property have contaminated soil and groundwater with phosphorous, VOCs, heavy metals, and PCBs. The contamination threatens sensitive wetland habitats in the area, and waterfowl deaths have been related to phosphorus poisoning.

Response actions included exhuming buried waste and contaminated soil (1993–1994). EPA studies were completed in 1995. Dredging is recommended to expose phosphorus-bearing soil to air, which will neutralize it through combustion.

Source: PNPLS-AK.

Fort Wainwright
Fairbanks North Star Borough
This U.S. Army installation near Fairbanks was built in 1947. Wastes from vehicle and aircraft maintenance operations, including oil, fuel, solvents, perchloroethylene, asbestos, fuel tank sludge, and heavy metal-bearing slag, were disposed of in a 74-acre

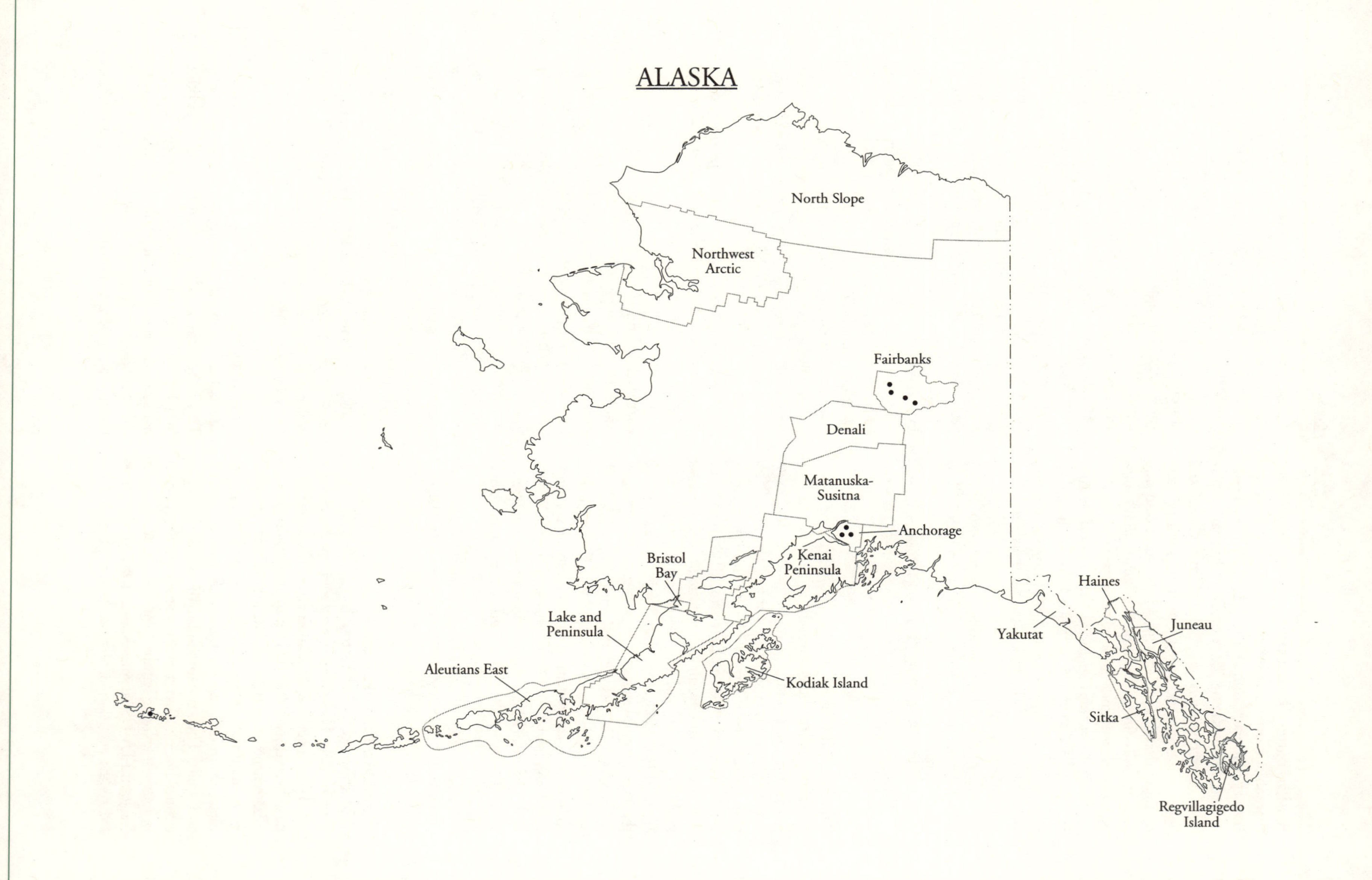
ALASKA
North Slope
Northwest Arctic
Fairbanks
Denali
Matanuska-Susitna
Anchorage
Kenai Peninsula
Bristol Bay
Lake and Peninsula
Aleutians East
Kodiak Island
Haines
Juneau
Yakutat
Sitka
Regvillagigedo Island

landfill. The groundwater and soil contain high levels of petroleum, solvents, pesticides, heavy metals (chromium and mercury), and PCBs. The contamination threatens the recreational Chena River and the drinking water of about 11,000 people within 3 miles of the fort.

Response actions have included site investigation studies by the U.S. Army and the EPA (1994–1996). A final remediation plan should be in place by 1997.

Sources: PNPLS-AK; Noel Wien Library, Fairbanks, AK, 99709.

Naval Air Station, Adak
Adak Island, Aleutian Islands
Located on Adak Island in the Aleutian Islands, this U.S. Navy base was built in 1942. Various hazardous materials such as drummed waste, oil, solvents, batteries, and transformer oils have been buried in landfills on the property. Tests in 1988 showed pollutants in the soil and groundwater containing petroleum, PCBs, and heavy metals (lead and silver). The contamination threatens the commercial fishing industry, wildlife habitats, and the water supplies of about 1,000 nearby residents.

Response actions included removing buried waste and drums (1988) and contaminated soil (1992). Site investigation studies are nearing completion and will recommend final clean-up remedies.

Sources: PNPLS-AK; Bureau of Land Management, Alaska Resources Library, 222 West 7th Avenue, No. 36, Federal Building, Anchorage, AK, 99513.

Standard Steel & Metals Salvage Yard
Anchorage Borough
Located in Anchorage, this site has been operated as a recycling yard since 1972. Processed material included transformers, batteries, and military equipment. EPA tests in 1985 found anomalous levels of PCBs, solvents, lead, dioxin, furans, and VOCs (tetrachloroethylene) in the soil and groundwater. The contaminants threaten Ship Creek, a recreational and salmon spawning stream, and the water supplies of over 120,000 people within 3 miles of the site.

Response actions included removing much of the hazardous waste, including transformers and batteries, sealing the soil cover, and fencing (1987–1988). Results and recommendations from the site investigation are pending.

Chugach Electric, one of the PRPs, signed a Consent Order (CO) (1992) with the EPA to finance the site investigation.

Sources: PNPLS-AK; Alaska Resources Library, Bureau of Land Management, 222 West 7th Avenue, No. 36, Federal Building, Anchorage, AK, 99513.

ARIZONA

Apache Powder Company
Cochise County
This 1,000-acre site near Benson is owned by Apache Nitrogen Products, Inc., and produces industrial chemicals and explosives, including nitric acid, ammonium nitrate, blasting powders, and nitrogen fertilizers. From 1922 to 1971 wastewater was discharged into dry washes that connected with the San Pedro River; since 1971 the wastewater has been discharged into unlined lagoons. Soil and groundwater are contaminated with arsenic, fluoride, nitrate, antimony, barium, beryllium, chromium, lead, manganese, and vanadium pentoxide. A number of nearby private wells are contaminated with nitrate.

Response actions included drilling deeper wells for affected residents and removing contaminated soil (1993–1994). Site investigations by the EPA recommended construction of wetlands to treat the contaminated groundwater and the excavation and capping of contaminated soil. Remedies were initiated in 1995.

Administrative Orders (1989, 1994) require Apache Nitrogen Products, Inc., to finance the site investigation and remediation work.

Sources: PNPLS-AZ; Benson Public Library, 300 South Huachuca, Benson, AZ, 85602.

Hassayampa Landfill
Maricopa County
This municipal landfill 40 miles west of Phoenix operated from 1961 to the 1990s, accepting both municipal and hazardous wastes in unlined trenches. Testing by the state in 1981 showed that the soil and groundwater are contaminated with VOCs, heavy metals, and pesticides. Private wells within 3 miles of the site are used for drinking water by about 400 residents and for irrigating farmland. The contamination also threatens the Hassayampa River about 4,000 feet away.

Site investigations were conducted between 1991 and 1995. The final remediation plan consists of groundwater extraction and treatment, groundwater monitoring, soil vapor extraction, and capping. Construction of these systems began in 1995.

The EPA identified over 100 PRPs in 1987. Seventy-seven parties responsible for small amounts of waste negotiated minimal settlements with the EPA;

ARIZONA

the others are required to finance the site investigation and clean-up work through a series of Consent Orders (1988) and Consent Decrees (1994).

Sources: PNPLS-AZ; Buckeye Public Library, 310 North Sixth Street, Buckeye, AZ, 85326.

Indian Bend Wash
Maricopa County
This area contains seven municipal supply wells for the cities of Scottsdale and Phoenix that are contaminated with VOCs. The contamination appears to be related to several industries in the area, including Motorola and Beckman Industries. Tests have shown that the groundwater and soil are contaminated with VOCs, boron, chloroform, lead, and zinc. Six of the wells have been shut down. The groundwater supplies about 70 percent of the water needs of Scottsdale, which has a population of 130,000, and is also used for irrigation and farming.

Response actions included the installation of an air-stripper on one of the contaminated wells, returning it to service. EPA site investigations recommended a final remediation plan of soil vapor extraction, groundwater extraction, and air-stripping. Construction of these remedies began in 1994.

The EPA has reached agreements with a number of PRPs to help finance the site investigations and remediation work.

Sources: PNPLS-AZ; Phoenix Public Library, 12 East McDowell Road, Phoenix, AZ, 85004.

Litchfield Area
Maricopa County
This 35-square-mile area in the city of Goodyear includes the Phoenix Airport. Testing in 1981 showed that the soil and groundwater are contaminated with VOCs and heavy metals (including chromium). Nearly half of the 90 municipal wells in the area contained VOCs. The contaminants were released by a number of manufacturing companies in the area.

Response actions included a number of site investigations. A chromium sludge bed was stabilized in 1993. Groundwater extraction and air-stripping began in 1989 and will operate for at least 10 years.

Agreements between the EPA and Goodyear Tire and Rubber, Unidynamics, and Loral Defense Systems (1988–1992) require these parties to finance the site investigations and remediation work.

Source: PNPLS-AZ.

Luke Air Force Base
Maricopa County
This U.S. Air Force base near the town of Glendale provides advanced training for fighter pilots. On-site disposal or leaks of waste products, including oils, radioactive waste, and ammunition derivatives, have contaminated the soil with VOCs and heavy metals. The contamination threatens the groundwater supplies for the towns of Glendale, Goodyear, and Phoenix.

Response actions included removing storage tanks and capping. Areas of soil contamination will be treated by vapor extraction, excavation, and biological treatment (1990–1995). Other site studies are still in progress.

Sources: PNPLS-AZ; Glendale Public Library, 5959 West Brown Avenue, Glendale, AZ, 85302.

Motorola Inc.
Maricopa County
This industrial site in the city of Phoenix manufactures semiconductors. Testing in 1983 showed that an underground storage tank was leaking VOCs into the soil and groundwater. The contamination plume extends at least 3 miles from the site and threatens the water supplies of about 1,000 residents within that area.

Monitoring wells were installed in 1983–1984. A test groundwater extraction and carbon adsorption treatment system was installed in 1986. In 1992 a soil vapor extraction system became operational.

Motorola, the PRP, has willingly financed the site investigations and remediation work.

Sources: PNPLS-AZ; Phoenix Public Library, 2802 North 46th Street, Phoenix, AZ, 85008.

Nineteenth Avenue Landfill
Maricopa County
This site operated as a sanitary landfill between 1957 and 1979. It received a variety of wastes, including municipal, radioactive, hospital, and industrial materials. The soil and groundwater have high levels of VOCs, heavy metals, pesticides, PCBs, beta radiation, and methane. Agricultural and industrial wells are located within 1,000 feet of the site.

Response actions included restricting access and installing a gas collection system and groundwater monitoring wells (1981). Construction of a new gas collection system and landfill cap was started in 1995.

The city of Phoenix, a PRP, has financed site investigations and the remediation work.

Sources: PNPLS-AZ; Ocotillo Branch Public Library, 102 West Southern Avenue, Phoenix, AZ, 85041.

Tucson International Airport
Pima County
In addition to the airport operations, a variety of industries formerly operated in the area, including aircraft and electronics companies. A variety of waste was disposed of in on-site landfills or discharged on the ground. Hughes Aircraft Company, for example, used VOC-bearing degreasers and chromium in electroplating. In 1950 high chromium levels were discovered in a nearby municipal well. Soil and groundwater are contaminated with VOCs and chromium that threaten the Santa Cruz River and the drinking water supplies of about 50,000 people.

Response actions included closing contaminated wells in 1981. Groundwater extraction and treatment systems were installed in 1987 and 1994. Sludges were removed in 1991. A soil remediation plan is being designed.

A series of legal agreements with PRPs requires them to finance the site investigations and remediation work.

Sources: PNPLS-AZ; Tucson Main Library, 101 North Stone Street, Tucson, AZ, 85701.

Williams Air Force Base
Maricopa County
U.S. Air Force activities at this 4,000-acre base near Chandler discharged wastes on-site, including or-

ganic solvents, petroleum, metal-plating solutions, pesticides, and radiological compounds. The groundwater and soil are contaminated with VOCs, nitrates, and heavy metals (cadmium, nickel, and chromium). About 5,000 people live or work in the area.

Response actions included removal of waste and drums (1988–1992), a landfill cap (1995), and groundwater extraction and treatment (1996). Some site investigation studies are still in progress and will recommend additional remedial actions.

Sources: PNPLS-AZ; Chandler Public Library, 75 East Commonwealth, Chandler, AZ, 85225.

Yuma Marine Corps Air Station
Yuma County
This U.S. Air Force station has discharged large volumes of waste fuels, warfare chemicals, and solvents on-site since the 1950s. Testing has shown high levels of VOCs and heavy metals in the groundwater and soil. The contamination threatens local drinking, industrial, and agricultural water supplies as well as the nearby Colorado River.

Site investigations are in progress and will recommend a final remediation plan.

Source: PNPLS-AZ.

ARKANSAS

Arkansas Inc.
Boone County
This site is about 3,000 feet southwest of the town of Omaha. Arkwood Industries and Mass Merchandiser, Inc., operated a pentachlorophenol (PCP) and creosote wood preserving business there from the 1960s to 1984. Waste was buried in drums or dumped in a sinkhole on the property. Investigations by the EPA (1986–1992) revealed soil contaminated with PAHs, PCP, and dioxins and nearby springs with anomalous levels of PCP. About 650 people live within 3 miles of the site; over 50 springs are within 2 miles of the site.

Response actions included fencing; supplying city water to threatened residents; removing structures and foundations, pumpable liquids, and drummed waste; on-site treatment of contaminated soil and sludge; and groundwater monitoring (1992–1995). Groundwater monitoring will continue through December 1977.

An AOC and AO with the EPA (1986–1987) and a Consent Decree with the Department of Justice (1988) and EPA (1992) forced the present owner, Mass Merchandiser, Inc., to finance the reclamation plan.

Sources: PNPLS-AR; Omaha Public School, College Street, Omaha, AR, 72662.

Frit Industries
Lawrence County
Frit Industries, near the town of Walnut Ridge, is a producer of micronutrients and trace element additives for fertilizer products. The groundwater and soil are contaminated with zinc sulfate, cadmium, chromium, lead, and other heavy metals. A spill of 81,000 gallons of contaminated water killed fish in nearby Coon Creek in 1979. Surface runoff from fertilizers, raw materials, and production waste stored on-site has also occurred. The site could contaminate the drinking water of about 15,000 nearby residents.

Response actions included excavating contaminated soil (1982) and constructing a surface water collection and treatment system (1984–1985). Plans were also made for constructing two hazardous waste storage buildings and a surge pond (1990–1991). The site is presently being monitored.

Three AOCs between the EPA and Frit Industries (1983, 1987, 1991) require Frit Industries to finance the remediation plan.

Sources: PNPLS-AR; Lawrence County Library, 1315 West Main Street, Walnut Ridge, AR, 72476.

Gurley Pit
Crittenden County
Used as a dump site by the Gurley Refinery from 1970 to 1975, the property received PCB-bearing sludge and oil and heavy metals (lead, zinc, and barium). Located 1 mile north of the town of Edmondson in an agricultural area, the site threatens Fifteen Mile Bayou and the drinking water of nearby homes. Leachate from the site has contaminated the bayou during periods of flooding.

Response actions have included draining the pit and pumping out contaminated storm water (1979), fencing (1984), installing surface runoff collection systems (1984–1989), and treating and discharging water (1989). Baseline air monitoring was started in 1992. An on-site sludge storage facility was built in 1994 to solidify the sludge material.

The Gurley Refining Company was sued by the EPA in 1990 as the potentially responsible party and is expected to finance the clean-up work.

Sources: PNPLS-AR; Edmondson City Hall, Edmondson, AR, 72332.

Industrial Waste Control
Sebastian County
This landfill site is a former coal strip pit near Fort Smith. The pit received a variety of municipal, con-

ARKANSAS

struction, and industrial waste materials from the 1960s to 1978, including hundreds of drums of liquid waste. The bottom of the pit is soft, fractured rock that transports groundwater. The site is within 200 feet of the nearest homes and water wells, and runoffs have killed fish in Prairie Creek. Soil and groundwater are contaminated with methylene chloride, toluene, PAHs, and heavy metals, including nickel, chromium, and lead.

Response actions included the removal of liquid-filled drums, treatment of contaminated soil, and construction of a slurry wall and multilayer cap (1988–1991). The property has now entered a 30-year monitoring phase.

The EPA has identified 19 PRPs for the recovery of clean-up costs. These parties purchased the site and surrounding property for the monitoring program.

Sources: PNPLS-AR; Fort Smith Public Library, Fort Smith, AR, 72901.

Jacksonville Municipal Landfill
Pulaski County
The city of Jacksonville operated this site as a landfill from 1960 to 1973. A variety of municipal and industrial wastes was received in unlined trenches, including drums of toxic industrial liquids. Soil and groundwater at the site are contaminated with 2,3,7,8-tetrachlorodibenzo (p) dioxin (2,3,7,8-TCDD) and herbicides 2,4-D, 2,4,5-T, and 2,4,5-TP. About 10,000 people and a number of wells are within 3 miles of the site.

Response actions included fencing (1986); excavation, transportation, and thermal treatment of exhumed wastes and capping (1990–1995); and long-term groundwater monitoring through 1999.

Three PRPs will contribute toward the cost of the remediation program.

Sources: PNPLS-AR; Jacksonville City Hall, Jacksonville, AR, 72076.

Midland Products
Yell County
Located less than 3,000 feet from the town of Ola, this is the former site of a sawmill and wood preservative treatment plant that operated from 1969 to 1979. During this time chemical solutions were held in processing lagoons, which occasionally overran

into nearby waterways. The groundwater, surface water, and soil are contaminated with creosote, pentachlorophenol (PCP), PAHs, chlorinated dibenzodioxins, and chlorinated dibenzo-furans. About 200 people live near the site; the closest water well and residence are within 400 feet. The Petit Jean State Wildlife Area is about a mile north of the site.

Response actions included thermal destruction of contaminated soil, sludges, and sediments (1991–1993); placement of a vegetated soil layer cover (1994); and groundwater extraction and treatment (1993–1998).

In 1988 the EPA identified Midland Products Company as the PRP for cost recovery.

Sources: PNPLS-AR; Ola City Hall, Ola, AR, 72853.

Mid-South Wood Products
Polk County
This commercial wood treatment site is located just west of the town of Mena. Since the 1930s four different companies have treated lumber with creosote and chromated copper arsenates. Storage ponds on the site held toxic industrial solutions, and residual sludges were mixed with soil and spread on two landfarmed areas. Groundwater, surface water, and soil are contaminated with creosote, arsenic, copper, chromium, and PCPs. About 6,000 people depend on private water wells within 1 mile of the site. Surface runoff flows into Prairie Creek and the East Fork of Moon Creek, threatening stream fisheries.

Response actions included connecting affected residents to municipal water (1984); excavating, stabilizing, and capping contaminated soil and sludges with clay; and installing a groundwater pump and treatment system (1986–1996).

The EPA has identified two PRPs. A Record of Decision (ROD) (1986) and two CDs (1987) were signed by the parties, the EPA, and Arkansas District Court to enforce private funding of the remediation plan.

Sources: PNPLS-AR; Polk County Library, Mena, AR, 71953.

Monroe Auto Equipment Company
Greene County
This rural site, a former gravel pit, is about 3 miles west of the town of Paragould. From 1973 to 1978 the Monroe Auto Equipment Company dumped alum and lime electroplating sludge in the pit. The groundwater and soil are contaminated with solvents and degreasing agents such as 1,1 dichloroethane, 1,2 dichloroethylene xylene, and heavy metals (chromium and lead). Although the area is lightly populated, a number of residences and wells are adjacent to the site.

The potentially responsible party, Monroe Auto Equipment Company, signed an AOC with the EPA (1991) to fund a detailed site investigation and feasibility study. A final decision regarding a remediation plan is expected by 1997.

Sources: PNPLS-AR; Northeast Arkansas Regional Library, Paragould, AR, 72450.

Popile Inc.
Union County
This site near El Dorado operated as a wood treatment facility from 1947 to 1982. Toxic wastewater containing creosote and PCPs was stored on-site in three surface ponds. Popile Inc. closed the ponds in 1984. The EPA identified soil contamination in 1989. The groundwater and soil are contaminated by PCP, acenaphthylene, benzo(a)pyrene, fluorene, and pyrene. The drinking water for the 26,000 residents of El Dorado is drawn from private and city wells within 4 miles of the site. Runoff from the site enters the recreational Bayou de Loutre.

Response actions included erosion and runoff control, fencing, and capping of the site (1991). Contaminated soil was excavated and stored on-site. Further investigations were done from 1991 to 1995. A final remediation plan consisting of groundwater extraction and treatment, excavation and biological treatment of contaminated soil and sludges, and final grading and seeding is under way.

Sources: PNPLS-AR; Barton Public Library, El Dorado, AR, 71730.

Rogers Road Municipal Landfill
Pulaski County
The Rogers Road Landfill site was used by the city of Jacksonville from 1953 to 1974. Municipal and industrial waste, including barrels of poisonous herbicides and pesticides, were buried in the dump. Soil and groundwater on the site are contaminated with 2,4-D, 2,4,5-T, 2,4,5-TP, 2,3,7,8-TCDD, and dieldrin. About 10,000 people live within 3 miles of the site, which is now a recreational woodland.

Response actions included the excavation and transport of contaminated soil and drums of hazardous waste for thermal treatment and capping the landfill (1990–1995). Long-term groundwater monitoring will continue through 1999.

Three PRPs will help finance the cost of the remediation plan.

Sources: PNPLS-AR; Jacksonville City Hall, Jacksonville, AR, 72076.

South 8th Street Landfill
Crittenden County
This waste facility is located on the Mississippi River flood plain near the town of West Memphis. Originally opened as a sand and gravel pit in

1957, the facility has received a variety of wastes, including a pit of oily sludge. The soil and groundwater are contaminated with PCBs, lead, PAHs, dioxin, and other pesticides. The material in the pit is highly corrosive with a pH of less than 2.0. The first EPA investigation was in 1981. Although the groundwater in the landfill flows away from West Memphis, the site is close to a trailer park and some private homes. Trespassers frequently cross the landfill to gain access to the Mississippi River.

Response actions included fencing the site (1992) and building a levee around the sludge pit (1992). An EPA study from 1992 to 1994 determined that the remediation plan would consist of the stabilization and removal of pit sludge and contaminated soil and building a new cover. Construction should be under way by 1997.

Eighty-five users of the landfill have been identified as PRPs. Through a Unilateral Administrative Order (UAO) (1992), an ROD (1994), and an AOC (1996) with the EPA, the parties will be financially responsible for the clean-up plan.

Sources: PNPLS-AR; West Memphis Public Library, West Memphis, AR, 72301.

Vertac, Inc.
Pulaski County
Located near the town of Jacksonville, this site was operated by four different companies that manufactured herbicides, including the defoliant Agent Orange, from 1948 to 1986. The site contains disposal areas that were filled with drums of herbicide production chemicals, sludges, and other wastes. The soil and groundwater are contaminated with these compounds, including 2,3,7,8-TCDD (dioxin), 2,4-D, chlorinated benzene, 2,4,5-T, chorinated phenols, herbicide production wastes, and 2,4,5-TP (silvex). Thirty thousand people live in nearby Jacksonville, and private homes, water wells, and Rocky Branch Creek are located beside the site. Fish in Rocky Branch Creek and Bayou Meto contain toxic levels of dioxin.

PRPs began disassembling the site in 1986. Response actions included removing drums of waste for incineration, pumping and treating liquid waste, repairing fences, and controlling runoff (1993–1996). Contaminated soil/sediments are also being removed and treated.

A series of AOs requires approximately eight PRPs to finance the clean-up activities.

Sources: PNPLS-AR; Jacksonville City Hall, Jacksonville, AR, 72076.

CALIFORNIA

Advanced Micro Devices, Building 915
Santa Clara County
Advanced Micro Devices builds semiconductor and microprocessor devices. Underground acid neutralization and waste organic solvent tanks are buried on this property near Sunnyvale. Tests in 1981 showed high levels of VOCs in the soil and groundwater. The contamination threatens the drinking water supplies of about 200,000 people in Santa Clara, Sunnyvale, and Mountain View.

Response actions included removing contaminated soil and leaking tanks (1982–1983). Seven additional wells were drilled between 1983 and 1985 to create a hydraulic barrier to contain the contamination plume. The extracted groundwater has been treated using air-stripping and carbon adsorption. Groundwater monitoring will continue for at least 5 years.

The PRP, Advanced Micro Devices, financed the site investigation and remediation work.

Sources: PNPLS-CA; Regional Water Control Board, 2101 Webster Street, Suite 500, Oakland, CA, 94612.

Advanced Micro Devices, Inc.
Santa Clara County
Advanced Micro Devices manufactures electronic equipment on this 6-acre site in Sunnyvale. Industrial solvents from spills and leaking tanks have contaminated the soil and groundwater with VOCs and freon. The contamination threatens the water supplies of about 300,000 people.

Response actions included removing leaking tanks. A groundwater extraction and air-stripping/carbon adsorption system was also installed. Contaminated soils were excavated and incinerated off-site (1992). New wells were installed in 1992 to contain the migration of the contamination plume.

Advanced Micro Devices, together with Signetics and TRW Microwave, signed agreements with the State of California to finance the site study and remediation work.

Sources: PNPLS-CA; Regional Water Control Board, 2101 Webster Street, Suite 500, Oakland, CA, 94612.

Aerojet General Corporation
Sacramento County
Located near Rancho Cordova, this company has manufactured liquid and solid rocket propellants and agricultural and pharmaceutical chemicals since 1953. Various hazardous wastes were disposed of on-site in landfills or deep injection wells. Testing has revealed high levels of VOCs, chloroform, freon, and heavy metals in the soil and groundwater. At least 16

CALIFORNIA

wells have been contaminated, and the chemicals threaten nearby recreational waterways and the water supplies of about 400,000 people. VOCs have been detected on the banks of the American River.

Response actions included the installation of groundwater extraction and treatment systems (1983–1986). Soil investigations are nearing completion. A final remediation plan, including upgraded groundwater treatment, will also be recommended.

Sources: PNPLS-CA; Carmichael Regional Library, 5605 Marconi Avenue, Carmichael, CA, 85608.

Applied Materials
Santa Clara County
This company manufactures semiconductor equipment on its 9-acre site in Santa Clara. A leaking underground acid neutralization system has contaminated the soil and groundwater with VOCs. The contamination threatens the water supplies of about 300,000 people within 3 miles of the site.

Response actions included installation of a temporary groundwater pump-and-treat system (1984), removal of underground tanks and contaminated soil (1985), and monitoring wells (1985). The EPA approved the temporary groundwater system as the permanent groundwater treatment system in 1990. Groundwater treatment and monitoring are continuing.

Applied Materials Inc. financed the site investigation and remediation work.

Sources: PNPLS-CA; Regional Water Quality Control Board, 2101 Webster Road, Suite 500, Oakland, CA, 94612.

Atlas Asbestos Mine
Fresno County
This mine near Coalinga, which recovered and processed asbestos ore, operated from 1963 to 1979. Various mining wastes were also stored within the city of Coalinga for later transportation. Surface runoff and flooding have contaminated nearby drainages and waterways. Tests have shown that the soils contain high levels of asbestos, chromium, and nickel.

Response actions included site investigations completed in 1991 that recommend surface runoff controls, sediment trapping dams to contain asbestos particles, and stabilizing waste piles. This work is nearing completion. Contaminated material and soil were removed from the site in Coalinga, which was later capped and graded (1993). Groundwater monitoring is in progress.

Sources: PNPLS-CA; San Jose Public Library, 180 West San Carlos Street, San Jose, CA, 95113.

Barstow Marine Corps Logistics Base
San Bernardino County
This 5,700-acre U.S. Marine Corps base near Barstow has been in operation since 1942. Activities have been primarily vehicle maintenance and war surplus storage; related wastes such as fuel, antifreeze, and acids were deposited on-site in landfills or unlined lagoons. Tests show that the soil and groundwater are contaminated with VOCs, pesticides, PAHs, heavy metals, and PCBs. The contamination threatens the water supplies of about 30,000 people in the Barstow area; the groundwater is also used for irrigation of farmland.

Response actions included supplying clean drinking water to the base and excavating sludges (1993). Ongoing site investigations should be completed in 1997 and will recommend a final remediation plan.

Sources: PNPLS-CA; San Bernardino Public Library, 304 East Buena Vista, Barstow, CA, 92311.

Beckman Instruments
Tulare County
Located in Porterville, this plant has built circuit boards and electronic equipment since 1968. From 1974 to 1983 various wastes, including solvents, acids, and heavy metals, were discharged into an evaporation pond on the 500-acre property. Tests in 1978 showed that the groundwater and soil are contaminated with VOCs, freon, and lead. The contamination threatens about 500 people who live within 1 mile of the site.

Response actions included deactivating and removing the pond, pond waste, and nearby soil (1983). Affected residents were connected to clean water supplies (1983–1985). A groundwater extraction and air-stripping system was installed in 1985 to control the migration of the contamination plume. Contaminated soils have also been excavated and removed. Groundwater monitoring is continuing.

A Consent Order (1987) and subsequent UAO with Beckman Instruments require the company to finance the cost of the site studies and reclamation work.

Sources: PNPLS-CA; Porterville Public Library, 41 West Thurman Avenue, Porterville, CA, 93257.

Brown and Bryant, Inc.
Kern County
This plant in the town of Arvin has processed agricultural chemicals, fertilizers, and poisons since 1960. Improper handling of hazardous waste from these activities has contaminated the soil and groundwater with pesticides and fumigants. The contamination threatens the water supply of about 8,000 nearby residents and 20,000 acres of farmland.

Response actions included excavating contaminated soil (1988). An EPA site investigation recommended consolidation of the contaminated soil, capping the soil, and groundwater extraction and treatment. Design of the remediation systems is nearing completion.

The PRP, Brown and Bryant, declared bankruptcy in 1989. The EPA and the State of California will finance the clean-up work.

Sources: PNPLS-CA; Kern County Library, 123 A Street, Arvin, CA, 93203.

Camp Pendelton Marine Corps Base
San Diego County
This 125,000-acre military installation near Oceanside provides training and support for the U.S. Marine Corps. Wastes generated from vehicle maintenance operations were discharged on-site into lagoons or landfills. Testing has shown that the soil and groundwater are contaminated with VOCs, oils, PCBs, pesticides, and heavy metals. The contamination threatens numerous nearby streams and wetland habitats as well as drinking water supplies for the base.

Response actions have consisted of a number of site investigation studies that began in 1990 and should be completed by 1997. A final remediation plan will then be selected.

Sources: PNPLS-CA; Oceanside Public Library, 330 North Hill Street, Oceanside, CA, 92054.

Castle Force Air Base

Merced County

This 3,000-acre military installation near Atwater began operation in 1941. Various wastes generated from aircraft maintenance and fire-fighting training, such as waste fuels, solvents, paints, and plating residues, were discharged in unlined pits and lagoons on the base. Tests have shown that the groundwater and soil are contaminated with VOCs (trichloroethylene and benzene). About 6,000 people depend on water wells within 3 miles of the site.

Response actions included drilling new replacement wells (1988) and installing a carbon filtration system for the contaminated groundwater. Toxic soils were excavated from a nearby residential area in 1991. Other site investigation studies are in progress.

Sources: PNPLS-CA; Merced County Library, 2100 O Street, Merced, CA, 95430.

Celtor Chemical Works

Humboldt County

This former chemical facility near Hoopa treated sulfide ore from the nearby Copper Bluff Mine from 1958 to 1962. Celtor Chemical Corporation abandoned the site in 1962. Testing has revealed that the soil and groundwater contain high levels of arsenic, copper, lead, cadmium, and zinc. These heavy metal solutions have penetrated and acidified local streams that drain into the recreational Trinity River, which is the only fishing source for the Hoopa Indian Reservation. About 1,000 people live within 3 miles of the site.

Response actions included removing contaminated material (1983). The area was fenced and the stream flow was diverted from the contaminated area. Additional contaminated soil was removed between 1987 and 1989. Monitoring is continuing.

Sources: PNPLS-CA; Bureau of Indian Affairs, Looped Road, Room 102, Hoopa, CA, 95546.

Coalinga Asbestos Mine

Fresno County

This former asbestos mine near Coalinga operated from 1962 to 1977. Asbestos and chromite were processed in the mill. Some of the mining products and materials were also stored in Coalinga. Mining and milling activities have polluted the soil and air with asbestos. Ten ranchers live within 5 miles of the site; Coalinga has a population of about 8,000 people.

Response actions included site investigations that were completed in 1991. The final remediation plan consisted of diverting surface flow, trapping asbestos behind sediment dams, dismantling the mill, and fencing the site. Contaminated soil and debris at the Coalinga site was removed in 1989. Monitoring will continue.

Sources: PNPLS-CA; Coalinga District Library, 305 North Fourth Street, Coalinga, CA, 93210.

Coast Wood Preserving Company

Mendocino County

This wood treatment facility near Ukiah has operated since 1971, using chromium, copper, and arsenic solutions to preserve timber. Spills and leaks have contaminated the groundwater and soil with these metals. The groundwater in the area is used for industrial, domestic, and agricultural purposes and flows into the Russian River.

Response actions included pumping out the contaminated groundwater and storing it on-site. Surface runoff was also contained through a system of berms. Soils were fixated with cement and capped. Groundwater extraction and treatment will continue for at least 5 more years.

The PRP, Coast Wood Preserving Company, has defied court injunctions to perform the remediation work.

Source: PNPLS-CA.

Concord Naval Weapons Station

Contra Costa County

This U.S. military site near Pleasant Hill is a major ammunition storage and shipping center. Wastes from these operations have been discharged into a tidal wetland area since 1942. Tests have shown that soil, groundwater, and marine sediments are contaminated with VOCs, heavy metals, and PCP. The contaminants have penetrated Suisun Bay, an important commerical and recreational fishing habitat. The salt marsh harvest mouse and black capper rail, two endangered species, are also at risk.

The U.S. Navy has excavated contaminated soils and restored wetlands (1983–1994). Ongoing site investigations will determine a final remediation plan.

Sources: PNPLS-CA; Contra Costa County Library, 1750 Oak Park Boulevard, Pleasant Hill, CA, 94523.

Cooper Drum Company

Los Angeles County

At this site near South Gate steel drums have been reconditioned since 1941. Caustic liquid waste from the Cooper Drum Company property seeped into a nearby school yard in 1987. The school was subsequently closed. Four South Gate water wells were found to be contaminated with VOCs and also closed. Tests have shown that the soil and groundwater contain high levels of VOCs (including vinyl chloride and benzene). The contamination threatens the drinking water supplies for about 340,000 residents in the area.

Response actions included closing Tweedy Elementary School and excavating the contaminated soil. Monitoring wells were installed to check groundwater contamination and flow (1987). An ongoing site investigation by the EPA will recommend a final remediation plan.

Source: PNPLS-CA.

Crazy Horse Sanitary Landfill
Monterey County
This landfill site near Salinas has been in operation since the 1930s. It received a variety of municipal and industrial wastes, including pesticides, vulcanizing by-products, oils, solvents, and VOCs. Tests in 1987 showed that nearby wells are contaminated with VOCs. About 6,000 people use private wells within 3 miles of the landfill.

Response actions included stabilization of sludge, control of runoff, and groundwater extraction and treatment. Seventeen new wells were drilled to control the migration of the contamination plume. A final recommendation on clean up is pending.

Sources: PNPLS-CA; State of California Water Quality Control Board, 82 Higuera Street, Suite 200, San Luis Obispo, CA, 93401.

CTS Printex, Inc.
Santa Clara County
This company built printed circuit boards near Mountain View from 1966 to 1985. Liquid wastes consisting of heavy metal and VOC solutions were discharged into an underground septic system. Testing indicates that the soil and groundwater are contaminated with VOCs, lead, and copper. About 200,000 people within 3 miles of the site use local groundwater, and Permanente Creek is only 400 feet from the site.

Response actions included razing the buildings and excavating contaminated soils and steel tanks (1985–1986). Groundwater extraction and treatment systems were installed in 1987 and 1989.

The PRP, CTS Printex, is participating in the site investigations and remediation work.

Sources: PNPLS-CA; Regional Water Control Board, 2101 Webster Street, Suite 500, Oakland, CA, 94612.

Del Amo Facility
Los Angeles County
Manufacturing at this site 12 miles south of Los Angeles produced synthetic rubber, butadiene, and styrene from 1943 to the 1960s. Liquid wastes were discharged into evaporation ponds on the site. The location has since been developed into an industrial park. In 1984 it was discovered that the groundwater and soil are contaminated with VOCs, PAHs, and free-product petroleum. The contamination threatens the drinking water of about 35,000 people located within 4 miles of the site.

Response actions including a series of site investigations are nearing completion. A final recommendation for a remediation plan will then be made.

Two major PRPs, Shell Oil Company and Dow Chemical Company, have signed an AOC to finance the site investigations.

Sources: PNPLS-CA; Carson Public Library, 151 East Carson Street, Carson, CA, 90745.

Del Norte Pesticide Storage
Del Norte County
This facility near Crescent City operated from 1970 to 1981 and stored pesticides for local agricultural and forestry needs. About 1,600 drums of waste and rinsewater were buried in an unlined pit. Tests showed that the soil and groundwater are contaminated with various pesticides and VOCs. The contamination threatens the water supplies for about 300 local residents.

Response actions included removing contaminated soil (1987) and air-stripping groundwater (1989). EPA site investigations recommended a remediation plan of excavating contaminated soil and treating groundwater via coagulation and sand filtration. In 1992–1993 the EPA installed a pumping well and air injection system to reduce contaminants in the groundwater further.

Sources: PNPLS-CA; Del Norte County Library, 190 Price Mall, Crescent City, CA, 95531.

Edwards Air Force Base
Kern County
Edwards Air Force Base covers over 300,000 acres near the city of Lancaster. Fuel spills and leaks have contaminated the soil and groundwater with VOCs; other areas contain heavy metals from dumped electroplating wastes. About 14,000 base personnel draw drinking water from wells within 3 miles of the site.

Response actions included removing drums and contaminated soil and capping (1984). A groundwater/fuel separation system was installed in 1989. Ongoing site investigations should be completed between 1997 and 1999 and will recommend a final remediation plan.

Sources: PNPLS-CA; Kern County Public Library, Rosamond Branch, 2645 Diamond Street, Rosamond, CA, 93560.

El Toro Marine Corps Air Station
Orange County
This military facility covers about 4,700 acres near the town of El Toro. Hazardous and solid waste from jet fighter maintenance operations have contaminated the soil and groundwater with VOCs, PCBs, acids, fuel, and heavy metals. The polluted groundwater is used for irrigating farmland and threatens the Upper Newport Bay Ecological Reserve.

Response actions have consisted of three site investigations that are nearing completion. A final remediation plan will then be recommended.

Sources: PNPLS-CA; Heritage Park Regional Library, 14361 Yale Street, Irvine, CA, 92714.

Fairchild Semiconductor Corporation (Mountain View)
Santa Clara County
This semiconductor manufacturing site is located in a heavy industrial section of the town of Mountain View. Chemical products, wastewater, and acid neutralization solutions are stored in both underground and aboveground tanks on the property. Leaks and spills have contaminated the soil and groundwater with VOCs, freon, isopropyl alcohol, and heavy metals. About 300,000 people depend on water from wells within 3 miles of the site.

Response actions included removing tanks and contaminated soil, sealing polluted wells, and installing 21 groundwater extraction wells to control the migration of the contamination plume. An EPA-recommended remediation plan in 1989 consisted of soil vapor extraction and groundwater treatment using air-stripping. The design of the remediation systems has recently been completed.

The PRP, through an AOC with the EPA (1985), is required to finance the site investigation and remediation work.

Sources: PNPLS-CA; Mountain View Public Library, 585 Franklin Street, Mountain View, CA, 94041.

Fairchild Semiconductor Corporation (South San Jose)
Santa Clara County
Semiconductors were built on this site from 1977 to 1983. In 1981 an underground solvent waste tank failed, contaminating the soil and groundwater with VOCs. Twenty-two of the 25 nearby private wells were also contaminated and later sealed. The contamination plume threatens the drinking water supplies of nearly 20,000 residents.

Response actions have included sealing contaminated wells, excavating the failed storage tank and contaminated soil, and installing a series of groundwater extraction and treatment systems to control the migration of the contamination plume (all construction was completed by 1992). Groundwater pumping will continue until at least 1999.

Fairchild Corporation has financed the site investigation and remediation work.

Sources: PNPLS-CA; Regional Water Quality Control Board, 2102 Webster Street, Suite 500, Oakland, CA, 94612.

Firestone Tire and Rubber Company
Monterey County
This former tire manufacturing plant near Salinas operated from 1965 to 1980. Production chemicals, including heavy metals and petroleum fuels, were frequently spilled or leaked while the plant was in operation. Liquid industrial wastes were treated on-site. In 1983 tests revealed that the soil and groundwater are contaminated with VOCs. The contamination threatens the drinking water supplies of about 15,000 people within 4 miles of the site.

Response actions included installing groundwater monitoring wells in 1983. Contaminated soil was removed in 1983–1984. The area was capped in 1984–1985. Groundwater treatment systems were installed in 1986–1987. In 1989 the EPA recommended that more groundwater extraction wells be installed. Initial results show that the size of the plume has been significantly reduced. Groundwater treatment will continue until at least 1999.

Firestone Tire and Rubber Company has undertaken the site investigation and remediation work.

Sources: PNPLS-CA; EPA Region 9 Office, 75 Hawthorne Street, San Francisco, CA, 94105.

Fort Ord
Monterey County
This former U.S. Army military base near Monterey was active between 1917 and 1994. Wastes generated from training operations, including oils, waste fuels, automotive chemicals, and ammunition byproducts, were buried in on-site landfills. Testing has shown that the groundwater and soil are contaminated with heavy metals, fuels, and VOCs. About 40,000 people depend on private drinking water wells within 3 miles of the site; the contamination also threatens the nearby Salinas River, Toro Creek, and Monterey Bay.

Response actions included installing soil vapor and groundwater treatment (carbon adsorption) systems in 1988. Contaminated soil and debris were excavated in 1994. Recommendations from ongoing site investigations are expected in 1997.

Sources: PNPLS-CA; Monterey County Free Library, 550 Harcourt Avenue, Seaside, CA, 93955.

Fresno Landfill

Fresno County

This landfill site in Fresno operated from 1935 to 1989 and received a variety of municipal and industrial wastes. Testing by the State of California in 1983 showed that the groundwater, soil, and air are contaminated with methane, carbon dioxide, and VOCs. Nearby off-site monitoring wells also detected the contamination. Nine city wells within 3 miles of the landfill serve about 300,000 people; agricultural wells are located within 3,000 feet of the site.

Response actions included the installation of monitoring wells (1984) and methane barrier walls (1984) and extraction of VOC-bearing gas through a vacuum system (1990–1991). An EPA decision in 1993 called for capping, a gas collection system, runoff controls, and possibly a leachate collection system. Construction will begin in 1997.

A number of PRPs identified by the EPA are participating in the site investigation and remediation work.

Sources: PNPLS-CA; Fresno County Library, 2420 Mariposa Street, Fresno, CA, 93721.

Frontier Fertilizer

Yolo County

This site near the city of Davis was used by two companies, Frontier Fertilizer and Barber and Rowland, Inc., from 1972 to 1987 for storing and processing pesticides, herbicides, and fertilizers. Waste products and rinsewater were discharged in on-site lagoons. Testing revealed high levels of pesticides and VOCs in the soil and groundwater. The contamination has penetrated 3 aquifers in the area. A new residential neighborhood has been built beside the site. The contamination plume threatens the drinking water supplies of about 44,000 people in the Davis region.

Response actions included removing contaminated soil (1985) and installing an emergency groundwater extraction and activated carbon treatment system (1994–1995). Pesticide-contaminated soil will be excavated and treated. Ongoing site investigations may determine that other remedies are warranted.

Sources: PNPLS-CA; Yolo County Library, 315 East Fourteenth Street, Davis, CA, 95616.

GBF Dump

Contra Costa County

This active landfill in the town of Antioch received hazardous waste from the 1960s to 1975, including heavy metal sludges, acids, petrochemicals, asbestos, formaldehyde, pesticides, and VOCs. Testing in 1990 showed that the soil and groundwater are contaminated with cyanide, VOCs, cadmium, nickel, and lead. Sediments from nearby Markley Creek contain very high levels of lead. The contamination threatens the water supplies of about 330,000 people in the area.

Detailed site investigations began in 1993. A final remediation plan should be recommended by 1997.

A number of PRPs identified by the EPA are financing the site investigation.

Source: PNPLS-CA.

George Air Force Base

San Bernardino County

This former U.S. Air Force installation near Victorville operated as a tactical fighter aircraft base until 1992. The soil and groundwater are contaminated with jet fuel, VOCs, and chlorinated solvents (trichloroethylene).

Response actions included installing a groundwater pump-and-treat system (1992), separating free-product jet fuel (1992), and removing contaminated soil (1992). Ongoing site investigations should be completed by 1997.

Sources: PNPLS-CA; San Bernardino Library, 11744 Bartlett Avenue, Adelanto, CA, 92301.

Hewlett-Packard Company

Santa Clara County

Located in Palo Alto, this optoelectronic manufacturer operated from 1962 to 1986. Leaks and spills of waste solvents contaminated the soil and groundwater with VOCs, threatening the water supplies of nearby residents.

Response actions have included removing leaking tanks and contaminated soil (1981) and pumping and treating contaminated groundwater (1982). A site investigation (1989–1995) also recommended soil vapor extraction and treatment.

Hewlett-Packard Company, the PRP, has financed the site investigation and remediation work.

Sources: PNPLS-CA; U.S. Geological Survey Library, 345 Middlefield Road, Menlo Park, CA, 94025.

Industrial Waste Processing

Fresno County

Located near Pinedale, this solvent recycling facility operated from 1967 to 1981. In addition to processing petroleum residues and nonchlorinated solvents, it also recovered lead and zinc from waste solder. Leaks and spills from operations and on-site storage have contaminated the soil and groundwater with solvents, heavy metals, asbestos, and acetone. The contamination threatens over 100 private and municipal wells within 3 miles of the site, which serve about 350,000 residents.

Response actions included removing stored wastes and contaminated soil in 1988. Ongoing site investigations (1992–1997) will recommend a final remediation plan.

The PRPs are financing the site investigation studies.

Sources: PNPLS-CA; Fresno County Library, Main Library, 2420 Mariposa Street, Fresno, CA, 93721.

Intel Corporation (Mountain View Plant)
Santa Clara County
Semiconductors were manufactured at this site in Mountain View from 1968 to 1981. Tests in 1981–1982 showed that the soil and groundwater are contaminated with VOCs, isopropyl alcohol, and freon. About 300,000 people depend on water drawn from within 3 miles of the site.

Response actions included the installation of a groundwater pump-and-treat system (1982) and removal of contaminated soil (1986). Tanks were also removed and contaminated wells sealed. An EPA investigation (1989–1995) recommended soil vapor extraction, slurry wall construction, groundwater extraction and air-stripping, and groundwater monitoring. The design of these remedies has recently been completed.

An AOC (1985) with the EPA requires three PRPs (Intel, Fairchild Semiconductor, and Raytheon) to conduct the site investigation.

Sources: PNPLS-CA; Mountain View Public Library, 585 Franklin Street, Mountain View, CA, 94041.

Intel Corporation (Santa Clara III)
Santa Clara County
Intel Corporation tests chemicals and microprocessors at this site in Santa Clara. The use of chlorinated solvents has contaminated the soil and groundwater with VOCs. About 300,000 people depend on groundwater drawn from within 3 miles of the site.

Response actions included the installation of a groundwater extraction and treatment system (1985). A site investigation in 1990 recommended expanding the system of groundwater extraction wells and using the granular-activated carbon treatment process. Groundwater pumping will continue at least until 1999, with longer-term monitoring.

The PRP will finance the site investigation and remediation work through agreements with the EPA.

Sources: PNPLS-CA; Regional Water Quality Control Board, 2101 Webster Street, Suite 500, Oakland, CA, 94612.

Intel Magnetics
Santa Clara County
Magnetic products were produced and tested at this manufacturing site in the 1980s. Surface spills and leaks from an underground storage tank have contaminated the soil and groundwater with VOCs. About 300,000 people depend on groundwater drawn from within 3 miles of the site.

Response actions included the installation of a groundwater pumping and treatment system (1990). After a site investigation, the series of extraction wells was expanded and the contaminated groundwater treated with carbon adsorption (1992). Groundwater treatment will continue at least until 1999, with longer-term monitoring.

Sources: PNPLS-CA; Santa Clara Library, 2635 Homestead Road, Santa Clara, CA, 95051.

Intersil/Siemens Components
Santa Clara County
Two companies have manufactured semiconductors at this site. Localized spills of waste solvents and leaking underground tanks have contaminated the soil and groundwater with VOCs. Part of a heavily industrialized corridor, the contamination threatens the water supply of about 300,000 people within 3 miles of the site. Calabazas Creek is about 1,500 feet from the site.

Response actions included removing leaking underground tanks and contaminated soil. Groundwater treatment and soil vapor extraction systems were also installed. The EPA recommended expanding the system of extraction wells and using carbon adsorption treatment technology. The soil and groundwater treatment systems will operate at least until 1999, followed by long-term monitoring.

The PRPs, through agreements with the EPA and the State of California, will help finance the site investigation and remediation work.

Sources: PNPLS-CA; Regional Water Quality Control Board, 2101 Webster Street, Suite 500, Oakland, CA, 94612.

Iron Mountain Mine
Shasta County
The Iron Mountain Mine near Redding was operated intermittently from the 1860s to 1963. Iron, silver, copper, gold, zinc, and pyrite were recovered from the mine. The remaining waste rock piles and mine tailings have contaminated the soil, surface water, and groundwater with heavy metals and sulfuric acid. The acidic drainages flow into Spring Creek Reservoir and the Sacramento River, killing aquatic life. Serious fish kills in the Sacramento River have led to the classification of the winter-run chinook salmon as an endangered species. About

50,000 people use the surface water within 3 miles of the site.

Response actions have included installing a lime neutralization system at the mine portal (1988–1989). An EPA decision in 1986 recommended capping selected areas and diverting streams from toxic areas, and capping was completed in 1989. The diversion of Slick Rock Creek, Upper Spring Creek, and the South Fork of Spring Creek are nearly completed (1990–1997). Acid mine drainage is being treated with a treatment plant (1993). Other site investigations are in progress.

Through a series of legal actions (1989–1993), the EPA has ordered the PRPs to conduct site investigations and remediation work.

Sources: PNPLS-CA; Shasta County Library, 1855 Shasta Street, Redding, CA, 96001.

J. H. Baxter & Co.
Siskiyou County
This wood treatment plant near the town of Weed has been in operation since the 1940s. Lumber is soaked in chemical solutions that contain petroleum, creosote, arsenic, copper, chromium, and zinc. In the earlier years of operation toxic sludges and wastewater were deposited on-site in unlined lagoons or sprayed on the ground. Testing has shown that the soil and groundwater contain high levels of PCP, creosote, polynuclear aromatic hydrocarbons (PNAs), arsenic, chromium, and dioxin. About 3,500 people within 3 miles of the site use the local groundwater. The contaminants have penetrated Beaughton Creek, destroying the trout population.

Response actions included fencing the site (1986). An EPA decision in 1989 recommended extracting the groundwater and using biotreatment and chemical precipitation. Contaminated soil is to be excavated and treated biologically or stabilized for reburial. The design of the remediation systems has been completed.

A series of legal actions by the EPA and the State of California requires the PRPs to finance the site investigations and remediation work.

Sources: PNPLS-CA; Weed Public Library, 780 South Davis Street, Weed, CA, 96094.

Jasco Chemical Corporation
Santa Clara County
The Jasco Chemical Corporation has been processing chemical products at this location in Mountain View since 1976. Bulk solvents, including PCP and paint thinners, were stored in underground storage tanks. Leakages and spills have contaminated the soil and groundwater with VOCs, PCP, and diesel fuel. The contamination threatens the water source for about 300,000 nearby residents.

Response actions have included removing contaminated soil and installing groundwater monitoring wells. A groundwater extraction system was installed in 1987. Site investigations from 1991 to 1994 recommended a soil vacuum extraction system to treat the soil contamination.

Jasco Chemical Corporation has worked with the State of California and the EPA in performing the site investigations and remediation work.

Sources: PNPLS-CA; Mountain View Public Library, 585 Franklin Street, Mountain View, CA, 94041.

Jet Propulsion Laboratory
Los Angeles County
Located in Pasadena, NASA's Jet Propulsion Laboratory began operations in 1945. Solid and liquid wastes generated from aeronautic and space technology research, including solvents, rocket fuels, cooling tower chemicals, acid, freon, and mercury, were deposited on-site in seepage pits. Testing in 1990 revealed the soil and groundwater to be contaminated with VOCs, which have also penetrated nearby municipal wells. About 70,000 people use water from domestic wells within 4 miles of the site.

Response actions included shutting down the affected municipal wells until treatment systems (air-stripping towers) could be installed (1987–1990). Recommendations for a final remediation plan from recently completed site investigations are expected by 1997.

Source: PNPLS-CA.

Koppers Company
Butte County
This wood treatment facility near Oroville began operations in 1948. Timbers are treated with solutions containing creosote and heavy metals. Poor handling, leaks, and spills have contaminated the soil and groundwater with creosote, PCP, dioxin, furans, PAHs, copper, chromium, and arsenic. About 11,000 people depend on local groundwater within 3 miles of the site.

Response actions included providing affected residents with clean water (1986) and temporary capping (1987–1988). EPA recommendations in 1989 included groundwater extraction and treatment with carbon adsorption, in-situ soil treatment, capping, and provision of permanent clean water supplies. Construction is in progress.

Consent Orders (1982, 1986) and Consent Decrees (1990) with the State of California and the EPA require the PRP to finance the site investigations and remediation work.

Sources: PNPLS-CA; Butte Public Library, 1820 Mitchell Avenue, Oroville, CA, 95965.

Lawrence Livermore National Laboratory
Alameda County
This 1-square-mile research laboratory managed by the U.S. Department of Energy has been in operation since the 1940s. Activities such as research on nuclear weapons and magnetic fusion generated hazardous wastes on-site. Soil and groundwater are contaminated with VOCs, chromium, lead, radioactive tritium, PCBs, and fuel derivatives. The contamination has penetrated nearby wells and threatens the water supplies of about 50,000 people within 2 miles of the site.

Response actions included supplying clean water to affected residents and excavating areas of contaminated soil. A groundwater pump-and-treat system using ultraviolet light/hydrogen peroxide became operational in 1989. Soil vapor extraction is also planned.

Sources: PNPLS-CA; Livermore Public Library, 1000 South Livermore Avenue, Livermore, CA, 94550.

Liquid Gold Oil Corporation
Contra Costra County
This was the former site of an asphalt plant and waste oil processing facility that operated from the 1940s to 1982. Spills and leaks have contaminated the soil with lead. About 90,000 people live within 4 miles of the site.

Response actions included removing surface oil, storage tanks, and contaminated soil (1974–1984). Plans include capping the site and long-term groundwater monitoring.

Sources: PNPLS-CA; Richmond Public Library, 325 Civic Center Plaza, Richmond, CA, 94804.

Lorentz Barrel and Drum Company
Santa Clara County
This drum recyling operation near San Jose was in business from 1950 to 1987. Many of the drums contained residual toxic waste that was rinsed out and discharged into ditches and ponds on the property. Tests have shown that the soil and groundwater are contaminated with heavy metals, VOCs, PCBs, and pesticides. Three municipal well fields and about 3,000 residents are within 1 mile of the site.

Response actions included removing over 26,000 drums and contaminated soil. Buildings and debris were removed in 1994, and the site was paved and fenced. In 1993 the EPA recommended soil vapor extraction, soil gas monitoring, and land use restrictions. Groundwater extraction and treatment using an ozone/ultraviolet process became operational in 1992.

A 1990 Consent Decree and a 1992 Administrative Order on Consent require approximately 20 PRPs to finance the site studies and remediation work.

Sources: PNPLS-CA; San Jose State University, One Washington Square, San Jose, CA, 95192.

Louisiana-Pacific Corporation
Butte County
A wood processing plant and landfill are located on this site near Oroville. Prior to 1980 wastewater, glue by-products, and other chemical solutions were discharged into an unlined pond. Tests in 1973 showed neighboring wells to be contaminated with PCPs. The groundwater and soil contain high levels of PCPs, VOCs, dioxins, arsenic, formaldehyde, PAHs, ether, boron, and copper. About 11,000 people within 3 miles of the site use the local groundwater for drinking purposes. The contamination also threatens the Feather River.

Response actions included fencing (1990) and a temporary groundwater treatment system (1996). Ongoing site investigations will result in a final remediation plan.

Sources: PNPLS-CA; Butte County Public Library, 1820 Mitchell Avenue, Oroville, CA, 95965.

March Air Force Base
Riverside County
This 7,000-acre U.S. Air Force base near Riverside has been used for aircraft maintenance and repair since 1918. Waste products were discharged on-site in pits, lagoons, and underground tanks. Testing in 1984 showed that the groundwater and soil contained high levels of VOCs and heavy metals. The contamination has penetrated neighboring wells. About 12,000 people use water from wells within 3 miles of the base.

Response actions included shutting down contaminated wells and providing affected residents with bottled water (1988). A groundwater extraction and treatment system was installed in 1992. Leaking tanks, contaminated soil, and free-product petroleum were also removed (1992–1993). A site investigation in progress will recommend a final remediation plan.

Sources: PNPLS-CA; Riverside Public Library, Moreno Valley Branch, 25480 Alessandro Boulevard, Moreno Valley, CA, 92388.

Mather Air Force Base
Sacramento County
This site operated as a U.S. Air Force base from 1918 to 1993. Toxic and hazardous chemicals such

as solvents and lubricants were used in maintaining aircraft. These chemicals and wastewaters were discharged on the ground or into unlined lagoons and pits. Testing in 1982 showed that the soil and groundwater are contaminated with VOCs, fuel, pesticides, and heavy metals. Nearby private wells have been contaminated with VOCs. About 60,000 people within 3 miles of the base use the local groundwater.

Response actions included connecting affected residents to clean water supplies (1984–1987). A groundwater extraction and air-stripping system was installed in 1995. Site investigations regarding soil contamination are still in progress.

Sources: PNPLS-CA; Sacramento Central Library, 828 I Street, Sacramento, CA, 95814.

McClellan Air Force Base
Sacramento County
This U.S. Air Force installation near Sacramento covers about 3,000 acres. Hazardous wastes, solvents, and radioactive materials from the maintenance of planes, electronics, and communications equipment were discharged on-site in unlined containment areas. The groundwater and soil are contaminated with VOCs. The contamination has penetrated a well on the base, which was shut down. About 23,000 people live within 3 miles of the site.

Response actions included removing contaminated soil, capping, and installing a groundwater extraction system to contain the plume of contaminants. Underground storage tanks were removed. A carbon filtration system was installed on the closed well, and about 550 nearby residents were connected to clean water supplies. Soil vacuum extraction has also been initiated. Site investigations by the EPA are nearing completion and will recommend a final remediation plan.

Sources: PNPLS-CA; Sacramento Public Library, 536 Downtown Plaza, Sacramento, CA, 95814.

McColl
Orange County
This site near Fullerton was once a refinery waste disposal facility that operated from the 1940s to the 1960s. Various wastes, including acidic sludges, were buried in unlined pits on the property. The area has since been developed into a golf course and residential neighborhood. The soil, air, and groundwater are contaminated with thiophene, sulfur dioxide, and VOCs. A tarlike leachate continues to seep from parts of the surface. About 7,000 people live within 3 miles of the site.

Response actions included a temporary cover (1983) and fencing. Toxic sludges were removed in 1984. EPA studies from 1984 to 1993 recommended that the contaminated soil be partially solidified and capped. An ongoing groundwater study is nearing completion.

Sources: PNPLS-CA; Fullerton Public Library, 353 West Commonwealth Avenue, Fullerton, CA, 92632.

McCormick and Baxter
San Joaquin County
This wood preservation facility operated near Stockton from 1942 to 1990. Timbers were treated with creosote, PCP, and arsenic and copper compounds. Waste oils and compounds were discharged on-site in unlined ponds. Runoff from the site caused a fish kill in the Mormon Slough and the Port of Stockton in 1977. Testing in 1983–1984 revealed that soil and groundwater contain high levels of arsenic, chromium, copper, PCP, and PAHs.

Response actions included installing runoff collection and groundwater extraction systems. The site was fenced and buildings demolished and removed between 1992 and 1994. EPA studies are nearing completion and will lead to a final remediation plan.

The owner, McCormick and Baxter, filed for bankruptcy in 1988; the clean up will be financed by federal funds.

Sources: PNPLS-CA; Stockton Public Library, 605 North El Dorado Street, Stockton, CA, 95203.

MGM Brakes
Sonoma County
Aluminum brake components have been manufactured at this site since 1965. Wastewater containing PCB-bearing hydraulic fluid and ethylene glycol were discharged into a field on-site. Testing in 1981 showed that the air, soil, and groundwater are contaminated with PCBs and VOCs. The contamination has penetrated Icaria Creek and the Russian River and threatens the water supplies of the 4,300 residents of Cloverdale.

Site investigations from 1981 to 1988 recommended excavating contaminated soil and groundwater monitoring. Soil clean up has been completed. The contamination plume in the groundwater is naturally attenuating and is expected to dissipate within 7 to 10 years.

A Consent Decree (1989) required MGM Brakes to fund the site investigations and remediation work.

Sources: PNPLS-CA; Cloverdale Regional Library, 401 North Cloverdale Boulevard, Cloverdale, CA, 95425.

Modesto Groundwater Contamination
Stanislaus County
This municipal well field in the city of Modesto has been contaminated from dry cleaning facilities that

have operated in the area for 50 years. Laundry compounds, including tetrachloroethylene, and naturally occurring uranium have contaminated the soil and groundwater. The contamination threatens the water supplies of the city of Modesto.

Response actions included applying carbon treatment systems to contaminated wells. The site was also paved to minimize human contact with the soil. A soil vapor extraction system was installed in 1990. Ongoing site investigations of the groundwater contamination should be completed by 1997.

Sources: PNPLS-CA; Stanislaus County Free Library, 15000 I Street, Modesto, CA, 95354.

Moffett Naval Air Station

Santa Clara County

This U.S. Navy facility near San Jose maintains aircraft operations that support submarine patrols. Solid and liquid wastes, including acids, PCBs, solvents, and fuels, have been discharged in landfills and sumps on the base. Testing has shown that the soil and groundwater are contaminated with VOCs. About 300,000 people depend on the local groundwater. The contamination also threatens the estuary habitats of San Francisco Bay.

Response actions included sealing contaminated wells, bioremediation of contaminated soil, and the extraction and carbon adsorption of groundwater. Ongoing site investigations are nearing completion and will recommend a remediation plan by 1997.

Sources: PNPLS-CA; Mountain View Public Library, 585 Franklin Street, Mountain View, CA, 94041.

Monolithic Memories

Santa Clara County

Integrated circuits and other computer parts have been manufactured at this site in Sunnyvale since the 1970s. Solvent wastes leaked from pipes and underground storage tanks. Testing in 1982 showed that the groundwater and soil are contaminated with VOCs. Part of a polluted industrial corridor, this site contributes to the endangerment of the water supply for over 300,000 people within 3 miles of the site.

Response actions included removing leaking tanks and extracting and treating contaminated groundwater (1986). A soil vapor extraction system was installed in 1995–1996.

Sources: PNPLS-CA; Santa Clara Library, 2635 Homestead Road, Santa Clara, CA, 95051.

Montrose Chemical Corporation

Los Angeles County

This location near Torrance was the site of a plant that manufactured, packaged, and distributed the pesticide DDT from 1947 to 1982. Liquid wastes and wastewater containing DDT and monochlorobenzene were discharged into a settling pond on the site. Testing indicated that the groundwater and soil are contaminated with DDT, monochlorobenzene, and VOCs. The contaminants have penetrated nearby wells and have been found in the soil in neighboring yards. The polluted groundwater threatens 4 aquifers in the area that supply water to the residents of Torrance and Carson.

Response actions included paving the site (1985). Montrose Chemical Corporation excavated contaminated backyards in the area in 1994. Site investigations are in progress that will recommend a final remediation plan.

Administrative Orders (1983, 1985, 1987, 1989) and other legal actions against Montrose Chemical Corporation require the PRP to finance the site investigations and remediation plan.

Sources: PNPLS-CA; Carson Public Library, 151 East Carson Street, Carson, CA, 90745.

National Semiconductor Corporation

Santa Clara County

This company makes electronic equipment at its facility in Santa Clara. Spills and leaks from underground storage tanks, sumps, and pipes have contaminated the soil and groundwater with VOCs. This site is located in a heavily industrialized corridor, contaminants from which threaten the water supplies of about 300,000 people within 3 miles of the site.

Response actions included excavating leaking tanks and removing contaminated soil as well as a long-term program of groundwater extraction and treatment (1982). The installation of a soil vapor extraction system is in progress.

Sources: PNPLS-CA; Santa Clara City Library, 2635 Homestead Road, Santa Clara, CA, 95051.

Newmark Groundwater Contamination

San Bernardino County

Two plumes of groundwater contamination (VOCs) were identified in 1980 in a part of the aquifer that underlies the city of San Bernardino. Twenty supply wells within 6 miles of the city were closed; 12 were reactivated with air-stripping towers to remove the VOCs. The contaminants have penetrated about 25 percent of the San Bernardino water supply. A former U.S. Army military base, the 1600-acre Camp Ono, and current light industry in the area are suspected sources of the VOCs. The contamination could ultimately affect the water supplies for the cities of San Bernardino, Riverside, Colton, Loma Linda, Fontanta, and Rialto (about 600,000 people combined).

Response actions included a number of EPA studies that tried to locate the sources of contamination (1990–1993). The EPA recommended an extensive groundwater extraction and carbon adsorption treatment system, which became operational in 1996. About 30 million tons of groundwater per day will be pumped and treated until the aquifer is clean. Investigations are continuing at the former military installation. A remediation plan for this site is expected in 1997.

Sources: PNPLS-CA; San Bernardino Library, 104 West Fourth Street, San Bernardino, CA, 92415.

Norton Air Force Base
San Bernardino County
This 2,000-acre U.S. Air Force base near San Bernardino overhauled and repaired aircraft from 1942 to 1994. Hazardous solid and liquid wastes were buried or discharged on the site, including waste oils, solvents, paint residues, gasoline, and acids. Low-level radioactive waste may also be buried on the site, which is being sold to the private sector for development. Testing has shown that groundwater and soil are contaminated with VOCS, PCBs, chromium, arsenic, and copper.

Response actions included removing sludge drying beds (1986) and underground storage tanks (1989). A groundwater pump-and-treat system was installed in 1992 to control the migration of the contamination plume. Site investigations will recommend a final remediation plan by 1997.

Sources: PNPLS-CA; Feldhym Central Library, 555 West Sixth Street, San Bernardino, CA, 92410.

Operating Industries Landfill
Los Angeles County
This landfill site near Monterey Park operated from 1948 to 1984, receiving a variety of municipal, industrial, and hazardous wastes. Testing has shown that the air, groundwater, and soil contain high levels of VOCs.

Response actions included fencing the site, rehabilitating the gas collection system, and controlling erosion. EPA site investigations (1989–1996) will recommend a final groundwater remediation plan. A leachate treatment system became operational in 1991. A new gas control system is also being installed.

The EPA has identified over 100 PRPs. Consent Decrees (1989, 1992) require the PRPs to finance a major portion of the remediation work.

Sources: PNPLS-CA; Montebello Library, 1550 West Beverly Boulevard, Montebello, CA, 90640.

Pacific Coast Pipelines
Ventura County
This oil refinery near the town of Fillmore operated from 1920 to 1952. Refinery wastes were discharged into unlined pits and lagoons on the property. Tests between 1980 and 1984 revealed that the soil and groundwater are contaminated with heavy metals and VOCs. About 10,000 people use local water wells within 3 miles of the site; the contamination also threatens the nearby Santa Clara River.

Response actions included excavating soil from the main pit (1986). A groundwater extraction and treatment system using solvent extraction was installed in 1994–1995. Groundwater monitoring is continuing.

Sources: PNPLS-CA; Fillmore City Hall, 524 Sespe Road, Fillmore, CA, 93015.

Purity Oil Sales
Fresno County
This oil recycling facility near Fresno operated from about 1934 to 1974. Petrochemical by-products and wastewater were discharged into sumps or pits on the property. Testing has shown that the groundwater and soil are contaminated with VOCs, zinc, iron, manganese, lead, phenols, PCBs, pesticides, oil, and grease. The contamination threatens the San Joaquin River drainage basin and the water supplies of hundreds of residents within a mile of the site.

Response actions included capping with concrete (1984), excavating waste and drums (1985), pumping out oil and gas, and fencing the site (1987). Affected residents were connected to clean water supplies. The remaining tanks were taken out in 1991. Soil vapor extraction technology was installed in 1993. The EPA recommended a groundwater extraction and treatment system using "greensand" technology to filter iron and manganese, which went into operation in 1995.

Sources: PNPLS-CA; Fresno County Library, 2420 Mariposa Street, Fresno, CA, 93721.

Ralph Gray Trucking
Orange County
Highly acidic petroleum refinery sludges were buried in unlined pits on this property near Westminster from the 1930s to the 1950s. Homes were later built on the site, and the new residents noticed an oozing, black substance in their yards and swimming pools; some became ill. Testing in 1986 indicated that the soil and groundwater had high levels of VOCs, petrochemicals, PAHs, and sulfuric acid.

Response actions included periodically removing the tarry leachate (1986–1992), excavating contaminated soil, and restoring the yards of 30 homes (1992). Additional contaminated soil was removed between 1994 and 1996. A soil vapor monitoring system is presently being installed. A groundwater site investigation is nearing completion and will recommend a groundwater treatment plan.

Sources: PNPLS-CA; Westminster Public Library, Westminster, CA, 92684.

Raytheon Corporation

Santa Clara County

Raytheon Corporation builds semiconductors at this 30-acre site near Mountain View. Tests by the State of California and the EPA in 1981 and 1982 showed that the soil and groundwater are contaminated with VOCs from leaks, spills, or direct waste discharge into a nearby creek. This site is part of a heavily industrialized corridor, contamination from which threatens the drinking water for about 300,000 people living within 3 miles of the site.

Response actions included sealing contaminated wells, removing tanks and contaminated soil, building slurry walls to control runoff, and installing a groundwater treatment system using air-stripping towers. A soil vapor extraction system will also be installed. All construction should be completed by 1997.

The EPA issued an AOC in 1985 requiring Raytheon Corporation to conduct the site investigation and remediation work.

Sources: PNPLS-CA; Mountain View Public Library, 585 Franklin Street, Mountain View, CA, 94041.

Riverbank Army Ammunition Plant

Stanislaus County

This U.S. Army facility near Modesto has manufactured ammunition since 1951. These operations generated corrosive wastes and toxic wastewater containing heavy metals that were discharged into unlined holding ponds on the property. Testing has shown that the soil and groundwater contain high levels of chromium, cyanide, zinc, and petroleum. The contaminants have penetrated the recreational Stanislaus River and threaten the drinking water for about 14,000 nearby residents.

Response actions included supplying affected residents with clean drinking water (1989) and installing a groundwater treatment system that removes chromium and cyanide (1991). By 1992 all affected residents were connected to permanent city water supplies. In 1994 the EPA recommended a comprehensive groundwater extraction and treatment system and the stabilizing and capping of contaminated soil. Design of the remediation systems has recently been completed.

The U.S. Army, the EPA, and the State of California signed a Federal Facility Agreement in 1990 that requires the army to conduct the site investigation and remediation work.

Sources: PNPLS-CA; Stanislaus County Library, 3442 Santa Fe Street, Riverbank, CA, 95367.

Sacramento Army Depot

Sacramento County

This 500-acre site in Sacramento has operated as a supply depot for the storage and maintenance of various electronic supplies since 1945. Wastes, including metal-plating sludges, degreasing solutions, and contaminated rinsewater, were discharged or buried on-site. Tests have shown that the soil and groundwater are contaminated with VOCs and heavy metals, including lead, chromium, and cadmium. The contamination has been detected in nearby Morrison Creek and threatens the drinking water of about 50,000 people who live within 3 miles of the site.

Response actions included a groundwater treatment operation (1990), soil vapor extraction (1994), soil washing (1995–1996), and solidification and capping of contaminated soil (1996–1998). Site investigations that will recommend additional groundwater extraction and treatment at least until 2001 are in progress.

Sources: PNPLS-CA; Sacramento Army Depot, 8350 Fruitridge Road, Sacramento, CA, 95813.

San Fernando Valley (Area 1)

Los Angeles County

This 9,400-acre area of groundwater contamination is located near North Hollywood in Los Angeles and Burbank. VOCs were detected in the water in 1980. Subsequent studies showed that nearly half of the supply wells were contaminated; these were later shut down or blended with clean water sources. About three million people live within 3 miles of the sprawling contamination plume, which is thought to have been caused by unregulated discharge of chemical wastes between the 1940s and 1960s.

Response actions included the installation of a groundwater extraction and treatment system using aeration and granulated carbon filters. This system became active in 1989 at a rate of about 15 million gallons of water per day and will continue for at least 10 years.

Sources: PNPLS-CA; Burbank Public Library, 110 North Glen Oaks Boulevard, Burbank, CA, 91502.

San Fernando Valley (Area 2)

Los Angeles County

This 6,700-acre area of groundwater contamination covers the Crystal Springs Well Field for the cities of Los Angeles and Glendale. VOCs were detected in the water in 1980. Subsequent studies showed that nearly half of the supply wells were contaminated; these were later shut down or blended with clean water sources. About three million people live within 3 miles of the contamination plume, which is thought to have been caused by unregulated discharge of chemical wastes between the 1940s and 1960s.

Response actions included the installation of a groundwater extraction and treatment system using aeration and granulated carbon filters. This system became active in 1989 and will operate for at least 10 years.

Sources: PNPLS-CA; Glendale Public Library, 222 East Harvard Street, Glendale, CA, 91205.

San Fernando Valley (Area 3)
Los Angeles County
This 5,200-acre area of groundwater contamination covers the Glorietta Well Field that services the city of Glendale. VOCs were detected in the water in 1980. Subsequent studies showed that nearly half of the supply wells were contaminated; these were later shut down or blended with clean water sources. About three million people live within 3 miles of the contamination plume, which is thought to have been caused by unregulated discharge of chemical wastes between the 1940s and 1960s.

Response actions included a 1993 EPA evaluation that recommended an interim groundwater extraction and treatment plan until a larger-scale study can be completed.

Sources: PNPLS-CA; Burbank Public Library, 110 North Glen Oaks Boulevard, Burbank, CA, 91502.

San Fernando Valley (Area 4)
Los Angeles County
This 5,800-acre area of groundwater contamination covers the Pollock Well Field that serves the city of Los Angeles. VOCs were detected in the water in 1980. Subsequent studies showed that nearly half of the supply wells were contaminated and later shut down or blended with clean water sources. About three million people live within 3 miles of the contamination plume. The contamination is thought to be caused by unregulated discharge of chemical wastes between the 1940s and 1960s.

Response actions included a 1993 EPA evaluation that recommended an interim groundwater extraction and treatment plan until a larger-scale study can be completed.

Sources: PNPLS-CA; Burbank Public Library, 110 North Glen Oaks Boulevard, Burbank, CA, 91502.

San Gabriel Valley (Area 1)
Los Angeles County
This 6-square-mile area of groundwater contamination affects numerous cities and towns of the San Gabriel Valley. Testing between 1979 and 1984 showed that the groundwater is contaminated with VOCs. The contaminated wells were later shut down or blended with clean water sources. The valley's aquifer supplies groundwater to about one million people, and over 400 supply wells are used for industrial and agricultural purposes. The VOCs have likely been discharged from hundreds of industrial sources that have operated in the San Gabriel Valley since the 1940s.

Response actions included groundwater monitoring and constructing a water treatment plant (1992). Recently completed EPA site investigations (1984–1996) will issue recommendations regarding soil and groundwater treatment plans.

Sources: PNPLS-CA; Whittier Central Library, 7344 South Washington Street, Whittier, CA, 90602.

San Gabriel Valley (Area 2)
Los Angeles County
This 1-square-mile area of groundwater contamination along Alhambra Creek affects numerous towns in the San Gabriel Valley. Testing between 1979 and 1984 showed that the groundwater is contaminated with VOCs. The contaminated wells were later shut down or blended with clean water sources. The valley's aquifer supplies groundwater to about one million people, and over 400 supply wells are used for industrial and agricultural purposes. The VOCs have likely been discharged from hundreds of industrial sources that have operated in the San Gabriel Valley since the 1940s.

Response actions included groundwater monitoring. EPA site investigations (1984–1996) have recommended the construction of a groundwater pump-and-treat system that will process about 27 million gallons of water per day.

From an original list of 1,600 PRPs, the EPA has identified 17 companies as being major polluters targeted for cost recovery.

Sources: PNPLS-CA; Whittier Central Library, 7344 South Washington Street, Whittier, CA, 90602.

San Gabriel Valley (Area 3)
Los Angeles County
This area of groundwater contamination along Alhambra Creek affects numerous towns in the San Gabriel Valley. Testing between 1979 and 1984 showed that the groundwater is contaminated with VOCs. The contaminated wells were later shut down or blended with clean water sources. The valley's aquifer supplies groundwater to about one million people, and over 400 supply wells are used for industrial and agricultural purposes. The VOCs have likely been discharged from hundreds of industrial sources that have operated in the San Gabriel Valley since the 1940s.

Response actions have consisted of recently completed EPA site investigations (1984–1996) that will recommend a final soil and groundwater remediation plan.

The EPA has identified at least 400 PRPs that have been targeted for cost recoveries.

Sources: PNPLS-CA; Whittier Central Library, 7344 South Washington Street, Whittier, CA, 90602.

San Gabriel Valley (Area 4)
Los Angeles County
This area of groundwater contamination along San Jose Creek in La Puente affects numerous towns in the San Gabriel Valley. Testing between 1979 and 1984 showed that the groundwater is contaminated with VOCs. The contaminated wells were later shut down or blended with clean water sources. The valley's aquifer supplies groundwater to about one million people, and over 400 supply wells are used for industrial and agricultural purposes. The VOCs have likely been discharged from hundreds of industrial sources that have operated in the San Gabriel Valley since the 1940s.

Response actions have consisted of recently completed EPA-PRP site investigations (1984–1996) that will recommend a final soil and groundwater remediation plan.

Sources: PNPLS-CA; Whittier Central Library, 7344 South Washington Street, Whittier, CA, 90602.

Selma Treating Company
Fresno County
This wood treatment plant near Selma has been in operation since 1936. Timber is soaked in various preservatives, including PCP and oil and copper- and arsenic-based solutions. Chemical wastes and sludges were discharged into ditches, dry wells, and an unlined pond on the property. Testing in 1971 showed that the soil and groundwater are contaminated with PCP, VOCs, and heavy metals (chromium, arsenic, and copper). The contamination threatens the drinking water of at least 10,000 people within 1 mile of the site.

Runoff controls were constructed in 1982. EPA studies conducted between 1988 and 1992 recommended groundwater extraction and treatment, excavation and solidification of contaminated soil, capping, restricted groundwater use, and long-term groundwater monitoring. Construction is nearing completion.

Sources: PNPLS-CA; Fresno County Library, Selma Beach, 2200 Selma Street, Selma, CA, 93662.

Sharpe Army Depot
San Joaquin County
This U.S. Army facility near Lathrop stored, shipped, and maintained general supplies from 1941 to 1975. Various wastes, including paints, sludges, solvents, phenols, and polychlorinated hydrocarbons were buried on-site. The soil and groundwater contain high levels of VOCs, including trichloroethylene. About 34,000 people within 3 miles of the site depend on the local groundwater.

Response actions included building interim groundwater extraction systems in 1987 and 1990. Final recommendations for clean up from site investigations conducted by the U.S. Army are expected by 1997.

Sources: PNPLS-CA; Manteca Library, 320 West Center Street, Manteca, CA, 95336.

Sola Optical USA
Sonoma County
This 35-acre property in Petaluma has been the site of an optical lens company since 1978. An on-site well and soil were identified in 1982 as being contaminated with VOCS and acetone leaked from underground storage tanks. Solvents were also discovered in a nearby city well, which was shut down. The contamination threatens the water supplies of nearby residents.

Response actions included the removal of the leaking tanks in 1985. Affected residents were connected to the city of Petaluma's water system. Groundwater extraction wells were installed in 1988. Site investigations (1989–1991) recommended additional groundwater extraction wells, which were built in 1992, and treatment by carbon adsorption. Groundwater treatment will continue for at least another 5 years.

Through agreements with the State of California (1985, 1987), Sola Optical USA has financed the site investigations and remediation work.

Sources: PNPLS-CA; Petaluma Public Library, 100 Fairgrounds Drive, Petaluma, CA, 94952.

South Bay Asbestos Area
Santa Clara County
This landfill site on the southern edge of San Francisco Bay operated from 1953 to the 1980s and received considerable asbestos waste. Asbestos-cement pipes and asbestos-contaminated soil taken from the dump were used by local residents in their yards. Flooding in 1983 spread the asbestos contaminants throughout the town of Alviso. Testing has shown that the soil and groundwater contain asbestos. A number of nearby residents draw water from private wells; the contamination also threatens wetlands in an adjacent wildlife refuge.

Response actions included removing asbestos waste (1983) and paving streets and parking lots that contained asbestos (1986–1987). An EPA site investigation in 1992 recommended additional paving, monthly wet-sweeping of Alviso streets, and deed restrictions (1993–1995). A wetlands restoration project is also planned.

Sources: PNPLS-CA; San Jose Public Library, Alviso Branch, 1060 Taylor Street, Alviso, CA, 95002.

Southern California Edison Company, Visalia Poleyard

Tulare County

A ultility pole treatment plant operated at this site in Visalia from the 1920s to 1980. The poles were soaked in creosote and PCP-bearing solutions. Leaking tanks, spills, and the on-site storage of treated poles on the ground have contaminated the groundwater and soil with PCP and creosote. About 40,000 people live within 5 miles of the site; 35 municipal drinking water wells within 3 miles of the site serve the residents of Visalia.

Response actions included construction of a slurry wall, removal of contaminated soil, and the installation of a groundwater pump-and-treat system (1977). Site investigations by the State of California and the EPA in 1994 recommended bioremediation of contaminated soil, deed restrictions, and soil capping. Construction of these remedies should be under way in 1997–1998.

Legal agreements between Southern California Edision and the State of California (1976, 1987) require the PRP to finance the site investigation and remediation work.

Source: PNPLS-CA.

Spectra-Physics

Santa Clara County

Electronic equipment and gas lasers have been manufactured on this site in Mountain View since 1961. Leaking tanks, spills, and improper storage of hazardous solutions and waste have contaminated the soil and groundwater with VOCs. The contaminants have penetrated Permanente Creek and threaten the groundwater supplies of about 200,000 people within 3 miles of the site. Nearly 200 private drinking water wells have been closed.

Response actions included connecting affected residents to clean water supplies (1982–1985), soil excavation (1987), and the installation of a 20-well groundwater pump-and-treat system (1988). An EPA site investigation in 1991 recommended soil vapor extraction, continued groundwater treatment, and groundwater monitoring. The soil vapor extraction has been completed.

The PRP, Spectra-Physics, is financing the site investigation and clean-up work.

Sources: PNPLS-CA; Mountain View Public Library, 585 Franklin Street, Mountain View, CA, 94041.

Stoker Company

Imperial County

Stoker Company has furnished agricultural pesticide services since 1966. Waste rinsewaters from cleaning equipment are disposed of by spraying them on this 20-acre site. Wildlife kills at a nearby pond were noted in 1988. Testing has shown that the air, surface water, soil, and groundwater are contaminated with pesticides. About 150 people live within 1 mile of the site; the contamination also threatens fisheries and wetlands in the area.

Response actions included closing the contaminated pond and removing contaminated soil (1988). An EPA site investigation is in progress.

Source: PNPLS-CA.

Stringfellow

Riverside County

This hazardous waste disposal facility near Glen Avon operated from 1956 to 1972. It received a variety of waste, including sludges from metal finishing and pesticide manufacturing. Testing has shown that the groundwater and soil are contaminated with VOCs, heavy metals, pesticides, PCBs, and sulfates. The wastes have penetrated nearby waterways. About 8,000 people live in Glen Avon; over 200 homes use private water wells.

Response actions included suppling bottled water to affected residents; removing millions of gallons of liquid waste (1975–1980); installing groundwater extraction wells (1980–1984); and soil capping, fencing, and installing a leachate collection system (1984). Affected residents were connected to the city water system in 1989. Groundwater treatment is through lime precipitation and carbon filtration. As of 1992 nearly 50 million gallons of groundwater had been treated. Soil vapor extraction is also being considered.

The Department of Justice and the EPA sued 23 PRPs in 1983. Administrative Orders on Consent (1988, 1992) and a Consent Decree (1992) require the parties to finance the remediation work.

Sources: PNPLS-CA; Riverside Public Library, 3581 Seventh Street, Riverside, CA, 92509.

Sulphur Bank Mercury Mine

Lake County

Mercury was mined at this abandoned site near Clear Lake intermittently from 1865 to 1957. Rock dumps and an open pit remain exposed on the site, and mine tailings extend into Clear Lake. Lake sediments and fish contain high concentrations of mercury and arsenic. The water of Clear Lake is part of the municipal drinking water supply for the 5,000 residents of the town of Clear Lake. The contamination also threatens nearby endangered species habitats, including the bald eagle and peregrine falcon.

Response actions included regrading waste slopes to minimize runoff into Clear Lake. EPA site investigations that will recommend a final remediation plan are nearing completion.

Sources: PNPLS-CA; Lake County Library, 1425 North High Street, Lakeport, CA, 95453; Redbud Library, 4700 Golf Avenue, Clearlake, CA, 95422.

Synertek Building One
Santa Clara County
Electronics were manufactured on this site in Santa Clara from 1978 to 1986. Underground acid neutralization tanks leaked VOCs into the soil and groundwater. About 300,000 people use water from wells within 3 miles of the site.

Response actions included removing the leaking tanks (1985) and installing a groundwater extraction and air-stripping system.

Synertek and Honeywell Corporation have undertaken the site investigations and remediation work.

Sources: PNPLS-CA; Regional Water Quality Control Board, 2101 Webster Street, Suite 500, Oakland, CA, 94612.

T. H. Agriculture and Nutrition Company
Fresno County
Companies occupying this site manufactured pesticides from 1950 to 1981. Pesticide-tainted waste solutions were discharged into landfill pits and trenches on the 5-acre property. Testing by the State of California has shown that the soil and groundwater are contaminated with chloroform and pesticides. About 35,000 people use the local groundwater for domestic purposes.

Response actions included excavating contaminated soil, connecting affected residents to the city water supply, and installing a soil vapor extraction system. A groundwater investigation in progress is expected to recommend a remediation plan by 1997.

One of the PRPs, T. H. Agriculture and Nutrition Company, has participated in the site studies and remediation work.

Sources: PNPLS-CA; Fresno County Library, 5547 East Kings Canyon Road, Fresno, CA, 93727.

Teledyne Semiconductor
Santa Clara County
Teledyne Corporation has built semiconductors on this 1-acre property near Mountain View since 1962. Testing in 1982 revealed that leaking underground storage tanks had contaminated the soil and groundwater with VOCs, which have penetrated about 50 nearby wells. About 200,000 people live within 3 miles of the site and use the local groundwater for domestic purposes.

Response actions included an EPA site investigation (1991) that recommended excavating contaminated soil, extracting and air-stripping groundwater, and long-term monitoring. All construction has been completed, and groundwater treatment and monitoring are ongoing.

Teledyne has been an active participant in the site investigations and remediation work.

Sources: PNPLS-CA; Mountain View Public Library, 585 Franklin Street, Mountain View, CA, 94041.

Tracy Defense Depot
San Joaquin County
This U.S. Army depot near the town of Tracy has been in operation since 1942. Activities have included processing chemicals, fumigating, maintaining vehicles, operating chemical and photographic laboratories, and storing chemicals, reagents, and pesticides. Wastes were discharged or buried on-site in pits, lagoons, or sewer systems. Testing in 1985 revealed that the soil and groundwater are contaminated with VOCs, heavy metals, and petroleum products. About 50,000 people depend on water pumped from the aquifer within 3 miles of the site.

Response actions included the installation of a temporary groundwater extraction system. In 1993 the EPA decided the groundwater extraction and treatment system should be expanded to process the entire contamination plume. A detailed site investigation, which began in 1991, is nearing completion.

Source: PNPLS-CA.

Travis Air Force Base
Solano County
This U.S. Air Force base near Fairfield has been in operation since 1943. Various waste products generated by military operations include cyanide, VOCs, solvents, paint strippers, pesticides, and jet fuel. Testing has shown high levels of VOCs, PAHs, and heavy metals in the soil and groundwater. The contaminants have caused fish kills in Union Creek. About 400 people within 1 mile of the base use the local groundwater; the contamination also threatens nearby Suisun Marsh, an important coastal wetland and migratory bird habitat.

Response actions included the removal of underground storage tanks in 1986. Ongoing site investigation studies (1990–1998) will recommend a final remediation plan.

Sources: PNPLS-CA; Fairfield Library, 1150 Kentucky Street, Fairfield, CA, 94533; Mitchell Memorial Library, 510 Travis Avenue, Travis Air Force Base, CA, 94535.

Treasure Island Naval Station, Hunters Point Annex
San Francisco County
This 900-acre military facility near San Francisco has been in operation since 1936, mostly as a shipyard. The U.S. Navy began leasing metal shops and other buildings to private tenants in 1976. Waste products from operations such as oil reclamation, transformer processing, metal plating, and battery refurbishing were discharged or dumped into unlined pits or lagoons on the property. Testing in 1987 revealed that the soil and groundwater are contaminated with fuels, heavy metals, PCBs, pesticides, asbestos, and VOCs. The contamination threatens the various water resources within a few miles of the site, including a bottled water operation.

Response actions included removing contaminated drums, debris, and soil (1988–1989). Asbestos-containing material was excavated in 1990. Additional contaminated soil was removed in 1993–1994. Bioremediation of soils and groundwater extraction and treatment systems are being considered by ongoing site investigations.

Source: PNPLS-CA.

TRW Microwave
Santa Clara County
Begun in 1968, operations at this site in Sunnyvale have consisted of building microwave components and semiconductors. Waste solvents, acids, and heavy metals have been spilled or leaked from underground storage tanks, contaminating the soil and groundwater with VOCs. The contamination threatens the deep drinking water aquifers that supply Santa Clara and Mountain View (affecting about 300,000 people altogether).

Response actions included removing tanks and contaminated soil (1973–1984). A groundwater extraction and air-stripping system has been operational since 1985. An EPA-California site investigation (1989–1991) recommended continuing the groundwater treatment until clean-up goals are achieved.

The operator of this property has been ordered by the State of California to undertake the remediation work.

Sources: PNPLS-CA; Regional Water Quality Control Board, 2101 Webster Street, Suite 500, Oakland, CA, 94612.

United Heckathorn Company
Contra Costa County
Various chemical production companies, including napalm and pesticide manufacturers, used this site near Richmond from 1948 to 1965. Improper handling procedures, discharges, leaks, and spills contaminated the air, soil, and groundwater with various pesticides, including DDT and dieldrin. The contaminants have been found in Lauritzen Canal, which flows to Richmond Inner Harbor, an important recreational and commerical fishing habitat. Fish in this area have the highest DDT levels in the state. About 11,000 people live within 1 mile of the site.

Response actions included dredging pesticide-contaminated shoreline (1990–1991). In 1994 the EPA recommended additional dredging of deeper shoreline sediments, which is expected to begin in 1996–1997.

Sources: PNPLS-CA; Richmond Public Library, 325 Civic Center Plaza, Richmond, CA, 94804.

U.S. DOE Laboratory/Old Campus Landfill
Solano County
This landfill site was leased by the U.S. Department of Energy from 1960 to 1989. Research on biological applications of radionuclides generated a variety of waste, including animal carcasses and radioactive and chemical solutions, which were buried on-site. Testing has shown that the groundwater and soil are contaminated with chloroform and radioactive wastes, including carbon-14 and tritium. The contaminants have penetrated nearby wells and threaten the water supplies of about 50,000 people within 4 miles of the site.

Site investigations that will determine the final remediation plan are in progress.

Source: PNPLS-CA.

Valley Wood Preserving Company
Stanislaus County
This wood treatment facility near Turlock operated from 1973 to 1979. Timber was preserved with an aqueous chromate-copper-arsenate solution. Active and spent solutions and contaminated rinsewater were stored in tanks or discharged into sumps and settling ponds. Testing in 1980 showed that the groundwater and soil are contaminated with chromium and arsenic. About 30,000 residents, including those in the town of Turlock, depend on groundwater from within 3 miles of the site. The local water is also used for irrigation and farming purposes.

Response actions included excavating contaminated soil and treating about 70 million gallons of groundwater (1979–1983). In 1991 site investigations led the EPA to recommend chemical fixation and on-site burial of contaminated soil and groundwater treatment through electrochemistry and alumina adsorption.

Valley Wood Preserving Company, the PRP, signed Consent Orders (1989, 1990) with the EPA to finance the site investigations and remediation work.

Sources: PNPLS-CA; Stanislaus County Library, 550 Minaret Avenue, Turlock, CA, 95380.

Waste Disposal Inc.
Los Angeles County
This landfill near Santa Fe Springs accepted a variety of industrial and municipal wastes from 1928 to 1965, including acetylene sludges, brewery waste, and cesspool fluids. Testing has revealed that the soil and groundwater are contaminated with heavy metals, PNAs, and VOCs. About 15,000 people in Santa Fe Springs depend on water from wells within 3 miles of the site.

Response actions included fencing the site in 1988. EPA studies, which began in 1988, are nearing completion and will recommend a final remediation plan consisting of soil vapor extraction, a landfill cap, and groundwater extraction and/or monitoring.

Sources: PNPLS-CA; Santa Fe Springs Library, 11700 East Telegraph Road, Santa Fe Springs, CA, 90670.

Watkins-Johnson Company
Santa Cruz County
This company has been manufacturing industrial furnaces and electrical parts on this site near Scotts Valley since 1963. Testing by the state in 1984 showed that the groundwater and soil are contaminated with VOCs from improper handling of hazardous waste. About 11,000 people depend on groundwater from wells within 3 miles of the site.

Response actions included the installation of an interim groundwater pump-and-treat system (1986). In 1991 the EPA recommended soil vapor extraction, an expanded groundwater extraction and treatment system using activated carbon, and long-term monitoring. The treatment systems became operational in 1994.

Sources: PNPLS-CA; Santa Cruz Library, 230-D Mount Hermon Road, Scotts Valley, CA, 95066.

Western Pacific Railroad Company
Butte County
This railyard near Oroville operated from 1920 to 1982. Wastes generated from sandblasting, welding, and fabricating were discharged into an unlined settling pond. Soil and groundwater are contaminated with VOCs and heavy metals, including arsenic, chromium, and lead. Water wells within 3 miles of the site serve about 10,000 nearby residents.

Response actions included removing contaminated soil, sludges, and storage tanks. A groundwater extraction and treatment system was also installed. Investigations regarding contaminated soil are still in progress.

The PRPs, Union Pacific and Solano Railcar, have agreed to undertake some of the site investigations and the remediation work.

Sources: PNPLS-CA; Butte County Public Library, Oroville, CA, 95965.

Westinghouse Electric Corporation
Santa Clara County
Electrical transformers were once manufactured at this site in Sunnyvale. The operation is presently making steam generators and missile launching systems for the U.S. government. Spills and leaking underground storage tanks have contaminated the groundwater and soil with PCBs, VOCs, and petrochemicals. Part of a heavily industrialized corridor, this site contributes to the overall contamination of a larger groundwater basin. About 300,000 people depend on water drawn from wells within 3 miles of the site.

Response actions included removing contaminated soil and tanks (1984–1986). Site investigations completed in 1991 recommended groundwater extraction and treatment, excavation and incineration of contaminated soil, and land-use restrictions. Construction of these remedies should be completed by 1997.

Westinghouse, the PRP, has cooperated with the State of California and the EPA in undertaking the site investigations and remediation work.

Sources: PNPLS-CA; Sunnyvale Public Library, 665 West Olive Avenue, Sunnyvale, CA, 94088.

Air Force Plant

Jefferson County

The U.S. Air Force, which owns this site near Waterton, has designed and tested engines and rockets here since 1957. Various wastes from these operations, including fuels, cooling water, and radioactive materials, were discharged or buried on the 464-acre property. Testing has shown that the groundwater and soil contain VOCs, freon, and thorium (as well as alpha, beta, and gamma radiation). The contamination threatens the recreational Chatfield Reservoir, Brush Creek, Roxbury State Park, and Waterton Canyon as well as the health and drinking water supplies of nearby residents.

Response actions included exhuming drums of radioactive magnesium-thorium waste and underground storage tanks (1988). The U.S. Air Force is conducting detailed site investigations.

Sources: PNPLS-CO; EPA Region 8 Office, 999 18th Street, Suite 500, Denver, CO, 80202.

ASARCO Globe Plant

Denver and Adams Counties

This metal-refining and smelter site in Denver began operations in 1896. The plant refined gold, copper, silver, and lead ores from nearby mines. In 1901 ASARCO Inc. converted the site into a lead smelter, which operated for about 25 years. The plant presently makes lead oxide. For most of the plant's history various liquid wastes and runoff flowed into the South Platte River. Testing from 1974 to 1992 confirmed that the soil and groundwater contain cadmium, lead, and zinc. The contamination threatens nearby wetlands and fisheries in the South Platte River. At least 30 homes in nearby Globeville are contaminated with heavy metals.

In 1992 the State of Colorado and the EPA recommended a remediation plan consisting of reducing smelter emissions, excavating contaminated soil, cleaning up contaminated residential neighborhoods, creating a medical monitoring program, and groundwater extraction and treatment. Construction began in 1993.

The State of Colorado sued ASARCO for damaged natural resources in 1983. A 1993 Consent Decree requires that ASARCO fund the clean-up work.

Sources: PNPLS-CO; EPA Region 8 Office, 999 18th Street, Suite 500, Denver, CO, 80202.

Broderick Wood Products

Adams County

This former wood treatment plant in Denver operated from the 1940s until 1981. Wooden timbers were preserved with creosote and PCP. Liquid wastes, including oils, grease, and sludge, were discharged into unlined ponds on the property. Testing has shown that the soil and groundwater contain PAHs, PCP, VOCs, and heavy metals. The contamination threatens Fisher Ditch, Clear Creek, Copeland Lake, and the water supplies for about 80,000 people within 3 miles of the site.

Response actions included excavating and storing contaminated sludges. In 1992 the EPA recommended a final remediation plan consisting of incinerating sludges, fencing, and bioremediation of contaminated soil and groundwater. Treatment of sludges and soil was completed in 1994. The groundwater treatment system should be operational by 1997.

A Consent Decree (1986) and Administrative Order (1990) require the PRPs, Broderick and its parent company, to conduct the clean-up work. The PRPs have not complied, and the EPA has filed a lawsuit against them.

Sources: PNPLS-CO; EPA Region 8 Office, 999 18th Street, Suite 500, Denver, CO, 80202.

California Gulch

Lake County

This former mining district near Leadville operated from 1859 to 1989. Various mine tunnels, dumps, and milling and smelting operations along California Gulch contaminated soil and groundwater over a 20-square-mile area with arsenic, cadmium, zinc, copper, and lead. Contaminated runoff and mine drainage flow into California Gulch and the recreational Arkansas River. The city of Leadville is contaminated with heavy metal residues. The contamination threatens the Arkansas River and the water supplies for about 4,000 nearby residents.

Response actions included connecting affected residents to clean water supplies in 1986. EPA site investigations are focusing on lead toxicity testing and on cleaning up acid mine drainage, mine tailings, waste rock piles, contaminated groundwater and surface water, slag piles, and private homes contaminated with fugitive dust. An acid drainage collection system was built in 1992. Most of these studies will be completed by 1997.

The EPA negotiated a settlement with ASARCO and Newmont Mining Company to remediate parts of the site. To date, the EPA has received about $10.6 million from the principal PRPs: ASARCO, Newmont, the Denver & Rio Grande Western Railroad, Hecla Mining Company, and the U.S. government.

Sources: PNPLS-CO; Lake County Public Library, 1115 Harrison Avenue, Leadville, CO, 80461.

Central City/Clear Creek

Clear Creek and Gilpin Counties

Gold mining from the 1800s contaminated the 400-square-mile drainage basin of Clear Creek.

COLORADO

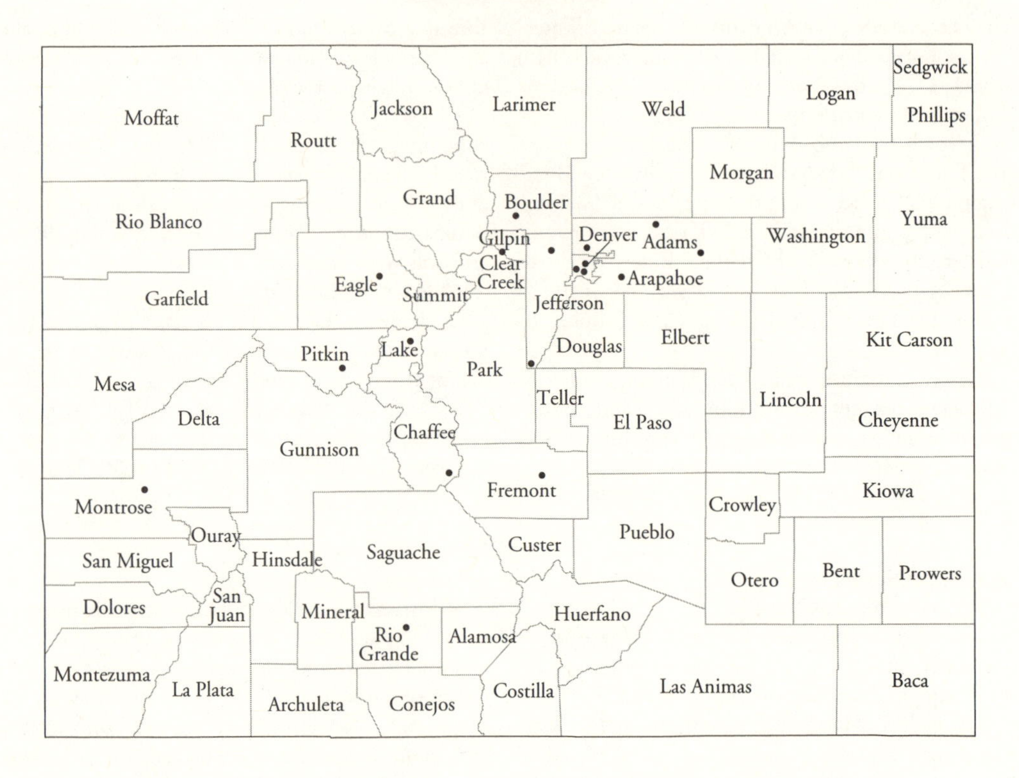

Smaller-scale mining is continuing today. Acid mine drainage from tailings piles and mine tunnels flows into Clear Creek and its tributaries. Soil, sediments, groundwater, and surface water contain arsenic, cadmium, chromium, copper, and lead. Clear Creek is used for agricultural and recreational purposes and supplies drinking water to the towns of Georgetown, Idaho Springs, Black Hawk, Golden, Westminster, Thorton, and Arvada. The contamination has devastated the ecostructure of Clear Creek.

Response actions included building barriers around tailings piles, identifying contaminated wells, and connecting affected residents to safe water supplies (1987–1991). Mine tunnel drainage will be treated through constructed wetlands or chemical precipitation treatment plants. Other tailings areas will be capped.

The EPA has signed Administrative Orders with new developers that require the companies to remediate their sites prior to the construction of new projects.

Sources: PNPLS-CO; Gilpin County Courthouse, 203 Eureka, P. O. Box 366, Central City, CO, 80427; Clear Creek Watershed Advisory Group, P. O. Box 847, Idaho Springs, CO, 80452.

Chemical Sales Company
Denver County
Located in northeastern Denver, this company has distributed industrial chemicals and detergents since 1976. Operations included storing and packing bulk chemicals. Testing in 1981 showed that the soil and groundwater contain high levels of VOCs resulting from leaks, spills, and improper handling of chemicals on the property. The contamination threatens the drinking water supplies of nearby residents.

Response actions included connecting affected residents to clean water supplies (1986), building a groundwater treatment plant (1989), and removing leaking drums. In 1991 the EPA recommended a final remediation plan of soil vapor extraction and groundwater extraction and air-stripping. Construction is under way.

Source: PNPLS-CO.

Denver Radium Site
Denver County
During World War I Denver was the site of an active radium processing industry. About 50 businesses and homes in the metropolitan area have since been built on radioactive wastes. Testing in the 1980s has shown

that the soil in these areas contain radium, thorium, arsenic, and lead. The contamination threatens the health of workers and residents in the areas.

Response actions included installing radon gas ventilation systems, excavating contaminated soil, demolishing contaminated buildings, and capping nonradioactive areas (1985–1995). Detailed site investigations are in progress.

Source: PNPLS-CO.

Eagle Mine
Eagle County
This former mining area along the Eagle River near the town of Gilman operated for at least 100 years. Waste rock and tailings were piled along the banks of Eagle River. The soil, groundwater, surface water, and runoff contain high levels of arsenic, chromium, copper, lead, cadminum, and zinc. The contamination has impacted the quality of the Eagle River and threatens the drinking water sources for about 2,500 nearby residents. Toxic dust particles can also be carried in the air.

Response actions included treating acid mine drainage and collecting surface runoff (1976–1979). In 1988 the State of Colorado recommended a final remediation plan of plugging mine portals, removing roaster waste, capping tailings piles, and treating groundwater. Soil studies are in progress.

The PRPs have conducted remediation work since 1976. A 1994 UAO and 1996 CD further define the scope of the remaining clean-up work.

Sources: PNPLS-CO; Minturn Town Hall, 302 Pine Street, Minturn, CO, 81645.

Lincoln Park
Fremont County
A former uranium mill operated on this site in Canon City from 1958 to the 1960s. Testing showed that the groundwater and soil in the mill area are contaminated with uranium, molybdenum, other heavy metals, and radionuclides. The contamination has penetrated the drinking water supplies for about 3,500 residents.

Response actions included connecting affected residents to clean water supplies (1988). The State of Colorado and the EPA recommended a final remediation plan of constructing hydraulic barriers to contain the contamination plume, groundwater treatment, and revegetating the site. Other site investigations are in progress.

Cotter Corporation, the PRP, signed a Consent Decree (1988) with the State of Colorado requiring it to fund the site investigations and clean-up work.

Sources: PNPLS-CO; Canon City Library, 516 Macon Avenue, Canon City, CO, 81212.

Lowry Landfill
Arapahoe County
This former landfill near Aurora operated from 1966 to 1980. It received a variety of municipal and industrial wastes, including liquid wastes, sludges, metal-plating solutions, petrochemicals, and insecticides. Testing has shown that the soil and groundwater contain organic and inorganic compounds. The contamination threatens the water supplies for about 5,000 residents within 3 miles of the site.

Response actions included an underground barrier wall and treatment system (1984). Solid and liquid wastes were removed between 1989 and 1992. A surface runoff control system was installed in 1993. Additional wastes were removed in 1995. In 1994 the EPA recommended gas collection, groundwater extraction and treatment, capping, and groundwater monitoring. Construction is expected to begin in 1997.

The EPA has recovered about $8.5 million from six negotiated settlements with the PRPs. The EPA issued UAOs to 34 remaining PRPs in 1995.

Sources: PNPLS-CO; Boulder Public Library, 1000 Canyon Blvd., Boulder, CO, 80302.

Marshall Landfill
Boulder County
This former landfill in the town of Marshall operated from the 1960s to 1991. It received a variety of municipal and industrial wastes, including sewage sludge and liquid wastes. In 1991 leachate was detected in a community waterway used for drinking and irrigation water. Testing showed that the groundwater and soil contain VOCs and inorganic compounds. The contamination threatens Marshall Lake, the water supply for the city of Louisville, and the drinking water wells of nearby residents.

The selected remediation plan includes fencing, regrading, installing a groundwater extraction and air-stripping system, and long-term groundwater monitoring. The groundwater treatment facility was completed in 1993.

The EPA negotiated Consent Decrees with a number of PRPs to conduct the site studies and reclamation work.

Source: PNPLS-CO.

Rocky Flats Plant
Jefferson County
This manufacturing facility on 6,500 acres of U.S. government-owned land near Denver has been in operation since 1951. Activities have included making plutonium triggers for nuclear weapons, recyling plutonium from older weapons, and conducting laboratory research. Wastes were stored or discharged on

the property. Testing in the late 1980s showed that the groundwater and soil contain VOCs, radionuclides, americium, plutonium, uranium, and other heavy metals. Air could also become contaminated. About 80 areas of contamination have been identified on the property; the contamination threatens about 300,000 people who live within 10 miles of the complex.

Response actions included sludge removal, installation of a surface water collection system and an interim groundwater extraction and treatment system, and soil vapor extraction. A number of ongoing studies should be completed by 1997 and will recommend a final remediation plan.

Sources: PNPLS-CO; Boulder Public Library, 1000 Canyon Road, Boulder, CO, 80302; Front Range Community College, 3645 West 112th Avenue, Westminster, CO, 80030.

Rocky Mountain Arsenal
Adams County
This 17,000-acre U.S. Army arsenal near Denver has been in operation since 1942, used by the federal government and by private industry. Various chemical products, including fuels, chemical warfare agents, and explosives, have been manufactured here. Solid and liquid wastes were deposited in numerous landfills on the site. The soil and groundwater are contaminated with VOCs, heavy metals, warfare chemicals, and pesticides. The contaminants can also be carried by dust in the air. The contamination threatens the water supplies for about 30,000 nearby residents.

Response actions included pumping out liquid wastes, plugging contaminated wells, connecting affected residents to clean water supplies, and installing a groundwater extraction system that treats about 1 billion gallons of water a year (1979–1990). Site investigations in progress are expected to recommend a final remediation plan in 1997.

The U.S. Army and Shell Chemical Company are the primary PRPs and have undertaken the response actions.

Sources: PNPLS-CO; Joint Administrative Document Facility, Room 14, Arsenal Security Building, 72nd and Quebec Streets, Commerce City, CO, 80022.

Sand Creek Industrial
Adams County
For over 25 years this site has been occupied by an oil refinery, a chemical company, acid pits, and a landfill. Spills, leaks, and improper waste handling procedures have led to high levels of VOCs, petrochemicals, pesticides, herbicides, and heavy metals in the soil and groundwater of this 550-acre site in Commerce City. The contamination has penetrated Sand Creek and threatens the health of hundreds of workers and residents in the area.

Response actions included removing acid waste, collecting seepage, and installing a methane gas collection system (1991). Contaminated soils were temporarily capped with a synthetic cover and later treated through soil washing and soil vapor extraction. Long-term groundwater monitoring is planned. Additional site studies are in progress.

PRPs conducted the early response actions.

Sources: PNPLS-CO; Adams County Public Library, 7185 Monaco, Commerce City, CO, 80022.

Smeltertown
Chaffee County
Metal smelting, wood treatment, and the production of zinc sulfate occurred on this site near Salida. Slag containing lead, zinc, arsenic, and other heavy metals was dumped along the Arkansas River in the early 1900s. Smelter emissions also contaminated the soil with heavy metals. Beazer East, Incorporated, treated lumber with creosote and PCP from about 1925 through the 1960s. Acid and galvanizing wastes from current zinc sulfate manufacturing operations on the site are discharged into lagoons and sludge pits. Testing has shown the groundwater and soils are contaminated with a variety of heavy metals, creosote, and PCP. The contamination threatens the Arkansas River and the drinking water supplies of nearby residents.

Creosote-contaminated soils were removed in 1986 and 1992. Other response actions included connecting affected residents to clean water supplies and removing contaminated soils from yards. Site investigations have been completed or are nearing completion. Clean-up actions are expected to begin in 1997 after a final remediation plan has been selected.

Beazer East, a PRP, has agreed to undertake site investigations according to an AOC (1996) with the EPA.

Source: PNPLS-CO.

Smuggler Mountain
Pitkin County
This former silver and lead mining district near Aspen produced ore from 1879 to 1918. Piles of waste rock and tailings were deposited along the Roaring Fork River. Some tailings were mixed with soil and spread over different areas on the 116-acre site. Mineral exploration still occurs in the area. The city of Aspen, a recreational community, has grown up around the mine waste. The soil contains high levels of lead, cadmium, and arsenic. Aspen has about 5,000 permanent residents and thousands of visitors every year.

Response actions included fencing (1985) and excavating and capping contaminated soil from residential properties (1990). In 1993 the EPA recommended capping and revegetating exposed mine waste, monitoring blood-lead levels in neighborhood children, and checking new construction projects for contaminated soil. A decision regarding a soil treatment plan is pending.

Three Administrative Orders in 1985 require the PRPs to fund the site studies and remediation work.

Sources: PNPLS-CO; Pitkin County Public Library, 120 East Main Street, Aspen, CO, 81611.

Summitville Mine
Rio Grande County
This remote mine in the San Juan Mountains operated sporadically from the late 1800s to 1990. In 1986 Summitville Consolidated Mining Corporation began using a cyanide heap-leach recovery system to mine gold. Illegal discharges containing cyanide and heavy metals have damaged nearby waterways, including Wightman Fork and Terrace Reservoir. The company declared bankruptcy in 1992 and abandoned the operation. Acid mine drainage containing heavy metals and cyanide continues to flow from the open pit and tailings.

Response actions include plugging mine openings and maintaining three water treatment plants on the site. Piles of waste rock and tailings were removed between 1993 and 1995. Other waste areas were capped. Heap-leach piles containing cyanide are being cleaned using bioremediation (bacteria) techniques that break down cyanide. Additional studies are in progress.

The EPA has funded the site studies and response actions.

Sources: PNPLS-CO; Del Norte Public Library, 790 Grand Avenue, Del Norte, CO, 81132; Conejos County ASCS Office, 15 Spruce Street, Box 255, La Jara, CO, 81140.

Uravan Uranium Project
Montrose County
This former uranium mining and milling complex in southwestern Colorado operated from 1915 to the 1980s. Radium, uranium, and thorium were processed or recovered from ore. Waste products including radioactive liquids were discharged on the property. Testing has shown that the soil and groundwater contain heavy metals, radionuclides, and radon gas. The contamination threatens the San Miguel River and a few nearby residents.

In 1986 the State of Colorado recommended a final remediation plan of capping and revegetating tailings, removing radioactive crystals, excavating mine waste from the San Miguel River, and processing pond and seepage water. Most of this work will be completed between 1995 and 2005.

Union Carbide, a PRP, has conducted the site remediation work.

Sources: PNPLS-CO; Colorado State Health Department, Radiation Control Division, 3773 Cherry Creek Drive North, Denver, CO, 80231.

CONNECTICUT

Barkhamsted–New Hartford Landfill
Litchfield County
This unlined landfill near Barkhamsted operated from the 1970s to the 1980s. It received a variety of municipal and industrial wastes, including sludges, oily metal grindings, and crushed barrels. The facility was cited for improper storage of hazardous solvents in 1983. Tests revealed that the soil and shallow and deep groundwater are contaminated with VOCs (xylene, toluene, and vinyl chloride). Adjacent streams flow into the Farmington River. About 5,000 people depend on wells that draw on groundwater within 3 miles of the site.

A site investigation (1991–1996) is expected to recommend a final remediation plan that includes gas and leachate control systems and a landfill cap.

Sources: PNPLS-CT; Beardsley Memorial Library, 690 Main Street, Winsted, CT, 06098.

Beacon Heights Landfill
New Haven County
The Beacon Heights Landfill operated from 1920 to 1979. It accepted a variety of city and industrial wastes, including spent chemical solvents, sludges, rubber, plastic, and oils. The soil and groundwater are contaminated with VOCs and heavy metals (lead). Nearby private wells were found to contain benzene in 1984. About 1,000 people within 1 mile of the site use local groundwater for drinking purposes. Surface runoff and the contaminated groundwater threaten the recreational Naugatuck River, Hockanum Brook, and related wetlands.

Response actions included connecting affected residents to the city water supply in 1989. The final plan consisted of excavating contaminated soil, gas venting, and restricting use of the groundwater (1985–1989). Leachate and surface runoff collection systems were installed between 1992 and 1996. Wetlands restoration and the landfill cap are expected to be under way by 1997.

The EPA identified about 70 PRPs. In 1987, 32 of these agreed to fund a large proportion of the clean-up work.

Sources: PNPLS-CT; Beacon Falls Town Hall, 10 Maple Avenue, Beacon Falls, CT, 06403.

CONNECTICUT

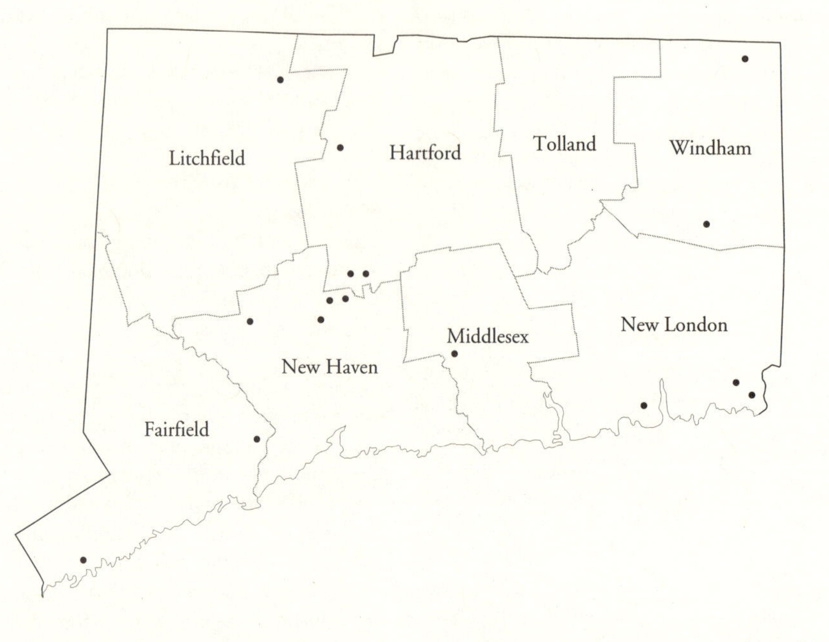

Cheshire Groundwater Contamination
New Haven County
Routine testing in Cheshire identified a 15-acre area of soil and groundwater contamination by VOCs, including trichloroethane, xylene, and tetrachloroethylene. The contamination is thought to be caused from the manufacture of plastic molding and electrochemical and electronic devices. Nearby wells are also contaminated. About 350 people within 1 mile of the site use the local groundwater. Cheshire's city wells are about 2 miles away. The contamination also threatens adjacents ponds and wetlands.

Response actions included removing contaminated soil and connecting affected residents to city water supplies (1983). A detailed site investigation initiated in 1994 is still in progress.

One of the PRPs, Cheshire Associates, signed a 1983 agreement with the State of Connecticut to remove the contaminated soil and monitor the contaminated wells.

Source: PNPLS-CT.

Durham Meadows
Middlesex County
Located on Main Street in the town of Durham, contamination at this site resulted from the manufacture of metal boxes by several different companies. Wastes such as sludge, wastewater, paints, and degreasing solvents were deposited on-site in unlined lagoons or stored in drums. Inspections in 1981 showed that soil and groundwater contain high levels of VOCs, including methylene chloride. Nearby private wells have been polluted. The contamination threatens the recreational Coginchaug and Connecticut Rivers and associated wetlands as well as the local water supplies for the 6,000 residents of Durham.

Drums of hazardous waste were removed (1983) and affected residents supplied with bottled water (1993). A site investigation was undertaken in 1996 and will recommend a final remediation plan.

The EPA is working to identify PRPs.

Sources: PNPLS-CT; Durham Public Library, 7 Maple Avenue, Durham, CT, 06422.

Gallup's Quarry
Windham County
This landfill site about 1 mile south of Plainfield was originally a gravel pit. It received a variety of municipal and industrial wastes, including illegal chemical wastes. Testing from 1978 to 1986 found the soil and groundwater to contain high levels of VOCs, heavy metals, and PCBs. The contamination has penetrated nearby drinking wells and recreational Mill Brook. About 6,500 people obtain drinking water from wells within 3 miles of the site.

Response actions included the removal of contaminated soil and waste drums (1978). A site investigation study is expected to be completed in

1996 and will recommend a final remediation plan.

An Administrative Order on Consent between the EPA and the PRPs (1993) requires the PRPs to finance the site investigation.

Sources: PNPLS-CT; Plainfield Public Library, 39 Railroad Avenue, Plainfield, CT, 06374.

Kellogg-Deering Well Field
Fairfield County
Testing of the 10-acre Kellogg-Deering Well Field on the bank of the Norwalk River showed that the soil and groundwater are contaminated with VOCs (trichloroethylene and perchloroethylene). Additional field work indicated that a major source of the contamination is the Elinco/Pitney Bowes/Matheis Court industrial complex. Water from this well field is blended with other municipal well water for the city of Norwalk (about 80,000 residents).

Response actions included installing air-strippers on some of the wells (1981–1988). The final remediation plan recommended by the EPA consists of soil vacuum extraction, extraction and treatment of groundwater, and long-term monitoring. Installation of the treatment systems should be completed by 1997. Additional studies regarding the size of the contamination plume are continuing.

The EPA signed an AO (1987) and CD (1990) with the PRPs to finance the site investigation and clean-up work.

Sources: PNPLS-CT; Norwalk Public Library, One Beldon Avenue, Norwalk, CT, 06770.

Laurel Park, Inc.
New Haven County
This landfill, located in a residential area near the town of Naugatuck, operated from the 1940s to 1987. It received a variety of municipal and industrial wastes, including solvents, sludge, tires, waste oils, and various hydrocarbons. Tests showed that the soil and groundwater are contaminated with VOCs (toluene and acetone) and heavy metals (calcium and magnesium). Contamination at the site threatens the private wells of nearly 100 homes within 3,000 feet of the landfill. Naugatuck's 26,500 residents are about 1 mile north of the site.

Response actions included installation of a court-ordered leachate collection system (1984) and fencing (1986). Affected residents were connected to the municipal water system (1989). EPA studies recommended a final remediation plan of capping, leachate and surface runoff collection, groundwater extraction and treatment, and long-term monitoring. Construction is expected to begin in 1997.

Over 20 PRPs have been identified by the EPA. Uniroyal Chemical Company, through a 1985 Administrative Consent Order, was the first PRP to begin working on the site. An AOC (1987) forced the PRPs to construct a new water line. A series of CDs and AOs (1989–1991) require 19 PRPs to fund the design and construction of the remediation plan.

Sources: PNPLS-CT; Howard Wittemore Library, 243 Church Street, Naugatuck, CT, 06770.

Linemaster Switch Corporation
Windham County
Operational since 1952, this company builds electrical foot switches and wiring harnesses near Woodstock. Wastes, including spent thinners and other solutions, are stored in barrels on-site. Tests showed that the soil and groundwater contain high levels of trichloroethylene. About 2,000 people within 3 miles of the site use wells that tap the contaminated groundwater.

Response actions included supplying affected residents with bottled water (1986). Nearby polluted wells have been equipped with carbon filters. A groundwater extraction and treatment system was also installed. In 1993 the EPA recommended a final remediation plan of soil vapor extraction and air-stripping groundwater. Construction of these systems is expected to begin in 1997.

Linemaster Switch Corporation is the main PRP. An Abatement Order (1986) and Consent Order (1987) required the company to undertake the initial response actions and fund the site investigation.

Sources: PNPLS-CT; Woodstock Town Hall, Route 169, Woodstock, CT, 06281.

New London Submarine Base
New London County
This military facility, which supports submarine operations, is located on 1,400 acres along the Thames River in Groton. Various wastes, including battery acid, VOCs, and pesticides, were buried in unlined landfills. Numerous spills and leaks have also contributed to surface contamination. Tests in 1988 indicated that the soil and groundwater contain high levels of pesticides, VOCs, PCBs, PAHs, and heavy metals (lead and cadmium). About 4,000 people live within 1 mile of the site.

Response actions included a variety of site investigations (1990–1997). Some areas of contaminated soil were removed and capped in 1994. A final remediation plan will be selected when the site investigations have been completed.

Sources: PNPLS-CT; Town of Groton Public Library, 52 Route 117 Newtown Road, Groton, CT, 06340; Public

Works Office, Naval Submarine Base, New London, Groton, CT, 06349.

Nutmeg Valley Road
New Haven County
This site is located around the Nutmeg Screw Machine Products Company, Waterbury Heat Treating Corporation, and Alpine Electronic Components, Inc., near Wolcott. Improper handling, leaks, and direct discharge of various degreasers, acids, oils, carbon tetrachloride, and solvents have contaminated the soil and groundwater with VOCs and heavy metals (lead and copper). About 11,000 residents depend on private wells within 3 miles of the site; the contamination also threatens the water supplies of Waterbury and Wolcott, about 3 miles away.

Response actions included supplying bottled water to affected residents (1987) and installing carbon filters on wells. Private homes were also been connected to city water. Some areas of toxic soil have been excavated and removed. An EPA site investigation is scheduled to begin in 1997. A final remediation plan will be recommended when the study is finished.

Source: PNPLS-CT.

Old Southington Landfill
Hartford County
Located in Southington, this landfill operated from the 1920s to 1967. It accepted a variety of municipal and industrial wastes. Closure consisted simply of covering the site with 2 feet of clean fill and reseeding. The site was later subdivided into small businesses and private homes. A nearby municipal well was closed in 1979 because it had high levels of VOCs. Tests showed that the soil and groundwater are contaminated with VOCs; the site is less than 2,000 feet from the recreational Quinnipiac River.

A site investigation was completed in 1993. An interim clean-up plan consists of capping, excavating contaminated soil, demolishing buildings on the site, collecting and treating soil gas, and long-term monitoring. Groundwater studies by the EPA will continue, and a final groundwater treatment plan may be recommended at a later time.

Three PRPs, through an AOC (1987) with the EPA, funded the site investigation study. In 1993 the EPA identified over 300 additional PRPs, most of which are targeted for cost recovery.

Sources: PNPLS-CT; Southington Public Library, 225 Main Street, Southington, CT, 06489.

Precision Plating Corporation
Tolland County
A chrome-plating facility has been located at this site near Vernon since 1970. Waste plating and etching solutions were discharged into a storm drain or stored on-site in drums. Chromium was detected in a nearby well in 1979. Subsequent testing showed that the soil and groundwater are contaminated with hexavalent and trivalent chromium. The site is located near wetlands, and about 11,000 people use groundwater from wells within 3 miles of the site.

Response actions included installing monitoring wells and removing contaminated soil (1986). Affected residents have been provided with clean water supplies. A site investigation is expected to start in 1997 and will recommend a final remediation plan.

Source: PNPLS-CT.

Raymark Industries
Fairfield County
Located in an industrial part of Stratford, this facility manufactured automotive brakes and clutches from 1919 to 1989. Waste by-products were used to fill former wetlands near the Housatonic River. PCBs, dioxin, VOCs, lead, asbestos, metals, and solvents were processed on-site, and liquid wastes were discharged into lagoons. The sludge was later used as landfill material for building projects in the city. Testing by the EPA in 1993 showed that the soil and groundwater at the site are contaminated with lead, asbestos, VOCs, chlorinated solvents, and PCBs. About 145,000 people live within 4 miles of the site, which also threatens recreational Ferry Creek and the Housatonic River.

Response actions included temporary capping of the site (1992), excavating soil from residential yards (1993–1996), capping the lagoons (1993), removing drummed toxic waste (1993), and fencing (1993). Soil removal at some of the other landfill sites in Stratford (including a junior high school) were completed in 1995. EPA site investigations should be completed by 1998 and will recommend a final remediation plan.

The main PRP, Raymark Industries, has filed for bankruptcy.

Sources: PNPLS-CT; Stratford Public Library, Reference Department, 2203 Main Street, Stratford, CT, 06497.

Revere Textile Prints Corporation
Windham County
This site is located in an industrial part of Sterling. Textiles were processed from the 1940s until 1980. Little is known about the former waste disposal practices, except that over 1,500 leaking drums of solvents, paints, and dyes were discovered on-site in 1980 after a fire destroyed the building. Sampling in 1984 showed that the soil and groundwater contain high levels of VOCs (toluene, ethyl benzene, and xy-

lene), barium, antimony, and methanol. The contamination threatens the local population's private drinking water wells.

Drums of toxic waste and some areas of contaminated soil were removed in 1983. Additional drums were hauled away in 1990. A site investigation completed by the EPA in 1992 only recommended monitoring the groundwater for 5 years.

The new owner of the site, Sterling Industrial Park Corporation, complied with an order made to the previous owner to remove the drummed waste and contaminated soil from the site.

Sources: PNPLS-CT; Sterling Public Library, 11110 Plainfield Pike, Oneco, CT, 06373.

Solvents Recovery Service of New England
Hartford County
Located in Southington, this hazardous waste treatment and storage facility operated from 1957 to 1991. It processed a variety of waste industrial solvents into new fuels. Waste sludges were discharged into unlined lagoons on-site. Testing in 1979 revealed that two city wells within 2,000 feet of the site were contaminated; these were later shut down. Further tests showed that the soil and groundwater contain high levels of isopropyl alcohol, toluene, lead, cadmium, and PCBs. The contamination threatens the water supplies of 38,000 people in Southington and is less than 500 feet from the Quinnipiac River.

Response actions included the emergency extraction and treatment of groundwater using an ultraviolet/oxidation system. A site investigation should be completed in 1997, with final recommendations for remediation to follow.

One of the PRPs, Solvents Recovery Service, signed a CD (1983) to implement the emergency groundwater recovery plan. A large group of PRPs that sent waste to the facility will also finance the site study and remediation work through a 1994 AO.

Sources: PNPLS-CT; Southington Public Library, 225 Main Street, Southington, CT, 07489.

Yaworski Waste Lagoon
Windham County
This landfill near Canterbury received a variety of waste from 1948 to 1973, including dyes, resins, acidic solutions, sludges, and solvents. Monitoring of wells installed in 1976 showed that the groundwater is contaminated with heavy metals, PCBs, PAHs, and VOCs. The site is located beside the Quinebaug River in a rural part of Canterbury township; the contaminants have been identified in river sediments.

Response actions included covering the lagoon (1990), improving the dike containment system, and groundwater monitoring. Additional site investigations are continuing. Groundwater monitoring will continue for at least 30 years.

The EPA has identified a number of PRPs, who subsequently undertook the response actions.

Sources: PNPLS-CT; Canterbury Public Library, 8 Library Road, Canterbury, CT, 06331.

DELAWARE

Army Creek Landfill
New Castle County
This former landfill near New Castle operated from 1960 to 1968. It received a variety of municipal and industrial wastes, including VOCs and acids. VOCs were discovered in nearby wells in 1972. Testing showed that the groundwater and soil contain VOCs, chromium, mercury, iron, zinc, and cadmium. Heavy metals are also present in Army Creek. The contamination threatens Army Creek and local groundwater supplies that are used by nearly 130,000 people.

Response actions included the installation of an interim groundwater pump-and-treat system. Other activities included capping (1990–1994). A larger-scale groundwater extraction and treatment system became operational in 1994. Other studies that will address wetland contamination are in progress.

A Consent Decree (1990) requires 18 PRPs to fund the site studies and remediation work.

Sources: PNPLS-DE; Delaware Department of Natural Resources, 715 Grantham Lane, New Castle, DE, 19720.

Chem-Solv
Kent County
Solvents were recycled on this site near Cheswold from 1982 to 1984. A fire and explosion destroyed the facility, and tanks and containers filled with solvents ruptured. Testing in 1984–1985 revealed that the groundwater and soils contain VOCs and heavy metals. The contamination threatens local wells that serve about 6,500 residents and agricultural needs.

Response actions included excavating and aerating contaminated soil in 1985. An interim groundwater extraction and air-stripping system operated from 1985–1988. Affected residents were also connected to clean water supplies. A site study (1989–1991) recommended only deed restrictions and long-term groundwater monitoring. The moni-

DELAWARE

toring is showing that VOC levels are approaching acceptable limits.

Various legal agreements, including a 1988 CO and 1993 UAO, require that the PRPs undertake the site investigations and remediation work.

Sources: PNPLS-DE; Jason Library, 1200 North DuPont Highway, Delaware State College, Dover, DE, 19901.

Coker's Sanitation Landfills
Kent County
These landfills near Cheswold received industrial and sanitation wastes from about 1962 to 1980. Waste materials included latex rubber sludges. Testing has shown that the groundwater, leachate, and soil contain heavy metals, and VOCs (styrene). The contamination threatens Willis Branch, Leipsic River, and associated wetlands as well as local groundwater supplies that serve about 4,000 residents and agricultural needs.

Drummed waste was excavated and removed in 1989. In 1990 the EPA recommended deed restrictions, fencing, capping, a leachate collection system, and groundwater and environmental monitoring. Construction was completed in 1993 and monitoring is in progress.

A 1988 AOC and 1992 CD require the PRPs to conduct the site investigations and clean-up actions.

Sources: PNPLS-DE; Clayton Post Office, Railroad Avenue, Clayton, DE, 19938.

Delaware City PVC Plant
New Castle County
Polyvinyl chloride (PVC) has been manufactured on this site near Delaware City since 1966. Various PVC waste solutions, including sludges, were held in unlined lagoons on the 400-acre property. Testing has indicated that the groundwater and soils are contaminated with VOCs, including vinyl chloride. The contamination threatens wells within 3 miles of the site, which serve over 100,000 residents.

Response actions included connecting affected residents and businesses to clean water supplies. PVC sludges were also removed and the areas capped. A groundwater extraction and air-stripping system has been in operation since 1992. Other studies are in progress.

Consent Orders (1984, 1987) require two PRPs to undertake the site studies and remediation work.

Sources: PNPLS-DE; Delaware Department of Natural Resources, 715 Grantham Lane, New Castle, DE, 19720.

Delaware Sand & Gravel Landfill
New Castle County
This former gravel pit near New Castle operated as an industrial waste landfill from about 1968 to 1976. It received a variety of municipal and industrial wastes, including drummed liquids and sludges from perfume, plastic, petroleum, and paint industries in the area. Testing has shown that the groundwater and soils contain VOCs, semi-volatile organic compounds (SVOCs), PCBs, and heavy metals (cadmium, mercury, iron, zinc, and chromium). The contamination threatens the drinking water supplies for about 2,000 nearby residents.

Response actions included installing an interim groundwater pump-and-treat system, removing drummed waste and flammable materials, fencing, capping, and installing surface runoff controls (1984–1991). In 1993 the EPA recommended soil vapor extraction and construction of a slurry wall. Soil treatment should be under way by 1997.

A 1992 AOC and 1995 CD require the PRPs to conduct the remediation work.

Sources: PNPLS-DE; Delaware Department of Natural Resources, 715 Grantham Lane, New Castle, DE, 19720.

Dover Air Force Base

Kent County

Wastes generated from activities on this active U.S. Air Force base near Dover were buried or discharged in 23 sites on the property from 1951 to 1970. Wastes included paints, solvents, fuels, and oil. Testing has shown that the groundwater and soils contain VOCs, arsenic, cadmium, mercury, and chromium. Heavy metals have also been detected in streams on the 3,700-acre base. The contamination threatens nearby streams and wetlands and aquatic life as well as the drinking water supplies for about 50,000 residents and base personnel.

Response actions included removing industrial and drummed waste and connecting affected residents to clean water supplies (1988). Contaminated soils were removed and the area capped in 1993. Other studies that will recommend a final remediation plan are in progress.

Sources: PNPLS-DE; Dover Air Force Base, 436 Military Airlift Wing, Dover, DE, 19902.

Dover Gas Company

Kent County

This former coal gasification plant near Dover operated from 1859 to 1948 and produced gas for street lamps. Waste coke, coal tar, and coal oil were buried on the site and discovered during construction in 1984. A site investigation revealed that the soils and groundwater contain PAHs and VOCs. The chemicals have also penetrated Tar Branch and the St. Jones River. The contamination could threaten nearby streams and wetlands as well as the private and public water wells in the area that serve about 45,000 residents.

A site investigation was conducted between 1990 and 1994. The EPA recommended the excavation and thermal treatment of contaminated soil, the extraction of free-product liquids, and groundwater monitoring to assure that natural attenuation of the contamination plume is occurring. Other studies are in progress.

A 1990 AOC requires the PRP to conduct the site investigations and evaluate different clean-up options.

Sources: PNPLS-DE; Dover Public Library, 45 South State Street, Dover, DE, 19901.

E. I. DuPont De Nemours Landfill

New Castle County

Two companies have manufactured paint pigments and chromium dioxide on this site near Newport since the early 1900s. Related solid and liquid wastes were buried or discharged in pits and lagoons on the property. Testing has shown that the groundwater and soil contain heavy metals, chlorinated solvents, oils, and petrochemicals. The chemicals have penetrated and threaten nearby wetlands and the Christina River. The contamination also threatens public and private wells that serve about 155,000 residents.

Response actions included driving sheet piles to control oil seepage from the landfill. In 1993 the EPA recommended capping, constructing groundwater barrier walls, connecting affected residents to clean water supplies, and dredging contaminated sediments from the wetlands and river. Construction should begin in 1997.

Various agreements (1988, 1993, 1994) require the PRPs to conduct the site studies and remediation work.

Sources: PNPLS-DE; Kirkwood Library, 600 Kirkwood Highway, Wilmington, DE, 19808.

Halby Chemical Company

New Castle County

Various inorganic and organic chemicals were manufactured on this site near Wilmington from 1948 to 1977. A variety of chemicals were stored on the site, and contaminated wastewaters were discharged into an unlined lagoon that drained into the Christina River. Testing has shown that the groundwater and soils contain VOCs, carbon disulfide, arsenic, zinc, PAHs, ammonia, and lead. Surface waters and river sediments are also polluted. The contamination threatens nearby wetlands, the Christina River and a small number of private and public wells that serve about 1,500 people in the area.

Response actions included removing chemicals, tanks, and other surface debris on the site. In 1991 the EPA recommended the consolidation and stabilization of contaminated debris and soil, capping, deed restrictions, and long-term monitoring to assure that natural attenuation of the contamination plume is occurring. A groundwater study is nearing completion.

A 1992 Consent Decree required the PRP to conduct the soil remediation work.

Sources: PNPLS-DE; Wilmington Institute Library, Tenth and Market Streets, Wilmington, DE, 19801.

Harvey & Knott Drum
New Castle County
This former dump near Kirkwood operated from 1963 to 1969. It received a variety of sanitary, industrial, and municipal wastes, including sludges, paints, solvents, acids, and heavy metals. An examination revealed that the groundwater and soils contain VOCs, arsenic, cadmium, lead, and PCBs. The contamination threatens nearby streams and wetlands and the local water supplies for a few hundred residents.

Response actions included connecting affected residents to clean water supplies (1981), fencing (1982), and removing drummed waste (1982–1984). All remaining drums of waste and contaminated soil were removed between 1988 and 1990. Lead-contaminated soils were capped. Groundwater monitoring is continuing.

Settlements in 1987 and 1988 required two PRPs to conduct the site studies and remediation work.

Sources: PNPLS-DE; Department of Natural Resources, Superfund Branch, 715 Grantham Lane, New Castle, DE, 19720.

Koppers Company, Newport Plant
New Castle County
This former wood preservation facility near Newport operated from 1929 to 1971. Lumber was treated with creosote and fuel oil-PCP solutions. Various liquid wastes were held in unlined ponds or discharged into an adjacent wetland. In 1984 the EPA discovered creosote residues in on-site soils and nearby stream sediments. Further testing showed that the soils and groundwater contain PAHs and other creosote by-products. The contamination threatens wetlands, Churchmans Marsh, Christina River, Hershey Run, White Clay Creek, and local wells that serve over 150,000 people.

A site investigation that will recommend a final remediation plan is nearing completion.

An AOC requires the PRPs to conduct the site investigations.

Source: PNPLS-DE.

NCR Corporation, Millsboro
Sussex County
Cash registers and other electronic equipment were manufactured on this site near Millsboro from 1967 to 1980. Various chemical wastes, including enamels, plating sludges, and degreasers, were discharged into concrete-lined lagoons on the property. Tests revealed that the groundwater and soil contain VOCs and heavy metals, including chromium, which have penetrated Iron Branch Creek. The contamination is a threat to the Indian River as well and to local drinking water supplies.

Sludges and contaminated sediments were removed in 1981. An interim groundwater extraction and air-stripping system was installed in 1988. A site investigation (1988–1991) recommended an expanded groundwater treatment system using precipitation and air-stripping as well as groundwater monitoring. The new system will be operational in 1997.

A 1988 CO and 1992 UAO require the PRPs to conduct the site studies and remediation work.

Sources: PNPLS-DE; Millsboro Town Hall, 322 Wilson Highway, Millsboro, DE, 19966.

New Castle Spill
New Castle County
A chemical company manufactured plastic foam on this site near New Castle from about 1954 to the 1980s. Various chemical components and spent solvents were stored on the property in drums that leaked or spilled. Tests by the State of Delaware in 1977 showed that the groundwater and soil contain VOCS, SVOCs (creosotes, phthalates, and trischloropropyl phosphate), and PCBs. The contamination threatens creeks, wetlands, and wells that supply about 7,000 people.

Response actions included connecting affected residents to clean water supplies (1977–1978) and pumping out contaminated groundwater. In 1989 the EPA recommended groundwater monitoring to ensure that the contamination plume is naturally attenuating. The level of contamination has fallen to well below clean-up goals.

A 1990 agreement requires the PRPs to conduct the environmental monitoring.

Sources: PNPLS-DE; Delaware Department of Natural Resources, 715 Grantham Lane, New Castle, DE, 19720.

Sealand Limited
New Castle County
Companies involved in animal waste rendering (1971 to 1979) and waste oil recycling (1982 to 1983) operated on this site near Mount Pleasant. Various wastes, including tars, inks, oils, creosote, and other solvents, were stored in drums and wooden tanks that leaked. Testing in 1984 showed that the soils and possibly the groundwater contain low levels of creosote, solvents, and PAHs. The contamination threatens the private wells in the area that serve about 1,000 people.

Response actions included removing toxic chemicals, drums, and solid waste, followed by capping (1983). A study in 1991 concluded that the response actions had made the site safe. Groundwater monitoring and a final evaluation are in progress. If these results are positive, the EPA will delete the site from the NPL.

Sources: PNPLS-DE; Appoquinimink Library, 218 North Broad Street, Middletown, DE, 19709.

Standard Chlorine
New Castle County
Chlorinated benzenes are manufactured on this site in Delaware City. Major benzene spills occurred on the property in 1981 and 1986, releasing nearly 1 million gallons of VOCs on the ground. Testing has shown that the groundwater, soil, surface water, and sediments in Red Lion Creek contain chlorobenzenes and other VOCs. The contamination threatens Red Lion Creek, wetlands, and wildlife habitats as well as the private and public water wells within 3 miles of the site, which serve about 150,000 people.

Response actions included building filter fences around wetlands, dikes, and a double-lined storage pond to hold recovered wastes (1986–1987). A site examination completed in 1995 is expected to recommend a final remediation plan by 1997–1998.

Standard Chlorine, the PRP, signed a 1988 agreement with the State of Delaware that requires the company to undertake the site studies and response actions.

Source: PNPLS-DE.

Sussex County Landfill
Sussex County
This former landfill near Laurel operated from 1970 to 1979. It received a variety of municipal and industrial wastes, including VOC solutions and heavy metal sludges. Testing by the EPA in 1986 revealed excessive levels of VOCs in the groundwater and soil that have penetrated nearby private wells. The contamination threatens the public and private wells in the area that serve about 6,000 residents and agricultural/irrigation needs.

Response actions included connecting affected residents to clean water supplies and installing a carbon treatment unit. A series of investigations from 1984 to 1992 concluded that VOC levels are low enough that they are within the EPA's acceptable risk range. Natural attenuation should further reduce VOC levels. Groundwater monitoring is continuing.

Sources: PNPLS-DE; Laurel Public Library, Six East Fourth Street, Laurel, DE, 19956.

Tybouts Corner Landfill
New Castle County
This former landfill near Wilmington operated during the 1960s and 1970s. It received a variety of municipal and industrial wastes, including degreasers and VOCs. Private wells were found to contain VOCs in 1976 and 1983. Further investigation showed that the groundwater and soils contain VOCs. The contamination threatens Pigeon Run Creek, Red Lion Creek, the Delaware River, and the local drinking water supplies.

Response actions included fencing (1982, 1987, 1993) and connecting affected residents to clean water supplies (1984–1986). In 1986 the EPA recommended excavating part of the landfill, capping, and constructing a slurry wall and a leachate collection system.

A 1988 Consent Decree requires the PRPs to fund some of the clean-up work.

Sources: PNPLS-DE; Delaware Department of Natural Resources, 715 Grantham Lane, New Castle, DE, 19720.

Tyler Refrigeration Pit
Kent County
Degreasing and cleaning solvents were used to prepare refrigeration equipment on this site near Smyrna from 1952 to 1969. Wastes solutions and sludges were discharged into unlined lagoons. The site was later converted into a metal fabrication facility and the lagoons paved with asphalt. In the late 1970s Smyrna's wells were discovered to contain VOCs. Additional studies showed that the groundwater and soil contain VOCs and pesticides. The contamination threatens the drinking water supplies for the 5,000 residents of Smyrna.

Response actions included installing an air-stripping and carbon filtration unit for the city water supply. Studies in progress indicate that another source of contamination may be the metal fabrication operations.

Consent Orders (1991, 1995) require the PRPs to conduct site investigations.

Sources: PNPLS-DE; Smyrna Public Library, 107 South Main Street, Smyrna, DE, 19977.

Wildcat Landfill
Kent County
This landfill near Dover operated from 1962 to 1973. It accepted a variety of municipal and industrial wastes, including heavy metals, pesticides, latex, sludges, and spent solvents. Testing has indicated that the groundwater and soil contain chlordane, VOCs, and PCBs. The contamination threatens the recreational St. Jones River and the drinking water sources for local residents.

The EPA recommended a final remediation plan of deed restrictions, removal of drummed waste, capping, drilling of new wells for affected residents, and groundwater monitoring. Construction was completed in 1992.

Sources: PNPLS-DE; Department of Natural Resources, Superfund Branch, 715 Grantham Lane, New Castle, DE, 19720.

Agrico Chemical Company
Escambia County
This chemical production site near Pensacola operated from 1889 to 1975. Sulfuric acid was manufactured until 1920, when production was shifted to fertilizers made from phosphate. A nearby city well was shut down in 1958 because of high acidity. Testing has shown that the groundwater and soil are contaminated with fluoride, lead, and arsenic. The contamination has penetrated Bayou Texar and threatens the water supplies of about 115,000 residents.

Response actions included connecting affected residents to city water supplies. In 1992 the EPA recommended the stabilization and reburial of contaminated soil, a multilayered cap, construction of a slurry wall to control runoff, and groundwater monitoring. These remedies are in progress.

The EPA has identified PRPs that are financing the site investigations and remediation work.

Sources: PNPLS-FL; Pensacola Public Library, 200 West Gregory Street, Pensacola, FL, 32501.

Airco Plating Company
Dade County
This electroplating shop near Miami has performed cadmium, chromium, copper, and zinc plating since 1957. Waste sludges were deposited in unlined lagoons on the property until 1973. EPA testing in 1985 revealed that the soil and groundwater are contaminated with perchloroethylene, cadmium, chromium, copper, and nickel. The contamination threatens the drinking water aquifer for Dade County; the well fields that supply 750,000 people are within 3 miles of the site. The chemicals may also be entering the Miami Canal and Miami River.

Response actions included outfitting the municipal wells with air-stripping towers. A recent EPA report recommends a remediation plan consisting of soil vapor extraction, capping, and a groundwater extraction and treatment system using air-stripping.

The EPA has identified PRPs that are working with the State of Florida and the EPA to conduct the site investigations and remediation work.

Sources: PNPLS-FL; John F. Kennedy Memorial Library, 190 West 49th Street, Hialeah, FL, 33012.

Alpha Chemical Corporation
Polk County
Unsaturated polyester resins for the production of fiberglass have been manufactured on this site since 1967. Wastewaters containing VOCs were discharged into unlined lagoons on-site. Testing in 1982 revealed that the soil and shallow groundwater are contaminated with VOCs (ethyl benzene and xylene).

Response actions included capping the site (1988–1989). Five years of groundwater monitoring indicated that the cap has reduced the contamination in the groundwater. The site will likely be deleted from the NPL by 1997.

A Consent Decree (1989) between the PRPs and the State of Florida requires the parties to finance the site investigation and remediation work.

Sources: PNPLS-FL; Lakeland Public Library, 100 Lake Morton Drive, Lakeland, FL, 33801.

American Creosote Works Pensacola Plant
Escambia County
A wood preservation plant operated on this site near Pensacola from the 1900s to 1981. Waste solutions generated from soaking lumber in creosote and PCP were discharged into unlined lagoons on the property, which occasionally overflowed into Bayou Chico and Pensacola Bay. Testing has revealed that the soil and groundwater are contaminated with VOCs, PAHs, PCP, and dioxin. The contaminants threaten the health of those in the surrounding commercial/residential area.

Response actions included draining the lagoons and solidifying and temporarily capping sludge (1983). Fencing and additional capping were completed in 1985–1986. In 1994 the EPA recommended separating free-product contaminants from the groundwater, followed by biological treatment. Construction of this system is expected to begin in 1997. The EPA is still studying the best way to treat PAH and PCP contamination in the soil.

American Creosote Works went bankrupt in 1981; site investigations and remediation work are being paid for with federal funds.

Sources: PNPLS-FL; Lakeland Public Library, 100 Lake Morton Drive, Lakeland, FL, 33801.

Anaconda/Milgo Electronics
Dade County
Two companies operated on this site in Miami from 1957 to 1984. Anaconda produced protective coatings on aluminum, and Milgo electroplated data processing equipment and built metal cabinets. Various hazardous wastes such as sulfuric acids, spent sodas and dyes, heavy metal sludges, and toxic wastewater were discharged into unlined pits, lagoons, and drain fields on the property. Testing from 1983 to 1987 indicated that the soil and groundwater are contaminated with cyanide, cadmium, lead, zinc, iron, selenium, chromium, copper, and mercury. The contaminated groundwater threatens the drinking water supplies for the 800,000 people in the Miami area.

Response actions included the removal of contaminated soil in 1993. EPA testing of the ground-

FLORIDA

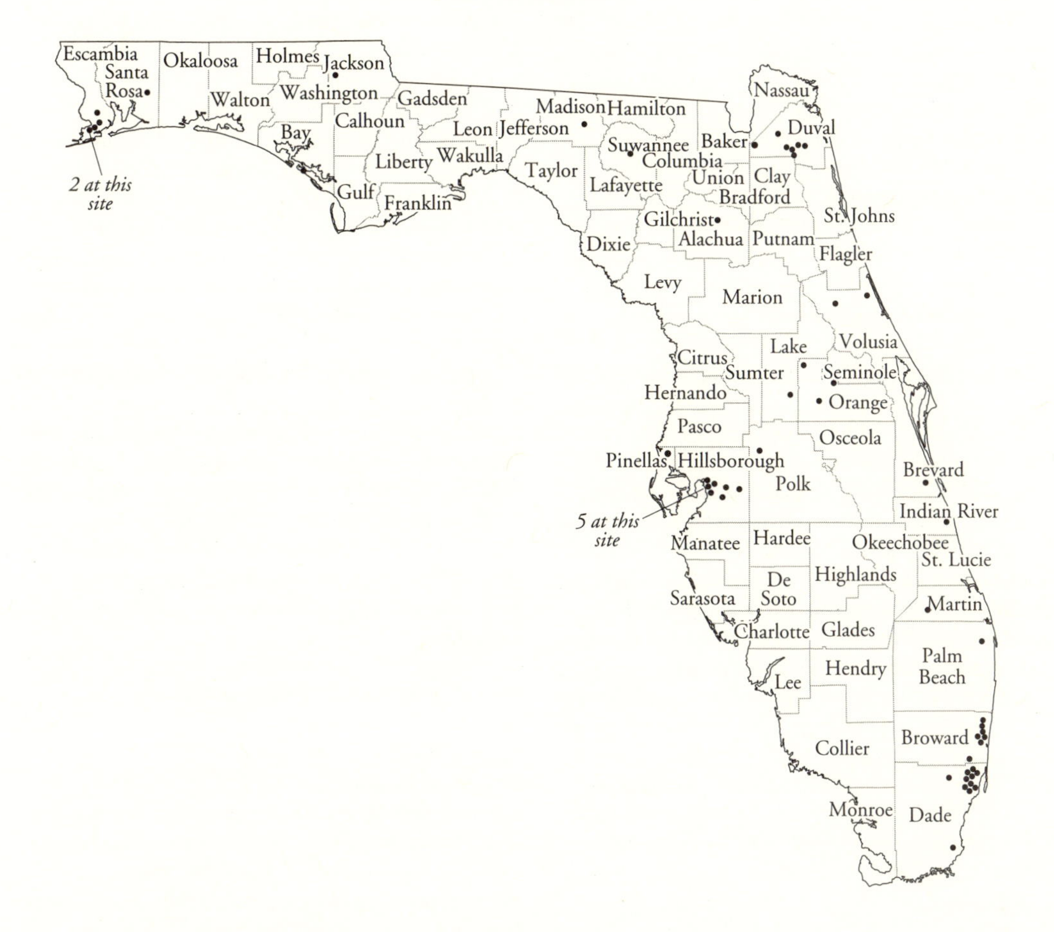

water between 1993 and 1995 indicated no further action was required, and the site has been recommended for deletion from the NPL.

Anaconda and Milgo Electronics have conducted the remediation activities.

Sources: PNPLS-FL; North Central Library, 9590 Northwest 27th Avenue, Miami, FL, 33172.

Anodyne
Dade County
Lithographs and silkscreen prints were produced at this North Miami Beach site from the 1960s to 1975. Liquid wastes were dumped on the ground or discharged into an injection well near the building. EPA testing in 1986 showed that the soil and groundwater are contaminated with VOCs and heavy metals. The contamination threatens the drinking water supplies of about 150,000 people within 3 miles of the site.

Response actions included EPA site investigations (completed in 1993). The final remediation plan consists of groundwater extraction and air-stripping and soil excavation. Construction is expected to be under way by 1997.

The EPA is researching PRPs for liability and cost recovery.

Sources: PNPLS-FL; North Dade Regional Library, 2455 Northwest 183rd Street, Miami, FL, 33056.

B & B Chemical Company
Dade County
Industrial cleaning compounds have been manufactured at this site in Hialeah since 1962. Before 1976 toxic washwater from mixing vats was discharged into unlined lagoons on the property. Testing in 1985 by the EPA identified high levels of VOCs (chlorobenzene and dichloroethylene) and chromium in soil and groundwater. The contamination threatens the drinking water of about 750,000 people; four municipal wells are within 3 miles of the site, and the Miami Canal is less than 1,000 feet away.

Response actions included installing a groundwater recovery and treatment system to control the contamination plume. In 1994 EPA studies concluded that natural attenuation of the groundwater contamination is occurring and will bring down levels to clean-up standards in two years. With the exception of groundwater monitoring, no additional action is required.

B & B Chemical Company, the PRP, installed the groundwater recovery system.

Sources: PNPLS-FL; John F. Kennedy Memorial Library, 190 West 49th Street, Hialeah, FL, 33102

Beulah Landfill

Escambia County

This landfill near Pensacola operated from 1950 to 1986. A variety of municipal and industrial wastes was accepted, including sludges, septic tank wastes, and demolition debris. Testing by the EPA and the State of Florida indicates that soil and groundwater contain low levels of anthracene, naphthalene, fluoroanthene, pyrene, PCP, PCBs, and zinc. The contamination threatens the private wells of local residents and the nearby recreational Eleven Mile Creek.

In 1993 EPA studies recommended that no action was required because the contamination was within an acceptable range for the protection of human health and the environment. The landfill is presently being closed according to Florida regulations.

The PRPs conducted the site investigations.

Sources: PNPLS-FL; George Stone Vocational School Media Center, 2400 Long Leaf Drive, Pensacola, FL, 32526.

BMI-Textron

Palm Beach County

Chrome-plated glass plates used in the production of electronic components were manufactured at this site in Lake Park from 1969 to 1986. Liquid wastes, including spent cyanide solutions, were discharged into ponds and drain fields on the property. Testing in 1983 revealed that the soil and groundwater are contaminated with cyanide, fluoride, barium, chromium, and nitrates. The contamination threatens nearby municipal well fields that supply water to about 110,000 residents.

Response actions included removing contaminated soil and capping (1984, 1986). Fencing was also installed. In 1994 the EPA recommended a remediation plan of natural attenuation and degradation of the contaminants.

The PRP, BMI, signed agreements with the State of Florida (1984, 1986) that require it to undertake the site investigation studies and remediation work.

Sources: PNPLS-FL; Lake Park Library, 529 Park Avenue, Lake Park, FL, 33403.

Broward County 21st Manor Dump

Broward County

This site was a former dump that operated from the 1950s to the 1960s beside Meadowbrook Elementary School near Fort Lauderdale. Owned and operated by the county school board, it received a variety of municipal wastes. Testing in 1987 indicated that the soil and groundwater are contaminated with low levels of heavy metals and VOCs; nearby private and municipal wells were also found to be contaminated, some of which were shut down. The contamination threatens the water supply of the elementary school and 13,000 people who live within 1 mile of the site.

Response actions included site investigations by the EPA. In 1991–1992 the EPA recommended that no action was necessary because the low level of contamination is not a threat to human health or the environment. The city of Fort Lauderdale has installed a treatment system for its municipal well that was shut down. The inclusion of the Broward County 21st Manor Dump on the NPL list is being reconsidered; the contamination may be related to other sources.

Sources: PNPLS-FL; Broward County Library, 1000 South Andrews Avenue, Fort Lauderdale, FL, 33301.

Brown Wood Preserving

Suwannee County

A wood preserving facility operated at this site near Live Oak from 1946 to 1978. Wood timbers were soaked in creosote or PCP. Contaminated wastewater was discharged into an unlined pond on the property. Testing in 1985–1986 revealed that soils are contaminated with PAHs. A number of private residences, water wells, and sinkholes are within 2 miles of the site.

Response actions included cleaning out the pond and removing sludges (1988). Contaminated soil was biologically treated on-site, and runoff controls constructed (1989). Well monitoring has indicated that contamination has not entered the water supply aquifer.

The PRPs have fulfilled their clean-up responsibilities under a CD (1988) with the EPA, which plans to delete the site from the NPL.

Sources: PNPLS-FL; Suwannee River Regional Library, 207 Pine Street, Live Oak, FL, 32060.

Cabot/Koppers

Alachua County

This site includes the Cabot and Koppers properties in Gainesville. Koppers Company has preserved wooden timbers and utility poles by soaking them in PCP, creosote, or chromate-arsenic-copper solutions.

Wastewater from earlier operations was discharged into unlined holding ponds. Cabot Carbon operated from 1945 to 1965 and made crude wood oil and pitch; waste solutions were discharged into an unlined pond. A developer drained the pond into nearby wetlands and Hogtown Creek in 1966. Remaining pine tar sludges were mixed with topsoil at the site, which created the release of a foul-smelling liquid. Testing has shown that the groundwater and soil are contaminated with arsenic, PAHs, and creosote compounds. The contamination could affect the water supplies of 11 schools and 4,000 residents within 1 mile of the site.

Response actions included collecting the foul-smelling leachate, which has been effective in reducing contaminant levels in Hogtown Creek. Investigation studies by the EPA and the State of Florida (1984–1990) recommended in-situ bioremediation of soil, soil excavation, soil washing, groundwater extraction and treatment, and free-product (creosote) separation. Construction should be under way by 1997.

The State of Florida won a suit for cost recovery in 1984 against Cabot Corporation. An AOC with the EPA required Cabot Corporation and Beazer East (formerly Koppers) to conduct the site investigation. A UAO with the EPA (1991) requires Beazer to clean up the Koppers site; Cabot Corporation is remediating its site under a Consent Decree.

Sources: PNPLS-FL; Alachua County Library, 401 East University Street, Gainesville, FL, 32601.

Cecil Field Naval Air Station
Duval County
This U.S. Navy facility near Jacksonville stores fuel and repairs aircraft and engines on this site. Various wastes, including oils, solvents, sludges, and paints were buried in landfills on the site. Testing has indicated that soil and groundwater are contaminated with heavy metals, solvents, VOCs, methylene chloride, and heavy metals. The chemicals have penetrated Rowell Creek, Yellow Water Creek, and Lake Fretwell. The contamination threatens the drinking water of about 6,000 people within 1 mile of the site and numerous wetland habitats. Local groundwater is also used for irrigation.

Response actions included excavating contaminated soil and sediment at recreational Lake Fretwell (1993). Site investigations (1989–1996) will recommend a final remediation plan that outlines how to remedy the groundwater, sludge pits, pesticide pits, and seepage pits. Soil treatment is already under way.

Sources: PNPLS-FL; Jacksonville Public Library, 6887 103rd Street, Jacksonville, FL, 32210.

Chemform
Broward County
This manufacturing facility near Pompano Beach built metal parts for the aerospace industry and electrochemical milling machines from 1967 to 1985. Toxic industrial wastes were discharged on the ground. Testing by the EPA in 1985 showed that the soil and groundwater contained high levels of chromium, copper, and nickel. About 95,000 people depend on groundwater drawn from within 3 miles of the site.

Response actions included removing drums and contaminated soil (1990, 1992). EPA studies recommended only groundwater monitoring, which indicated the levels of pollution in the groundwater had dropped as a result of the surface clean up. The site is presently being deleted from the NPL.

The PRPs, through an AO with the EPA (1989), agreed to finance the site investigation and remediation work.

Sources: PNPLS-FL; Broward County Library, 100 South Andrews Avenue, Fort Lauderdale, FL, 33301.

Chevron Chemical Company
Orange County
Two companies have operated on this site in Orlando. Chevron Chemicals made pesticide sprays until 1976. Wastewater and pesticide-contaminated rinsewater were discharged into unlined ponds on the site and later backfilled. From 1978 to 1986 Central Florida Mack Trucks Service Center overhauled truck engines and generators on the property. A tanker truck leaked a significant amount of hydrochloric acid and nitric acid in 1984. Testing by the EPA in 1989 revealed that the soil and groundwater contain anomalous levels of VOCs, pesticides, and heavy metals. The contamination threatens the water supplies of nearby residents.

Response actions included the excavation of the acid spill (1984), site dewatering, and removal of contaminated soil (1990–1992). Soil from yards in a neighboring trailer park contaminated by storm runoff were excavated in 1994. Detailed site investigations that will recommend a final remediation plan should be completed by 1997.

Chevron Chemical Company and the Central Florida Mack Trucks Service Center signed AOCs (1990, 1993) with the EPA that require the parties to finance the site investigation and remediation work.

Sources: PNPLS-FL; Orlando Public Library, Edgewater Branch, 6250 Edgewater Drive, Orlando, FL, 32810.

City Industries
Orange County
This site near Winter Park operated from 1971 to 1983 as a reprocessing facility for a variety of waste

chemicals, including solvents, plating solutions, PCBs, and inks. The company abandoned the site in 1983, leaving behind thousands of drums of chemical waste and sludge. Testing revealed that the soil and groundwater are contaminated with VOCs, phthalates, and heavy metals. The contamination threatens the water supplies for about 300,000 people living within 3 miles of the site.

Response actions included removing drums of liquid wastes and incinerating contaminated soil (1983–1984). The EPA recommended a final remediation plan of groundwater extraction and air-stripping that became operational in 1994. It will continue for at least 10 years.

City Industries ignored an Administrative Order from the EPA in 1984. The State of Florida sued City Industries and participating companies in 1984. In 1991 the EPA signed a Consent Decree with the PRPs that requires the parties to finance the remediation work.

Sources: PNPLS-FL; Winter Park Public Library, 460 East New England Avenue, Winter Park, FL, 32789.

Coleman-Evans Wood Preserving Company
Duval County
This company has been preserving wood at this site in Whitehouse since 1954. The timber is treated with PCP, and wastewater and sludges are discharged into unlined disposal pits and ditches. Testing in 1980 showed that the soil and groundwater contain high levels of PCP, VOCs, heavy metals, dioxin, and fuel oil. About 1,000 local residents within 1 mile of the site use private water wells.

Response actions included excavating sludge pits (1986), PCP- and dioxin-contaminated soil (1991–1993), and contaminated soil from nearby residences (1994). EPA studies (1986–1995) recommended a final remediation plan of soil washing, soil solidification, and groundwater treatment through carbon adsorption and chemical precipitation.

Coleman-Evans has not obeyed an Administrative Consent Order with the State of Florida (1984) that requires it to clean up the site.

Sources: PNPLS-FL; Whitehouse Elementary School, 11160 General Avenue, Whitehouse, FL, 32220.

Davie Landfill
Broward County
This landfill facility near Fort Lauderdale operated from 1964 to 1987. It accepted a variety of municipal and industrial wastes, including ash, sludges, tires, and demolition debris. Testing has shown that the groundwater and soil are contaminated with sulfate, chloride, lead, ammonia, antimony, VOCs (benzene and vinyl chloride), and cyanide. The contamination threatens numerous ranches, dairy farms, and horse stables and the water supplies of 10,000 people in the area.

Response actions included connecting affected residents to clean water supplies. The EPA recommended stabilizing and capping sludge and letting the groundwater contaminants biodegrade through natural attenuation. Long-term groundwater monitoring will continue.

Sources: PNPLS-FL; Broward County Library, 100 South Andrews Avenue, Fort Lauderdale, FL, 33301.

Dubose Oil Products
Escambia County
This oil recovery facility near Cantonment operated from 1979 to 1982. It processed waste oils, petroleum refining waste, spent solvents, and paint byproducts. Sampling by the EPA revealed that the groundwater and soil contain low levels of VOCs, manganese, iron, aluminum, and PCP. The contamination threatens the headwaters of Jack Branch and the drinking water supplies of about 2,500 people within 3 miles of the area.

Response actions included excavating contaminated soil (1984–1985) and collecting leachate. In 1990 the EPA recommended a final remediation plan of soil excavation and bioremediation, runoff controls, and groundwater monitoring. Construction should be completed by 1997.

Consent Decrees between the EPA and the PRPs require the parties to finance the site studies and remediation work.

Sources: PNPLS-FL; Tate High School, Tate High School Road, Gonzalez, FL, 32560.

Florida Steel Corporation
Martin County
This former steel mill near Indiantown operated from 1970 to 1982. Waste products from steel fabrication, including contaminated cooling water, heavy metals, and oils, were captured in concrete drains and sent to a concrete recirculating reservoir. Scrap was stacked in waste piles on the site. EPA testing in the early 1980s revealed that the soil and groundwater are contaminated with arsenic, cadmium, lead, sodium chloride, radium, PCBs, and zinc. About 5,000 people depend on water pumped from wells within 3 miles of the site. The contamination also threatens nearby wetlands.

Response actions included removing and incinerating contaminated soil (1985). EPA studies from 1987 to 1992 recommended a final remediation plan of excavation, solidification, and removal of contaminated soil and groundwater pumping and treatment.

Legal agreements with PRPs were completed in 1987. Additional CDs between the EPA and the parties signed in 1993 and 1994 require the parties to finance the site investigations and remediation work. A total of about $330,000 has been recovered from the PRPs for past costs.

Sources: PNPLS-FL; Indiantown Public Library, 15200 Southwest Adams Avenue, Indiantown, FL, 34956.

Gold Coast Oil Corporation

Dade County

This solvent reclamation plant near Miami recycled and stored various waste oils and paint by-products. Wastewaters were sprayed on the ground or stored on-site in drums and tanks. Testing has shown that the groundwater and soils have anomalous concentrations of VOCs and metals. The contaminated groundwater is part of the aquifer that supplies drinking water to the city of Miami.

Response actions included removing waste liquids, sludges, soils, and leaking drums (1982–1990). In 1989 the EPA recommended a final remediation plan of removing contaminated soil and sludge, groundwater extraction and treatment with air-stripping, and removing the remaining structures on the site. To date, 80 million gallons of groundwater have been treated.

The EPA has signed agreements with 14 PRPs that require the parties to finance the site studies and the remediation work.

Sources: PNPLS-FL; West Dade Regional Library, 9445 Coral Way, Miami, FL, 33165.

Harris Corporation

Brevard County

Harris Corporation manufactures a variety of electronic devices on this property in Palm Bay. In 1982 the EPA discovered that operations at Harris Corporation contaminated Palm Bay's nearby municipal wells that supply drinking water to about 33,000 residents. The groundwater contains elevated amounts of VOCs and heavy metals (chromium and lead).

Response actions included shutting down contaminated wells and the later installment of air-strippers. Groundwater extraction and air-stripping systems were installed in 1985 and 1991 and will operate until at least 2029.

Harris Corporation has been an active participant in cleaning up the site. Consent Decrees (1983, 1990, 1991) and an AO (1992) required Harris to finance the site investigations and remediation work.

Sources: PNPLS-FL; Palm Bay Public Library, 1520 Port Malabar Boulevard Northeast, Palm Bay, FL, 32905.

Helena Chemical Company, Tampa Plant

Hillsborough County

Located in an industrial part of Tampa, Helena Chemical Company has operated at this site since 1967. Activities included processing liquid pesticides and formulating liquid fertilizers. Raw materials stored at the site include zinc, manganese, toluene, and xylene. Liquid wastes and runoff were discharged into a holding pond and later neutralized in underground holding tanks. EPA tests in 1989–1990 revealed that the soil and groundwater have high levels of pesticides, VOCs, and heavy metals. About 7,000 people use water pumped from wells within 4 miles of the site.

Detailed site investigations should be completed by 1997 and will recommend a final remediation plan.

The EPA has identified Helena Chemical Company as the PRP and is negotiating an agreement for the clean up of the site.

Sources: PNPLS-FL; Tampa-Hillsborough Public Library, 5701 East Hillsborough, Tampa, FL, 33610.

Hipps Road Landfill

Duval County

This landfill near Jacksonville Heights operated from the 1960s to 1970. A cypress swamp was filled with various municipal and industrial waste, including cans of trichloroethylene and old U.S. Army artillery shells. The site was later developed into a residential area. Leachate seepage killed vegetation and fish in a nearby pond in the 1970s. Testing in 1983 showed that the groundwater is contaminated with VOCs (vinyl chloride and benzene). The contamination has penetrated nearby private wells and also threatens the recreational surface waters in the area.

Response actions included connecting affected residents to municipal water supplies and removing private residences on the site (1988). Fencing and a clay cap were installed in 1990. EPA studies in the 1980s recommended a final remediation plan of groundwater extraction and air-stripping. Built in 1994, the treatment plant will operate at least until 1999, followed by long-term monitoring.

One of the PRPs, Wastecontrol Inc., has conducted the reclamation work according to Consent Decrees (1989, 1992) with the EPA.

Sources: PNPLS-FL; Jacksonville Public Library, 6681 103rd Street, Jacksonville, FL, 32210.

Hollingsworth Solderless Terminal

Broward County

A solderless terminal plant operated on this site in Fort Lauderdale from 1968 to 1982. Toxic rinsewater, degreasing agents, and electroplating sludges were discharged on the surface or in drain fields and injection wells. Testing has shown that the soils and groundwater contain high levels of VOCs and heavy metals (copper and tin). The contamination threatens the drinking water supplies of local residents.

Response actions including pumping out the injection well (1982), and excavating and treating contaminated soil (1987). EPA recommendations also included installing a groundwater pump-and-treat system. Soil excavation, aeration, and replacement were completed by 1993. Groundwater treatment and monitoring are continuing.

Hollingsworth Company, the PRP, has participated in the site investigations and remediation work.

Sources: PNPLS-FL; Broward County Library, 1000 South Andrews Avenue, Fort Lauderdale, FL, 33301.

Homestead Air Force Base
Dade County
This U.S. Air Force installation 25 miles southwest of Miami stores fuel and supplies and maintains equipment and vehicles. Waste products, including fuels, pesticides, solvents, and heavy metals, were discharged on-site through early disposal practices. Testing has revealed that the groundwater and soils are contaminated with petroleum by-products and pesticides. The contamination threatens the waterways and wetland habitats of Military Canal and Biscayne Bay as well as the drinking water supplies for about 1,700 local residents.

Response actions included excavating contaminated soil and storage tanks (1992–1994). Site investigations in progress will recommend a final remediation plan.

Sources: PNPLS-FL; Dade Community College Library, Miami, FL, 33132.

Jacksonville Naval Air Station
Duval County
This U.S. Navy air station maintains aircraft and naval weapons and equipment. Maintenance activities, especially the repair of engines and aircraft, generated toxic wastes that were discharged in a landfill on the base. Some painting wastes were dumped in the St. Johns River. Testing has revealed that the soil and groundwater contain high levels of VOCs, PCBs, and heavy metals. The contaminants have penetrated the St. Johns River and associated wetlands, including bald eagle and Florida manatee habitats. About 400 people depend on groundwater pumped from within 3 miles of the base.

Response actions included removing contaminated soil (1992). Detailed site investigations (1992–1997) will recommend a final remediation plan, including the separation of free-product petroleum.

Sources: PNPLS-FL; Webb Wesconnet Public Library, 6887 103rd Street, Jacksonville, FL, 32210.

Kassouf-Kimmerling Battery Disposal
Hillsborough County
This former landfill site in Tampa operated in the 1960s and 1970s. Originally a peat mine, it was filled with discarded battery casings. Testing has shown that the groundwater and soil are contaminated with heavy metals (lead, arsenic, and cadmium). The contamination threatens abundant wetlands in the area; there are about 1,500 water wells within 3 miles of the site.

Response actions included excavating the contaminated soil, solidifying the soil through chemical fixation, and reburial (1994). Contaminated sediments were also removed from the marsh. Existing woodlands around the site will be expanded. Groundwater monitoring is continuing.

PRPs identified by the EPA have funded the site investigations and remediation work.

Sources: PNPLS-FL; Tampa-Hillsborough Library, 900 North Ashley Drive, Tampa, FL, 33602.

Madison County Landfill
Madison County
This landfill near Madison has been in operation since 1971. It has received a variety of municipal and industrial wastes, including solvents and VOCs. Testing in 1984 identified VOCs in the groundwater and soil. The contamination has penetrated some of the 98 wells within 3 miles of the site that service about 5,000 people.

Response actions included removing drums of VOC waste (1984–1985). Bottled water, filtration systems, and city water connections were provided to affected residents. In 1992 the EPA recommended a final remediation plan of runoff controls, a clay cap, a groundwater extraction and treatment system, and a groundwater monitoring program. Construction is expected to begin by 1997.

A series of Consent Agreements, Consent Orders, Administrative Orders on Consent, and Unilateral Administrative Orders (1986–1994) requires the PRPs to finance the site investigations and remediation work.

Sources: PNPLS-FL; North Florida Junior College Library, Turner Davis Drive, Madison, FL, 32340.

Miami Drum Services
Dade County
A drum recycling plant was operated at this site in industrial Miami from 1957 to 1982. Various wastes from the drums, including phenols, solvents, and acids, were spilled or dumped on the ground. Testing

revealed that the groundwater and soil contain high levels of VOCs, phenols, heavy metals, petrochemicals, and pesticides. The contaminants penetrated the nearby Medley Well Field, which was shut down in 1982, and threaten the regional water supply.

Response actions included removing contaminated soil and pumping and treating groundwater (1982). Air-stripping towers were added in 1992. Groundwater treatment and monitoring is ongoing and processes about 150 million gallons of water per day.

Lawsuits by Dade County and the EPA against the PRPs recovered most of the costs for the site investigations and remediation work.

Sources: PNPLS-FL; Miami-Dade County Public Library, 101 West Flagler Street, Miami, FL, 33130.

MRI Corporation

Hillsborough County

Metal recycling and chemical detinning were conducted at this site in Tampa from 1979 to 1986. Scrap metals was treated in an alkali solution, washed, and the tin removed through an electrochemical procedure. Wastewaters containing compound acids, alkalis, chlorine, degreasers, and cyanides were discharged into a drainage ditch. This effluent exceeded permit limits for zinc, mercury, grease, cyanide, cadmium, and oil in 1984. Recent testing has shown that the groundwater and soils contain high levels of mercury, zinc, and cyanide. The contamination threatens wells that supply about 16,000 residents, and nearby streams.

This site was proposed to the NPL in 1996. A final decision is expected in 1997–1998. Additional studies are planned.

Source: EPA Publication 9320.7-071.

Munisport Landfill

Dade County

This former landfill operated at this site in North Miami from the 1970s to the 1980s. It received a variety of municipal and industrial wastes, including demolition debris and hospital pathological wastes. Sampling by the EPA in the 1980s revealed that the groundwater and soil are contaminated with ammonia, heavy metals, pesticides, and VOCs. The contamination threatens a mangrove wetland and Biscayne Bay.

Response actions included EPA studies (1987–1990) that recommended a final remediation plan of treating leachate in air-stripping ponds, tidal restoration in the mangrove wetland, and groundwater extraction and treatment. The groundwater system is expected to be operational by 1997.

The city of North Miami, identified as the only PRP, signed a Consent Decree with the EPA in 1991 to perform the remediation work.

Sources: PNPLS-FL; Florida International University Library, North Miami Campus Library, North Miami, FL, 33181.

Northwest 58th Street Landfill

Dade County

This landfill near Medley operated from 1952 to the 1990s. It received a variety of municipal wastes, including paints and paint sludge, pesticides, and solvents. Testing has shown that the soil and groundwater are contaminated with heavy metals (arsenic and lead) and VOCs. The contaminants threaten municipal wells that serve about 750,000 people in the area. About 60 private wells have been contaminated.

Response actions included selectively pumping municipal wells and drilling extraction and injection wells to control the plume. Dade County installed an alternate water supply and leachate collection system in 1988–1989. The construction of erosion controls and a landfill cap is expected to be under way by 1997.

In 1989 Dade County, the major PRP, signed a Consent Decree with the Department of Justice that required it to close the landfill. The EPA has recovered past remediation costs from the county.

Sources: PNPLS-FL; Miami-Dade County Public Library, 101 West Flagler Street, Miami, FL, 33130.

Peak Oil Company/Bay Drum Company

Hillsborough County

These two companies have occupied this site near Tampa since the 1950s. Peak Oil Company processed waste oils, and Bay Drum reconditioned steel drums that had to be rinsed of their often-toxic contents. Waste products and contaminated wastewater and rinsewater were indiscriminantly discharged on-site. Testing in the 1980s found that the soil and groundwater contain high levels of PCBs, VOCs, and heavy metals. The contamination threatens bordering wetlands and the water supplies of nearby neighborhoods.

Response actions included excavating and removing contaminated soil, sludge, old tanks, and used oil (1987–1990). EPA site investigations (1989–1993) recommended in-situ soil treatment, groundwater pumping and air-stripping, excavation of contaminated sediments from the wetlands, and construction of a clay cap. Construction of these remedies is expected to be under way by 1997.

PRPs identified by the EPA signed an AOC in 1989 that requires the parties to pay for the site investigations and remediation work.

Sources: PNPLS-FL; Brandon Public Library, 135 West Robertson Street, Brandon, FL, 33511.

Pensacola Naval Air Station
Escambia County
Operations on this 6,000-acre U.S. Navy base near Pensacola include storing fuel and maintaining aircraft, vehicles, and engines. Various waste products such as insecticides, oils, solvents, paints, and electroplating sludges were discharged in sewer systems until 1973. Testing has shown that the soils and groundwater contain high levels of VOCs, arsenic, and pesticides. The contamination threatens about 50,000 people who draw water from wells within 3 miles of the site; numerous wetlands and recreational waterways are also endangered.

Detailed site investigations are being conducted by the U.S. Navy, and a final remediation plan should be recommended by 1997.

Sources: PNPLS-FL; Pensacola Regional Library, 200 West Gregory Street, Pensacola, FL, 32501.

Pepper Steel and Alloys
Dade County
A number of businesses have operated on this site in an industrial strip northwest of Miami since the 1960s. Activities included manufacture of batteries and concrete products, painting, metal recycling, and truck service and repair. Pepper Steel recycled transformers and dumped the PCB-laden waste oils directly on the ground. Solid waste, including old batteries and transformers, were stored in piles on-site. Testing has revealed that the soil and groundwater are contaminated with VOCs, PCBs, and heavy metals. The contamination threatens the water supplies that serve the local businesses and residents.

Response actions included removing contaminated soil and free-product petroleum in 1983. Remaining soil was either removed or fixated and reburied. Groundwater monitoring since 1987 has indicated that the soil treatment has been effective in cleaning the groundwater. The site will be deleted from the NPL.

A Consent Decree between the EPA and Florida Power and Light (1987) required the PRP to finance the site investigation and remediation work.

Sources: PNPLS-FL; Miami-Dade Public Library, 101 West Flager Street, Miami, FL, 33130.

Petroleum Products Corporation
Broward County
This oil recyling facility near Pembroke Park operated from 1952 to 1972. Waste products were discharged into a holding pond on the property that frequently overflowed. The site has since been developed into a small industrial park. Sampling by the State of Florida and the EPA has shown that the soil and groundwater are contaminated with petrochemicals, VOCs, lead, chromium, and PCBs. Municipal water wells that serve about 150,000 people are located within 3 miles of the site.

Response actions included removing drums, tanks, and contaminated sludge and installing an oil recovery system (1985). A temporary groundwater extraction and treatment system was installed by the EPA in 1990. Detailed site investigations (1990–1996) will recommend a final remediation plan.

Petroleum Products signed a Consent Order with the EPA in 1985 that required it to take immediate actions on the property. A 1991 Consent Decree with PRPs requires the parties to finance the site investigations and remediation work.

Sources: PNPLS-FL; Broward County Library, 100 South Andrews Avenue, Fort Lauderdale, FL, 33301.

Pickettville Road Landfill
Duval County
This landfill site near Jacksonville operated from the 1940s until 1977. It received a variety of hazardous wastes, including battery acids and PCBs. EPA tests in 1981 showed that the groundwater and soils are contaminated with VOCs, heavy metals, and PCBs. Runoff and seeping leachate have penetrated Little Sixmile Creek. The contamination has also been detected in private wells. Agricultural wells in the area are used to irrigate farmlands and water livestock. Over 300 residences are located within 1 mile of the landfill.

Response actions included backfilling, grading, and seeding the site (1977, 1983). Affected residents were connected to city water supplies in 1993. A landfill cap to limit infiltration of rainwater was built in 1995.

An Administrative Order (1988) and Consent Decree (1991) require the PRPs to finance the site investigations and remediation work.

Sources: PNPLS-FL; Jacksonville Public Library, 1826 Dunn Avenue, Jacksonville, FL, 32218.

Pioneer Sand Company
Escambia County
This abandoned quarry near Pensacola was operated as a landfill from 1974 to 1981. It received a variety of municipal and industrial wastes, including sludges, scrap metal, resins, phenols, and electroplating wastes. In 1981 the EPA found that the soil and groundwater contained high levels of heavy metals, VOCs, PCBs, and PCP. The contamination threatens the water supplies of about 70,000 people.

Response actions included excavating contaminated soil (1986). EPA recommendations included a leachate collection system and groundwater monitoring. Construction was completed in 1991, and the site was deleted from the NPL in 1993. Monitoring is continuing.

A Consent Decree (1988) between the EPA and PRPs required the PRPs to finance the site investigations and remediation work.

Sources: PNPLS-FL; John C. Pace Library, University of West Florida, 11000 University Parkway, Pensacola, FL, 32514.

Piper Aircraft/Vero Beach
Indian River County

Piper Aircraft Company has built and painted small airplanes at this site in Vero Beach since 1957. Chemicals and solvents were stored in underground tanks. Leaks from the tanks contaminated a Vero Beach municipal well in 1978. Subsequent testing showed that the groundwater, surface water, and soil are contaminated with VOCs. The contaminants have been detected in oysters and fish from nearby Main Canal and threaten the water supply for Vero Beach.

Response actions included the installation of a groundwater extraction and air-stripping system (1981). EPA site investigations (1992–1998) have recommended expanding the groundwater treatment system.

A Consent Agreement (1981) with the EPA forced Piper Aircraft Company to repair the site and treat the contaminated groundwater.

Sources: PNPLS-FL; Indian River Library, 1600 21st Street, Vero Beach, FL, 32960.

Plymouth Avenue Landfill
Volusia County

This landfill near De Land has been operating since the 1940s. It received a variety of municipal and industrial wastes, including nitric acid slurry, sulfuric acid, and heavy metal sludges. Testing in 1990 indicated excessive amounts of VOCs, phthalates, and heavy metals in the monitoring wells. The chemicals have penetrated nearby private wells. The contamination threatens two aquifers that supply water to about 26,000 people within 4 miles of the site. The groundwater is also used for irrigation.

Detailed site investigations are in progress and will recommend a final remediation plan.

Source: PNPLS-FL.

Reeves Southeast Galvanizing Corporation
Hillsborough County

Reeves Southeast Galvanizing Corporation has galvanized steel on this site near Tampa since the 1960s. Various waste solutions, including spent acids, were neutralized and discharged into ponds. The soil and groundwater are contaminated with heavy metals, especially zinc. The contamination threatens the water supplies of about 56,000 people within 3 miles of the site.

Detailed site investigations (1989–1995) will recommend a final remediation plan, including stabilizing and capping contaminated soil, excavating and treating groundwater, and monitoring nearby wetlands. Construction is expected to be under way by 1997.

Two AOCs (1989, 1993) require the PRPs to undertake the site investigations and remediation work.

Sources: PNPLS-FL; Brandon Public Library, 135 West Robertson Street, Brandon, FL, 33511.

Sapp Battery Salvage
Jackson County

A battery recycling company operated at this site near Alford from 1970 to 1980. Batteries were crushed to recover lead. Acid was dumped on the ground, and the casings were thrown in a pond on-site. In 1977 acidic leachate killed nearby cypress trees. The groundwater and soil are contaminated with lead, antimony, and cadmium. The contamination threatens the drinking water for about 3,500 residents within 3 miles of the site.

Response actions included neutralizing the acid discharges (1980) and excavating contaminated soil (1984). In 1986 the EPA recommended excavating, stabilizing, and capping contaminated soil; installing a groundwater pump-and-treat system; and groundwater monitoring. Wetland studies should be completed by 1997.

The EPA negotiated minimum settlements with a number of small-volume PRPs. A 1993 CD with major PRPs requires the parties to undertake the remediation work.

Sources: PNPLS-FL; Jackson County Public Library, 413 North Green Street, Marianna, FL, 32446.

Schuylkill Metals Corporation
Hillsborough County

This was the site of a battery recycling operation near Plant City from 1972 to 1986. Waste acids were poured into an unlined pond on the property. Testing in the 1980s revealed that the soil and groundwater contain excessive amounts of lead, chromium, sulfate, and ammonia. The contaminants have been detected in adjacent wetlands and threaten the water supplies of about 25,000 residents and workers within 3 miles of the site.

Detailed site investigations were completed in 1990. Recommendations included the excavation and solidification of contaminated soil and wetland sediments, groundwater pumping and treatment, and creation of additional wetlands. Construction is expected to be under way by 1997.

A Consent Order (1986) and Consent Decree (1991) require Schuylkill Metals Corporation to finance the site investigations and remediation work.

Sources: PNPLS-FL; Plant City Library, 501 North Wheeler Street, Plant City, FL, 33566.

Sherwood Medical Industries
Volusia County
This property has been the site of a biological and medical laboratory and manufacturing plant near De Land since 1959. The main operation is the making of stainless steel and aluminum hypodermic needles. Since 1983 industrial waste products and wastewater have been processed in a wastewater treatment facility on-site. Prior to 1983, liquid wastes and sludges were held in unlined ponds. Testing by the State of Florida in 1982 showed excessive amounts of VOCs in the groundwater and soils. The contamination threatens the drinking water of nearby residences and neighborhoods.

Response actions included detailed site investigations and water well monitoring from 1985 to 1991. In 1992 the EPA recommended groundwater extraction and air-stripping to control the migration of the contamination plume. Construction is under way.

Sherwood Medical Industries has been an active and willing participant in the site investigations. Sherwood signed a Consent Decree with the EPA in 1992 that requires the company to finance the remediation work.

Sources: PNPLS-FL; De Land Public Library, 212 West Rich Avenue, De Land, FL, 32720.

Sixty-Second Street Dump
Hillsborough County
This site was originally a sand mine. Located in Tampa, it was later used as an industrial waste dump until 1976. Various wastes were buried there, including automobile scrap, batteries, and kiln residues. Leachate killed fish in a neighboring fish hatchery in 1976. An examination showed that the soil and groundwater are contaminated with heavy metals (arsenic, chromium, cadmium, and lead) and PCBs. The contamination threatens wetland habitats and the water source for about 6,000 nearby residents.

The State of Florida conducted studies between 1984 and 1990 that recommended stabilizing the waste, building slurry walls, and capping the site. The remediation work is in progress.

Consent Decrees with the EPA require the PRPs to finance the remediation work.

Sources: PNPLS-FL; Tampa-Hillsborough Library, 900 North Ashley, Tampa, FL, 33602.

Standard Auto Bumper Corporation
Dade County
This company chrome-plated car bumpers and metal furnishings at this site in Hialeah from 1959 to 1993. Waste electroplating and stripping solutions were poured into a ditch on the property until 1972. EPA tests from 1985 to 1987 discovered high amounts of cadmium, copper, lead, chromium, and nickel in the soil and groundwater. Drinking water wells that supply about 750,000 people in Dade County are within 3 miles of the site.

Response actions included the removal of contaminated soil (1989, 1993). In 1993 the EPA concluded that groundwater attenuation and monitoring was the best remediation strategy. Additional soil was excavated in 1994. Groundwater monitoring is continuing.

Standard Auto Bumper Corporation signed Administrative Orders with the EPA in 1989 and 1990 to clean up the site. The PRP declared bankruptcy and abandoned the site in 1993. Federal funds will be used to finance the remediation work.

Sources: PNPLS-FL; John F. Kennedy Memorial Library, 190 West 49th Street, Hialeah, FL, 33012.

Stauffer Chemical Company, Tampa Plant
Hillsborough County
This company manufactured pesticides on this site in Tampa from 1951 to 1986. Toxic chemical ingredients that were used to formulate insecticides and herbicides were stored on-site. Wastewaters were discharged into an unlined pond; leaks and spills have been documented. Testing by the EPA in 1987–1988 showed that excessive levels of pesticides (DDT, DDD, and lindane) occur in soil and groundwater. Water wells within 4 miles of the site supply about 7,000 people. The contamination also threatens the Tampa Bypass Canal, wetlands, and the endangered West Indian manatee.

Response actions included removing contaminated soil and treating it in a low-temperature desorption unit on-site (1993, 1994). Site investigations that will recommend a final remediation plan are in progress.

Sources: PNPLS-FL; Tampa-Hillsborough Library, 5701 East Hillsborough, Tampa, FL, 33610.

Stauffer Chemical Company, Tarpon Springs Plant
Pinellas County
Chemicals have been manufactured on this site near Tarpon Springs since 1950; mainly elemental phosphorus is produced from phosphate ore. Various liquid wastes, including calcium, silica, fluoride, and phosphate slags, were dumped in unlined pits near

the Anclote River. Solid wastes such as calcined phosphate dust were buried on the property. EPA tests in 1988–1989 showed that the soil and groundwater are contaminated with heavy metals (barium, chromium, lead, vanadium, zinc, copper, and arsenic). The contaminants have penetrated the recreational Anclote River and associated wetlands and threaten water wells that supply about 10,000 people.

Detailed site investigations that will recommend a final remediation plan should be completed by 1997.

The EPA has signed an AOC with PRP Atkemix Thirty-Seven Incorporated.

Sources: PNPLS-FL; Tarpon Springs Library, Spring Boulevard, Tarpon Springs, FL, 34689.

Sydney Mine Sludge Ponds
Hillsborough County
Originally a phosphate strip mine, this site operated as a landfill from 1973 to 1981. A variety of municipal and industrial wastes were received, including oils and grease, septic system wastes, and sludges. Testing in 1979 revealed that the soil and groundwater are contaminated with VOCs. The contamination threatens water wells within 3 miles of the site, which supply about 4,200 residents.

Response actions included building a slurry wall and extracting and treating groundwater. Contaminated soil and sludge have been excavated and incinerated. By 1990 over 40 million gallons of groundwater had been cleaned. A deeper aquifer was found to be contaminated in 1991, and the groundwater recovery system was expanded to include this new area.

The EPA signed agreements with the PRPs that require the parties to finance the site investigations and remediation work.

Sources: PNPLS-FL; Brandon Public Library, 135 West Robertson Street, Brandon, FL, 33511.

Taylor Road Landfill
Hillsborough County
This former gravel pit was operated as a landfill from 1975 to 1984. A variety of municipal and industrial wastes were received, including sludges. Testing revealed that the soil and groundwater are contaminated with VOCs (benzene and vinyl chloride), methane gas, and lead. The contamination has penetrated some of the nearly 600 drinking water wells within 1 mile of the site. It threatens nearby dairy farms and the water supplies of about 30,000 residents.

Response actions included installing a cap, runoff controls, and a gas collection system (1983). Clean water supplies for affected residents were also provided. Ongoing site investigations are expected to recommend a final remediation plan by 1997.

A Consent Decree (1983) signed by the Hillsborough County Utilities Department required it to carry out the initial response actions and a 30-year groundwater monitoring program.

Sources: PNPLS-FL; Thonotosassa Library, 10715 Main Street, Thonotosassa, FL, 33592.

Tower Chemical Company
Lake County
This chemical manufacturing company 15 miles west of Orlando made pesticides from 1957 to 1981. Waste solutions and spent acids were discharged into an unlined evaporation pond. The pond occasionally overflowed into nearby wetlands and Lake Apopka, killing vegetation and wildlife. Testing has shown that the soil and groundwater contain high levels of copper, VOCs, and pesticides. The contamination threatens the drinking water supplies of about 1,000 people who live in the area.

Response actions included excavating drums of waste and contaminated soil and sediments (1983). Contaminated wastewater was also pumped from the pond. The EPA recommended installing a groundwater pump-and-treat system, incinerating contaminated soil, and connecting affected residents to clean water supplies. Over 100 million gallons of groundwater have been extracted. Other site investigations and monitoring are continuing.

Tower Chemical Company has refused to obey an EPA order in 1983 to clean up the site. Work to date has been financed by federal funds.

Sources: PNPLS-FL; Cooper Memorial Library, 620 West Montrose Street, Clermont, FL, 32711.

Tyndall Air Force Base
Bay County
This U.S. Air Force base near Panama City has been active since 1941. Activities have included gunnery, pilot, and weapons training and equipment testing. Testing in 1985, 1990, 1992, and 1993 revealed that the groundwater and soil contain the pesticides DDD, DDE, chlordane, and DDT. The chemicals have also been detected in recreational Shoal Point Bayou. The contamination is a threat to local drinking water wells, Shoal Point Bayou, and its associated ecosystems and fisheries.

This site was proposed to the NPL in 1996; a final decision is expected in 1997–1998. Additional studies are planned.

Source: EPA Publication 9320.7-071.

Whitehouse Oil Pits
Duval County
This was the site of seven unlined pits that held acidic waste oil sludges from 1958 to 1968. Two spills in

1976 poured over 200,000 gallons of waste into McGirts Creek. Testing has shown high levels of heavy metals, VOCs, and SVOCs in the soil and groundwater. About 6,000 people in the area use water from private wells.

Response actions included draining and stabilizing the pits (1986). The EPA recommended a remediation plan consisting of a slurry wall construction; groundwater extraction and treatment; dredging sediments from McGirts Creek; soil treatment through washing, biotreatment, and solidification; and capping. These systems should be operational by 1997.

The PRP, Allied Petroleum, went bankrupt in 1968. Federal funding has been used for the site studies and remediation work.

Sources: PNPLS-FL; Whitehouse Elementary School, 11160 General Avenue, Whitehouse, FL, 32220.

Whiting Field Naval Air Station
Santa Rosa County
This military installation near Milton has been in operation since 1943. Various wastes, including demolition debris, waste solvents, oil, fuel, machine fluids, sludges, solvents, and paint by-products, have been buried on-site. Testing by the State of Florida in 1986 showed that the groundwater and soil are contaminated with VOCs (benzene and trichloroethylene). About 7,000 people obtain their water from wells located within 4 miles of the base.

Response actions included installing carbon adsorption filters on the water supply wells on the base. Detailed site investigations are in progress and will recommend a final remediation plan.

Sources: PNPLS-FL; West Florida Regional Library, Milton Branch, 805 Alabama Street, Milton, FL, 32583.

Wilson Concepts
Broward County
This property has been the site of three metal machinery operations since 1967. Wastes from degreasing, deburring, steam cleaning, and spray coating activities were often discharged on the ground or overflowed from underground tanks. Testing between 1985 and 1987 revealed that the soil and groundwater are contaminated with heavy metals and VOCs. The contamination threatens the drinking water of about 100,000 people who use groundwater pumped from within 4 miles of the site.

Detailed site investigations were completed by the EPA in 1992. The EPA recommended only groundwater monitoring. The results indicate there is no threat to human health or the environment, and the site will be deleted from the NPL.

Two PRPs signed AOCs in 1989 with the EPA to conduct the site investigations. The parties did not comply, and the EPA took over the work in 1991.

Sources: PNPLS-FL; Broward County Library, 100 South Andrews Avenue, Fort Lauderdale, FL, 33301.

Wingate Road Dump
Broward County
This landfill in Fort Lauderdale operated from 1955 to 1978. It received a variety of residential, municipal, commercial, and incinerator wastes. One hundred steel drums filled with pesticides were discovered in 1985. Testing by the EPA identified high levels of pesticides (DDT, aldrin, dieldrin, and chlordane) and dioxins in the soil and groundwater. About 350,000 people depend on groundwater pumped from wells within 3 miles of the site. The contamination has penetrated Rock Pit Lake and threatens the Biscayne Aquifer.

EPA site investigations (1991–1996) will recommend a final remediation plan. The EPA is also conducting a search for PRPs.

Sources: PNPLS-FL; Broward County Library, 1000 South Andrews Avenue, Fort Lauderdale, FL, 33301.

Yellow Water Road Dump
Duval County
This landfill near the town of Baldwin operated from 1981 to 1984. PCB-contaminated fluids and oils from recycled transformers were stored at the site in drums and tanks, which leaked about 63,000 gallons into the ground. Testing has shown that the soil and groundwater are contaminated with PCBs, iron, lead, hexachlorobenzene, and arochlor. About 500 residents within 1 mile of the site use private water wells.

Response actions included removing drums and transformers (1984), cleaning a contaminated pond (1985), and stockpiling contaminated soil (1988). In 1990 the EPA recommended excavating, stabilizing, and capping contaminated soil along with long-term groundwater monitoring.

A Unilateral Administrative Order (1991) requires the PRPs to finance the remediation work.

Sources: PNPLS-FL; Baldwin Town Hall, 10 U.S. Highway 90, West Baldwin, FL, 32234.

Zellwood Groundwater Contamination
Orange County
This site in Zellwood is occupied by four businesses that have operated since the 1960s. Activities include recycling and crushing steel drums, manufacturing chemicals and fertilizers, and washing and packing vegetables. Various contaminated rinsewaters and wastewaters were discharged into ditches or unlined ponds. Testing in 1982 showed that the groundwater and soil

have high levels of PAHs, pesticides, and heavy metals (chromium and lead). The contamination threatens the local drinking water of about 7,000 residents.

Response actions included the removal of buried drums (1982). EPA studies (1982–1990) recommend a final remediation plan of excavation, solidification, and capping of contaminated soil. A groundwater decision is expected by 1997.

Sources: PNPLS-FL; Zellwood Elementary School, 3551 East Washington Street, Zellwood, FL, 32798.

GEORGIA

Cedartown Industries
Polk County
This industrial facility in Cedartown operated from 1978 to 1980, originally as a machine shop that recycled lead from automobile batteries. Smelter residues and waste were left behind when the company sold the property in 1980. Tests show that the groundwater and soil contain lead. The contamination threatens recreational Cedar Creek and the drinking water source for about 26,000 local residents.

Response actions included removing the slag pile and contaminated soil (1990). In 1993 the EPA recommended excavating and stabilizing contaminated soil, monitoring groundwater, and installing a groundwater pump-and-treat system if the contamination plume did not naturally attenuate. Construction is expected to be under way by 1997.

Sources: PNPLS-GA; Cedartown Public Library, 245 East Avenue, Cedartown, GA, 30125.

Cedartown Landfill
Polk County
This abandoned iron ore mine near Cedartown was used as a landfill from the 1960s to 1980. It received a variety of municipal and industrial wastes, including heavy metals, solvents, and degreasers. Testing in the 1980s revealed excessive levels of heavy metals (beryllium, cadmium, chromium, lead, and manganese) in the soil and groundwater. The contamination threatens the local drinking water supplies of about 34,000 people in Polk County.

Detailed site investigations by the EPA recommended leachate collection and groundwater monitoring. If the monitoring shows that the contamination is not naturally attenuating, a groundwater extraction and treatment system will be installed. Monitoring is under way.

A Consent Order (1990) requires 15 PRPs to undertake the site investigations and remediation work.

Sources: PNPLS-GA; Cedartown Public Library, 245 East Avenue, Cedartown, GA, 30125.

Diamond Shamrock Landfill
Polk County
This former landfill in Cedartown operated in the 1970s. A variety of industrial wastes, including fungicides, oil sludges, alcohols, and metallic salts, were buried in unlined trenches. Testing showed that the soil and groundwater contain heavy metals and VOCs, including toluene. The contamination threatens Cedar Creek, the Coosa River, and the local groundwater supply for Cedartown.

Drummed waste and contaminated soil were removed in 1990. Other soil areas were biotreated. About 9,000 gallons of liquid waste were also recovered. In 1994 the EPA recommended a groundwater program of natural attenuation and long-term monitoring.

PRPs identified by the EPA are required to conduct the site investigations and remediation work.

Sources: PNPLS-GA; Cedartown Public Library, 245 East Avenue, Cedartown, GA, 30125.

Firestone Tire and Rubber Company, Albany
Dougherty County
Tires have been manufactured on this site near Albany since 1968. Industrial wastes were stored or buried on the property. Testing has shown that the groundwater and soil contain VOCs and heavy metals. The contamination threatens the local drinking water supply for about 500 residents. The groundwater is also used to irrigate crops.

In 1993 the EPA recommended installing a groundwater extraction and treatment system and excavating contaminated soil. Additional groundwater studies should be completed by 1997.

Firestone Company, the PRP, has undertaken the site investigations and remediation work.

Sources: PNPLS-GA; Dougherty Public Library, 300 Pine Avenue, Albany, GA, 31701.

Hercules 009 Landfill
Glynn County
This former landfill site near Brunswick operated in the 1970s. It received about 20,000 tons of waste from a company that produced pesticides. Testing by the state and the EPA indicated that the shallow and deep groundwater and soil are contaminated with the insecticide toxaphene. The contamination threatens coastal wetlands and numerous private wells in the area.

Response actions included connecting affected residents to clean water supplies (1992). Detailed site

GEORGIA

investigations were conducted between 1988 and 1992. Excavation of contaminated residential soil is in progress. An in-situ soil solidification program should be operational by 1997.

Hercules Corporation, the PRP, signed a Consent Order (1988) and Consent Decree (1991) with the EPA, accepting responsibility to fund the site investigations and remediation work.

Sources: PNPLS-GA; Brunswick-Glynn County Regional Library, 208 Gloucester Street, Brunswick, GA, 31523.

LCP Chemicals

Glynn County

Petroleum refining and chemical manufacturing have been conducted on this site in Brunswick from the 1920s to the present. Products include caustic soda, hydrochloric acid, and chlorine. Various chemical wastes, including mercury sludges, were stored in lagoons and pits near a tidal marsh. Testing by the EPA in 1989 revealed that the soils and groundwater contained dioxin, mercury, PCBs, and lead. Shellfish offshore also contained high levels of these compounds. The contamination threatens the private and municipal water supplies for about 35,000 residents as well as Purvis Creek, Turtle River, and surrounding wetlands.

Response actions included removing contaminated soils (1995). Three PRPs—AlliedSignal, Georgia Power, and ARCO—are building surface run-off controls and removing contaminated soil and buildings. Site investigations are under way.

Source: EPA Publication 9320.7-071, Volume 3, No. 1.

Marine Corps Logistics Base

Dougherty County

This U.S. Marine Corps facility near Albany repairs and maintains military equipment, vehicles, and fuel storage areas. Related wastes such as fuels, oil, solvents, sludges, and paints have been deposited on the base. Testing in 1986 revealed that the soil and groundwater contain DDE, DDT, PCBs, heavy metals, and VOCs. About 4,500 people on the base use the local groundwater for domestic purposes.

Response actions included excavating sludge beds, capping, and installing an interim groundwater extraction and treatment system. Site investigations are nearing completion and will recommend a final remediation plan by 1997.

Sources: PNPLS-GA; Dougherty County Public Library, 300 Pine Street, Albany, GA, 31701.

Marzone/Chevron Chemical
Tift County
Marzone Incorporated manufactured pesticides on this site near Tifton from 1950 to 1964. Chevron Chemical Company also produced and stored chemicals, including solvents and toxaphene, on the property during that time. A succession of other agricultural chemical companies has used the facility since 1970. Liquid wastes were discharged into an unlined lagoon on the property. Testing in the 1980s by the State of Georgia indicated that the groundwater and soils contain toxaphene, DDT, and VOCs. The contamination threatens about 35 private drinking water wells in the area.

Response actions included removing stored wastes, decontaminating abandoned buildings, removing contaminated soil, and digging out the wastewater lagoon (1984–1985). EPA site investigations from 1990 to 1993 recommended low-temperature thermal desorption of soil and a groundwater extraction and treatment system.

PRPs Chevron Chemical Company and Kova Fertilizer participated in the 1984–1985 clean-up efforts. A Consent Order (1990) with four PRPs requires them to conduct the site investigations and remediation work.

Sources: PNPLS-GA; Tifton-Tift County Public Library, One Library Lane, Tifton, GA, 31794.

Mathis Brothers Landfill
Walker County
This former landfill in the town of La Fayette operated from 1974 to 1980. It received a variety of municipal and industrial wastes, including latex, benzonitrile, dicama, and dichlorobenzene. Testing has shown that the soil contains latex and herbicide residues. Fish kills have been reported in the area. The contamination threatens nearby recreational waterways and water wells that supply about 5,000 local residents.

Site investigations that recommended capping and the excavation and incineration or biodegradation of contaminated soil were completed in 1992. Construction should be finished in 1997.

A number of PRPs are contributing to the cost of the site studies and clean-up work.

Sources: PNPLS-GA; La Fayette County Commissioner's Office, Highway 136, La Fayette, GA, 30728.

Monsanto, Augusta Plant
Richmond County
Various wastes from Monsanto Corporation's plant in Augusta were buried in on-site landfills from 1966 to 1977. The hazardous wastes include arsenic trisulfide and phosphoric acid sludge. Testing in the 1980s discovered that the soil and groundwater are contaminated with arsenic. The contamination threatens Butler Creek, Phinizy Swamp, and the drinking water supplies of about 2,000 residents.

Response actions included removing arsenic wastes in 1983. In 1990 the EPA recommended a final remediation plan of installing a groundwater pump-and-treat system and quarterly groundwater monitoring.

Monsanto Corporation, the PRP, signed AOCs (1989, 1990) with the EPA that require the company to fund the site investigation and remediation work.

Sources: PNPLS-GA; Augusta-Richmond County Public Library, 902 Green Street, Augusta, GA, 30901.

Powersville Landfill
Peach County
This former gravel pit in Powersville operated as a landfill from the 1940s to 1979. It received a variety of waste, including pesticides. In response to complaints about the taste of drinking water, in 1983 the State of Georgia conducted tests that indicated the groundwater and soil are contaminated with VOCs, heavy metals, and pesticides. The contamination threatens the local groundwater, which is used for domestic and agricultural purposes.

In 1987 the EPA recommended a remediation plan of deed restrictions, capping, connecting affected residents to city water supplies, and long-term groundwater monitoring. Construction was completed in 1993 and monitoring is continuing.

A 1988 Consent Decree requires PRPs to fund the site investigations and remediation work.

Source: PNPLS-GA.

Robins Air Force Base
Houston County
This 9,000-acre U.S. Air Force base is located near the town of Warner Robins. Various industrial wastes from activities on the base were buried or discharged in pits and lagoons. Testing has shown that the groundwater and soils are contaminated with heavy

metals (cadmium, cyanide, and lead), VOCs, DDT, and PCBs. The contamination threatens wetlands, the Ocmulgee River, and the drinking water supplies of 10,000 nearby residents.

Detailed site investigations by the Air Force recommended the solidification of sludges, collection of leachate, capping, and construction of sludge lagoon groundwater recovery wells. Additional studies on groundwater and wetlands restoration are in progress.

Sources: Nola Brantly Memorial Library, 721 Watson Boulevard, Warner Robins, GA, 31093.

T. H. Agriculture and Nutrition Company
Dougherty County
Pesticides were stored and processed on this site in Albany from the 1960s until 1982. Testing has shown that the groundwater and soil contain toxaphene, lindane, DDT, and methyl parathion. The contamination threatens recreational Lake Worth, Flint River, Muckalee Creek, Kinchafoonee Creek, and the local groundwater that is used for drinking water and for agricultural purposes.

Response actions included removing contaminated soil and debris (1984). Additional soil was removed in 1992. About 3,000 tons of soil were thermally treated on-site (1992–1993). A groundwater pump-and-treat system is being installed.

An AOC (1990) with T. H. Agriculture and Nutrition and UAOs with other PRPs (1992) require the companies to fund the site investigation and remediation work.

Sources: PNPLS-GA; Dougherty County Public Library, 300 Pine Avenue, Albany, GA, 31701.

Woolfolk Chemical Works
Peach County
Woolfolk Chemical Works Incorporated has produced agricultural chemicals and pesticides at this site in Fort Valley since 1910. Wastewater was being discharged into on-site sewers or nearby creeks as late as 1979. Testing has shown that the groundwater and soil contain heavy metals, VOCs, dioxins, chlordane, DDT, lindane, and toxaphene. Nearby residential yards contain arsenic. About 10,000 people depend on water supplies drawn from within 3 miles of the site. The contamination has also penetrated Bay Creek and Big Indian Creek.

Response actions included removing contaminated soil and capping (1986–1987). Contaminated soil was also removed from residential yards, and other contaminated properties have been bought by a PRP. A groundwater extraction and treatment system is expected to be operational by 1997.

The EPA has targeted PRPs for cost recovery and site remediation.

Sources: PNPLS-GA; Thomas Public Library, 213 Persons Street, Fort Valley, GA, 31030.

GUAM

Anderson Air Force Base
This U.S. Air Force base is located on 20,000 acres in the city of Yigo. Since 1940 the base has supported Strategic Air Command operations. Waste products from aircraft maintenance activities, such as trichloroethane, paint thinners, fuels, pesticides, antifreeze, PCBs, and cleaning compounds, were discharged into unlined landfills or stored in tanks and drums. The soil and groundwater are contaminated with VOCs, lead, chromium, SVOCs, PCBs, fuel, and pesticides. About 40,000 people are serviced by private water wells within 4 miles of the site.

Response actions included closing and capping the landfill (1994). A number of site investigations were initiated between 1993 and 1996 and will continue for at least 5 to 8 years. A final remediation plan will be recommended after the data have been evaluated.

Sources: PNPLS-UST; Robert F. Kennedy Memorial Library, Mangilao, Guam, 96923; Nieres M. Flores Library, Agana, Guam, 96910.

Ordot Landfill
This landfill near the villages of Ordot and Chalan Pago has been in operation since World War II. It has received a variety of municipal and industrial wastes, including explosives and PCB-contaminated oils from transformers. Testing showed that the groundwater and soil contain heavy metals, phthalates, and VOCs. The contamination could threaten Lonfit River, Pago Bay, and the water supplies of island residents.

EPA site investigations concluded that the threat to human health and the environment was minimal and recommended no further action. Groundwater monitoring wells were installed in 1992 and will be monitored for at least 5 years to determine if any remedial actions are needed.

Source: PNPLS-UST.

GUAM

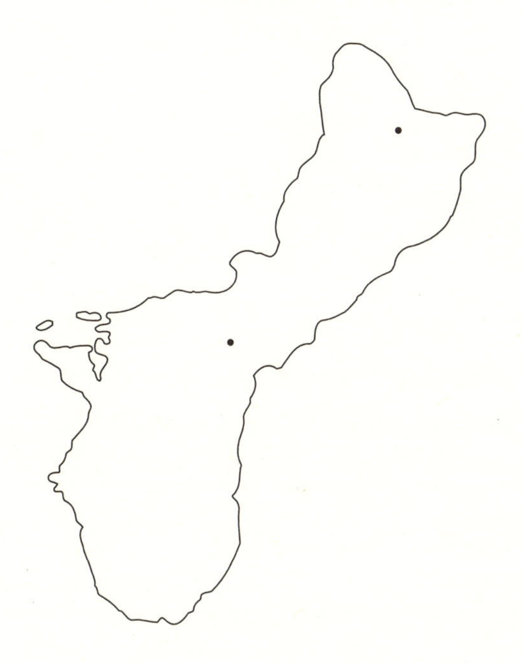

HAWAII

Del Monte Corporation
Honolulu County
The Del Monte Corporation began growing pineapple on this 6,000-acre site on the island of Oahu in the 1940s, using various fumigants, such as ethylene dibromide, to control mematodes. Testing by the state in 1983 showed that local wells and soil were contaminated with fumigants ethylene dibromide and 1,2-dibromo-3-chloropropane.

Response actions included removing 18,000 tons of soil contaminated by fumigant spills or leaking storage containers. A site investigation that will identify other sources of contamination and recommend a clean-up plan is in progress.

Sources: PNPLS-HI; Wahiawa Public Library, 820 California Avenue, Wahiawa, HI, 96786.

Naval Computer and Telecommunications Area
Honolulu County
This sprawling U.S. naval area consists of the facilities of the Naval Computer and Telecommunications Area Master Station in Wahiawa, Lualualei, Opana, Kokekole Pass, and Pearl Harbor. Fourteen areas of hazardous waste sites—mostly old landfills, storage areas, and transformer locations—have been identified in Lualualei and Wahiawa. The soil and groundwater are contaminated with PCBs, chlorinated solvents, petrochemicals, VOCs, creosote, and heavy metals. The contamination threatens various wildlife habitats, nearby residents, and certain Pacific Ocean fishing beds.

Response actions included excavating PCB-contaminated soil (1990–1991). A detailed site investigation to recommend a final remediation is planned.

Source: PNPLS-HI.

Pearl Harbor Naval Complex
Honolulu County
Since World War II this U.S. naval base has been a military industrial center. On-site disposal areas have received a variety of wastes, including pesticides, chromic acid, PCBs, waste oils and fuel, and heavy metals (mercury). Groundwater and soil are contaminated by mercury, chromium, pesticides, VOCs, and PCBs. The contamination threatens the National Wildlife Refuge, numerous ocean fisheries, and various private, municipal, and industrial water resources.

A site investigation and feasibility study is under way; a final remediation plan will be recommended upon completion.

Source: PNPLS-HI.

HAWAII

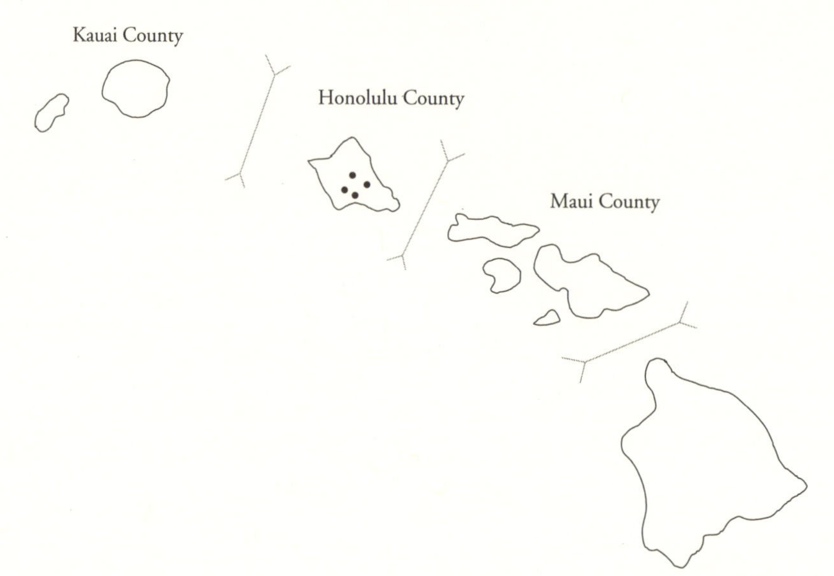

Schofield Barracks
Honolulu County
This 17,000-acre site near Wahiawa and Oahu is a U.S. Army base that maintains and repairs various vehicles and equipment. Various spent organic solvents, paints, and degreasers were discharged on-site. Testing in 1985 showed that the soil and groundwater are contaminated with VOCs (trichloroethylene). About 60,000 people depend on well water from within 3 miles of the site; the recreational Wahiawa Reservoir about 3 miles away is also used to irrigate pineapple crops.

Response actions included the installation of an air-stripping system on four existing wells (1986). A site investigation started in 1991 should be completed by 1997 and will recommend additional remedies.

Source: PNPLS-HI.

IDAHO

Arrcom (Drexler Enterprises)
Kootenai County
This facility near Rathdrum operated as a recycling plant from 1960 to 1982. Recycled materials included oils, solvents, lead, and PCBs. Waste by-products, including sludges, were stored on-site. Tests showed high levels of VOCs (xylene and methyl ethyl ketone), lead, mercury, acid, PCBs, and PCP in soils on the property. The contamination threatened the aquifer that supplies groundwater to about 6,000 local residents. Farmers also use the water for irrigating fields.

Response actions included removing hazardous waste and debris from the property, including contaminated soil (1983). Buildings were razed and more material removed between 1987 and 1990. The site was deleted from the NPL in 1992. Monitoring will continue.

Sources: PNPLS-ID; Rathdrum Library, 731 South First Street, Rathdrum, ID, 83858.

Blackbird Mine
Lemhi County
Located in the Salmon National Forest, this area of about 11,000 mining claims has been mined by a number of companies for cobalt and copper. An open pit and underground shafts are on the property. Noranda Mining Company mined the last ore in 1982. Tailings, waste rock piles, and a open pit generate acid drainage and leachate that have contaminated the local drainages that flow into the Panther Creek basin with acid, copper, cobalt, arsenic, and

IDAHO

Boundary
Bonner
Kootenai
Shoshone
Benewah
Latah
Clearwater
Nez Perce
Lewis
Idaho
Adams
Valley
Lemhi
Washington
Custer
Clark
Fremont
Payette
Boise
Gem
Butte
Jefferson
Madison
Teton
Canyon
Ada
Elmore
Camas
Blaine
Bonneville
Bingham
Gooding
Lincoln
Caribou
Jerome
Minidoka
Power
Bannock
Owyhee
Twin Falls
Cassia
Oneida
Franklin
Bear Lake

nickel. The contamination threatens the endangered chinook and sockeye salmon.

Response actions have included stabilizing a tailings dam and diverting West Fork Creek (1993). Waste rock piles and tailings were removed from the site in 1995–1996. A site investigation will be completed in 1997 with a recommendation for a final clean-up plan.

The State of Idaho filed a lawsuit against Noranda Mining Company and two previous operators in 1983. An Administrative Order (1994) between the EPA and the PRPs requires the companies to finance the site investigation and remediation plan.

Sources: PNPLS-ID; Salmon Public Library, 204 Main Street, Salmon, ID, 83467.

Bunker Hill Mining and Metallurgical Complex
Shoshone County
This area covering 21 square miles includes the towns of Kellogg, Wardner, Page, Smelterville, and Pinehurst. The Bunker Hill Mining and Metallurgical Company was the main employer in the area from 1889 to the 1970s. The facility mined lead and zinc and included smelters and phosphoric acid, sulfuric acid, and fertilizer processing plants. Heavy metal contamination is widespread from unregulated early mining, including discharge of mine tailings into the Coeur d'Alene River. Tailings ponds built in the 1970s to control the problem leaked or seeped because they were unlined. Sulfuric emissions from the smelter stacks have killed extensive areas of vegetation. Groundwater, surface water, soil, and stream sediments are contaminated with heavy metals (lead, zinc, cadmium, arsenic, and mercury). Dangerous levels of lead were discovered in children in Kellogg in the 1970s.

Response actions included excavation of contaminated soil from the mine sites and neighborhoods (1986–1991). Over 400 neighborhood yards have been detoxified. Fencing, erosion control, and revegetation was started in 1991. Sludges were removed in 1991–1992. Other clean-up phases in unpopulated parts of the complex are continuing.

An Administrative Order with Gulf Resources (1987), a Unilateral Order with Gulf Resources and Bunker Hill (1991), and a Consent Decree with other PRPs (1994) have required the PRPs to contribute to the clean-up work.

Sources: PNPLS-ID; Kellogg Public Library, 16 West Market Avenue, Kellogg, ID, 83837.

Eastern Michaud Flats Contamination
Power and Bannock Counties
FMC Corporation and J. R. Simplot Company have operated phosphate processing facilities on this site near Pocatello. Crushed waste rock, slag, and slurry stored on-site in piles or unlined holding ponds contain heavy metals. A drinking well near the site was condemned in 1976 because of high arsenic levels. The groundwater and soils are contaminated with arsenic, chromium, copper, nickel, phosphorus, zinc, cadmium, fluoride, nitrate, selenium, sodium, and sulfate. Groundwater from within 3 miles of the site supplies about 55,000 people and is used to irrigate thousands of acres of farmland. The contamination also threatens important fishing and recreation waterways, such as the Portneuf River.

Response actions included closing an unlined pond in 1993. A study begun in 1991 is expected to recommend a clean-up plan by 1997.

An Administrative Order on Consent with the two PRPs requires them to finance the site investigation and remediation plan.

Sources: PNPLS-ID; Idaho State University Library, Document Department, P.O. Box 8089, Pocatello, ID, 83209.

Idaho National Engineering Laboratory (INEL)
Butte County
This site was established as a nuclear reactor testing station in 1949 by the Atomic Energy Commission. It is now an engineering research center for the U.S. Department of Energy. Waste materials deposited on-site in landfills and wells include cooling tower water, radionuclides, and sulfuric acid. The site has contaminated the soils and groundwater with chromium, lead, mercury, acetone, sodium hydroxide, sulfuric acid, and VOCs. The contaminants have penetrated the underlying aquifer, a major water source in southeastern Idaho.

Response actions included removal of debris, solid waste, and contaminated soil (1993); treatment of radioactive soil (1993–1995); groundwater extraction and treatment (1994); and gas treatment and landfill caps (1994–1995). Other investigations to address other aspects of the clean-up plan will continue until 1998.

A Consent Order (1987) and Interagency Agreement (1991) between the EPA and Idaho National Engineering Laboratory have assured clean up of the site.

Sources: PNPLS-ID; INEL Technical Library, 1776 Science Center Drive, Idaho Falls, ID, 83415.

Kerr-McGee Chemical Corporation
Caribou County
This manufacturing plant located 1 mile north of Soda Springs has been in operation since 1963.

Kerr-McGee Company manufactures vanadium pentoxide from ferrous-phosphate raw material. Liquid wastes containing vanadium, arsenic, copper, and silver are stored in on-site ponds; the chemicals have leaked into the soils and underlying groundwater. About 3,000 people live within 3 miles of the site; a number of private wells also irrigate farmland.

Response actions included a site investigation (1991–1995). A final remediation plan has yet to be selected.

An Administrative Order on Consent between Kerr-McGee and the EPA calls for the company to finance the site investigation and remediation plan.

Sources: PNPLS-ID; Soda Springs Public Library, 149 South Main Street, Soda Springs, ID, 83276.

Monsanto Chemical Company
Caribou County
This facility located about 1 mile north of Soda Springs is owned by Monsanto Chemical Company and processes phosphate ore. Soils and groundwater have high levels of cadmium, selenium, vanadium, and fluoride from stockpiles of ore, ore waste, slag piles, and holding ponds containing wastewater. About 4,000 people live within 3 miles of the site.

Response actions included lining wastewater ponds and removing pockets of contaminated soil. Over 50 groundwater monitoring wells supply data on the migration of the contamination plume. A detailed EPA study (1991–1995) has yet to select a final remediation plan.

Sources: PNPLS-ID; Soda Springs Public Library, 149 South Main Street, Soda Springs, ID, 83276.

Mountain Home Air Force Base
Elmore County
Established by the U.S. Air Force in 1943 as a tactical command base, this site has served as an aircraft maintenance facility. Various waste materials, including oils, solvents, pesticides, and solid waste, have been deposited in landfills and sewage systems on the property. Drinking water wells on the base are contaminated with VOCs, benzene, PAHs, bromoform, DDT, lindane, and dieldrin. About 14,000 people use groundwater derived from within 3 miles of the site for drinking and irrigation purposes.

Response actions included extracting radioactive soil and solid waste. EPA investigations (1991–1996) will result in the selection of a final remediation plan.

Sources: PNPLS-ID; Mountain Home Public Library, 790 North Tenth Street East, Mountain Home, ID, 84647.

Pacific Hide and Fur Recycling Company
Bannock County
Located in Pocatello, this metal salvage yard was operated from the 1950s to 1983 by McCartys, Pacific Hide and Fur Depot, and Union Pacific Railroad. Pits on the property are filled with waste sludge and automotive scrap, including oil filters and battery and transformer casings. The soil is contaminated with PCBs, lead, and other inorganic compounds. The 45,000 residents of Pocatello draw their water supplies from groundwater sources within 3 miles of the site; the recreational Portneuf River is less than 1,500 feet from the site.

Response actions included the removal of scrap waste and contaminated soil, fencing, installation of groundwater monitoring wells, and decontamination of large scrap metal (1983–1989). Soil treatments, including excavation, fixation, and capping, were completed from 1988 to 1993. Lead-contaminated soil was excavated and treated in 1995–1996. Groundwater monitoring will continue.

Sources: PNPLS-ID; Pocatello Public Library, 812 East Clarke Street, Pocatello, ID, 83201.

Triumph Mine Tailings Piles
Blaine County
The Triumph Mine produced silver, zinc, and lead ore from 1882 to 1957. The ore was processed on-site through crushing, grinding, and flotation operations. The waste slurries were pumped into two large tailings piles. EPA tests in 1991–1992 showed that the groundwater, soils, surface water, and nearby stream sediments have high concentrations of heavy metals, including lead, arsenic, zinc, and silver. Nearby wetlands, residential yards, and a municipal well are also contaminated. The recreational Wood River is located at the edge of tailings piles.

A detailed site investigation study by the EPA and State of Idaho is under way to determine a clean-up plan.

Source: PNPLS-ID.

Union Pacific Railroad Company
Bannock County
Located in Pocatello, this site is an unlined pit in which the Union Pacific Railroad Company dumped wastewater treatment plant sludge from 1961 to 1983. EPA tests in 1983 discovered seepage from the pit that was contaminating the groundwater and soil with heavy metals, PAHs, and organic compounds. About 45,000 people live within 4 miles of the site; a number of wells, including Pocatello's municipal supply system, are within 3 miles of the site.

Response actions included fencing the site and restricting public access. Removal of sludge and contaminated soil, capping, and installation of groundwater extraction and treatment systems were completed between 1991 and 1995. Groundwater monitoring is continuing.

An Administrative Order (1988) and Consent Decree (1992) between the EPA and Union Pacific Railroad Company required the company to finance the site investigation and remediation plan.

Sources: PNPLS-ID; Pocatello Public Library, 812 East Clarke Street, Pocatello, ID, 83201.

ILLINOIS

A & F Material
Cumberland County
This waste reclamation facility near Greenup operated from 1977 to 1980. It received a variety of industrial wastes, including oil, sludge, and acids. In 1978 storage lagoons overflowed into the Embarras River. Testing has shown that the groundwater and soils contain low levels of VOCs, sulfates, petroleum, PNAs, and heavy metals. The contamination threatens the county fairgrounds, the Embarras River, and the drinking water supply for about 2,000 residents.

Lagoon dikes were fixed in 1980. Lagoon wastes were pumped and treated in 1982–1983. The lagoons were excavated, consolidated sludges capped, and bulk and drummed wastes exhumed (1983–1985). In 1986 the EPA recommended razing the buildings, removing all contaminated soils, and monitoring groundwater to make sure contamination was naturally attenuating. Groundwater monitoring was started in 1990.

Four PRPs have agreed to clean up the property through Consent Decrees (1984, 1989).

Sources: PNPLS-IL; Greenup City Clerk's Office, Greenup Municipal Building, 115 Cumberland Road, Greenup, IL, 62428.

Acme Solvent Reclamation Plant
Winnebago County
This solvent reclamation facility near Rockford operated from 1960 to 1973. Solvents, oils, paints, and sludges were discharged into lagoons on the property or stored in tanks and drums. After being cited for illegal operations, the company failed to close the facility properly according to state regulations. Testing in the 1980s showed that the groundwater and soils contain VOCs, PCBs, chromium, lead, and phthalates. Methane gas was also detected in nearby homes. About 6,000 people within 3 miles of the site use private well water. The contamination also threatens recreational Killbuck Creek and the Kishwaukee and Rock Rivers.

Response actions included removing contaminated soil and debris (1987) and connecting affected residents to clean water supplies or installing carbon filtration units. In 1995 the EPA recommended low-temperature treatment of contaminated soil, soil vapor extraction, groundwater extraction and air-stripping, and long-term groundwater monitoring. Soil treatment is completed, and the groundwater system should be operational by 1997.

Acme Solvent Company signed a 1986 Consent Order with the EPA to conduct site investigations and clean up the site.

Sources: PNPLS-IL; Rockford Public Library, 215 North Wyman Road, Rockford, IL, 61101.

Amoco Chemicals
Will County
From 1958 to 1976 this industrial landfill site near Joliet received industrial wastes from Amoco Chemical Company, including spent solvents, acids, and heavy metal sludges. Leachate regularly seeped into the Des Plaines River until 1976. Testing in 1982 confirmed that the groundwater and soils contain benzene, xylene, other VOCs, cadmium, copper, chromium, and lead. The contamination threatens the recreational Des Plaines River and the private water wells of about 1,000 people in the area.

A detailed site investigation begun in 1994 will recommend a final remediation plan.

Amoco Chemicals Company has been identified as the principal PRP.

Source: PNPLS-IL.

Beloit Corporation
Winnebago County
Beloit Corporation researches and manufactures paper-making machines at this plant in Rockford. Wastewaters were held in three unlined lagoons on the property, and the contaminated sediment was later spread on the ground. Testing in 1985 revealed that the soil and groundwater contain VOCs. About 15,000 people use private or public water wells within 3 miles of the site, and the Rock River is less than 50 feet away.

Carbon treatment units were installed in affected homes. A site investigation that will recommend a final remediation plan is in progress.

Sources: PNPLS-IL; Talcott Free Library, 101 East Main Street, Rockton, IL, 61072.

ILLINOIS

Jo Daviess
Stephenson
Winnebago
Boone
McHenry
Lake
Carroll
Ogle
DeKalb
Kane
DuPage
Cook
Whiteside
Lee
Kendall
Will
Rock Island
Henry
Bureau
La Salle
Grundy
Mercer
Putnam
Kankakee
Stark
Marshall
Knox
Warren
Peoria
Woodford
Livingston
Iroquois
Henderson
McDonough
Fulton
Tazewell
McLean
Ford
Hancock
Mason
Schuyler
Logan
De Witt
Champaign
Adams
Brown
Cass
Menard
Piatt
Vermilion
Macon
Sangamon
Morgan
Douglas
Edgar
Pike
Scott
Moultrie
Christian
Coles
Greene
Shelby
Cumberland
Clark
Calhoun
Macoupin
Montgomery
Jersey
Fayette
Effingham
Jasper
Crawford
Madison
Bond
Clay
Lawrence
Marion
Richland
Clinton
St. Clair
Wayne
Wabash
Edwards
Washington
Jefferson
Monroe
White
Randolph
Perry
Franklin
Hamilton
Saline
Jackson
Williamson
Gallatin
Union
Johnson
Pope
Hardin
Alexander
Massac
Pulaski

Belvidere Landfill
Boone County
This former landfill site near Belvidere operated from 1939 to 1973. It received a variety of municipal and industrial wastes, including paint, heavy metal, and sewage sludges. Testing by the State of Illinois showed that the groundwater and soil contain VOCs, PAHs, PCBs, nitrite, and heavy metals. The contamination threatens the Kishwaukee River, West and East Ponds, and the water sources for about 14,000 nearby residents

Response actions included removing debris and contaminated soil and sludge in 1986. In 1988 the EPA recommended capping, a groundwater treatment system, fencing, erosion controls, and long-term groundwater monitoring. The groundwater extraction/treatment system became operational in 1992.

Sources: PNPLS-IL; Ida Public Library, 320 North State Road, Belvidere, IL, 61008.

Byron Salvage Yard
Ogle County
This salvage yard near Byron operated in the 1960s and 1970s, processing a variety of wastes, including paint and heavy metal solutions, oils, solvents, and scrap metal. Hazardous liquid wastes held in unlined ponds frequently overflowed. Testing in 1976 discovered that the soil and groundwater contain cyanide, heavy metals, and VOCs. The contamination has penetrated nearby Meyer's Spring and the private water supplies of nearly 130 homes in the area.

Response actions included fencing and supplying bottled water to affected residents (1984). Free-product petroleum was removed from the groundwater in 1988, and affected residents were connected to the city water supply system (1988–1989, 1993). Contaminated soils and buried drums were removed by 1986. Groundwater monitoring is under way to check the natural attenuation of the contamination plume.

Sources: PNPLS-IL; Byron Public Library, 109 North Franklin Street, Byron, IL, 61010.

Central Illinois Public Service Company
Christian County
This former coal gasification plant near Taylorville operated from 1892 to 1932. Residual coal tars and buried tanks were discovered during construction in 1985. Testing in 1986 revealed that the soil and groundwater contain PAHs and VOCs. Runoff has contaminated nearby properties. Water wells that serve about 13,000 people are located within 4 miles of the site.

Response actions included removing underground tanks and contaminated soil in 1987. Affected residents were also connected to the municipal water supply system. A groundwater extraction and treatment system was constructed in 1994 and is operational.

PRPs have been involved in the site investigations and remediation work.

Sources: PNPLS-IL; Taylorville Public Library, 121 West Vine Street, Taylorville, IL, 62568.

Circle Smelting Corporation
Clinton County
Zinc was recovered from ore and scrap metal at this site near Beckemeyer from about 1904 to the 1960s. Various smelting waste, including a 17-acre slag pile and coal cinders, were left on the property. Smelter emissions have contaminated surrounding properties with heavy metals. Testing in 1992 has shown that the groundwater, surface water, and soil contain zinc, lead, nickel, copper, and cadmium. The metals have been detected in Beaver Creek and on the premises of the Beckmeyer Elementary Public School. The contamination threatens Beaver Creek, associated wetlands, and the drinking water supplies of about local 800 residents, businesses, and schools.

This site was proposed as an NPL site in 1996; a final decision is expected in 1997–1998. Response actions have included first-phase studies that have estimated that about 21 million square feet of soil are contaminated. Other studies are under way.

Source: EPA Publication 9320.1-071.

Cross Brothers Recycling
Kankakee County
A former drum and pail reclamation facility operated on this site near Kankakee from 1961 to 1980. Residues in the containers were stripped using toxic solvents and discharged on the ground. Containers were buried on the property. Testing has revealed that the groundwater and soil contain excessive amounts of VOCs, PCBs, and heavy metals. The contamination threatens agricultural wells and the water supplies for about 3,500 nearby residents.

Contaminated wells were deepened to provide clean water to affected residents. Drums of waste and contaminated soils were removed in 1985. In 1989 the EPA recommended soil removal, soil washing, groundwater extraction and treatment, fencing, and capping. Construction began in 1994.

The PRPs have been involved in the site studies and remediation work.

Sources: PNPLS-IL; Kankakee Public Library, 304 South Indiana Road, Kankakee, IL, 60901.

DuPage County Landfill
DuPage County
This former landfill in the Blackwell Forest Preserve near Warrenville operated from 1965 to 1970. It received a variety of demolition, municipal, and hazardous/industrial wastes. About 45,000 people depend on water drawn from wells within 3 miles of the site. Testing has indicated anomalous levels of VOCs in the soil and groundwater.

Leachate collection began in 1987. Detailed site investigations (1989–1995) will recommend a final remediation plan.

The Forest Preserve District, a co-operator of the landfill and PRP, signed a 1989 AOC with the EPA to conduct the site investigation.

Sources: PNPLS-IL; Warrenville Public Library, 28 West 751 Stafford Place, Warrenville, IL, 60555.

Galesburg/Koppers Company
Knox County
This former wood treatment plant near Galesburg operated from 1907 to the 1980s. Railroad ties were treated with PCP, creosote, coal tar, or fuel oil. Wastewaters were stored in unlined slurry ponds and lagoons or sprayed on fields. The soils and groundwater contain VOCs, phenols, asbestos, heavy metals, PCBs, and PAHs. The contamination threatens agricultural wells, recreational Brush Creek, Lake Bracken, and the drinking water supplies for about 60,000 people.

Response actions included removing liquid wastes and contaminated soils from the ponds and lagoons (1983). In 1989 the State of Illinois recommended bioremediation of excavated soils, groundwater extraction and treatment, and long-term monitoring. The remedies are presently being designed.

Koppers Company, a PRP, signed a Consent Decree (1994) with the EPA and State of Illinois that requires the company to fund the remediation work.

Source: PNPLS-IL.

HOD Landfill
Lake County
This former landfill near the town of Antioch operated from 1963 to 1984. It received a variety of municipal and industrial wastes, including a tanker load of PCBs. Soils and groundwater contain VOCs, cadmium, manganese, zinc, and lead. About 5,000 people in the area use private water wells. The contamination also threatens wetlands and Sequoit Creek.

Site investigations were completed in 1995 and will recommend a final remediation plan.

Sources: PNPLS-IL; Antioch Public Library, 757 Main Street, Antioch, IL, 60002.

Ilada Energy Company
Alexander County
This site in East Cape Girardeau was operated as a fuel oil processing center from 1942 to the 1950s and as a waste oil recycling center from 1981 to 1983. Various PCB-contaminated oils were frequently spilled or leaked from storage tanks. Testing has shown that the groundwater and soils contain VOCs, PCBs, and heavy metals. The contamination threatens agricultural wells, the Mississippi River, and the drinking water wells of about 500 residents.

Response actions included installing monitoring wells in 1986. Tanks and pipes were removed in 1989. Site investigations that will recommend a final remediation plan are in progress.

A 1993 Consent Decree, 1989 Consent Order, and 1989 Unilateral Administrative Order require the PRPs to conduct the site investigations and remediation work.

Sources: PNPLS-IL; Cape Girardeau Library, 711 North Clark Road, Cape Girardeau, MO, 63701.

Interstate Pollution Control
Winnebago County
Located in industrial Rockford, this site was a hazardous waste storage facility from 1974 to 1982. Wastes included oils, solvents, and heavy metal–plating solutions that were often stored in tanks that leaked or unlined lagoons. Testing in 1986 by the EPA showed that the groundwater and soil contain VOCs, cadmium, and copper. The contamination has penetrated nearby private wells and threatens the municipal water supply system for Rockford.

Response actions included removing leaking drums and tanks from the property and capping (1993). A site investigation completed in 1994–1995 will recommend a final remediation plan.

Interstate Pollution Control Incorporated has conducted removal actions on the site.

Sources: PNPLS-IL; Rockford Public Library, 215 North Wyman Road, Rockford, IL, 61101.

Jennison-Wright Corporation
Madison County
Wood treatment operations were conducted on this site in Granite City from about 1910 to 1989. Lumber was treated with creosote, PCP, and zinc compounds. Asphalt sealers were also produced. Various liquid wastes were stored in lagoons and storage tanks on the property. Testing in 1988 and 1991 revealed that the soil contained high levels of heavy metals, creosote, and PCP. The contamination threatens the drinking water supplies for about 30,000 residents of Granite City.

Response actions included removing and stabilizing contaminated material and securing drummed waste (1992). Site investigations are planned.

The PRP, Jennison-Wright Corporation, filed for bankruptcy in 1989.

Source: EPA Publication 9320.7-071, Volume 3, No. 1.

Johns-Manville Corporation
Lake County
This manufacturing plant near Waukegan has produced building materials since 1922. Various wastes, including heavy metals and asbestos, have been buried in pits on the property. Groundwater and soil contain asbestos, arsenic, chromium, and VOCs. Asbestos fibers were also detected in the air. The contamination threatens the health of the 68,000 residents of Waukegan, as well as recreational Lake Michigan and Illinois Beach State Park.

Response actions included closing and capping the asbestos pit in 1989. In 1987 the EPA recommended removing asbestos waste and debris, closing the asbestos pit, and long-term groundwater and air monitoring (1989–1991). Monitoring is continuing.

Johns-Manville, the PRP, participated in the site clean up.

Sources: PNPLS-IL; Waukegan Public Library, 128 North County Road, Waukegan, IL, 60085.

Joliet Army Ammunition Plant, Load Assembly
Will County
This U.S. Army base near Joliet operated from the 1940s to 1977. The army produced artillery shells, bombs, and ammunition. Ammunition was also tested, and waste materials were exploded or burned on the site. Various solid and liquid wastes were deposited in landfills. The groundwater and soil contain excessive levels of TNT, lead, mercury, chromium, and cadmium. The contamination threatens the local water supply for 250 people as well as recreational Kemery Lake, the Des Plaines River, and the Kankakee River. Local groundwater is also used for agricultural purposes.

Site investigations between 1989 and 1995 will recommend a final remediation plan.

Sources: PNPLS-IL; Joliet Public Library, 150 North Ottawa Road, Joliet, IL, 60431.

Joliet Army Ammunition Plant, Manufacturing Area
Will County
This U.S. Army base near Joliet operated from the 1940s to 1977. It produced artillery shells, bombs, and ammunition. Ammunition was also tested, and waste materials were exploded or burned on the site. Various solid and liquid wastes were deposited in landfills. Process waters from the manufacturing buildings were discharged into Jackson and Grant Creeks. The groundwater and soils contain excessive levels of TNT, lead, chromium, arsenic, and VOCs. The contamination threatens the local water supply for 1,250 people and has penetrated Jackson and Grant Creeks.

Response actions included pumping out contaminated water from the lagoons (1985). Sludges were also removed and the area capped (1985). Site investigations between 1989 and 1995 will recommend a final remediation plan.

Sources: PNPLS-IL; Joliet Public Library, 150 North Ottawa Road, Joliet, IL, 60431.

Kerr-McGee (Kress Creek)
DuPage County
This site covers approximately a 2-mile stretch of Kress Creek and a 2-mile stretch of the west branch of the DuPage River in West Chicago. Wastes from the processing of uranium and thorium ores by three different companies between 1931 and 1973 reached these streams via surface runoff or direct discharges. The stream sediments are radioactive. About 20,000 residents live within a few miles of the sites.

Detailed site investigations between 1992 and 1996 will recommend a final remediation plan.

In 1985 Kerr-McGee Corporation, a PRP, signed a Consent Decree to excavate contaminated sediments from Kress Creek. No excavation has occurred. The EPA has since ruled that the site does not pose an immediate threat to the public, and no excavation will be performed until the EPA recommends a final remediation plan.

Sources: PNPLS-IL; West Chicago Public Library, 118 West Washington Street, West Chicago, IL, 60185.

Kerr-McGee (Reed Keppler Park)
DuPage County
A mill that processed radioactive ores operated near this site in West Chicago from 1931 to 1973. Wastes from these processes were buried in an abandoned gravel pit that later became Reed Keppler Park. Soil and groundwater contain radioactive compounds and heavy metals. Radioactive waste has also been detected in air emissions. About 15,000 people live within 3 miles of the park.

Response actions included the removal of contaminated waste tailings from the park (1986). Detailed site investigations between 1993 and 1996 will recommend a final remediation plan.

Kerr-McGee, the PRP, signed a Consent Decree in 1985 to excavate and decontaminate this site.

Sources: PNPLS-IL; West Chicago Public Library, 118 West Washington Street, West Chicago, IL, 60185; Warrenville Public Library, 28 West 751 Stafford Place, Warrenville, IL, 60555.

Kerr-McGee (Residential Areas)
DuPage County
A mill that processed radioactive ores operated near this site in West Chicago from 1931 to 1973. Mill tailings from these processes were used to fill numerous areas in West Chicago that were later developed into residential neighborhoods. About 2,000 homes built on radioactive waste have been identified.

Contaminated soils were removed from numerous yards in the 1980s. Additional yard clean ups will be conducted between 1995 and 1997. Field studies by the EPA (1993–1996) will recommend a final remediation plan.

Kerr-McGee, a PRP, signed a Consent Decree in 1985 to excavate and decontaminate these residential sites. Kerr-McGee has carried out the clean-up work completed to date.

Sources: PNPLS-IL; West Chicago Public Library, 118 West Washington Street, West Chicago, IL, 60185.

Kerr-McGee Sewage Treatment Plant
DuPage County
A mill that processed radioactive ores operated near this site in West Chicago from 1931 to 1973. Some of the radioactive waste was dumped in septic tanks at the city sewage treatment plant or used for landfill on the property. The sewage plant is still in operation, and in the process of upgrading the facility many areas of radioactive soil were located. The contamination threatens about 15,000 people and the nearby DuPage River.

In 1986 Kerr-McGee excavated the contaminated soils and stored them in a temporary facility. An EPA investigation (1993–1996) indicated that tailings are still on the property; a final remediation plan will be recommended.

Kerr-McGee, the PRP, signed a 1985 Consent Decree to clean up the contamination.

Sources: PNPLS-IL; West Chicago Public Library, 118 West Washington Street, West Chicago, IL, 60185; Warrenville Public Library, 28 West 751 Stafford Place, Warrenville, IL, 60555.

La Salle Electric Utilities Company
La Salle County
La Salle Electric Utilities Company produced PCB-containing capacitors on this site in LaSalle from 1940 to 1981. Numerous spills, leaks, and discharges contaminated the surface with PCB wastes. Waste oil was used to control dust in the parking lot. PCB-contaminated wastes and trichloroethylene were stored in tanks on the property. The groundwater, soils, and structures contain PCBs. It is likely that PCBs have penetrated the Illinois River. About 20,000 people live within a few miles of the site.

Response actions included fencing, capping, removing solid and liquid wastes, and excavating contaminated soil (1982–1985). Additional soil was removed between 1987 and 1991 and the buildings steam cleaned. Private homes were also decontaminated. A groundwater extraction and treatment system will also be installed.

The EPA has performed all the site investigations and clean-up work.

Sources: PNPLS-IL; City Clerk's Office, La Salle City Hall, 745 2nd Street, La Salle, IL, 61301.

Lenz Oil Service
Cook County
Various companies recycled and stored solvents and oils at this site in Lemont from the 1960s to 1986. Hazardous wastes were stored in lagoons, drums, underground tanks, and aboveground tanks. Lenz Oil Service filed for bankruptcy in 1986. Testing by the State of Illinois showed that the groundwater and soils contain VOCs, PAHs, and PCBs. The contamination has penetrated nearby drinking wells and threatens the water supplies for about 12,000 nearby residents.

Affected residents were connected to clean water supplies in 1986. Contaminated soil was removed and incinerated in 1989. Detailed site investigations between 1989 and 1992 will recommend a final remediation plan.

Over 200 PRPs conducted the site investigations.

Sources: PNPLS-IL; Lemont Town Hall, 418 Main Street, Lemont, IL, 60439; Burr Ridge Village Hall, 7660 South County Line Road, Burr Ridge, IL, 60521.

MIG/Dewane Landfill
Boone County
This former landfill site near Belvidere operated from 1969 to 1988. It received a variety of municipal and industrial wastes, including paint sludges and spent organic solvents. Testing indicated that the groundwater and soil contain VOCs, arsenic, cyanide, lead, and zinc. Wells within 3 miles of the landfill supply water to about 16,000 residents. The contamination also threatens the Kishwaukee River.

Leachate was removed and treated from 1989 to 1992, and a temporary cap was built. Detailed site investigations between 1991 and 1996 will recommend a final remediation plan.

PRPs identified by the EPA signed a 1991 Consent Order to conduct the site investigations.

Sources: PNPLS-IL; Ida Public Library, 320 North State Street, Belvidere, IL, 61008.

NL Industries
Madison County
Operations at this site in Granite City since 1900 have consisted of metal refining, smelting, and fabrication. Wastes from lead smelting, including slag, smelter

dust, and battery casings, were stored on-site and later used for landfill in Granite City. Testing from 1978 to 1981 indicated that the soil and groundwater contain lead and other heavy metals. Lead was also detected in the air. The contamination threatens the health and water supplies of the 15,000 residents of Granite City.

Site investigations recommended removing contaminated soil from residential areas, consolidating lead-contaminated soil and solid waste into the slag pile and capping it, and long-term groundwater and air monitoring.

The PRPs identified by the EPA have refused to participate in the clean-up work, despite a 1985 Consent Order.

Sources: PNPLS-IL; Granite City Public Library, 2001 Delmar Street, Granite City, IL, 62040.

Ottawa Radiation Areas

La Salle County

This area of radioactive contamination is in a residential section of Ottawa. Radioactive wastes from businesses that made luminous clock dials were processed or stored here from 1918 to 1978. Soils were found to contain radioactive radium, lead, and bismuth. Site access is unrestricted. About 50 residences have been built on radioactive materials, and other persons are exposed through recreational activities in the area.

Response actions included dismantling buildings and removing radioactive debris and waste (1985). Residential homes were decontaminated or purchased. Additional soil removal was completed in 1996. A site investigation will make final recommendations for clean up by 1997.

Sources: PNPLS-IL; Reddick Public Library, 1010 Canal, Ottawa, IL, 61350.

Outboard Marine Corporation

Lake County

This extensive area of PCB contamination includes Waukegan Harbor and a 37-acre stretch of Lake Michigan. Outboard Marine Corporation discharged liquid wastes from its aluminum die cast machines directly into Waukegan Harbor and North Ditch, both of which flow into Lake Michigan. Testing has shown that the groundwater, sediments, and soil contain PCBs and that fish may be contaminated. The contamination threatens those who work in the area or use the harbor for recreational activities.

Site investigations by the EPA recommended excavating and removing contaminated soils and sediment and constructing surface runoff controls. Waste cells were constructed on-site to treat contaminated water, sediments, and soils. The waste cells will be capped by 1997. Long-term groundwater monitoring is under way.

Consent Decrees (1985, 1989) required the PRPs to conduct the remediation work.

Sources: PNPLS-IL; Waukegan Public Library, 128 North County Road, Waukegan, IL, 60085.

Pagel's Pit

Winnebago County

This landfill site near Rockford began operations in 1972. It has received a variety of municipal and industrial wastes, including sewage treatment sludge. Testing has shown that shallow groundwater contains arsenic and VOCs. The contamination has penetrated nearby private wells and threatens Killbuck Creek. The area is sparsely populated.

Affected residents were connected to water filtration systems. Site investigations between 1986 and 1991 recommended groundwater extraction and treatment, leachate collection and treatment, upgrading the gas extraction unit, and capping. Construction is expected to begin by 1997.

PRPs have funded the site investigations, and a 1992 Consent Decree requires the operator to undertake the clean-up work.

Sources: PNPLS-IL; Rockford Public Library, 215 North Wyman Road, Rockford, IL, 61101.

Parsons Casket Hardware Company

Boone County

This electroplating facility in Belvidere operated from the 1920s to 1982. Wastes, including heavy metal sludges and cyanide-plating solutions, were stored in drums, storage tanks, and an unlined lagoon on the property. Testing in 1987 revealed that the groundwater and soil contain VOCs, cyanide, and heavy metals. The contamination threatens the private and municipal water supplies for the 20,000 nearby residents as well as the recreational Kishwaukee River.

Response actions included removing solid and liquid wastes in 1985. A detailed site investigation completed by the EPA in 1992 will recommend a final remediation plan.

The two most recent owners, Parsons Casket Hardware Company and Filter Systems Inc., conducted the removal action.

Sources: PNPLS-IL; Ida Public Library, 320 North State Street, Belvidere, IL, 61008.

Quincy Landfills

Adams County

These former landfill sites near Quincy operated during the 1970s. They received a variety of municipal and industrial wastes, including solvents, degreasers, heavy metal sludges, oils, coolants, acetone, paint byproducts, and drummed hazardous waste. Testing by the state in 1985–1986 revealed that the groundwater and soil contain VOCs, PCBs, and heavy metals (including selenium). Two nearby wells were also contaminated. The contamination threatens about 100 private wells within 3 miles of the sites.

Response actions included connecting affected residents to clean water supplies. A site investigation between 1987 and 1993 recommended a final remediation plan of fencing, constructing a leachate collection and treatment system, improving the cap, and monitoring groundwater.

PRPs identified by the EPA funded the site investigation and will help in the remediation work.

Sources: PNPLS-IL; Quincy Public Library, 526 Jersey Street, Quincy, IL, 62301.

Sangama Electric Dump
Williamson County
This site is located beside Crab Orchard Lake near Marion. Originally owned by the U.S. Army, the site housed manufacturing operations that produced explosives, munitions, boats, plated metals, and electrical equipment during World War II. Afterward the land was transferred to the Department of the Interior, and former military buildings were used by private businesses. All wastes were discharged or buried on the site. Testing in the 1980s indicated at least 40 areas of groundwater and soil contaminated by heavy metals, PCBs, and VOCs. Fish also contain PCBs and are dangerous to eat. These areas are now part of the 42,000-acre Crab Orchard National Wildlife Refuge. The contamination has penetrated Crab Orchard Lake, a local drinking water source.

Response actions included a site investigation that recommended excavating, removing, or stabilizing contaminated soil and sediment; fencing; and groundwater monitoring. The soil remediation work is expected to be finished in 1997. Other studies are continuing.

Sources: PNPLS-IL; Marion Carnegie Public Library, 206 South Market Street, Marion, IL, 62959.

Sauget Area 1
St. Clair County
This site in and around the village of Sauget has been contaminated by various industrial waste disposals since the 1930s. Waggoner Trucking Company, Monsanto Corporation, Ruan Trucking Company, H. H. Hall Construction Company, and others have buried or discharged solid and liquid industrial wastes in gravel pits, dumps, wetlands, and streams in the area. Dead Creek has been damaged by repeated illegal discharges of liquid wastes. Testing has shown that the groundwater and soil in nine point sources contain SVOCs, VOCs, PCB, PAHs, chlorophenols, and numerous heavy metals. The chemicals have entered Dead Creek and associated wetlands. The contamination threatens Dead Creek, the Mississippi River, Old Prairie duPont Creek, Cahokia Chute, associated wetlands, and the air and drinking water supplies for over 100,000 people.

This site was proposed as an NPL site in 1996; a final decision is expected in 1997–1998. Site investigations are planned.

Source: EPA Publication 9320.7-071.

Savanna Army Depot
Carroll and Jo Daviess Counties
This 13,000-acre U.S. military base on the Mississippi River near Savanna has been in operation since 1918. Activities included processing and storing ammunition and explosives, loading explosives, and detonating waste materials and old munitions. Groundwater and soil over 70 areas contain explosives, VOCs (tetrachloroethylene and chloroform), PAHs, and heavy metals. The contamination threatens the private water wells of about 500 people near the site as well as a population of bald eagles.

Response actions have consisted of detailed site investigations (1992–1996) that will recommend a final remediation plan.

Sources: PNPLS-IL; Savanna Public Library, 326 3rd Street, Savanna, IL, 61074.

Southeast Rockford Groundwater Contamination
Winnebago County
The groundwater in this 2-square-mile area in southeast Rockford is contaminated with VOCs. The source areas have yet to be identified. The contamination threatens the Rock River and the drinking water supplies of thousands of Rockford residents.

Response actions included connecting 283 affected residents to bottled water and carbon filtration systems and later to clean water supplies (1989– 1990). Additional homes were connected in 1991. An ongoing site investigation should be completed by 1997.

Sources: PNPLS-IL; Rockford Public Library, 215 North Wyman Street, Rockford, IL, 61101.

Tri-County Landfill
Kane County
This former gravel pit near South Elgin was operated as a landfill from 1968 to 1977. It received a variety of municipal and industrial wastes. Testing in 1984 showed that the groundwater and soil contain VOCs, cyanide, and heavy metals. The contamination threatens a wetland, the recreational Fox River, and the private water wells of about 10,000 people within 3 miles of the site,

In 1992 the EPA recommended a final remediation plan of a landfill cap, gas venting, soil excavation, and a groundwater extraction and treatment system. Construction is expected to begin in 1997.

PRPs have been involved in the site studies and design of the remediation plan.

Sources: PNPLS-IL; Gail Borden Library, 200 North Grove Avenue, Elgin, IL, 60120.

Velsicol Chemical Corporation
Clark County
This former chemical plant near Marshall operated from the 1930s to 1987. It manufactured or processed petroleum resins, solvents, and pesticides, including chlordane. Liquid wastes were stored in surface lagoons, which occasionally overflowed into East Mill Creek. Groundwater and soil contain VOCs, pesticides, and cadmium. Fish contain toxic levels of chlordane. The contamination threatens East Mill Creek and the drinking water of the 17,000 residents of Marshall.

Response actions included excavating and stabilizing contaminated soil under a temporary clay cap in the 1980s. In 1988 the EPA recommended a remediation plan of excavating and treating contaminated soil and extracting and treating groundwater using activated carbon. Construction activities were completed in 1994, and the groundwater treatment plant is operational.

The EPA reached a financial settlement with PRPs in 1989.

Sources: PNPLS-IL; Marshall Public Library, 612 Archer Avenue, Marshall, IL, 62441.

Wauconda Sand & Gravel
Lake County
This former gravel pit near Wauconda operated as a landfill from 1941 to 1978. It received a variety of municipal and industrial wastes. Leachate seeped into Mutton Creek in the 1970s. Testing in the 1980s revealed that the groundwater and soil contain heavy metals, cyanide, pesticides, and VOCs. The contamination threatens the water supplies of nearby residents as well as Mutton Creek.

Response actions included installing erosion controls and a leachate collection system (1985). In 1989 the EPA recommended a final remediation plan of long-term groundwater monitoring, gas controls, leachate and surface runoff collection systems, and cap upgrading.

An AOC (1986) and UAO (1989) require the PRPs to conduct the site investigations and remediation work.

Sources: PNPLS-IL; Wauconda Area Library, 801 North Main Street, Wauconda, IL, 60084.

Woodstock Municipal Landfill
McHenry County
This former landfill near Woodstock was operated from 1935 to 1975. It received a variety of wastes, including heavy metal sludges and drummed VOC and PCB solutions. In 1985 leachate was observed in nearby wetlands. Testing indicated that the groundwater and soil contain VOCs (vinyl chloride), naphthalene, cadmium, mercury, arsenic, and cobalt. About 13,000 people use water wells within 3 miles of the site.

Response actions included site investigations between 1989 and 1993. EPA recommendations consist of a landfill cap and a groundwater extraction and treatment system. Construction is expected to be under way by 1997.

Sources: PNPLS-IL; Woodstock Public Library, 414 West Judd Street, Woodstock, IL, 60098.

Yeoman Creek Landfill
Lake County
This former landfill near Waukegan operated from 1959 to 1969. It received a variety of municipal and industrial wastes, including sludges. Leachate has penetrated Yeoman Creek. Testing has indicated that the soil and groundwater contain VOCs, PCBs, phthalates, lead, manganese, iron, and ammonia. VOC-bearing gases have also been detected in nearby buildings. The contamination threatens the water supplies of about 70,000 Waukegan residents in addition to Yeoman Creek.

Response actions included upgrading the cap (1980), fencing (1990), and installing a gas venting system (1994). Recommendations from detailed site studies are expected by 1997.

The EPA has identified PRPs who are participating in the site investigations and clean-up work.

Sources: PNPLS-IL; Waukegan Public Library, 128 North County Road, Waukegan, IL, 60085.

INDIANA

American Chemical Service
Lake County
This chemical recycling facility near Griffith operated from 1958 to 1975. Various waste chemicals, byproducts, solutions, sludges, transformer oils, and wastewaters were buried in drums on the site or discharged on the 21-acre property. Testing in the 1970s determined that the groundwater and soils contain VOCs, PCP, phthalates, PCBs, SVOCs, heavy metals, and coal tar derivatives. The compounds have

INDIANA

Lake
Porter
La Porte
St. Joseph
Elkhart
La Grange
Steuben
Noble
De Kalb
Starke
Marshall
Kosciusko
Whitley
Allen
Newton
Jasper
Pulaski
Fulton
Wabash
Huntington
Wells
Adams
White
Cass
Miami
Benton
Carroll
Howard
Grant
Blackford
Jay
Warren
Tippecanoe
Clinton
Tipton
Delaware
Madison
Randolph
Fountain
Montgomery
Boone
Hamilton
Henry
Vermilion
Parke
Hendricks
Marion
Hancock
Wayne
Putnam
Rush
Fayette
Union
Shelby
Vigo
Clay
Morgan
Johnson
Franklin
Decatur
Owen
Monroe
Brown
Bartholomew
Dearborn
Sullivan
Greene
Ripley
Jennings
Jackson
Ohio
Switzerland
Lawrence
Jefferson
Knox
Daviess
Martin
Scott
Washington
Orange
Clark
Pike
Gibson
Dubois
Crawford
Floyd
Harrison
Warrick
Perry
Posey
Spencer
Vanderburgh

penetrated nearby wetlands and streams. The contamination is a threat to the local drinking water supplies that serve about 10,000 residents.

Response actions included constructing groundwater barriers and surface runoff controls and removing drummed waste and liquid chemicals. A site investigation was completed in 1992. The EPA recommended groundwater extraction and treatment, removal and incineration of contaminated soil, soil vapor extraction, environmental monitoring, deed restrictions, and possibly wetland reconstruction. Construction work should be under way by 1997.

Sources: PNPLS-IN; Griffith Public Library, 940 North Broad Street, Griffith, IN, 46319.

Bennett Stone Quarry
Monroe County
A former limestone quarry, this site near Bloomington was used as an illegal landfill and industrial dump in the 1960s and 1970s. A variety of waste was received, including electrical parts and PCB-bearing capacitors. Testing revealed that the groundwater and soil contain PCBs, which have also penetrated nearby ponds and Stout Creek. The contamination threatens private wells in the area, surface water used for agricultural purposes, and nearby streams, wetlands, and wildlife habitats.

Response actions included removing the capacitors and contaminated soil, capping, and fencing (1983). Contaminated sediments were hydro-vacuumed from Stouts Creek in 1987. Other surface debris and contaminated soil were also removed. Environmental monitoring is in progress.

The PRPs are remediating the site according to a 1985 Consent Decree with the EPA.

Sources: PNPLS-IN; Monroe County Public Library, 303 East Kirkwood Avenue, Bloomington, IN, 47491.

Carter Lee Lumber Company
Marion County
Land purchased in 1971 by the Carter Lee Lumber Company in Indianapolis was previously used by railroad companies. Various liquid wastes from tank cars were dumped on the property. Testing by the EPA in 1985 showed that the soil and groundwater contain heavy metals, cyanide, VOCs, and PNAs. The contamination threatens the White River and private and public water wells that serve about 750,000 local residents.

A detailed site investigation is under way and should be completed by 1997. The owner, Carter Lee Lumber Company, is conducting the site investigation.

Sources: PNPLS-IN; Hawthorn Community Center, 2440 West Ohio Street, Indianapolis, IN, 46222.

Columbus Old Municipal Landfill No. One
Bartholomew County
This municipal landfill in the city of Columbus was in operation from 1938 to 1983. It received a variety of municipal and industrial wastes, including solvents, bases, heavy metal sludges, paints, and acids. Testing has shown that the groundwater and soil contain a variety of contaminants, including VOCs and heavy metals. The contamination threaten the White River and private and public water wells that supply about 35,000 residents.

A detailed site investigation, conducted between 1987 and 1992, recommended no further action. The site was later fenced and capped. Groundwater monitoring has been in progress since 1992. It is probable that the site will be deleted from the NPL within 5 years.

A 1987 Consent Order required three PRPs to conduct the site investigation.

Sources: PNPLS-IN; Bartholomew County Public Library, Columbus, IN, 46901.

Conrail Rail Yard, Elkhart
Elkhart County
This railroad yard in Elkhart began operation in 1956. Various activities have polluted the groundwater and soil with oil, diesel fuel, acid, petrochemicals, degreasers, and cleaning fluids. In 1986 the EPA discovered excessive levels of VOCs in the groundwater, soil, and private drinking water wells. The contamination threatens the public and private wells that service about 50,000 nearby residents.

Response actions included providing clean water supplies to affected residents (1986–1987). A site investigation was conducted from 1988 to 1991. A groundwater extraction and air-stripping system became operational in 1996.

Sources: PNPLS-IN; Elkhart Public Library, 300 South Second Street, Elkhart, IN, 46516.

Continental Steel Corporation
Howard County
Nails, wire, and fencing were manufactured from scrap metal at this steel mill in Kokomo from 1914 to 1986. Various solid and liquid wastes, including slags, were deposited or discharged on the 140-acre property. Testing has shown that the groundwater, surface water, and soil contain VOCs, heavy metals, PCBs, phenols, and phthalates that have penetrated Kokomo and Wildcat Creeks. The contamination threatens nearby streams and wetlands as well as water wells that serve about 2,000 residents.

Response actions included removing liquid wastes from on-site lagoons from 1989 to 1993. Drummed waste and contaminated soil were also removed. Detailed site investigations are nearing completion and will recommend a final remediation plan.

The PRP, Continental Steel Corporation, filed for bankruptcy in 1985.

Sources: PNPLS-IN; Kokomo-Howard County Public Library, 220 West Union Street, Kokomo, IN, 46901.

Douglass Road/Uniroyal Landfill
St. Joseph County

This industrial landfill in Mishawaka operated from 1954 to 1979. It received various solid and liquid wastes from Uniroyal Company, including spent solvents, petrochemicals, fly ash, and rubber and plastic products. Testing has shown that the groundwater and soils contain various hydrocarbons. The contamination threatens nearby Juday Creek and private and public wells that serve about 125,000 residents.

Response actions included a site investigation that should be completed by 1997.

Uniroyal, operating under a 1989 Consent Order, conducted the site study until it filed for bankruptcy in 1992.

Sources: PNPLS-IN; Mishawaka-Penn Public Library, 209 Lincoln Way East, Mishawaka, IN, 46544.

Envirochem
Boone County

A former solvent recycling facility, Envirochem operated on this site near Indianapolis from 1977 to 1982. Various wastes such as resins, thinners, sludges, oils, chlorinated hydrocarbons, and other flammables were stored or processed on the property. Illegal discharges, leaks, and spills contaminated the groundwater and soil with VOCs, heavy metals (barium, lead, and nickel), PCBs, phenols, and phthalates. The pollutants have also been detected in the sediments of Eagle Creek Reservoir.

Response actions included pumping out storage tanks, removing contaminated soil, and erecting fencing (1983–1984). Underground tanks were removed and a temporary clay cap installed. A groundwater collection sump was installed by the EPA in 1985. In 1987 the EPA recommended a permanent cap, a groundwater extraction and treatment system, and soil vapor extraction. Most of the construction was completed by 1993.

Consent Decrees between the EPA, the State of Indiana, and about 250 PRPs require the companies to perform the clean-up work.

Sources: PNPLS-IN; Hussey Memorial Library, 225 West Hawthorne, Zionsville, IN, 46077.

Fisher-Calco
La Porte County

This former industrial chemical facility operated on this site near Kingsbury Heights from the 1950s to the 1970s. Products included weapons and explosives, sodium hypochlorite, sulfur dioxide, chloride, ammonia, and solvents. Paints and solvents were also recycled. Processing and waste solutions included cleaning solvents, degreasers, acids, metal-plating sludges, and cyanide. A fire destroyed the operation in 1978. Testing revealed that the groundwater and soil contain VOCs, SVOCs, and PCBs. The contamination threatens public and private water wells that supply nearby residents.

Response actions included removing contaminated drums and other solid waste and fencing the site (1989). About 3,400 drums have been removed. In 1990 the EPA recommended the excavation and incineration of PCB-contaminated soil, soil vapor extraction of VOC-contaminated soil, and groundwater extraction and air-stripping. Affected residents will also be connected to a new water supply system.

A 1982 Consent Agreement and 1988 Unilateral Order require the PRPs to conduct groundwater monitoring and the remediation work.

Sources: PNPLS-IN; La Porte County Library, 904 Indiana Avenue, La Porte, IN, 46350.

Fort Wayne Reduction Dump
Allen County

This former municipal and industrial landfill near Fort Wayne operated from the 1960s to 1976. Buried wastes included sewage, solvents, and industrial liquids. Testing in the 1980s showed that the groundwater and soil contain VOCs, heavy metals, PCBs, PAHs, and phenols. The contamination threatens the Maumee River and associated wetlands as well as water wells that serve about 1,200 people. In 1988 the EPA recommended groundwater monitoring, a groundwater pump-and-treat system, removal of drummed waste, fencing, erosion controls, capping, and deed restrictions. Construction was completed between 1991 and 1994. The groundwater treatment plant has been operational since 1994.

Sources: PNPLS-IN; Allen County Public Library, 900 Webster Street, Fort Wayne, IN, 46801.

Galen Myers Dump
St. Joseph County

Various local industrial waste was buried, stored, or recycled at this site in Osceola from 1960 to 1982. Testing has shown that the groundwater and soil contain VOCs, phthalates, PCBs, and pesticides that have penetrated nearby private wells. The contamination threatens the St. Joseph River and its wetland habitats as well as the drinking water source for about 20,000 residents.

Response actions included removing drummed waste (1985) and connecting affected residents to clean water supplies (1987, 1992, 1993). A detailed site investigation begun in 1989 is nearing completion and will recommend a final remediation plan.

Sources: PNPLS-IN; Indiana Department of Environmental Management, 100 North Senate Avenue, 12th Floor, Indianapolis, IN, 46206.

Himco Dump

Elkhart County

This town dump in the city of Elkhart operated from 1960 to 1976. It received a variety of municipal, industrial, medical, and pharmaceutical wastes, including heavy metal sludges. Inspections in 1984 discovered leachate streams leaving the property and strong sulfur and methane odors. Testing indicated that the groundwater and soils contain heavy metals and VOCs that have penetrated nearby wells. The contamination threatens wetland habitats and nearby water wells that supply about 20,000 people.

Response actions included drilling deeper wells for affected residents. Drummed waste and contaminated soil were removed in 1992. An EPA study conducted from 1989 to 1993 recommended capping, gas collection, and groundwater monitoring.

A 1975 Consent Agreement required the PRPs to drill the new residential wells.

Sources: PNPLS-IN; Elkhart Public Library, 2400 Benham Avenue, Elkhart, IN, 46517.

Lake Sandy Jo

Lake County

Located in the city of Gary, this former gravel pit was used as a landfill from 1971 to 1980. It received a variety of municipal and industrial wastes, including drummed industrial waste and demolition debris. Testing has shown that the groundwater, surface water, and soil contain heavy metals, VOCs, PCBs, PAHs, phthalates, and DDT. The contamination threatens the water supplies of about 5,500 local residents.

Response actions included fencing (1986). In 1986 the EPA recommended removal of contaminated soil, deed restrictions, and groundwater monitoring. Affected residents are also being connected to clean water supplies. All construction has been completed, and groundwater monitoring is in progress.

Sources: PNPLS-IN; Gary Public Library, 220 West Fifth Avenue, Gary, IN, 46402.

Lakeland Disposal Service

Kosciusko County

This former landfill near Claypool operated from 1974 to 1978. It received a variety of municipal and industrial wastes, including cyanide, paint sludges, and heavy metal sludges. The property was later developed into a mobile home park. Testing in 1982 showed that the groundwater and soil contain high levels of arsenic, cadmium, barium, other heavy metals, and VOCs. Methane gas was also detected in the air. The contamination threatens about 300 private wells and the municipal wells for the town of Claypool. Runoff from the site also enters nearby wetlands.

Response actions included completing a site investigation in 1992. The EPA recommended a remediation plan of fencing, deed restrictions, removal of contaminated soils, capping, gas collection, a slurry wall, and groundwater extraction and treatment. Construction is expected to be under way by 1997.

Sources: PNPLS-IN; Kosciusko County Health Department, 100 West Center Street, Third Floor, Warsaw, IN, 46580.

Lemon Lane Landfill

Monroe County

This former landfill near Bloomington operated from 1950 to the 1960s. It received a variety of municipal and industrial wastes, including drummed industrial chemicals and leaking capacitors. Testing has shown that the groundwater, soil, and local springs contain PCBs. The contamination threatens the local drinking water supplies.

Response actions included fencing, removing PCB-contaminated capacitors, and constructing erosion controls (1983). Affected residents were connected to clean water supplies. The landfill was capped in 1988. Contaminated soils were removed and incinerated, and groundwater monitoring is continuing.

The PRP, Westinghouse Corporation, has conducted the site studies and remediation work according to a 1985 Consent Decree.

Sources: PNPLS-IN; Monroe County Public Library, 303 East Kirkwood Avenue, Bloomington, IN, 47491.

Main Street Well Field

Elkhart County

In 1981 VOCs were detected in the groundwater of a well field that supplies about 30,000 residents of Elkhart. Further investigations revealed that two industries near the well field were likely point sources. The contamination continues to threaten these drinking water supplies and nearby streams.

Response actions included supplying affected residents and businesses with clean water supplies in 1987. A groundwater extraction and air-stripping system began to treat the contaminated groundwater in 1987. A soil vapor extraction system became operational in 1994.

Sources: PNPLS-IN; Elkhart Public Library, 300 South Second Street, Elkhart, IN, 46516.

Marion Dump

Grant County

This former gravel pit near Marion operated as a landfill in the 1960s and 1970s. It received a variety of municipal and industrial wastes, including solvents, plasticizers, and heavy metals. Testing has shown that the groundwater and soils contain VOCs, heavy metals (arsenic), and PAHs. The contamination threatens the local drinking water supplies, nearby ponds, and the Mississinewa River.

Response actions included a site investigation. In 1987 the EPA recommended erosion controls, capping, fencing, connecting affected residents to clean water supplies, and groundwater monitoring. A groundwater decision is expected by 1997.

Sources: PNPLS-IN; Marion Public Library, 600 South Washington Street, Marion, IN, 46953.

Midco No. 1
Lake County
This abandoned industrial waste recycling facility in Gary operated in the 1970s. Various liquid wastes, including solvents and oils, were recycled, stored, buried, or discharged on the 4-acre property. Testing has shown that the groundwater, soils, and nearby stream sediments contain VOCs, SVOCs, PCBs, heavy metals, cyanide, and pesticides. The contamination threatens the local drinking water supplies and nearby streams and wetlands.

Response actions included fencing (1981); removing contaminated debris, tanks, and soil; and capping (1982). In 1989 the EPA recommended the extraction and solidification of contaminated soil and wetland sediment, soil vapor extraction, capping, installation of a groundwater extraction and treatment system, deed restrictions, and groundwater monitoring. Construction of these remedies began in 1994.

A 1985 Consent Decree and Administrative Order (1989) require the PRPs to fund the site investigations and remediation work.

Sources: PNPLS-IN; Gary Public Library, 220 West Fifth Avenue, Gary, IN, 46402.

Midco No. 2
Lake County
This abandoned industrial waste recycling facility in Gary operated from 1976 to 1978. Various chemical wastes, including solvents, acids, and caustics, were processed on the 7-acre property. Wastewaters and spent solutions were discharged into pits or into the Grand Calumet River. Testing revealed that the groundwater and soil contain VOCs, isophorone, cyanide, sodium, potassium chloride, PCBs, and heavy metals. The contamination threatens wetlands, the Grand Calumet River, and the drinking water supplies for about 250,000 residents.

Response actions included fencing, removing drums and tanks, and excavating sludge pits (1984–1989). In 1989 the EPA recommended solidification and reburial of contaminated soil, a groundwater pump-and-treat system, deed restrictions, and long-term monitoring. Contaminated sediments in the neighboring wetlands will also be treated. Construction work began in 1993.

A Consent Decree (1985), Administrative Order (1989), and Consent Decree (1992) require the PRPs to fund the site investigations and remediation work.

Sources: PNPLS-IN; Gary Public Library, 220 West Fifth Avenue, Gary, IN, 46402.

Neal's Dump
Owen County
This former landfill in Spencer operated from 1966 to 1971. It received a variety of industrial wastes, including capacitors, waste oils, and sawdust. Testing has indicated that the groundwater and soil contain PCBs. The contamination is a threat to nearby springs, wetlands, the White River, and the drinking water supplies for about 1,000 residents.

Response actions included capping and fencing (1983). The EPA recommended a final remediation plan of removing debris, excavating contaminated soil, and monitoring groundwater. This work should be completed by 1997.

The PRP, Westinghouse Corporation, is conducting the remediation work.

Sources: PNPLS-IN; Monroe County Public Library, 303 East Kirkwood Avenue, Bloomington, IN, 47491.

Neal's Landfill
Monroe County
This former landfill in Bloomington received a variety of municipal and industrial wastes from 1950 to 1972, including PCB-containing capacitors and arrestors. Testing has shown that the soils and groundwater contain PCBs. The contamination threatens wetlands, Conard's Branch, recreational Richland Creek, and the drinking water source for about 1,500 local residents.

Response actions included installing a cap and erosion controls and fencing (1983). Capacitors were also excavated. PCB-contaminated sediments from Richland Creek and Conard's Branch were excavated in 1989. Other contaminated soils were incinerated, and carbon treatment units were installed in contaminated springs (1988–1992). Groundwater monitoring is continuing.

The PRP, Westinghouse Company, is conducting the remediation work according to a 1985 Consent Decree.

Sources: PNPLS-IN; Monroe County Library, 303 East Kirkwood Avenue, Bloomington, IN, 47491.

Ninth Avenue Dump
Lake County
This 17-acre property in Gary was used as a chemical and industrial landfill in the 1970s. It received a variety

of wastes, including demolition debris, tires, batteries, oils, sludges, resins, acids, and solvents. Testing has shown that the groundwater and soil contain VOCs, PAHs, petrochemicals, heavy metals, PCBs, and pesticides. The contamination is a threat to local drinking water supplies, ponds, and associated wetlands.

Response actions included a detailed site investigation. The EPA recommended a slurry wall, oil extraction, thermal treatment of oil-contaminated soil, groundwater extraction and treatment, and soil vapor extraction. These remedies should be completed by 1997.

Various legal agreements (1975, 1980, 1983, 1988, 1989) require the PRPs to conduct the site investigations and clean-up work.

Sources: PNPLS-IN; Gary Public Library, 220 West Fifth Avenue, Gary, IN, 46402.

Northside Sanitary Landfill
Boone County
This former landfill near Zionsville operated from the 1950s to the 1980s. It received a variety of municipal, industrial, and hazardous wastes, including solvents, fuels, and sludges. Underground fires plagued the landfill in the 1970s. Testing has shown that the groundwater, soil, surface water, and nearby stream sediments contain high amounts of acids, petrochemicals, pesticides, and VOCs. The contamination threatens the drinking water supplies for about 5,000 local residents, part of the Indianapolis water supply system, Finley Creek, Eagle Creek, Eagle Creek Reservoir, and related wetlands.

Response actions included a site investigation. In 1987 the EPA recommended deed restrictions, capping, erosion controls, a leachate collection system, groundwater extraction and treatment, and long-term monitoring. Construction is expected to begin in 1997.

The EPA has identified PRPs who have agreed to undertake the clean-up actions.

Sources: PNPLS-IN; Hussey Memorial Library, 225 West Hawthorne, Zionsville, IN, 46077.

Prestolite Battery Division
Knox County
Lead-acid batteries were produced on this site near Vincennes from 1945 to 1985. Various wastewaters and spent acids were discharged into the sewer system and on-site lagoons. Testing revealed that the groundwater, surface water, soil, and air are contaminated with lead. PCBs were also found in the soil, and lead was detected in air emissions from the plant. The contamination threatens Kelso Creek, Snapp Creek, the Wabash River, and private and public wells that serve about 1,000 local residents and the city of Vincennes.

Response actions included removing lead-contaminated soil, draining and fencing lagoons, and capping (1989). A site investigation (1988–1994) recommended natural attenuation of the contamination plume, to be confirmed by long-term groundwater monitoring.

The PRP, Allied Signal Corporation, will conduct the remediation work according to a 1992 Administrative Order of Consent.

Sources: PNPLS-IN; Knox County Public Library, 502 North Seventh Street, Vincennes, IN, 47591.

Reilly Tar & Chemical Company
Marion County
Specialty chemicals have been manufactured on this site in Indianapolis since the 1950s. Other activities included refining coal tar and treating lumber with creosote. Various wastewaters and spent solutions were discharged into a pond on the 120-acre property. Testing has shown that the groundwater and soils contain VOCs, ammonia, and PAHs. The contamination threatens the local drinking water supplies for about 6,000 residents as well as nearby ponds and streams.

Response actions included a site investigation that was completed in 1992. The EPA recommended a groundwater extraction and treatment system, low-temperature thermal desorption of contaminated soil, solidification of sludge, capping, and monitoring. Additional studies are in progress.

Legal agreements between the EPA and Reilly Industries, the PRP, require the company to conduct the site studies and remediation work.

Sources: PNPLS-IN; Indianapolis-Marion County Public Library, 40 East St. Clair Street, Indianapolis, IN, 46206.

Seymour Recycling Corporation
Jackson County
This recycling facility near Seymour operated in the 1970s. Various waste chemicals were processed, including solvents, degreasers, chloroform, and heavy metal sludges. These compounds were stored on the property in drums and tanks that frequently leaked. Testing has shown that the groundwater and soil contain VOCs, phenols, chloroform, arsenic, iron, barium, beryllium, and manganese. The pollution has been detected in nearby stream sediments and aquatic life. Fish kills were also recorded. The contamination threatens the White River, the Ohio River, and the local drinking water supplies for about 1,000 residents.

Response actions included removing drummed waste (1980–1984), erecting fencing (1982), removing tanks and contaminated soil (1982–1984), and capping (1984). Affected residents were connected to clean water supplies in 1984–1985. The EPA recommended the installation of a groundwater extraction and treatment system, soil bioremediation, soil vapor extraction, capping, and long-term monitoring. Construction is in progress. The groundwater treatment system is expected to be in operation for about 30 years.

PRPs are funding the remediation work according to a 1988 Consent Decree.

Sources: PNPLS-IN; Jackson County Public Library, 2nd and Walnut Streets, Seymour, IN, 47274.

Southside Sanitary Landfill
Marion County
This landfill near Indianapolis began operations in 1971. It has received a variety of municipal and industrial wastes, including coal tar, asbestos, heavy metal sludges, and paints. Testing has shown that the groundwater and soil contain arsenic, cadmium, nickel, chromium, and PAHs. The contamination threatens the local drinking water supplies for about 7,500 residents as well as nearby Eagle Creek, White River, and recreational Fall River.

Response actions included building a leachate collection system and slurry wall in 1988. A detailed site investigation is nearing completion and will recommend a final remediation plan.

A 1986 agreement between the EPA and the PRP required the company to undertake the response actions.

Sources: PNPLS-IN; Indianapolis Public Library, 40 East St. Clair Street, Indianapolis, IN, 46204.

Tippecanoe Landfill
Tippecanoe County
This former landfill near Lafayette operated from 1971 to 1989. It received a variety of industrial and municipal wastes, including sludges and out-of-state wastes. Testing in the 1980s revealed that the groundwater and soils contain various contaminants, including VOCs and PCBs. Gases have also been released into the air. The contamination threatens local drinking water supplies and nearby streams.

Response actions included the installation of alarms that alert local residents to high levels of gas coming from the landfill. A site investigation (1991–1996) will recommend a final remediation plan.

A Consent Decree (1988) and Consent Order (1990) require the PRPs to fund the site investigations.

Sources: PNPLS-IN; Tippecanoe Public Library, 627 South Street, Lafayette, IN, 47901.

Tri-State Plating
Bartholomew County
This former metal-plating facility in Columbus operated from the 1940s to 1984. Various violations and illegal discharges of wastewater led to the company's demise in 1984. State examinations revealed that the groundwater and soil contain excessive levels of cadmium, chromium, copper, lead, zinc, nickel, and arsenic. The contamination threatens Haw Creek, the White River, and the drinking water supplies for the 30,000 residents of Columbus.

Response actions included fencing (1987), waste removal (1987), and excavating contaminated soil. In 1989 the EPA recommended razing structures, removing debris, excavating contamination soil, and backfilling. A groundwater extraction and treatment system became operational in 1992.

Sources: PNPLS-IN; Bartholomew County Health Department, 440 Third Street, Suite 303, Columbus, IN, 47201.

U.S. Smelter and Lead Refinery
Lake County
This former lead and copper smelter in East Chicago operated from 1906 to 1985. Lead was recovered from old batteries from 1973 to 1985. Various solid and liquid wastes, including slag and smelter dust, were stored on the site. Stockpiled slag was dumped in nearby wetlands. Contaminated wastewaters that violated permit levels for heavy metals were discharged into the Grand Calumet River. Testing has shown that groundwater, soil, surface water, and nearby wetlands contain lead, cadmium, copper, arsenic, and zinc. The air also carries lead particles. The contamination threatens the Grand Calumet River, Lake Michigan, and the health and drinking water supplies of up to four million residents.

A detailed site investigation is planned.

The PRPs, U.S. Smelter and Lead and its parent company, Sharon Steel, have both filed for bankruptcy.

Source: PNPLS-IN.

Waste Landfill
La Porte County
This former wetland and landfill near Michigan City operated from 1966 to 1982. It received a variety of municipal and industrial wastes, including heavy metals and PCB-contaminated oils. Testing has shown that the groundwater and soils contain VOCs, PCBs, PAHs, phthalates, and heavy metals. The contamination threatens recreational Trail Creek, Lake Michigan, and private and public wells that serve about 33,000 people.

Response actions included a site investigation that was completed in 1993. In 1994 the EPA recommended capping, a leachate collection system, and a groundwater extraction and treatment system. Construction should be under way by 1997.

A 1987 Consent Order requires nine PRPs to undertake the site investigation.

Sources: PNPLS-IN; Michigan City Public Library, 100 East Fourth Street, Michigan City, IN, 46360; La Porte County Health Department, Michigan City Branch, 104 Brinckmann Avenue, Michigan City, IN, 46360.

Wayne Waste Oil

Whitley County

This former waste oil recycling plant near Columbia City operated from 1975 to 1980. Oil wastes were stored in drums, tanks, and trenches or dumped on the ground. Testing by the State of Indiana in 1980 indicated that the groundwater and soil contain VOCs, heavy metals, cyanide, PAHs, phenols, and phthalates. These pollutants have entered nearby ponds and the Blue River. The contamination threatens the drinking water supplies for about 6,000 local residents as well as nearby streams and wetlands.

Response actions included removing contaminated soil, sludges, and tars (1986, 1988). Drummed waste was removed from the site in 1988–1989. An EPA site investigation (1985–1990) recommended soil vapor extraction, groundwater extraction and air-stripping, capping, soil consolidation and reburial, a slurry wall, and deed restrictions. Construction began in 1994.

An AOC (1986) and UAO (1988) require the PRPs to conduct the removal actions.

Sources: PNPLS-IN; Peabody Library, 203 North Main Street, Columbia City, IN, 46725.

Whiteford Sales and Service

St. Joseph County

A trucking service operated on this site in South Bend from 1960 to 1980. Various wastewaters and solvents from cleaning trucks and trailers were illegally buried on the property. The contamination was discovered during construction in 1983. Testing showed that the groundwater and soils contain VOCs, including trichloroethylene and vinyl chloride, and a variety of heavy metals, which have also penetrated city wells. The contamination threatens public and private wells within 1 mile of the site, which serve about 250,000 residents.

Response actions included the excavation and removal of contaminated soils. A recently completed EPA site investigation will recommend a final remediation plan.

The PRPs conducted the removal actions.

Sources: PNPLS-IN; St. Joseph County Public Library, 122 West Wayne Street, South Bend, IN, 46601.

IOWA

Des Moines TCE

Polk County

This area of contaminated groundwater in Des Moines was discovered in 1976. Sampling indicated that the city's water supply system contains trichloroethylene (TCE). The source area was identified as Dico Company, a nearby business that used TCE-bearing solvents and also made pesticides and herbicides. Contaminated wastewater was discharged on the property. Testing has shown that the groundwater and soil contain VOCs (vinyl chloride) and pesticides. The contamination threatens the Raccoon River and the water supply for about 260,000 city residents.

Response actions included shutting down the contaminated wells, decontaminating buildings, and building an asphalt cap. The EPA recommended a program of groundwater extraction and air-stripping, which became operational in 1987. Soil treatments are still being considered.

Dico Company has been identified as the PRP. Administrative Orders on Consent (1986, 1989) and a Unilateral Administrative Order (1994) require Dico Company to perform most of the remedial work. A group of smaller PRPs is working on other areas of the site.

Sources: PNPLS-IA; Des Moines City Library, 100 Locust, Des Moines, IA, 50308.

Electro-Coatings Inc.

Linn County

This electroplating shop in Cedar Rapids has been in operation since 1947. Testing in 1976 showed high chromium levels in a nearby industrial well. The point source was a leaking underground tank containing chromic acid. The groundwater contamination is a threat to Cedar Lake, Cedar River, and the drinking water source for about 110,000 people.

Response actions included removing the leaking tank in 1976 and installing groundwater monitoring wells. Contaminated soil was discovered in 1992 and excavated. An EPA investigation conducted between 1990 and 1994 recommended groundwater extraction and treatment.

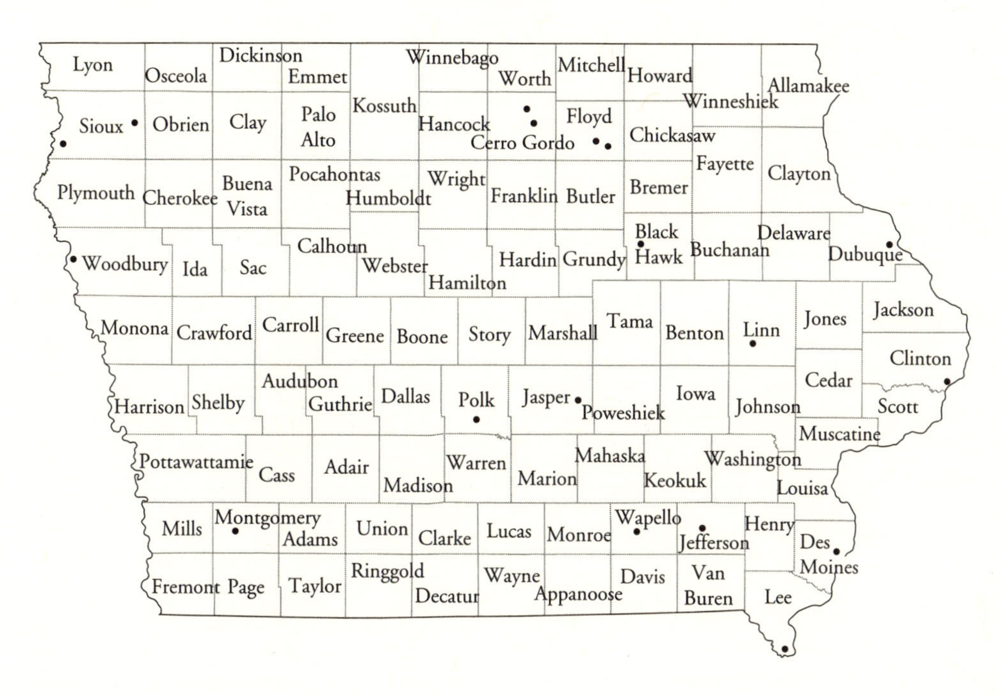

Electro-Coatings received an Executive Order (1977) to install the monitoring wells. A 1990 Consent Order required the PRP to conduct the site investigation. Another Consent Order is being negotiated for additional clean-up work.

Sources: PNPLS-IA; Cedar Rapids Library, 500 First Street Southeast, Cedar Rapids, IA, 52401.

Fairfield Coal Gasification Plant
Jefferson County
This plant in Fairfield produced natural gas from coal from 1878 to 1950. Wastes include PAH- and VOC-containing coal tar and iron oxide slag containing cyanide salts. The majority of the waste was buried on-site. Testing from 1985 to 1987 indicated that the groundwater and soil contain excessive levels of PAHs, anthracene, pyrene, lead, mercury, and VOCs (benzene and xylene). The contamination threatens private and public wells that serve 11,000 people within 3 miles of the area. The recreational Cedar River is also nearby.

Response actions included removing solid waste and debris and installing a groundwater pump-and-treat system. Site studies (1990–1992) recommended continued groundwater extraction and the excavation of coal tar and contaminated soil (1993–1995). Groundwater extraction has achieved the EPA's health-based standards, and only groundwater monitoring is continuing.

Iowa Electric Utilities Company signed an AOC (1989) and CD (1991) to conduct the remediation work.

Sources: PNPLS-IA; Fairfield Public Library, Court and Washington Streets, Fairfield, IA, 52556.

Farmers' Mutual Cooperative
Sioux County
This agricultural supply business has operated on this site near Hospers since 1908. Fertilizers and pesticides were stored and sold on the property until 1992. Testing by the State of Iowa in 1984 detected herbicides and VOCs in nearby municipal wells. The groundwater and soils contain a variety of herbicides, including atrazine. Carbon tetrachloride (a VOC) is also present and was used as a grain fumigant. The contamination threatens the Floyd River and the groundwater drinking supplies of nearby residents.

Response actions included shutting down the contaminated wells. Site investigations from 1984 to 1992 indicate that contamination levels are decreasing through natural attenuation. Monitoring is continuing.

The Farmers' Cooperative is funding the site investigation and remediation work through two Consent Orders.

Sources: PNPLS-IA; Hospers City Hall, Hospers, IA, 51238.

Iowa Army Ammunition Plant
Des Moines County
This 18,000-acre U.S. Army base near Danville has assembled fuses, ammunition, and explosives since 1941. Various liquid and solid wastes, including explosives, heavy metal sludges, pesticides, radionuclides, and laboratory products, have been discharged into lagoons or buried on-site. Testing from 1981 to 1993 showed that the soil and groundwater contain lead and explosives. The contamination threatens private and agricultural wells in the area and recreational waterways. The contamination has already penetrated Brush Creek.

Site investigations are in progress that will recommend a final remediation plan. A possible remedy is creating wetlands as part of an in-situ phytoremediation program for explosive-contaminated groundwater.

Sources: PNPLS-IA; Main Administrative Building, Iowa Army Ammunition Plant, Middletown, IA, 52638; Danville City Hall, Danville, IA, 52623.

John Deere
Wapello County
The John Deere plant near Ottumwa has manufactured farm equipment since 1946. Various wastes, including solvents, sludges, paints, acids, cyanide, and heavy metals, were buried in pits on the site. Testing has revealed low levels of VOCs and heavy metals in the soil and groundwater. The contamination threatens agricultural land, the recreational Black Lake, the Des Moines River, and the private water wells of about 1,000 nearby residents.

EPA recommendations from a 1990–1991 site investigation included fencing, deed restrictions, and groundwater monitoring. The EPA will likely remove the site from the NPL list in the next few years.

A 1989 Administrative Order on Consent and a 1992 Consent Decree require the John Deere Corporation to conduct the site investigation and clean-up work.

Sources: PNPLS-IA; Ottumwa Public Library, 129 North Court Street, Ottumwa, IA, 52501.

Lawrence Todtz Farm
Clinton County
This former landfill site near Clinton operated from 1958 to 1975. It received a variety of municipal and industrial wastes, including spent acids, plastics, resins, paints, and alcohols. Testing revealed that groundwater and soil contain heavy metals (arsenic, barium, lead, and sodium) and VOCs. The contamination threatens about 100 private water wells and recreational Murphy's Lake and Bandixen Lake.

Response actions including connecting affected residents to a new well system (1989). Erosion controls and a groundwater monitoring system were installed in 1991. A groundwater extraction and treatment system will be considered if the groundwater does not naturally attenuate.

The PRPs have conducted all actions at the site. A 1990 Consent Decree requires that the parties conduct the remediation work.

Sources: PNPLS-IA; Clinton Main Library, 306 Eighth Avenue, South Clinton, IA, 52732; Camanche Public Library, 102 12th Avenue, Camanche, IA, 52730.

Mason City Coal Gasification Plant
Cerro Gordo County
This former plant in Mason City produced natural gas from coal from 1900 to 1951. The buildings were razed in 1952. Digging in 1984 discovered oily sludges on the property. Later testing showed that the soil and groundwater contain PAHs. The contamination threatens the municipal wells that serve the 30,000 residents of Mason City and has already penetrated recreational Willow Creek.

Response actions included the removal of underground storage tanks and contaminated soils (1988). The soils were stored on-site under a membrane cap. Testing from 1991 to 1994 resulted in recommendations for removal and treatment of the contaminated soil, followed possibly by a groundwater treatment program.

Interstate Power Company, the PRP, signed a 1991 AOC with the EPA that requires the company to undertake the site investigation and clean-up work.

Sources: PNPLS-IA; Mason City Public Library, 225 Second Street Southeast, Mason City, IA, 50401.

Mid-America Tanning Company
Woodbury County
The Mid-America Tanning Company near Sergeant Bluff has processed animal hides since 1969. Waste sludges containing heavy metals were discharged into unlined lagoons. Treated waste solutions were also poured into a lake on the 100-acre property. Testing has shown that the soil, groundwater, and lake sediments contain arsenic, barium, chromium, lead, cadmium, and hydrogen sulfide gas. The contamination threatens the Missouri River, a wetland habitat for the endangered piping plover, and the private drinking wells for about 900 residents.

Response actions included excavating contaminated soil and sludge that were stored on-site (1990–1991). EPA recommendations in 1991 included solidifying and capping the contaminated soil and sludge and capping other contaminated areas.

This work should be under way by 1997. A groundwater program may also be considered.

The main PRP could not comply with a 1989 UAO for financial reasons. The site investigations were completed with EPA funds. Another PRP performed the clean-up work.

Sources: PNPLS-IA; Sergeant Bluff City Hall, 401 Fourth Street, Sergeant Bluff, IA, 54054.

Midwest Manufacturing Site
Jasper County
Manufacturing companies have operated on this site near Kellogg since 1973. Products have included electroplated metal pieces and high-speed flywheel ring gears for automobiles. Electroplating wastes were poured directly into the North Skunk River from 1973 to 1977. Untreated and treated sludges were discharged into settling ponds from 1977 to 1981. Testing in 1982 showed that the groundwater and soil contain excessive levels of VOCs, cadmium, and nickel. Site examinations discovered a black oily leachate in a wetland beside the property in 1987. The contamination threatens the private drinking water wells of about 700 people. It has already penetrated a city well and the North Skunk River.

Response actions included fencing, deed restrictions, removing contaminated soil, capping the lagoons, and installing a groundwater extraction and treatment system. Construction should be under way by 1997.

Smith and Jones, the PRP, signed a 1994 Consent Decree that requires the company to conduct all site investigations and clean-up work.

Sources: PNPLS-IA; Kellogg City Library, Kellogg City Hall, Kellogg, IA, 50135.

Northwestern States
Portland Cement Company
Cerro Gordo County
This former limestone quarry near Mason City was used as a landfill from 1969 to 1982. It received a variety of municipal and industrial wastes, including kiln dust containing chromium and sulfates. Runoff from the landfill contaminated the Winnebago River, turning it white. Testing showed that the groundwater and soil are contaminated with sulfates, sodium, and heavy metals. The contamination is a threat to about 100 private wells in the area and to the municipal water supply that serves 30,000 residents in Mason City. Calmus Creek and the recreational Winnebago River are also threatened.

Runoff controls were constructed in the 1980s to protect Calmus Creek. Water was also pumped from the quarry and treated. In 1990 an EPA investigation recommended constructing a permanent runoff control system, capping the quarry, and collecting and treating contaminated groundwater. Construction was completed in 1994. The site is expected to be removed from the NPL.

Administrative Orders (1985, 1989) required the PRPs to conduct the site investigations and clean-up work.

Sources: PNPLS-IA; Mason City Public Library, 225 Second Street, Mason City, IA, 50401.

Peoples Natural Gas Company
Dubuque County
This coal gasification plant in East Dubuque operated from 1890 to 1954. It was later used as a gas storage area and maintenance facility. Coal tar sludges and iron oxide waste from the gasification process were either stored in holding tanks or buried on the property. Testing in 1983 revealed that the soil and groundwater contain PAHs, phenols, and inorganic compounds. About 60,000 people depend on water drawn from wells within 3 miles of the site. Recreational wetlands, a wildlife refuge, and the Mississippi River are also nearby.

Response actions included removing and incinerating coal tar sludges and contaminated soils. In 1991 the EPA recommended additional soil removal and the construction of a groundwater extraction and treatment system. These are expected to be operational by 1997.

An AOC (1989) and Consent Decree (1992) require four PRPs to conduct the clean-up operations.

Sources: PNPLS-IA; Carnegie Stout Public Library, Eleventh and Bluff Streets, Dubuque, IA, 52001.

Red Oak Landfill
Montgomery County
This former limestone quarry near Red Oak operated as a landfill from 1962 to 1974. It received a variety of municipal, industrial, and hazardous wastes, including heavy metal sludges, degreasers, and demolition debris. Testing revealed that the groundwater and soil contain toluene, xylene, chromium, lead, and barium. About 7,000 people depend on local wells for drinking water. The contamination also threatens the recreational East Nishnabotna River.

Site investigations completed in 1992 recommended capping the landfill and groundwater monitoring. Construction is expected to begin in 1997.

EPA-identified PRPs have undertaken the site investigations and remediation work.

Sources: PNPLS-IA; Red Oak Public Library, Second and Washington Streets, Red Oak, IA, 51566.

Shaw Avenue Dump
Floyd County
This landfill in Charles City began operations in 1949. It has received a variety of municipal and industrial wastes, including arsenic wastes from animal pharmaceutical production and lime sludges. Open burning has also been carried out. Testing has shown that the groundwater and soil contain arsenic. Nearby water wells serve about 9,000 people. The contamination also threatens the recreational Cedar River.

Site investigations from 1987 to 1993 recommended excavating contaminated soil and sludge, which was completed in 1994. Groundwater studies are in progress and will recommend a clean-up plan in 1997.

A 1988 Consent Order requires the PRPs to conduct the site investigations and remediation work.

Sources: PNPLS-IA; Charles City Public Library, 106 Milwaukee Street, Charles City, IA, 50616.

Sheller-Globe Corporation
Lee County
This industrial landfill and solvent burning site near Keokuk operated from 1947 to 1970. Wastes included rubber products, weather stripping, sludges, paint wastes, resins, and heavy metals. Testing in 1987 detected heavy metals (arsenic, chromium, lead, nickel, and zinc) and VOCs. The contamination threatens nearby waterways and about 300 private water wells in the area.

Detailed site investigations (1990–1993) recommended only a clean soil cover and deed restrictions.

An AOC (1990) requires the PRPs to undertake the site studies.

Sources: PNPLS-IA; Keokuk Public Library, 210 North 5th Street, Keokuk, IA, 52632.

Vogel Paint and Wax Company
Sioux County
This company disposed of paint, wax liquid, and solid wastes at this site near Orange City from 1971 to 1979. Testing in 1979 revealed that the groundwater and soil contain VOCs, chromium, and lead. The contamination threatens the rural water system for about 3,000 people.

Response actions included a clay cap and removing free-product VOCs from the water table. A final remediation plan in 1991 recommended excavating and treating contaminated soil and extracting and air-stripping groundwater. These systems became operational in 1992.

A 1990 Consent Order required the PRPs to conduct the site investigations and clean-up work.

Sources: PNPLS-IA; Orange City Public Library, 112 Albany Avenue Southeast, Orange City, IA, 51041.

Waterloo Coal Gasification Plant
Black Hawk County
This former gasification plant near Waterloo produced natural gas from coal from 1901 to 1956. The plant was later dismantled. Coal tar and cyanide byproducts remain on the site. Testing has shown that the soils and groundwater contain PAHs, VOCs, and cyanide. About 75,000 people depend on well water from within 4 miles of the site. The recreational Cedar River is also nearby.

Site studies identified the areas of contamination. Coal tars and contaminated soil were excavated and removed in 1994–1995. Additional studies in progress will address groundwater remediation.

An AOC (1993) between the EPA and the PRPs require the companies to finance the remediation work.

Source: PNPLS-IA.

White Farm Dump
Floyd County
This former gravel pit near Charles City was used as a dump by the White Farm Equipment Company. Wastes included sludges and baghouse dust from manufacturing operations. Testing has revealed that the soil contains heavy metals and VOCs. The contamination threatens the recreational Cedar River and local wells that serve about 10,000 people.

Response actions included site investigations that recommended a soil cap, which was constructed in 1994–1995.

PRPs have conducted the site investigations and clean-up work according to a 1989 AOC and 1991 CD.

Sources: PNPLS-IA; Charles City Public Library, 106 Milwaukee Street, Charles City, IA, 50616.

Ace Services
Thomas County
This former chrome-plating shop in Colby operated from 1969 to 1990. For 6 years chromium-bearing wastewaters were discharged on the ground. After 1975 waste solutions were held in unlined lagoons. A nearby city well was shut down in 1980 because of high chromium levels. Tests have shown that the soil, groundwater, and sludge on the property are contaminated with chromium, which threatens the drinking water of about 6,200 local residents.

Response actions included excavating contaminated soil and sludge (1981–1982). Drummed wastes and razed buildings were removed in 1992–1994. Investigations that will lead to a final groundwater remediation plan are in progress.

Ace Services, the PRP, has been involved in the remediation work.

Source: PNPLS-KS.

Arkansas City Dump
Cowley County
This former landfill site near Arkansas City began operations in 1916. It received a variety of municipal and industrial wastes, including acidic sludges from oil refineries, residual oil products, and demolition debris. Groundwater and soil contain sulfuric acid, PAHs, heavy metals, ammonia, sulfur, and oil. About 7,000 people live within 4 miles of the site.

Response actions included neutralizing acidic sludges with high pH materials (1991–1992). The EPA considers the contamination to be of little threat to the surrounding communities or to the Arkansas River. The site will probably be removed from the NPL in the next few years.

Sources: PNPLS-KS; Arkansas City Public Library, 120 East Fifth Avenue, Arkansas City, KS, 67005.

Chemical Commodities
Johnson County
This chemical recycling facility in Olathe operated from 1951 to 1989. Poor management and waste handling practices led to frequent spills and illegal discharges of chemical wastes. Testing has shown that the groundwater and soil contain heavy metals (chromium), VOCs, SVOCs, pesticides, and halogenated organic compounds. The contamination threatens the drinking water supplies of about 7,000 residents who live within 3 miles of the site.

Response actions included removing stored chemicals and contaminated soil and decontaminating structures (1989). Runoff controls were also installed. Detailed site investigations are under way and will recommend a final treatment plan.

Sources: PNPLS-KS; Olathe City Library, 201 East Park Street, Olathe, KS, 66061.

Cherokee County
Cherokee County
This 110-square-mile mining district near Galena is part of the larger Tri-State Mining District. Over 100 years of mining created 4,000 acres of sulfidic mine tailings and waste dumps. Acidic runoff has contaminated nearby streams. Testing has revealed the presence of radon gas in the air around Galena. Groundwater and soil contain cadmium, lead, selenium, zinc, and chromium. The heavy metals can also be transported in dust particles. The acidic, heavy metal mine drainage has impacted Spring River, Short Creek, Shoal Creek, and Tar Creek. About 22,000 people live in Cherokee County, and many of them depend on the local groundwater for domestic purposes.

Response actions included installing water treatment units on contaminated wells in Galena and municipal water hook-ups. Contaminated soil was removed from around day care centers and about 200 private homes. Two deeper municipal wells for Galena were completed in 1992–1993. Other strategies being considered are diverting streams from tailings areas and constructing runoff barriers. A number of studies that deal with different areas of contamination elsewhere in the county are in progress.

The PRPs have failed to comply with a 1990 UAO ordering them to undertake the site investigations and clean-up work. To date the work has been funded by the EPA.

Sources: PNPLS-KS; Galena Public Library, 315 West Seventh Street, Galena, KS; 66739; Johnston Public Library, 210 West Tenth Street, Baxter Springs, KS, 66713.

57th and North Broadway Streets
Sedgwick County
The groundwater in this residential/commercial area in Wichita Heights was discovered to be contaminated with VOCs and heavy metals in 1983. The exact point sources have yet to be identified. The EPA has identified oil refining, trucking, and paint manufacturing businesses as being possible PRPs. The contamination has penetrated private and municipal wells in the area.

Response actions included supplying bottled water and city water hook-ups to affected residents (1992). Detailed site investigations that will recommend a final remediation plan should be completed by 1997.

Sources: PNPLS-KS; Amelia Earhart Environmental Magnet School, 4401 North Arkansas, Wichita, KS, 67209.

KANSAS

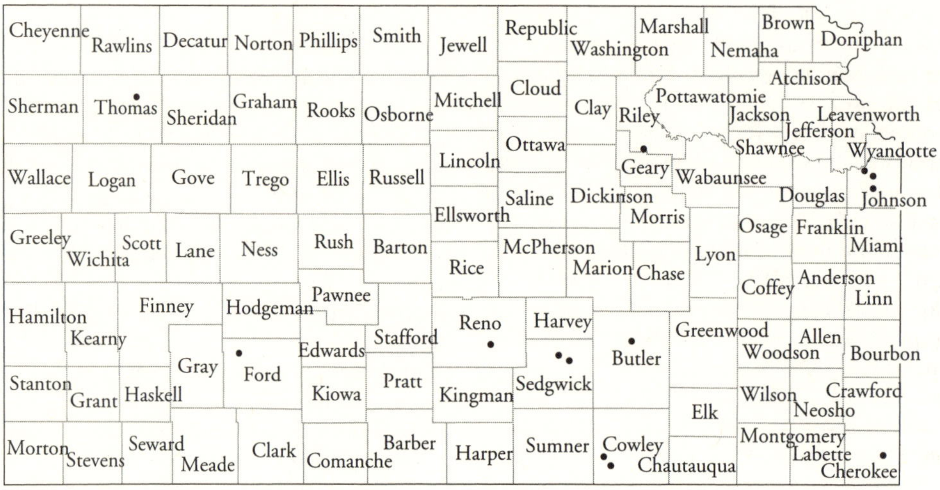

Fort Riley
Geary and Riley Counties
This 152-square-mile U.S. Army base near Junction City was established in 1853. Various liquid and solid wastes from motor pools, burn areas, hospitals, chemical mixing and storage buildings, and dry cleaners were dumped on the ground or buried in landfills. Testing in 1992 detected VOCs in the groundwater at different sites on the base. The contamination threatens the Kansas and Republican Rivers, private and public wells that serve about 50,000 people in the area, and agricultural wells.

Response actions included site studies that were completed in 1994. Contaminated soils were removed in 1994–1995. A pilot soil vapor extraction system and groundwater treatment system were also installed. Some studies are still ongoing, and a final remediation plan is expected by 1997–1998.

Sources: PNPLS-KS; Manhattan Public Library, Juliette and Pyntz Streets, Manhattan, KS, 66502.

Holliday Disposal Site
Johnson County
This former landfill near Merriam operated from 1963 to 1970. It accepted a variety of municipal and industrial wastes, including paint and metal sludges, pesticides, resins, and solvents. Leachates seeped into the nearby Kansas River. Liquid wastes were later stored in unlined ponds on the site. Testing has shown that the groundwater and soil contain VOCs, pesticides, PCBs, PAHs, and heavy metals. The contamination threatens the Kansas River and water wells that serve about 205,000 residents.

In 1989 the EPA recommended a final remediation plan of pumping out the liquid waste, building a cap, installing a leachate collection system, groundwater monitoring, and deed restrictions. Construction should be under way by 1997.

Deffenbaugh Industries, the PRP, signed a Consent Agreement (1987), Administrative Order on Consent (1990), and UAO (1995) that require the company to undertake the site studies and remediation work.

Sources: PNPLS-KS; Johnson County Public Library, 8700 West 63rd Street, Merriam, KS, 66201.

Obee Road
Reno County
This area of groundwater contamination in the town of Obeeville was discovered by state officials in 1983 following local complaints about water quality. VOCs were detected. The source of the contamination is thought to be the old city landfill near the airport that operated in the 1960s and 1970s and received a variety of liquid wastes and sludges, including heavy metals. The contamination threatens the water supply wells of about 2,500 residents within 2 miles of the site.

Response actions included drilling deeper city wells and connecting affected residents to the new municipal water supply. Site investigations from 1990 to 1994 recommended deed restrictions, long-term groundwater monitoring, and possibly groundwater treatment at a later time.

A group of PRPs signed a Consent Agreement (1990) and Administrative Order (1993) requiring them to undertake the site studies and clean-up work.

Sources: PNPLS-KS; Hutchinson Public Library, 901 North Main Street, Hutchinson, KS, 67504.

Pester Refinery Company
Butler County
This oil refinery plant near El Dorado has been in operation since 1917. Various wastes were stored in a pond or burned on-site until about 1975. Testing by the State of Kansas in 1987 discovered leachate seeping into the recreational Walnut River. Since then, seepage and contaminated runoff have occasionally overflowed into the river during heavy rains. The soil and groundwater contain heavy metals (lead), VOCs, and petrochemicals. The contamination threatens the private water wells of about 200 people within 3 miles of the site and has impacted the Walnut River.

Site investigations were conducted from 1990 to 1992. The EPA recommended a final remediation plan of removing toxic sludges and bioremediation of contaminated soils. A groundwater treatment plan will be recommended by 1997.

The Pester Refinery Company declared bankruptcy when it was issued an Administrative Order in 1986. Fina Corporation, a past owner, signed Consent Orders in 1990 and 1993 that require the company to undertake the site studies and clean-up work.

Sources: PNPLS-KS; Bradford Memorial Library, 611 South Washington, El Dorado, KS, 67042.

Strother Field Industrial Park
Cowley County
This industrial park near Winfield was a former military facility and waste disposal site that received a variety of municipal and industrial wastes. Testing by the State of Kansas has shown that the groundwater contains VOCs. About 4,200 local residents and workers use water from public or private wells within 3 miles of the site.

Response actions included supplying affected residents with bottled water and connections to new municipal wells. Groundwater extraction and air-stripping systems were installed in 1985 to remove the VOCs. A final remediation plan for other point sources is pending.

General Electric, a PRP, installed the groundwater extraction system. Five PRPs signed a variety of agreements between 1986 and 1990 that require the companies to undertake the site investigations and clean-up work.

Sources: PNPLS-KS; Strother Field Commission, Terminal Building, Fourth and A Streets, Cowley County, KS, 67156.

Sunflower Army Ammunition Plant
Johnson County
This 9,000-acre U.S. Army military base and ammunition plant near De Soto operated from 1941 to 1992. It manufactured weapons, cannons, rockets, nitric acid, sulfuric acid, and nitroguanidine. Explosives were stored on-site. Various solid and liquid wastes generated during the manufacturing operations were buried in 70 pits, ditches, or sumps. Testing has shown that the groundwater and soil contain nitrates, mercury, arsenic, ammonia, and other explosives derivatives. The contamination threatens private and agricultural wells in the area and has already penetrated Captain Creek and Kill Creek, causing fish kills.

Detailed site investigations that will recommend a final remediation plan are in progress.

Sources: PNPLS-KS; De Soto Public Library, 33145 West 83rd Street, De Soto, KS, 66018.

29th and Mead Groundwater Contamination
Sedgwick County
Groundwater in this industrialized part of Wichita was discovered to be contaminated with VOCs over the period 1983–1986. The contamination plume covers about 1,500 acres and is thought to be related to discharges from several industries in the area. The contamination threatens water wells that supply about 3,500 local residents.

Detailed site investigations that will recommend a final remediation plan should be completed by 1997. The point source area is already being treated by soil vapor extraction and groundwater extraction and air-stripping.

About 70 PRPs have been identified. The PRPs formed a steering committee to address clean-up issues. Administrative Orders on Consent (1989) that require the PRPs to undertake the site investigations and remediation work have been signed.

Sources: PNPLS-KS; Kansas Department of Health, District Office, 1919 Amidon Street, Wichita, KS, 67203.

Wright Groundwater Contamination
Ford County
Testing in 1988–1989 revealed that groundwater in the city of Wright contained high levels of VOCs, pesticides, and heavy metals. A source has yet to be identified. The contamination threatens the private water supplies for about 225 residents.

Response actions included connecting affected residents to clean water supplies (1991–present). Detailed site investigations are in progress that will possibly identify the source of contamination.

Source: EPA Publication 9320.7-071, Volume 3, No. 1.

KENTUCKY

Airco
Marshall County
This former landfill near Calvert City operated from the 1950s to 1980. It received a variety of municipal and industrial wastes, including PVCs, sludges, spent acids, VOCs, mercury, and other heavy metals. Testing has shown that the groundwater and soil contain excessive levels of PCBs, PAHs, and VOCs. The contamination threatens the nearby Tennessee River and the drinking water source for about 4,000 people who live in the area.

Response actions included building a clay cap in 1981. EPA site investigations from 1982 to 1988 recommended deed restrictions, a groundwater pump-and-treat system, excavation of contaminated soil, and a soil vapor extraction system. Construction should be under way by 1997.

The EPA has identified a number of PRPs, including Airco Carbide, who are funding the site investigations and remediation work.

Sources: PNPLS-KY; Marshall County Public Library, Calvert City Branch, City Hall, Calvert City, KY, 42029.

B. F. Goodrich
Marshall County
This former landfill near Calvert City operated from 1969 to 1972. Located beside the Airco NPL site, the B. F. Goodrich Company buried various industrial wastes on the 2-acre property, including sludges and chlorinated solvents. In 1980 state officials observed leachate seeping into the Tennessee River. Testing showed that the groundwater and soil contain VOCs. The contamination threatens the Tennessee River and the drinking water source for about 4,000 nearby residents.

Response actions included installing a clay cap (1980). In 1988 EPA site investigations recommended a final remediation plan of deed restrictions; excavation of contaminated soil; runoff, leachate, and flood controls; and a groundwater pump-and-treat system. Construction should be under way by 1997.

B. F. Goodrich, the PRP, has agreed to fund the site investigations and remediation work.

Sources: PNPLS-KY; Marshall County Library, 1003 Poplar Street, Benton, KY, 42025.

Brantley Landfill
McLean County
This former coal strip pit near the town of Island operated as a landfill from the 1960s to 1980. It received a variety of municipal and industrial wastes, including residues from an aluminum recycling operation. Testing in 1986 discovered ammonia in air samples near the landfill. Soil and groundwater contain high levels of heavy metals (including chromium, copper, and aluminum. Other gases were also detected (acetylene, methane, and hydrogen sulfide). The contamination threatens the water supply of about 500 nearby residents. The local groundwater is also used for farming purposes.

Response actions included fencing the site (1990). Site investigations from 1990 to 1994 recommended capping the landfill, removing contaminated soil, installing a groundwater extraction and treatment system, and long-term groundwater monitoring. The systems are presently being designed.

A Consent Order (1990) between the EPA and identified PRPs requires the companies to fund the site investigations and remediation work.

Sources: PNPLS-KY; Island City Hall, 16 South First Street, Island, KY, 42350.

Caldwell Lace Leather Company
Logan County
This former tannery near Auburn operated from 1972 to 1985. Liquid wastes from the leather tanning process, including chromium solutions and sludges, were poured into unlined ponds. In 1983 the state officials found high chromium levels in a neighboring private well. Further testing showed that the soil and groundwater on the site are contaminated with hexavalent chromium. About 600 people within 3 miles of the site use the local groundwater for drinking purposes.

Response actions included capping the site, fencing, and connecting affected residents to the city water system. An EPA site investigation (1990–1993) concluded that soils and surface water were safe but that the groundwater risks were impossible to determine because of the underlying karst (cave) structures; monitoring was recommended.

The PRP has agreed to finance the site investigations and remediation work.

Sources: PNPLS-KY; Logan Public Library, 201 West Sixth Street, Russellville, KY, 42276.

Distler Brickyard
Hardin County
This brick manufacturing facility near West Point operated from the late 1800s to 1975. The site was later used as a waste dump from 1976 to 1979 and received various liquids and sludges. Testing has determined that the soil and groundwater contain VOCs and heavy metals, including lead. Surrounding vegetation and wildlife have already been killed by the contaminants. The contamination threatens nearby Bee Branch and Ohio River and the water supplies of about 70,000 people.

Response actions included removing areas of contaminated soil and about 2,000 drums of liquid

KENTUCKY

Kenton
Boone
Campbell
Gallatin
Pendleton
Trimble
Carroll
Grant
Bracken
Mason
Lewis
Greenup
Owen
Robertson
Henry
Harrison
Oldham
Fleming
Boyd
Nicholas
Carter
Scott
Shelby
Franklin
Bourbon
Jefferson
Bath
Rowan
Woodford
Elliott
Lawrence
Hancock
Bullitt
Spencer
Fayette
Montgomery
Anderson
Clark
Menifee
Morgan
Meade
Jessamine
Johnson
Henderson
Nelson
Mercer
Powell
Martin
Breckinridge
Madison
Wolfe
Daviess
Washington
Magoffin
Hardin
Estill
Union
Boyle
Garrard
Lee
Floyd
Marion
Breathitt
Larue
Pike
Webster
McLean
Lincoln
Ohio
Grayson
Jackson
Owsley
Hopkins
Crittenden
Taylor
Rockcastle
Knott
McCracken
Hart
Casey
Livingston
Butler
Green
Perry
Edmonson
Clay
Muhlenberg
Caldwell
Adair
Pulaski
Laurel
Leslie
Letcher
Ballard
Lyon
Warren
Barren
Russell
Christian
Metcalfe
Knox
Logan
Marshall
Carlisle
Harlan
Wayne
Trigg
Cumberland
Todd
Whitley
Bell
Simpson
Allen
Hickman
Graves
Monroe
Clinton
McCreary
Calloway
Fulton

waste (1979–1982). The remaining contaminated soil was removed in 1989. A groundwater extraction and treatment system became operational in 1994.

Sources: PNPLS-KY; West Point City Hall, 509 Elm Street, West Point, KY, 40177.

Distler Farm

Jefferson County

This property near West Point was used to store industrial waste illegally from the 1960s to the 1970s. A flood in 1978 exposed numerous drums of waste along the banks of Stump Gap Creek. Waste chemicals included paint and varnish by-products. Further investigations discovered four burial sites. The soil and groundwater contain high levels of VOCs and heavy metals. The contamination threatens Stump Gap Creek, the Ohio River, and the drinking water of about 3,000 people within 4 miles of the site.

Response actions included the emergency removal of the exposed drums of waste (1978). More drums were excavated and removed from 1979 to 1984. In 1988 the EPA recommended a final remediation plan of excavating and removing contaminated soil, installing a groundwater treatment system, and groundwater monitoring until 2020.

Sources: PNPLS-KY; West Point City Hall, 509 Elm Street, West Point, KY, 40177.

Fort Hartford Coal Company Quarry

Ohio County

This stone quarry near Olaton operated from the 1950s to 1970s. Compacted dust residues (salt cake fines) from Barmet Aluminum Company were stored in the underground tunnels on the property. In 1986 the EPA discovered ammonia gas in the air and ammonia, methane, acetylene, and hydrogen sulfide gas in the underground waste areas. About 700 people within 3 miles of the site depend on private water wells. The contamination also threatens the recreational Rough River and the town of Hartford's water supply.

Response actions included efforts to divert runoff from the waste piles underground and pumping out mine water. Data from studies conducted between 1991 and 1994 are being evaluated and will lead to the selection of a final remediation plan.

An AOC with Barmet Aluminum Corporation (1989) requires the company to perform the site investigations and remediation work.

Sources: PNPLS-KY; Court House Square, Main Street, Hartford, KY, 42347.

Green River Disposal

Daviess County

This former landfill site near Maceo operated from 1970 to 1984. It received a variety of municipal and industrial wastes, including sludges. An examination by the State of Kentucky in 1985 discovered high amounts of heavy metals in a nearby well. Soil and groundwater contain heavy metals (arsenic and barium) and ammonia gas. The contamination threatens the private water wells of about 500 nearby residents and has penetrated nearby Blackford Creek.

Response actions included fencing and a temporary leachate collection system (1990). In 1994 the EPA recommended a final remediation plan of capping, leachate collection and treatment, and removing contaminated sediments. Construction should be under way by 1997.

Green River Disposal Company was ordered by the State of Kentucky to clean up the site in 1983. The company filed for bankruptcy 3 years later. The EPA has identified other PRPs who have signed an AOC that requires them to undertake the site studies and remediation work.

Sources: PNPLS-MI; Owensboro Library, 450 Griffith Avenue, Owensboro, KY, 42301.

Howe Valley Landfill

Hardin County

This former landfill site near Howe Valley operated from 1967 to 1976. It received a variety of municipal and industrial wastes, including heavy metal sludges and insulation materials. Testing by the state in 1979 showed that soils contain low levels of VOCs. The contamination threatens the local drinking water of about 100 residents within 2 miles of the site.

Response actions included removing drummed waste in 1988. Detailed site investigations from 1990 to 1994 recommended the excavation or in-situ aeration of the contaminated soil and 5 years of groundwater monitoring. The soil work was completed in 1994.

The PRPs signed an AO (1988) and Consent Decree (1991) with the EPA to fund the site investigations and remediation work.

Sources: PNPLS-KY; Hardin County Public Library, 201 West Dixie Highway, Elizabethtown, KY, 42701.

Maxey Flats Nuclear Disposal

Fleming County

This disposal site near Hillsboro received low-level radioactive wastes (uranium, thorium, and plutonium) from 1963 to 1977. Radioactive leachates have penetrated surrounding property. Testing by the EPA has shown that the groundwater, soil, and leachate contain radioactive materials, VOCs, petrochemicals, and heavy metals. The contamination threatens the local groundwater supplies for about 300 local residents as well as Rock Lick Creek and the Licking River. The groundwater is also used for agricultural purposes.

Response actions included collecting and treating radioactive leachate (1989). In 1991 the EPA recommended the excavation and solidification of contaminated soil and leachate, runoff controls, capping, a groundwater flow barrier, and long-term groundwater monitoring. Construction should be under way by 1997.

Legal agreements signed with PRPs in 1987 call for the PRPs to fund the site investigations and remediation work.

Sources: PNPLS-KY; Rowan County Public Library, 129 Trumbo Street, Morehead, KY, 40351; Fleming County Public Library, 303 South Main Cross Street, Flemingsburg, KY, 41041.

National Electric Coil
Harlan County

Electric motors and transformers were rebuilt on this site near Harlan from 1951 to 1987. Trichloroethylene was used to clean equipment. Liquid wastes, sludges, and oils were discharged into the Cumberland River. Testing by the state in 1989 revealed that the groundwater and soil contain excessive levels of VOCs and PCBs. Nearby water wells are also contaminated. The contamination threatens the water supply of about 2,000 people within 3 miles of the site as well as the Cumberland River.

Response actions included connecting affected residents to the public water supply system (1989). Contaminated soil was removed in 1991. In 1992 the EPA recommended a final remediation plan of groundwater extraction and treatment. The system should be operational by 1997.

Cooper Industries, the PRP, signed UAOs (1991, 1992) and a Consent Order (1992) that require the company to fund the site investigation and clean-up work.

Sources: PNPLS-KY; Harlan County Public Library, 107 North Third Street, Harlan, KY, 40831.

National Southwire Aluminum
Hancock County

National Southwire Aluminum Company near Hawesville refines aluminum metal. Various production wastes, including pot linings, calcium fluoride, and cyanide, were buried on-site or discharged into unlined lagoons. Testing in the late 1970s revealed cyanide in the company's drinking water. Groundwater and soil contain excessive amounts of fluoride, arsenic, cyanide, lead, and nickel. The contamination threatens the Ohio River and the drinking water of about 16,000 local residents.

Response actions included closing the older, unlined lagoons on the site. The plant's contaminated wells were closed. The plant was later connected to the city water supply (1989). Site investigations are nearing completion and have recommended an interim groundwater pump-and-treat system until a final remedy is selected.

National Southwire Aluminum Company has violated effluent limits twice since 1987. Under a 1992 agreement with the EPA the company has agreed to fund the site investigations.

Sources: PNPLS-KY; Hancock County Public Library, Court Street, Hawesville, KY, 42348.

Paducah Gaseous Diffusion Plant
McCracken County

This U.S. Department of Energy complex near the town of Kevil has operated since 1952 as a uranium-enrichment center. A variety of wastes, including PCBs, VOCs, technetium-99, uranium isotopes, fuels and lubricants, chemical solvents, heavy metals, rinsewater, and cooling tower water, were stored or discharged on-site. Testing in 1988 discovered that the groundwater and soil are contaminated with trichloroethane and technetium. The contaminants have penetrated nearby private wells, recreational Little Bayou Creek, Big Bayou Creek, and the West Kentucky Wildlife Management Area.

Response actions included connecting affected residents to city water and building runoff/erosion control systems. An interim groundwater pump-and-treat system to remove VOCs and technetium became operational in 1995. Contaminated sediments from ditches on the property were removed (1995). Other studies are in progress and should be completed by 1997.

Sources: PNPLS-KY; U.S. Department of Energy, Environmental Information Center, 175 Freedom Boulevard, Kevil, KY, 42053.

Red Penn Sanitation Company Landfill
Oldham County

This landfill site near Pewee Valley operated from 1954 to 1986. It received a variety of municipal and industrial wastes, including phenol, acids, VOCs, paint waste, and various sludges. Recent testing has shown that the soil contains low levels of VOCs, PCBs, lead, chromium, and selenium. The groundwater may also be contaminated. The testing indicates that the contamination has not migrated beyond the site. It could later threaten the private water wells of about 1,000 nearby residents as well as recreational Floyd's Fork. Other creeks in the area are used for swimming and agricultural purposes.

Response actions included removing sealed drums and contaminated soil in 1986. The EPA concluded in 1993 that contamination at the site was not serious enough to warrant long-term clean up. The EPA did, however, order proper closure of the landfill.

PRPs identified by the EPA are closing the site.

Sources: PNPLS-KY; South Oldham Library, 6720 West Highway 46, Crestwood, KY, 40014.

Smith's Farm
Bullitt County
This former unregulated landfill near Shepherdsville operated for over 30 years. It received a variety of municipal and industrial wastes, including over 100,000 drums of hazardous waste. Leachate has migrated into Bluelick Creek. Testing has shown that the soil and groundwater are contaminated with VOCs, PCBs, arsenic, lead, nickel, chromium, creosote, and phenols. The contamination is a threat to about 500 people who live within 1 mile of the site.

Response actions included removing drums and erecting fencing (1984–1988). In 1991 the EPA recommended chemical treatment of the contaminated soils to remove PCBs and PAHs. This procedure should be completed by 1997. Construction of a leachate collection system and landfill cap should be under way by 1997.

A Consent Order (1989) and UAOs (1990, 1994) require the PRPs to fund the site investigations and clean-up work.

Sources: PNPLS-KY; Bullitt County Public Library and Ridgeway Memorial Library, Second and Walnut Streets, Shepherdsville, KY, 40165.

Tri-City Industrial Disposal Site
Bullitt County
This former landfill site near Shepherdsville operated in the 1960s. It accepted a variety of municipal and industrial wastes, including fiberglass and liquid wastes such as paint thinners. Testing by the state in 1987 showed that the soil and groundwater contain VOCs, lead, chromium, mercury, PCBs, and creosote. Two nearby springs are also polluted. About 1,600 people use local water wells or springs within 3 miles of the site. The contamination also threatens recreational Knob Creek.

Response actions included removing drummed waste and connecting affected residents to a clean water supply (1988). In 1991 the EPA recommended a program of groundwater extraction and treatment and groundwater monitoring. Soil treatment may also be considered.

PRPs identified by the EPA are funding the site investigations and remediation work.

Sources: PNPLS-KY; Bullitt County Public Library and Ridgeway Memorial Library, Second and Walnut Streets, Shepherdsville, KY, 40165.

LOUISIANA

Agriculture Street Landfill
Orleans Parish
Located in the city of New Orleans, this former city landfill operated from 1910 to 1966 and took in a variety of municipal and construction wastes. It has been partially redeveloped with new neighborhoods and a school. At the request of local community leaders, the EPA conducted tests in 1993–1994 that found anomalous concentrations of heavy metals, including lead, in the shallow groundwater and surface soil. The groundwater is not used as a water source, but lead levels in the soil (1,000–4,000 ppm) are considered dangerous.

Response actions included fencing the undeveloped portion of the landfill site and closing the local school (1994). A site investigation/feasibility study and proposed plan of action was completed in 1996 and has been distributed for public comment.

PRPs identified by the EPA have been reluctant to participate in the site study, and legal action is being considered.

Sources: PNPLS-LA; Helen Edwards Elementary School Library, New Orleans, LA, 70126; Community Outreach Office, 3221 Press Street, New Orleans, LA, 70126.

American Creosote Works Inc.
Winn Parish
This site in residential Winnfield (population 7,000) operated as a wood treatment plant from 1901 to the 1980s. A succession of owners discharged liquid wastes into unlined ponds and pits. Additional contamination came from on-site storage areas, leaking storage tanks, and drainage ditches. The soil and groundwater contain high levels of creosote, PCP, and PAHs. The groundwater flows into nearby streams and wetlands and threatens residential drinking water supplies.

Response actions included excavation of oil and sludge from a spill in 1988. The PRP fenced the site in 1988. A site investigation in 1993 recommended incineration of the "tar mat," in-situ biotreatment of contaminated soil, and pumping and disposal of liquid wastes, including 1 million gallons of subsurface creosote product and 24 million gallons of contaminated groundwater. The clean up, which should be under way by 1997, will boost local employment.

Sources: PNPLS-LA; Winn Parish Public Library, Winnfield, LA, 71483.

LOUISIANA

Bayou Bonfouca
St. Tammany Parish
This former wood treatment plant is located near the town of Slidell. A number of operators treated wood with creosote products from 1882 to 1970. Creosote leaks and spills happened frequently during operations. The soil and groundwater are contaminated with creosote compounds, including PNAs. The contamination threatens the water supplies of local residents and the Bayou Bonfouca wetland. Further studies have indicated that the bayou is biologically sterile, and exploratory divers received second-degree burns from creosote.

Response actions included fencing of the site by the PRPs (1985). A site investigation (1987) recommended a clean-up plan that called for on-site incineration of contaminated soil, installation of a groundwater extraction and treatment system, and dredging of contaminated sediments from the bayou. Groundwater treatment began in 1991. Incineration activities were conducted from 1993 to 1995. Bayou dredging has also been conducted.

Three PRPs have been targeted for the recovery of site investigation and clean-up costs.

Sources: PNPLS-LA; St. Tammany Parish Library, Slidell, LA, 70458.

Bayou D'Inde
Calcasieu Parish
This industrial site is located near the towns Westlake and Lake Charles on Bayou D'Inde, part of the Calcasieu River system. Petrochemical companies produce petroleum, sodium hydroxide, teflon, butadiene, and trichloroethylene. EPA tests in 1987 showed that the groundwater, soil, and river sediments contained high levels of PCBs, phenanthrene,

xylene, hexachlorobenzene, hexachlorobutadiene, mercury, and chromium. The fish and shellfish in the bayou, an important recreational waterway, are also contaminated.

Response actions included an EPA site investigation (1991–1995). The recommended clean-up plan includes dredging the bayou and treating the contaminated sediment. The communities view the dredging and deeper channel as an economic benefit that will attract more industry to the bayou.

Thirty-eight PRPs have been identified. Pittsburgh Paint and Glass, one of the larger PRPs, will finance an ecological risk assessment for Bayou D'Inde (1995–1999).

Source: PNPLS-LA.

Bayou Sorrel
Iberville Parish
Located in a remote backswamp environment, this disposal area received wastes from a variety of petrochemical companies from 1977 to 1978. It was closed in 1979. EPA studies in 1981–1982 indicated that the soil, groundwater, and surface water were contaminated with hydrocarbon and pesticide/herbicide wastes and by-products. Less than 100 people live within 5 miles of the site.

A final site clean-up plan was recommended in 1987, and consent decrees with PRPs were signed in 1988–1989. Multilayer caps, slurry walls, and gas venting systems were completed between 1988 and 1990. Groundwater monitoring is continuing. The site is expected to be removed from the NPL by 1997.

Twenty-three major PRPs were identified in 1983; a number of smaller companies also contributed minimal settlements to the clean-up fund. The EPA has recovered all past costs from the PRPs with the exception of $150,000 from former operator Cyril Hinds.

Sources: PNPLS-LA; Iberville Parish Public Library, Iberville, LA, 70776.

Cleve Reber
Ascension Parish
This landfill accepted both municipal and industrial wastes, including about 6,400 drums of hazardous material. It closed in 1974. EPA tests in the mid-1980s indicated that the soil and groundwater at the site contain high levels of hexachlorobenzene and hexachlorobutadiene. The contamination threatens the surrounding swamps and the drinking water wells of nearby residences.

Response actions included fencing the site, removing drums and waste piles, and temporary capping (1983). The EPA finished the site investigation and designed a clean-up plan from 1984 to 1990. Contaminated soil and bulk waste were incinerated on-site in 1995, and capping and revegetation was completed in 1996. Groundwater monitoring is continuing.

A UAO between the EPA and 23 PRPs requires the parties to finance the site investigation and remediation plan.

Sources: PNPLS-LA; Ascension Parish Public Library, Gonzales, LA, 70737.

Combustion
Livingston Parish
Operated as a waste oil recycling operation, this site near Denham Springs was closed in 1982. Tests in the mid-1980s showed that the soil and groundwater contained high levels of lead, silver, nickel, benzene, toluene, mercury, toluene diamine, toluene disocyanate, and 1,2 dichloroethane. About 500 people reside within 1 mile of the site. The contaminated groundwater and surface runoff flow into West Colyell Creek and the Amite River.

Response actions included removing oil from surface ponds, taking down buildings, and pulling out storage tanks (1993). Contaminated sludges and soil were also excavated. An advanced site investigation begun in 1995 is studying a groundwater treatment plan.

A series of compliance orders and settlement agreements (1987–1992) have been signed with PRPs. Eighty-seven PRPs have been identified, 28 of which have formed a steering committee that will help oversee and fund the clean-up plan.

Sources: PNPLS-LA; Livingston Parish Library, 10095 Florida Boulevard, Denham Springs, LA, 70726.

D. L. Mud, Inc.
Vermilion Parish
This site is a former drilling mud facility near the town of Abbeville and adjacent to the Gulf Coast Vacuum Services Superfund site. From 1969 to 1980 it stored and mixed barium sulfate mud. EPA tests in the early 1990s revealed anomalous amounts of petroleum compounds, mercury, chromium, lead, zinc, barium, and arsenic in the soil and groundwater. The contamination threatens the drinking water of about 2,600 within 3 miles of the site as well as waterways that flow to the nearby Vermilion River. The groundwater is also used for irrigating crops.

Response actions included the removal by former owners of contaminated soil and liquid wastes from tanks and dismantling tanks in the mid-1980s. An EPA site investigation that studied the remaining risks was completed between 1990 and 1994. The recommended clean-up plan called for fencing, deed restrictions, groundwater monitoring, and removal and treatment of additional barium-contaminated soil.

The EPA has identified 97 PRPs who will finance the clean-up operation.

Sources: PNPLS-LA; Vermilion Parish Library, 200 North Street, Abbeville, LA, 70511.

Dutchtown Treatment Plant

Ascension Parish

Located near Dutchtown, this site was operated as an oil refinery and reclamation plant from 1965 to 1982. Wastes were disposed of or stored on-site, including in sludge ponds and an oil pit. State-administered tests in 1984 revealed that the soil and shallow groundwater contained anomalous levels of benzene, ethylbenzene, toluene, and lead. About 4,000 people live within 3 miles of the site; Grand Goudine Bayou is less than 2,000 feet from the site.

Response actions included excavating an oil spill (1987), removing additional waste materials (1990–1991), and temporary capping. A final remediation plan consisting of removal and incineration of contaminated soils and waste (approved in 1994) has yet to be started. Groundwater monitoring is continuing.

Of the 85 PRPs identified by the EPA, 20 have signed Consent Decrees (1989) and AOCs (1989) to contribute toward the site investigation and remediation plan.

Sources: PNPLS-LA; Ascension Parish Public Library, Gonzales, LA, 70737.

Gulf Coast Vacuum Services

Vermilion Parish

This site is located beside the D. L. Mud, Inc., Superfund site near Abbeville and the Vermilion River. The main operation processed primary wastes from oil and gas exploration from 1969 to 1984. Sludge and liquid wastes were illegally dumped in ponds, tanks, and a buried pit. EPA tests in the mid-1980s found elevated levels of benzene, toluene, mercury, lead, chromium, arsenic, barium, and organic compounds in the soil and groundwater. About 2,600 residents rely on groundwater from within 3 miles of the site. The water is also used for irrigation and flows into LeBoeuf Canal and Coulee Galleque.

Response actions included fencing the site and constructing surface runoff containment systems (1990–1991). Sludges were capped in 1994. The final remediation plan calls for on-site biological treatment of sludges and soils and stabilizing inorganic soils. Water will be removed from pits and the groundwater monitored.

Over 400 PRPs have been identified by the EPA, 150 of whom can contribute funds to the cost of the study and clean up. Three million dollars have been recovered from minimal settlements with 54 PRPs.

Sources: PNPLS-LA; Vermilion Parish Library, Abbeville, LA, 70510.

Gulf States Utilities

Calcasieu Parish

Gulf States Utilities Company is located along the Calcasieu River in the town of Lake Charles and has been active since 1916. Gas operations (including coal gasification), landfill operations, and storage tanks have contaminated the soil and groundwater with PAHs, PCBs, and volatile organic compounds (VOCs). Transformers, oil, and coal tar from the gas operations were dumped in wetlands on the property. About 60,000 people live in Lake Charles and nearby Westlake. The contamination also threatens the recreational Calcasieu River, Lake Charles, and Prien Lake.

Response actions included an EPA site investigation (1990–1992). The site was proposed as an NPL site in 1995. A final remediation plan has yet to be recommended.

Gulf States Utilities has issued a good faith offer (1995) to finance the site investigation and clean-up plan. Other PRPs are being contacted.

Sources: PNPLS-LA; Calcasieu Parish Library, 1160 Ryan Street, Sulphur, LA, 70663.

Highway 71/71 Refinery

Bossier Parish

This former refinery site is located in downtown Bossier City (population 52,000) and operated from 1923 to 1967. Refinery activities and on-site landfills contaminated the soil and groundwater with lead, petroleum, hydrocarbon gas, and PAHs. The site has since been redeveloped with private homes, apartments, and businesses. Residents complained of gas fumes and headaches in 1990. The Red River is less than 2,000 feet from the site.

Site investigations by the EPA and the PRP have been ongoing since 1991. A final remediation plan has yet to be selected.

An agreement was signed in 1995 by the EPA, the State of Louisiana, and the PRP, Canadian OXY Offshore Production Company, to remediate the site.

Sources: PNPLS-LA; Bossier Parish Library, 2206 Beckett Street, Bossier City, LA, 71111.

Lincoln Creosote

Bossier Parish

Located in Bossier City, this site is an abandoned wood treatment facility that was operated by a variety of owners from 1935 to 1969. Lumber was preserved using creosote, PCP, and chromated copper arsenate. Wastewater was discharged onto the ground. EPA tests in 1985 found the soil to be contaminated with creosote, PAHs, SVOCs, PCP, chromium, copper, and arsenic. The chemicals have been found on neighboring properties and threaten local groundwater, which flows into the nearby Red River.

Response actions included excavation and treatment of contaminated soil and fencing (1992). An EPA study conducted from 1992 to 1996 will recommend a final remediation plan, including additional soil removal.

Four PRPs have been identified. An AOC with Josyln Corporation (1995) calls for the company to remove contaminated soil from the site.

Sources: PNPLS-LA; Bossier Parish Library, 2206 Beckett Street, Bossier City, LA, 71111; EPA Region 6, 1445 Ross Avenue, Dallas, TX, 75202.

Louisiana Army Ammunition Plant

Webster Parish

Located in an agricultural area, this plant began manufacturing explosives for the U.S. Army in 1942. A series of contractors has worked with the U.S. Army on this property. Various production wastes have been discharged into landfills, pits, and sludge lagoons. The soils and shallow groundwater are contaminated with chemical compounds derived from explosives, including cyclonite (RDX) and TNT. The contamination threatens the drinking water of about 10,000 people living within 2 miles of the site.

Response actions included soil removal and capping contaminated areas (1990). EPA and U.S. Army investigations from 1978 to 1993 recommended the incineration of contaminated waste and soil. The U.S. Army is presently conducting additional site studies.

The costs of the site studies and remediation plans will be paid for by the U.S. Army.

Sources: PNPLS-LA; Webster Parish Public Library, Minden, LA, 71058.

Madisonville Creosote Works

St. Tammany Parish

A wood treatment facility operated on this site near Madisonville from the 1950s to 1994. Wood products were soaked in creosote. Contaminated sludges and wastewaters were discharged into lagoons or sprinkled on the surface from the 1960s to 1984. Testing in 1995 revealed that the soil and groundwater contain creosote and creosote derivatives such as anthracene, benzo(a)pyrene, furans, and naphthalene. The chemicals have also been detected in nearby streams and wetlands. The contamination threatens local drinking water supplies, Lake Ponchartrain, streams, wetlands, and habitats for the endangered gulf sturgeon and American bald eagle.

Response actions included removing contaminated soils and sludges from lagoons and ditches, backfilling, and capping (1986). This site was proposed as an NPL site in 1996; a final decision is expected in 1997–1998. Additional studies are planned to determine interim and final remediation actions.

The PRP declared bankruptcy in 1994.

Source: EPA Publication 9320.7-071.

Old Inger Oil Refinery

Ascension Parish

This former oil refinery on the Mississippi River near the town of Darrow was operated from 1967 to 1980. Used oil was recycled on the site, and wastes were discharged into lagoons on-site. A major spill occurred in 1978. EPA tests in the mid-1980s indicated that the soil and groundwater are contaminated with petroleum, PNAs (phenanthrene), and heavy metals (zinc). About 20,000 people live within 10 miles of the site; the contamination plume threatens local water sources used for drinking and irrigation.

Response actions included fencing, runoff containment, and excavation of contaminated soil (1983–1988). The final remediation plan to excavate and treat soils and sludge, extract and treat groundwater via carbon adsorption, and cap is under way.

Although 17 PRPs have been identified, none can contribute funds toward the clean-up remedies.

Sources: PNPLS-LA; Ascension Parish Public Library, Gonzales, LA, 70737.

Pab Oil & Chemical Service, Inc.

Vermilion Parish

Located near the town of Abbeville (population 13,000), this former disposal facility for oil field waste processed oil-based drilling muds from 1979 to 1982. Contaminated sludges were dumped in pits. Testing in the 1980s revealed that the soil and surface water contain barium, chromium, lead, manganese, ethylbenzene, acetone, toluene, xylenes, and PAHs. The contamination threatens the Vermilion River and the water supplies of about 15,000 people who use private and municipal wells within 3 miles of the site.

Response actions included removing a damaged storage tank in 1991. An ongoing site investigation will call for bioremedation of sludges, surface water treatment and disposal, and groundwater monitoring (to be completed by 1999).

The EPA has identified over 100 PRPs. Some minimum settlements have been negotiated with smaller parties. A PRP group is financing the site investigation.

Sources: PNPLS-LA; Vermilion Parish Public Library, Abbeville, LA, 70510.

Petro-Processors of Louisiana
East Baton Rouge Parish
Four former petrochemical dump sites in this rural area are located near Scenic Highway 694. These landfills and disposal lagoons received drilling chips, muds, and other liquid drilling waste from 1961 to 1980. Tests showed that the soil and groundwater are contaminated with chlorinated hydrocarbons (including hexachlorobutadiene), PAHs, heavy metals, and oils. Cases of spontaneous combustion in lagoons and spills that have killed livestock are well documented. The contamination threatens local drinking water supplies, Bayou Baton Rouge, and the Mississippi River.

Site studies from 1984 to 1989 recommended a clean-up plan consisting of groundwater extraction and treatment (air-stripping and carbon adsorption), and capping the lagoons. Response actions included soil caps, erosion controls, and a groundwater treatment system (1994). Over 100 sumps and 170 extraction wells have been installed. Bayou Baton Rouge was rerouted so as not to flow through the sites. Studies on other portions of the sites should be concluded by 1997.

Lawsuits filed by the U.S. Department of Justice and the State of Louisiana against a number of PRPs resulted in a CD in 1984 that requires them to fund the site studies and remediation plans. Notable PRPs include U.S. Steel Corporation, Uniroyal, Dow Chemical, Shell Oil, and Exxon.

Sources: PNPLS-LA; Alsen Community Library, 303 Old Rafe Mayer Road, Baton Rouge, LA, 70801.

Southern Shipbuilding
St. Tammany Parish
Operated from 1919 to 1993 as a shipbuilding and repair plant, this facility on the outskirts of Slidell deposited wastes in pits and sludge lagoons on the site. Investigations by the EPA in 1992–1993 found that the soil and groundwater were contaminated with PAHs, tributyltin (TBT), explosives by-products, paints, lead, copper, and PCBs. The contamination threatens the water supplies of the town of Slidell and the recreational Bayou Bonfouca.

Response actions included pumping and treating contaminated surface water from the pits and lagoons (1993–1995). Debris was removed and levees along the bayou were reinforced in 1995. A final remediation plan (decreed in 1995) will consist of off-site incineration of pit waste, capping soils, and excavating TBT-contaminated soil.

Although one PRP has been identified, its financial situation will not allow it to contribute toward the clean-up costs.

Sources: PNPLS-LA; St. Tammany Parish Public Library, Slidell, LA, 70458.

MAINE

Brunswick Naval Air Station
Cumberland County
This U.S. Navy base near Bath began operations in the 1940s. Various solid and liquid wastes from maintenance activities, including acids, asbestos, caustics, solvents, and heavy metals, were buried in landfills on the base. Testing has shown that the groundwater and soil contain VOCs. About 5,000 people live or work within a few miles of the site. The contamination threatens their drinking water supplies as well as Harpswell Cove, an important commercial fishery.

Response actions included removing contaminated soil (1994), installing a groundwater extraction and UV/oxidation treatment system (1995), and long-term groundwater monitoring (1995). Other site investigations are still in progress to evaluate the potential for capping, building a slurry wall, and soil treatment.

Sources: PNPLS-ME; Curtis Memorial Library, 23 Pleasant Street, Brunswick, ME, 04011.

Eastern Surplus Company
Washington County
This former army surplus and salvage company near Meddybemps was in business from 1946 to the 1980s. Hazardous materials, including leaking transformers, compressed gas, calcium carbide, and other chemicals, were spilled or improperly stored on-site. The soil and groundwater contain PCBs, solvents, acids, paints, heavy metals, oils, pesticides, and asbestos. The contamination threatens the private water wells of about 200 people in the area as well as the recreational Meddybemps Lake, Dennys River, and a bald eagle habitat.

Response actions included erecting a fence and removing designated wastes. The State of Maine and the EPA are conducting detailed site investigations that will recommend a final remediation plan.

Sources: PNPLS-ME; Calais Free Library, Union Street, Calais, ME, 04619.

MAINE

Aroostook
Piscataquis
Penobscot
Somerset
Washington
Franklin
Hancock
Waldo
Oxford
Kennebec
Androscoggin
Knox
Lincoln
Sagadahoc
Cumberland
York

Loring Air Force Base
Aroostook County
This 9,000-acre U.S. Air Force base near Limestone operated from 1952 to 1994. Aircraft and vehicle maintenance and repair generated a variety of wastes that were buried on-site, including waste oil and fuel, chlorinated organic solvents, PCBs, and pesticides. Testing has revealed that the soil and groundwater contain VOCs, petrochemicals, PAHs, PCBs, and heavy metals. The contamination threatens the private water wells of about 1,200 nearby residents, Greenlaw Brook, and recreational wetlands and lakes.

Response actions included the removal of contaminated soil and tanks, a pilot groundwater treatment project, and capping contaminated soil. A number of site investigations (1989–1997) that will address other aspects of the clean up are nearing completion.

Sources: PNPLS-ME; Robert Frost Memorial Library, 238 Main Street, Limestone, ME, 04750.

McKin Company
Cumberland County
This former gravel pit near Gray operated as a landfill from 1965 to 1978. It received a variety of municipal and industrial wastes, including fuels, solvents, oils, and sludges. Complaints from residents in 1973 led to an investigation that revealed that the soil and groundwater contain VOCs, petrochemicals, arsenic, mercury, and lead. The contamination threatens the water supplies of about 1,200 nearby residents.

Response actions including supplying affected residents with clean water supplies (1977–1978), removing liquid chemical waste (1979–1985), and excavating and treating contaminated soil (1985–1987). A groundwater extraction and treatment system and long-term monitoring were initiated in 1990.

Agreements between the EPA and about 320 PRPs (1988) require the companies to finance the site investigations and remediation work.

Sources: PNPLS-ME; Gray Public Library, 5 Skilling Street, Gray, ME, 04039.

O'Connor Company
Kennebec County
This former scrap and recycling facility near Augusta operated from the 1950s to the 1970s. Activities included stripping PCB-laden oil from transformers. Testing revealed that the soil and groundwater contain PCBs, VOCs, lead, aluminum, and PAHs. The contamination threatens the drinking water of about 200 nearby residents, wetlands, and Riggs Brook.

Response actions included erecting a fence in 1984 and removing drums of liquid waste. An EPA decision in 1995 required the excavation and removal of contaminated soils. Wetlands will also be restored. A study on groundwater remediation is in progress.

Administrative Orders (1984, 1986, 1987) and a Consent Decree (1990) with the PRPs, O'Connor Company and Central Maine Power, require the companies to fund the site studies and remediation work.

Sources: PNPLS-ME; Lithgow Public Library, Winthrop Street, Augusta, ME, 04330.

Pinette's Salvage Yard
Aroostook County
This former auto repair shop and salvage yard near Washburn operated in the 1970s and 1980s. Major spills of PCB-containing fluids occurred on the site. The groundwater and soil contain PCBs and VOCs. The contamination threatens the water supply of local residents, wetland habitats, and the Aroostook River.

Response actions included removing contaminated soil in 1983. The EPA recommended a final remediation plan consisting of soil excavation, incineration, on-site solvent extraction, and groundwater extraction and treatment using carbon adsorption. The groundwater treatment plant should be operational by 1997.

Sources: PNPLS-ME; Washburn Town Office, Main Street, Washburn, ME, 04286.

Portsmouth Naval Shipyard
York County
This 300-acre site near Kittery was first established as a shipyard in 1690 and became a U.S. Navy facility in 1800. The complex builds ships and submarines and overhauls nuclear propulsion vessels. Various wastes, including acids, battery acids, and electroplating sludges, were discharged into the Piscataqua River. Tidal flats were filled with heavy metal wastes, drummed paint waste, mercury, and asbestos. Testing has revealed that the soil and groundwater contain heavy metals, PCBs, asbestos, and VOCs. The contamination threatens the water supplies of about 10,000 people as well as wetlands, the Great Bay estuary, and the Piscataqua River.

Response actions included removing underground storage tanks in 1994. A temporary cap was also built. Detailed site investigations are in progress and will recommend a final remediation plan.

Sources: PNPLS-ME; Rice Public Library, 8 Wentworth Street, Kittery, ME, 03904; Portsmouth Library, 8 Islington Street, Portsmouth, ME, 03801.

Saco Municipal Landfill
York County
This landfill near Saco has been in operation since 1960. It has received a variety of municipal and in-

dustrial wastes, including tires, tannery sludge, heavy metal sludge, and VOCs. Soils and groundwater contain phthalates, VOCs, heavy metals, and VOCs. The contamination threatens water wells that serve about 800 residents. The contamination has penetrated nearby Sandy Brook.

Response actions included removing solid wastes in 1991. Ongoing site investigations begun in 1992 are nearing completion and will recommend a final remediation plan.

PRPs identified by the EPA are participating in the site investigations.

Sources: PNPLS-ME; City Hall, 300 Main Street, Saco, ME, 04072.

Saco Tannery Waste Pits
York County
This landfill near Saco operated from 1959 to 1981. Until its bankruptcy in 1981, Saco Tannery Corporation dumped processing wastes, including acids, sludges, methylene chloride, and heavy metal sludges, at the site. Testing by the State of Maine and the EPA confirmed that groundwater and soil contain excessive levels of VOCs, arsenic, antimony, and zinc. The contamination threatens nearby recreational wetlands and the drinking water source for about 3,000 people in the area.

Response actions including pumping out acidic waste, fencing, and capping (1983). The EPA recommended a final remediation plan of capping, groundwater monitoring, and excavating contaminated soil. Construction was completed in 1993, and groundwater monitoring is ongoing. The EPA purchased about 250 acres of wildlife habitat to compensate for the loss of 10 acres of wetland during the remediation work.

Sources: PNPLS-ME; Dyer Library, 371 Main Street, Saco, ME, 04072.

Union Chemical Company
Knox County
This chemical company near South Hope manufactured paint and coating strippers and other solvents from 1967 to 1984. Various wastes, including sludges, were buried or stored on-site in leaking tanks and drums. The company was frequently cited for violating state laws. Testing has revealed that the groundwater and soil contain VOCs, including toluene and dichloroethane. The contamination has penetrated Quiggle Brook. The contamination threatens the drinking water supplies for about 23,000 people in the towns of Camden, Rockport, Rockland, and Thomaston as well as Quiggle Brook and recreational Grassy Pond.

Response actions included removing drummed waste, pumping out storage tanks, and removing contaminated soils (1984). In 1990 the EPA recommended a treatment plan of soil excavation and low-temperature aeration, groundwater extraction and treatment, and long-term monitoring. Buildings on the property were decontaminated in 1993–1994. Soil vapor and groundwater extraction systems became operational in 1996.

Administrative Orders with about 300 PRPs (1987–1988) and a Consent Decree (1989) with 9 other PRPs require the companies to fund the site investigations and clean-up work.

Sources: PNPLS-ME; Hope Town Office, Route 105, Hope, ME, 04847.

West Site/Howes Corners
Penobscot County
A waste oil storage and processing facility operated on this site near Plymouth from 1965 to 1980. Waste oil was stored in tanks on the property and used for controlling dust on local roads. EPA testing in 1988 revealed that the soil and groundwater contain PCBs, chlorinated organic compounds, and VOCs. The contamination has penetrated nearby private water wells and threatens the water supply for the town of Plymouth.

Response actions included removing the storage tanks in 1980. Fencing was erected and some contaminated soil removed between 1988 and 1991. A detailed site investigation that will select a final remediation plan is planned.

The PRP removed the tanks in 1980.

Source: PNPLS-ME.

Winthrop Landfill
Kennebec County
This former gravel pit near Winthrop operated as a landfill from the 1930s to 1982. It received a variety of municipal and industrial wastes, including liquid resins, plastics, solvents, and other processing chemicals. Testing has shown that the soil and groundwater contain VOCs and inorganic chemicals. The contamination has migrated off the site and threatens the private wells of about 500 people in the area as well as Lake Annabessacook and nearby wetlands.

Response actions included connecting affected residents to clean water supplies. The landfill was fenced and capped in 1992. Soil vapor and groundwater extraction and treatment systems were installed in 1994–1995. Groundwater monitoring is continuing.

A number of PRPs signed a Consent Decree with the EPA in 1986 that requires the parties to contribute to the clean-up plan.

Sources: PNPLS-ME; Winthrop Town Hall, 574 Main Street, Winthrop, ME, 04364.

Aberdeen Proving Ground, Edgewood Area
Harford and Baltimore Counties
Part of a 79,000-acre tract of land and water near Chesapeake Bay, this 13,000-acre section was used by the U.S. military to develop and test chemical agent munitions. Chemical research is still being done on the site, and various toxic wastes are stored or discharged on the property. Waste compounds include napalm, mustard gas, and phosphorus. The groundwater and soils contain heavy metals, phosphorus, and VOCs. The contamination threatens wells that serve about 3,000 residents as well as numerous wildlife and aquatic habitats.

Response actions have included removing leaking storage tanks and contaminated soil, capping, dismantling missile silos, and excavating solid wastes (1991–1995). Studies that address remedies for contaminated groundwater and soils are in progress. A groundwater extraction and air-stripping/ultraviolet oxidation treatment system became operational in one area in 1995.

Sources: PNPLS-MD; Harford Public Library, 21 Franklin Street, Aberdeen, MD, 21001.

Aberdeen Proving Ground, Landfill
Harford County
Part of a 79,000-acre tract of land and water near Chesapeake Bay, this 17,000-acre section contains numerous landfills and burial sites. The main Michaelsville landfill operated during the 1970s and received a variety of municipal and industrial wastes, including pesticides, oils, solvents, transformer oils, and waste munitions from U.S. military operations on the site. Runoff flows into seven creeks and the Bush River. The groundwater and soil contain heavy metals, phosphorus, VOCs, pesticides, petroleum, and PCBs. The contamination threatens the drinking water supplies of about 40,000 residents as well as nearby wildlife and aquatic habitats.

Response actions included excavating contaminated soil and storage tanks, installing interim groundwater treatment plants, and building a landfill cap (1991–1994). Other site investigations in progress will be completed between 1997–2002.

Sources: PNPLS-MD; Harford Public Library, 21 Franklin Street, Aberdeen, MD, 21001.

Beltsville Agricultural Research Center
Prince Georges County
This U.S. Department of Agriculture research center in Washington, D.C., has conducted agricultural research and testing since 1910. The complex covers 6,500 acres. Disposal areas on the property received a variety of wastes, including transformers, laboratory wastes, pesticides, and heavy metals. Testing has shown that the groundwater and soil contain PAHs, pesticides, PCBs, VOCs, and heavy metals. The contamination threatens the water supplies of nearby workers and residents as well as Paint Branch, Beaver Dam Creek, Indian Creek, the Anacostia River, and the Potomac River.

Response actions have consisted of initiating a series of site investigations.

Sources: PNPLS-MD; National Agricultural Library, 10301 Baltimore Boulevard, Beltsville, MD, 20705.

Bush Valley Landfill
Harford County
This former landfill near Abingdon operated from 1974 to 1983. It received a variety of municipal and industrial wastes, including heavy metals, sludges, and construction debris. Examinations in 1983 revealed that nearby seeps and wells were contaminated with VOCs. The groundwater and soil contain VOCs, beryllium, manganese, and arsenic. VOCs have also been detected in air samples. The contamination threatens streams, tidal wetlands, and wells that supply local residents.

Response actions have consisted of site investigations (1992–1997) that are nearing completion. A final remediation plan will then be issued.

PRPs are conducting the site investigations according to a 1990 Administrative Order.

Sources: PNPLS-MD; Harford Library, 100 Pennsylvania Avenue, Bel Air, MD, 21014.

Central Chemical
Washington County
Pesticides and fertilizers were processed on this site in Hagerstown from the 1930s to 1984. The buildings are currently used for storage and auto repair. Various wastes, including DDT and chlordane, were buried in pits, trenches, and sinkholes on the 19-acre property. Testing from 1976 to 1979 by the state revealed that the groundwater and soil contain excessive amounts of pesticides and heavy metals. After another landfill was discovered on the property in 1987, additional sampling revealed high DDT levels in surface runoff and in nearby Antietam Creek. The contamination threatens local drinking water supplies, streams, and wetlands.

Response actions included capping the landfill areas in 1979. This property was proposed as an NPL site in 1996; a final decision is expected in 1997–1998. More site investigations are planned.

The PRP, Central Chemical Company, has yet to comply with a Consent Order from the State of Maryland to remediate the newly discovered dump site.

Source: EPA Publication 9320.7-071.

MARYLAND

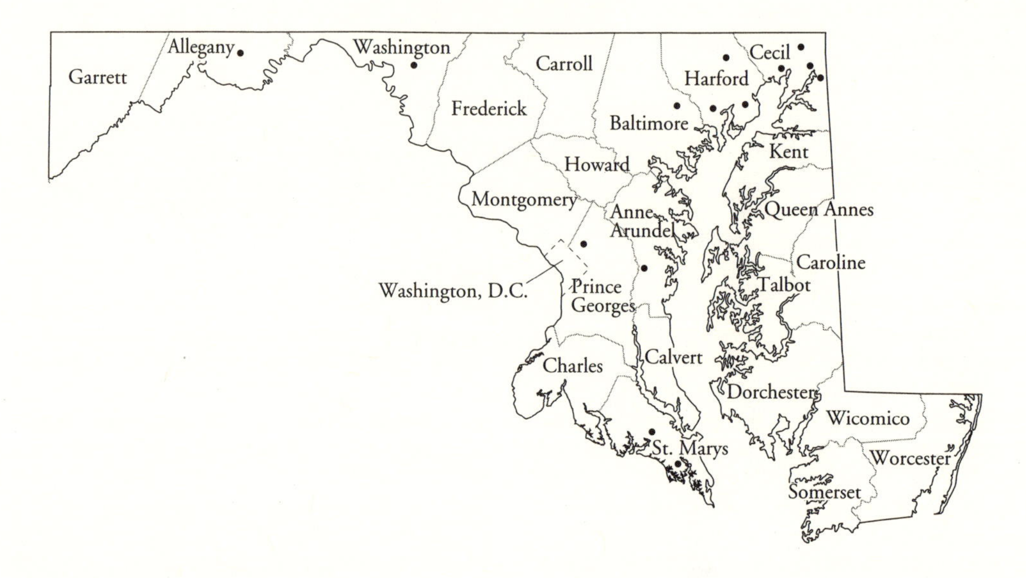

Kane & Lombard Street Drums
Baltimore County
This former dump site in Baltimore was in operation from 1962 to 1984. Municipal, demolition, and industrial wastes were buried on the property, which is now in a highly developed area. The groundwater and soil contain VOCs, cadmium, mercury, lead, PCBs, PAHs, and phthalates. The contamination threatens the health of nearby residents and workers as well as Herring Run, Back River, and the Chesapeake Bay.

Response actions included the removal of drummed waste, construction of fencing, and erosion controls (1984). Building of slurry walls and a cap was completed in 1990. Groundwater studies will make final recommendations in 1997–1998.

PRPs are participating in the clean-up work and site studies according to a 1992 UAO and 1993 AOC.

Sources: PNPLS-MD; Highland Multi-Purpose Center, 3411 Bank Street, Baltimore, MD, 21224.

Limestone Road
Allegany County
This former landfill near Cumberland operated in the 1970s. It received a variety of commercial, municipal, and demolition wastes, including heavy metal sludges. The groundwater and soils contain heavy metals. The contamination threatens the private water wells that supply about 500 nearby residents.

Response actions included constructing erosion controls, capping, and fencing (1992–1995). Affected residents were also connected to clean water supplies. A groundwater study (1987–1997) will recommend a remediation plan shortly.

A 1989 Consent Decree with the EPA requires the PRPs to finance the remediation work on the property.

Sources: PNPLS-MD; Alleghany County Library, 31 Washington Street, Cumberland, MD, 21502.

Mid-Atlantic Wood Preservers
Anne Arundel County
This former wood treating plant near Harmans ceased operations in 1993. Timbers were treated with chromated copper-arsenate solutions and drip-dried on the property. Testing by the state in 1978 revealed that the groundwater contain chromium from the drippings and from overflows from storage tanks. The soil also contains arsenic and chromium. The contamination threatens nearby drinking water wells as well as Stoney Run, associated wetlands, and the Patapsco River.

Response actions included capturing drippings and removing contaminated soils and sludges (1980). Other contaminated soils were left on-site and capped (1993).

A Unilateral Order (1993) requires the PRPs to undertake the remediation work.

Sources: PNPLS-MD; Provinces Library, Severn Square Shopping Center, 2624 Annapolis Road, Severn, MD, 21144.

Ordnance Products
Cecil County
Grenade fuses and other explosives were manufactured on this site near North East from the 1960s

to 1972. The site was later developed into an industrial park. In 1987 testing by the state revealed that the soil and groundwater contain arsenic, selenium, barium, and VOCs. The chemicals have also been found in nearby wells. The contamination threatens the water supplies of nearby workers and residents.

Response actions included supplying affected residents with carbon filtration units and removing contaminated soils and buried ordnance (1988). A groundwater extraction and air-stripping system was also installed. A detailed site investigation is under way.

A 1988 UAO required the PRPs to undertake site investigations and response actions.

Source: PNPLS-MD.

Patuxent River Naval Station
St. Mary's County
This 6,400-acre U.S. Navy facility near Cedar Point has three landfills that have contaminated the waters of Chesapeake Bay. The landfills have received a variety of wastes from military operations, including sewage sludge, spent oils, paints, thinners, antifreeze, pesticides, and photolab wastes. The groundwater and soil contain VOCs, asbestos, petrochemicals, heavy metals, DDT, and chlordane. The contamination threatens local drinking water supplies as well as streams and wetlands in the Patuxent River/Chesapeake Bay area.

Erosion controls were built in 1994. Detailed site investigations are under way.

Sources: PNPLS-MD; Lexington Park Library and St. Mary's County Library, Lexington Park, MD, 20653.

Sand, Gravel, and Stone
Cecil County
This former quarry near Elkton was used as a landfill from 1964 to 1974. It received a variety of bulk wastes, including wastewater, sludges, and drummed chemicals. Testing has shown that the groundwater and soil contain heavy metals, VOCs, and pesticides. The contamination threatens the Elk River and the local drinking water supplies of about 600 residents.

Response actions included removing liquid waste (1974) and erecting fencing (1985). The EPA conducted studies from 1983 to 1985 that recommended a groundwater pump-and-treat system and removal of drummed waste. The groundwater plant is being constructed. A final decision on soil remediation is pending.

The PRPs signed Consent Agreements (1986) and Consent Decrees (1987, 1991) that require the companies to undertake the site investigations and clean-up work.

Sources: PNPLS-MD; Cecil Public Library, 301 Newark Avenue, Elkton, MD, 21921

Southern Maryland Wood Treating
St. Mary's County
Wooden timbers were treated with creosote and PCP at this site near Hollywood from 1965 to 1978. Waste solutions were stored in unlined lagoons on the property. Lagoon sludge and wastewater were sprayed or landfarmed on the property by order of the State of Maryland in 1982. Wood treatment operations and improper landfarming have contaminated the soil and groundwater with VOCs, PCP, PAHs, and creosote. The contamination threatens the private drinking water wells of about 1,000 residents as well as nearby wetlands.

Response actions included removal of sludge and contaminated soil (1985, 1986), erosion controls, removal of tanks and drums, and building demolition (1993). Groundwater treatment began in 1995 using carbon adsorption; contaminated soils will be thermally treated.

Sources: PNPLS-MD; St. Mary's County Memorial Library, Route 245, Leonardstown, MD, 20650.

Spectron Inc.
Cecil County
A succession of paper manufacturers and chemical producers occupied this site in Elkton from the 1800s to 1988. Galaxy Chemicals/Solvent Distillers recycled waste organic solvents from various industries from 1961 to 1988, when the owner went bankrupt. Inspections revealed rusting or leaking drums filled with waste chemicals on the property. Sampling indicated that the groundwater and soils contain VOCs. Leachate has also seeped into recreational Little Elk Creek and the Elk River. Private wells within 4 miles of the site supply water to about 5,300 residents. The contamination also threatens waterways and wetlands.

Response actions included removing wastes and tanks (1989–1991). A groundwater study that will recommend a final remediation plan is in progress.

AOCs (1989, 1990, 1991) require the PRPs to undertake the response actions and site investigations.

Sources: PNPLS-MD; Cecil County Library, 301 Newark Avenue, Elkton, MD, 21921.

Woodlawn County Landfill
Cecil County
This former gravel pit in Woodlawn operated as a landfill from 1965 to 1979. It received a variety of waste, including PVC sludges from Firestone Tire and Rubber Company. Testing by the EPA and the

State of Maryland revealed that the groundwater and soil contain VOCs, PAHs, and heavy metals. About 6,000 residents draw drinking water from wells within 4 miles of the site. The contamination also threatens recreational Basin Run, a trout stream.

Response actions included capping the sludge and supplying affected residents with a carbon water treatment system. An EPA groundwater study from 1988 to 1993 recommended excavating contaminated soil, gas monitoring, capping the landfill, and building a groundwater extraction and treatment system. Construction will begin in 1997–1998.

PRPs identified by the EPA signed a Consent Order (1988) that requires the companies to conduct the site investigation.

Sources: PNPLS-MD; Cecil County Library, 301 Newark Avenue, Elkton, MD, 21921.

MASSACHUSETTS

Atlas Tack Corporation
Bristol County
Located in Fairhaven, the Atlas Tack Corporation manufactured wire tacks and nails from the 1940s to the 1970s. Various wastes, including heavy metals and cyanide solutions, were poured into an unlined lagoon beside the tidal marshes in Buzzards Bay. Testing revealed high levels of cyanide, VOCs (toluene and ethylbenzene), cadmium, arsenic, lead, nickel, chromium, pesticides, PCBs, and PAHs in the soil, groundwater, and marine sediments in the bay. About 10,000 people live within 2 miles of the site.

Response actions included fencing (1992). An EPA site investigation is presently under way, and a final remediation plan will be chosen in 1997.

Sources: PNPLS-MA; Fairhaven Public Library, Center Street, Fairhaven, MA, 02719.

Baird & McGuire
Norfolk County
Baird & McGuire mixed and batched chemicals at this facility in Holbrook from 1912 to 1983. Pesticides, soaps, disinfectants, and other solvents were also processed and stored. Wastes were poured into the soil, an on-site lagoon, and a nearby creek, gravel pit, and wetland. Testing has shown that the soil and groundwater contain high levels of pesticides (DDT and chlordane), VOCs, arsenic, lead, and PAHs. Dioxin has been found in wetland soils. A municipal well field was closed in 1982 because of high VOC levels. The nearby Cochato River was a water source for the towns of Holbrook, Randolph, and Braintree until it was also determined to be contaminated.

Response actions included razing buildings, removing contaminated waste and soil, temporary capping, fencing, and installing a groundwater recirculation system. A groundwater extraction and treatment system was installed in 1993 and will operate for at least 10 years. Contaminated soil will be removed and incinerated on-site (1995–1998). Toxic sediments in the Cochato River were dredged from 1991 to 1995.

Four PRPs have been identified. Baird & McGuire Company was fined at least 35 times for various past violations. A CD requiring them to contribute to the cost of the site investigations and cleanup work has been signed by the parties.

Sources: PNPLS-MA; Holbrook Public Library, 2 Plymouth Street, Holbrook, MA, 02343.

Blackburn and Union Privileges
Norfolk County
Manufacturing at this site dates back to the seventeenth century when ironworks, textile factories, and tanneries operated along the Neponset River. Later industries in the eighteenth century used arsenic and mercury. More recent activities include the manufacturing of asbestos brake linings and textiles. Various wastewaters from these industries were discharged into either sewers or settling lagoons. Testing has shown that the soil and groundwater are contaminated with asbestos, lead, arsenic, nickel, and VOCs. City wells within 4 miles of the site service about 20,000 people in the town of Walpole.

Response actions included the removal of buried tanks (1988). Asbestos-contaminated soils were removed and the area capped in 1992. An ongoing site investigation will be concluded in 1997 with recommendations for a final remediation plan.

The two new owners of the site were issued UAOs (1988, 1992) to undertake the response actions and finance the site investigation.

Sources: PNPLS-MA; Walpole Public Library, 275 Common Street, Walpole, MA, 02081.

Cannon Engineering Corporation
Plymouth County
This site in Bridgewater operated as a hazardous waste processing and handling facility from 1974 to 1980. Some of these wastes included oils, lacquers, plating solutions, solvents, old filters, sludges, and pesticides. Testing in 1980–1981 showed the soil and groundwater to be contaminated with VOCs, PAHs, dioxin and other pesticides, heavy metals (including

MASSACHUSETTS

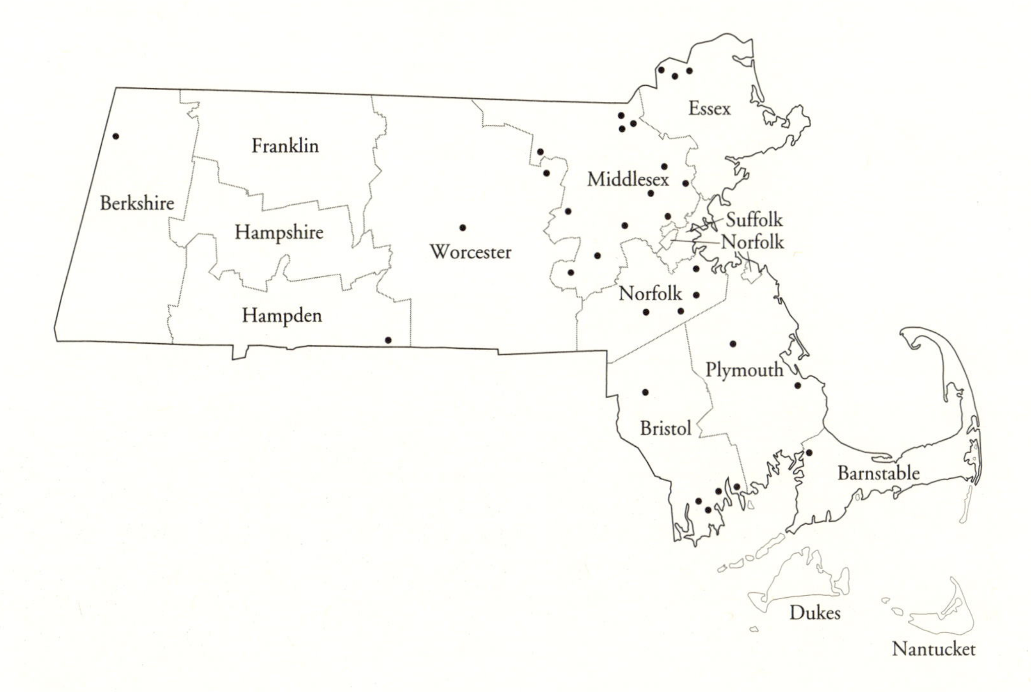

selenium, lead, manganese, and silver), and PCBs. Private and municipal wells that service Raynham and Bridgewater are within 3 miles of the site. The contamination also threatens Lake Nippenicket and nearby wetlands.

Response actions included removing drummed and lagoon waste (1982–1988) and fencing (1989). The final remediation plan consisted of thermally treating contaminated soil, groundwater monitoring, razing buildings, and restoring wetlands. The soils were treated from 1990 to 1991. Removal of the contaminated soil improved the quality of the groundwater. Long-term groundwater monitoring will continue.

The EPA has identified a number of PRPs who have contributed to the site investigation and remediation work.

Sources: PNPLS-MA; Bridgewater Public Library, 15 South Street, Bridgewater, MA, 02324.

Charles-George Reclamation Trust Landfill
Middlesex County
Located about a mile from the town of Tyngsboro, this landfill operated from the 1950s to 1983. It received a variety of municipal and industrial wastes, including drummed chemical waste and sludges, including mercury. Testing revealed that the soil and groundwater contain tetrahydrofuran, arsenic, and butanone. The contamination has migrated into nearby wells. The site is bordered by wetlands and streams, and about 2,000 people live within 3 miles of the site.

Response actions included fencing, supplying affected residents with clean water, and installing a gas control system (1983–1984). A permanent water supply was installed by the U.S. Army Corps of Engineers in 1988. A landfill cap was constructed in 1990. A gas collection and flare system became operational in 1995. Leachate and groundwater extraction and treatment systems will be installed in 1997.

The owner filed for bankruptcy in 1983. Two AOs with other PRPs have recovered about $40 million of the site investigation and clean-up costs.

Sources: PNPLS-MA; Littlefield Public Library, 25 Middlesex Road, Tyngsboro, MA, 01879.

Fort Devens
Worcester and Middlesex Counties
This U.S. Army base about 35 miles west of Boston covers about 9,500 acres. Seventy-six hazardous waste sites identified on the site (mostly unlined pits and trenches) contain ammunition waste, oils, asbestos, construction debris, and petroleum. Soil and groundwater are contaminated with heavy metals (arsenic, cadmium, chromium, lead, iron, magnesium, and mercury), PAHs, perchloroethylene, and explosive

residues. The site threatens the drinking water supplies of about 30,000 people as well as the Nashua River and the Oxbow National Wildlife Refuge. Fish and sediments in Plow Shop Pond and Grove Pond are contaminated with mercury. The base is expected to be closed by 1997.

U.S. Army site investigations begun in 1991 are still ongoing for some disposal areas. PCB-contaminated scrap was removed in 1993. A few selected areas of PAH-contaminated soils have been excavated and treated (1995–1996). Final clean-up remedies will be selected in 1997–1998.

Sources: PNPLS-MA; Fort Devens Installation Library, Fort Devens, MA, 01433; Shirley Public Library, Shirley, MA, 01464.

Fort Devens–Sudbury Training Annex
Middlesex County
This U.S. Army training facility covers about 2,750 acres near the towns of Maynard, Stow, Hudson, and Sudbury. Army operations included training troops, storing ammunition, and disposing of laboratory waste. Storage areas and tanks, a demolition area, and on-site landfills have contaminated the soil and groundwater with VOCs, pesticides, and inorganic compounds.

Response actions included removing PCB-contaminated soil (1985–1988), excavating underground storage tanks (1993), and fencing portions of the site (1993). The laboratory waste landfill should be capped by 1997. Other site investigations are continuing, and a groundwater treatment plan will be recommended at a later date.

Sources: PNPLS-MA; Goodnow Library, 21 Concord Street, Sudbury, MA, 01776; Hudson Public Library, Wood Square, Hudson, MA, 01749; Randall Library, P.O. Box 263, Common Road, Stow, MA, 01775.

Groveland Wells
Essex County
An 850-acre area near Groveland that contains that city's municipal water wells is polluted with VOCs. The contamination was first detected in wells in 1979, and they were shut down. Three companies in the area are known to have contributed pollutants to the soil and groundwater. Valley Manufacturing Company discharged toxic wastes through underground disposal systems or simply by dumping them on the ground. About 5,000 people live within 3 miles of the site.

Response actions included shutting down the contaminated wells (1979), drilling a new well (early 1980s), and building a groundwater treatment system for one of the contaminated wells (1989). Groundwater investigations (1983–1995) recommended ultraviolet light/oxidation treatment of the contaminated groundwater. Contaminated soils will be vacuum-extracted to remove VOCs. The final clean up has yet to be initiated.

A lawsuit by the city of Groveland against the PRPs led to one of the parties funding a site investigation. Despite a 1992 UAO requiring the PRPs to finance the technical design and implementation of the remediation plan, the PRPs have refused to participate.

Sources: PNPLS-MA; Langley-Adams Library, Main Street, Groveland, MA, 01834.

Hanscom Field/Hanscom Air Force Base
Middlesex County
This 1,100-acre U.S. Air Force base is located near the towns of Bedford, Lexington, Lincoln, and Concord. Military operations ended in 1973, and the site is now a public airport. Areas of soil and groundwater contamination include old burning sites, landfills, and underground storage tank sites. Pollutants include VOCs, PAHs, and jet fuel. Three Bedford drinking water wells were shut down in 1983 because of high VOC levels.

Response actions included removing contaminated soil and drums (1988), capping landfills (1988), installing groundwater pump-and-treat systems (1991), and soil vapor extraction and groundwater collection (1995). Site investigations for other areas of contamination on the base are in progress.

Sources: PNPLS-MA; Hanscom Air Force, Environmental Flight Building 1810, Bedford, MA, 01730.

Haverhill Municipal Landfill
Essex County
This former landfill site near the town of Haverhill operated from the late 1930s to 1981. A variety of municipal and industrial wastes was accepted, including sludges, dyes, paints, oils, lacquers, tires, and drummed wastes. Sludges and liquid wastes flowed into the adjacent Merrimack River. The soil and groundwater are contaminated with benz(a)anthracene, dibenzofuran, VOCs, heavy metals (chromium, arsenic, lead, mercury, and manganese), toluene, and xylene. The contamination has already polluted the Merrimack River and threatens the groundwater supplies of the 52,000 residents of Haverhill and Groveland.

Response actions included the removal of drummed waste in 1990. A site investigation and feasibility study is expected to begin in 1997.

Source: PNPLS-MA.

Hocomonco Pond
Worcester County
A wood treatment facility operated beside Hocomonco Pond from 1928 to 1946. Lumber was sat-

urated with creosote, and wastewater was discharged into pit lagoons on the property and Kettle Pond. Tests have shown that the soil and groundwater are contaminated with creosote and creosote by-products (heavy metals, including arsenic and chromium). The contamination threatens the water supplies for about 17,000 people within 3 miles of the site.

Response actions including closing contaminated Hocomonco Pond to the public in 1980. A site investigation study completed in 1989 recommended capping and runoff collection (completed in 1990). Contaminated sediments in Hocomonco Pond were dredged in 1994. A groundwater treatment system was installed in 1994; recovering free-product creosote began in 1995.

A Consent Decree (1987) requires the PRPs to finance the design and construction of the remediation plan.

Sources: PNPLS-MA; Westborough Public Library, West Main Street, Westborough, MA, 01581.

Industri-Plex
Middlesex County
Located near North Woburn, this has been the site for various manufacturing companies over the past 130 years. Hazardous products and wastes that have been manufactured or disposed of on-site include arsenic, insecticides, acetic acid, sulfuric acid, benzene, toluene, phenol, and glue products from animal hides. Soil and groundwater are contaminated with VOCs, benzene, and arsenic; the air is fouled by hydrogen sulfide from the decay of buried animal waste. Runoff has polluted nearby streams and wetlands. About 34,000 people live within 3 miles of the site.

Response actions included fencing (1986). The final remediation plan consists of a soil cap, groundwater extraction and treatment, groundwater monitoring, and gas collection and treatment using carbon or thermal oxidation. Construction of the treatment systems began in 1992 and is still continuing.

A Consent Order (1982) and Consent Decree (1989) with the PRPs require the parties to finance the site investigations and remediation work.

Sources: PNPLS-MA; Reading Public Library, 45 Pleasant Street, Woburn, MA, 01801.

Iron Horse Park
Middlesex County
This 500-acre site near North Billerica has operated as an industrial complex since 1913. Various wastes, including asbestos, have been deposited in landfills, lagoons, and open storage areas. Tests have shown that the soil and groundwater contain high levels of VOCs, asbestos, cadmium, arsenic, lead, selenium, PCBs, and petrochemicals. About 60,000 residents, six schools and day care centers, and sensitive wetland habitats within 3 miles of the site are threatened by the contamination.

Response actions included removing asbestos (1984) and placing temporary caps on the landfill sites. A gas collection and vent/flare system was also installed to control odor. Lagoon sludges will be treated on-site through bioremediation (1991–1997). Other site investigations are under way and will recommend final treatment plans for contaminated groundwater and soil.

Consent Decrees (1984, 1989) require the PRPs to finance the closure of the landfills, site investigations, and remediation work at the site.

Sources: PNPLS-MA; Billerica Public Library, 25 Concord Road, Billerica, MA, 01821.

Materials Technology Laboratory
Middlesex County
Owned by the U.S. Army, this facility has been in existence since 1816. Ammunition was manufactured until about 1970, when it became an atomic research, armor research, and materials testing center. The site closed in 1995. These military operations have contaminated the soil and groundwater with PAHs, PCBs, pesticides, VOCs, radiologic compounds, and oil. The contamination threatens nearby residents and the recreational Charles River, which flows through the site.

Response actions included removing leaking underground storage tanks and razing the nuclear reactor (1994). Ongoing site investigations will determine a final remediation plan.

Sources: PNPLS-MA; Watertown Public Library, 123 Main Street, Watertown, MA, 02172.

Natick Laboratory Army Research Development
Middlesex County
This U.S. Army facility is located on Lake Cochituate near the town of Natick and is used as an industrial, research, and storage area. Military activities have used a variety of hazardous materials, including VOCs, paints, thinners, gasoline, lead, fuels, pesticides, and ink. Tests have shown that the soil and groundwater are contaminated with petroleum, VOCs, heavy metals, and chlorinated solvents. The contamination threatens the drinking water supplies for about 38,000 people within 4 miles of the site as well as recreational Lake Cochituate.

Site investigations were begun in 1993; a final remediation plan has yet to be selected.

Sources: PNPLS-MA; Natick Library, Natick, MA, 01760.

Naval Weapons Industrial Reserve Plant
Middlesex County
This U.S. military base has been in operation since 1952 as a missile and radar research, flight testing,

and storage facility. Hazardous wastes were discharged into an on-site septic system or stored in barrels. Wastes included VOCs, oils, coolants, lacquers, and various acids. Testing indicates that the groundwater and soils are contaminated with VOCs and heavy metals. Three Bedford municipal wells were closed in 1984 because of VOC contamination. About 11,000 people rely on drinking water wells within 4 miles of the site; special wetland habitats along the Shawsheen River are also threatened.

Response actions included a short-term groundwater extraction and treatment system to contain the contamination plume. Ongoing site investigations are exploring the extent of soil and groundwater contamination.

Sources: PNPLS-MA; Bedford Public Library, 7 Mudge Way, Bedford, MA, 01730.

New Bedford Site

Bristol County

This site is an urban harbor and bay in an industrial part of New Bedford. Industries that built electric capacitors from 1940 to 1978 discharged PCB-contaminated waste into the estuary. The sediments and aquatic life have high levels of cadmium, lead, copper, and chromium. Extensive contamination has resulted in the closure of over a 6-mile length of the bay to lobster and other fishing and a restriction on further development. About 100,000 people live within 3 miles of the site.

Response actions included posting warning signs and fencing (1982). About 14,000 cubic yards of highly contaminated harbor sediments were removed by 1995 and have been stored on-site for further treatment. Additional site investigations are continuing, and it may take another 10 years to install remaining clean-up systems.

Five PRPs have reached settlements with the EPA regarding financing the site investigations and clean-up costs.

Sources: PNPLS-MA; Wilkes Branch New Bedford Free Library, 1911 Acushnet Avenue, New Bedford, MA, 02746.

Norwood PCBs

Norfolk County

Located in a commercial/industrial part of Norwood, this site was the former location for a number of manufacturers, including the Grant Gear Company, which manufactured electrical components. Manufacturing operations have contaminated the soil and groundwater with PCBs, VOCs, PAHs, and heavy metals. The contamination has penetrated nearby Meadow Brook. About 8,000 residents are within 1 mile of the contamination plume.

Response actions included the emergency removal of PCB-contaminated soil in 1983. Fencing was installed in 1986. In 1989 the EPA recommended excavating contaminated soil, collecting and air-filtering the groundwater, and restoring wetlands after dredging Meadow Brook. Construction of the groundwater treatment system began in 1996.

Sources: PNPLS-MA; Morrill Memorial Library, Walpole Street, Norwood, MA, 02062.

Nyanza Chemical Waste Dump

Middlesex County

This active industrial complex in Ashland was the site of textile operations from 1917 to 1978. Various wastes such as acids and mercury were stored on-site or discharged into nearby streams; sludges and spent solvents were buried on-site. The groundwater and soil are contaminated with chlorinated organic compounds and heavy metals. About 10,000 people live within 3 miles of the site. The contaminants have also been detected in sediments of the Sudbury River, associated wetlands, and fish.

Response actions included excavating tanks and contaminated soil in 1987. Contaminated soil and sludge were excavated, treated, reburied, and capped (1988–1992). Groundwater treatment systems will be pilot-tested by 1997. Contaminated sediments are being dredged from the Sudbury River and associated wetlands (1994–1997).

Sources: PNPLS-MA; Ashland Public Library, 66 Front Street, Ashland, MA, 01721.

Otis Air National Guard

Barnstable County

This site covers about 21,000 acres near the town of Falmouth. Various research, maintenance, and training facilities produced wastes such as fly ash, solvents, fuels, herbicides, sludges, and emulsifiers. The liquid wastes were discharged on the surface in sand beds. Testing in 1984 detected high VOC levels in the groundwater and soil. About 36,000 people live within 3 miles of the site. Ashumet Pond is also nearby.

Response actions included connecting affected residents to city water (1986) and removing contaminated soil (1990). A groundwater extraction and treatment system was installed in 1993. A landfill cap should be completed by 1997. Additional remedial recommendations from ongoing studies are expected by 1998.

Sources: PNPLS-MA; Jonathon Bourne Library, 19 Sandwich Road, Bourne, MA, 02532.

Plymouth Harbor

Plymouth County

Located near the town of Plymouth, this abandoned manufacturing site used old storage tanks that held fuel, oil, solvents, pesticides, and plating sludge. Soil and groundwater are contaminated with PAHs, pesticides, and heavy metals., About 100,000 people live within 3 miles of this popular recreational area.

Response actions included removing and decontaminating tanks (1983). Contaminated soil was excavated in 1986–1987. After 5 years of monitoring, the site was removed from the NPL list in 1992. Monitoring is continuing.

A Consent Decree with Salt Water Trust, a PRP, required the party to remediate part of the site.

Sources: PNPLS-MA; Plymouth Public Library, 132 South Street, Plymouth, MA, 02360.

PSC Resources

Hampden County

This site operated in the 1970s as a processing facility for waste oils and solvents. Millions of gallons of contaminated wastewater were left in leaking tanks and lagoons when the plant closed in 1978. Soil and groundwater are contaminated with VOCs, PCBs, heavy metals, and oil. The contamination threatens the recreational Quaboag River and the water supplies of about 5,000 people within 3 miles of the site.

Response actions included pumping out the tanks (1979–1984) and fencing (1986). A site investigation recommended a final remediation plan of on-site stabilization and treatment of contaminated soils and sediments, followed by capping. This procedure should be under way by 1997.

The EPA, Department of Justice, and the State of Massachusetts settled with about 200 PRPs, who will contribute about $6 million toward the clean-up work and site investigations.

Sources: PNPLS-MA; Palmer Public Library, 455 North Main Street, Palmer, MA, 01069.

Re-Solve Inc.

Bristol County

Re-Solve Inc. operated this waste chemical processing facility from 1956 to 1980. Liquid wastes, including sludges, were discharged into unlined lagoons on the site. Oil waste was sprayed across the site to control dust. Soil and groundwater are contaminated with VOCs, PCBs, lead, arsenic, zinc, and mercury. Nearby waterways and ponds, including fish, are also contaminated. About 300 people within 3 miles of the site use the local groundwater for drinking purposes.

Response actions included excavating sludges and contaminated soil (1985) and capping (1987). PCB-contaminated sediments were treated by dechlorination or low-temperature desorption, replaced, and capped (1992–1995). A groundwater treatment system using carbon filters and air-stripping should be operational by 1997.

A Consent Decree (1989) requires the PRPs to reimburse past costs and finance the clean-up work.

Sources: PNPLS-MA; Southworth Public Library, 732 Dartmouth Street, Dartmouth, MA, 02748.

Rose Disposal Pit

Berkshire County

From 1951 to 1959 this site accepted waste oils and solvents from General Electric Company in Pittsfield. Testing by the state in 1980 discovered high levels of PCBs and VOCs in the soil and groundwater. The contamination has migrated off-site, and has penetrated nearby wetlands and private wells. About 100 people within 1 mile of the site use the contaminated groundwater for drinking purposes.

Response actions included fencing and temporary capping (1984). Affected residents were connected to clean water supplies. Contaminated soils were excavated and incinerated. A groundwater treatment system using air-stripping and carbon adsorption was also installed. Wetland sediments were removed and treated in 1993–1994. Groundwater monitoring is continuing.

A Consent Decree (1989) with the EPA requires General Electric Company to finance all site investigation and clean-up costs.

Sources: PNPLS-MA; Lanesboro Public Library, Main Street, Lanesboro, MA, 01970.

Salem Acres

Essex County

This landfill site near Salem accepted a variety of waste from 1946 to 1969, including sludges, grease, fly ash, and tannery waste. The soil is contaminated with PCBs, VOCs, arsenic, and chromium. Nearly 130,000 people live within 3 miles of the site, which borders Strongwater Brook.

Response actions included capping the sludge pits and building runoff walls (1988). A site investigation completed in 1993 recommended sludge stabilization and removal, followed by capping (1996).

A Consent Order (1987), Consent Decree (1993), and Consent Order (1994) require the PRPs to finance the site investigation and remediation work.

Sources: PNPLS-MA; Salem Public Library, 370 Essex Street, Salem, MA, 01970.

Shpack

Bristol County

This landfill outside the town of Attleboro operated from 1946 to 1965. It received a variety of municipal and industrial wastes, including radioactive materials.

The soil and groundwater are contaminated with VOCs (vinyl chloride and trichloroethylene), uranium, radium, and a number of heavy metals (arsenic and lead). The contaminants have penetrated the nearby wetlands. About 40,000 people within 3 miles of the site use the local groundwater for drinking water.

Response actions included fencing. Site investigation studies were initiated in 1990 and will recommend a remediation plan upon completion.

A number of PRPs identified by the EPA are funding the site investigation study.

Sources: PNPLS-MA; Norton Public Library, East Main Street, Norton, MA, 01237.

Silresim Chemical Corporation
Middlesex County

Located in an industrial part of Lowell, this recycling facility dealt with a variety of wastes from 1971 to 1977, including heavy metal sludges and petrochemicals. Silresim went bankrupt in 1977 and left behind about 30,000 leaking drums and tanks of liquid waste. The soil and groundwater contain high levels of VOCs, pesticides, PCBs, dioxin, and heavy metals. The contamination threatens Meadow Brook and Merrimack River as well as the water supplies of about 24,000 people who reside within 3 miles of the site.

The state began cleaning up the site in 1978. The EPA performed additional work in 1983–1984. A detailed site investigation (1985–1991) recommended a remediation plan of soil vapor extraction, soil stabilization and capping, and air-stripping groundwater. The groundwater system was installed in 1995, and construction of the soil vapor extraction system is in progress.

The EPA negotiated with a number of PRPs to help finance the site investigation and remediation work.

Sources: PNPLS-MA; Pollard Memorial Library, 401 Merrimack Street, Lowell, MA, 01850.

South Weymouth Naval Air Station
Norfolk and Plymouth Counties

This U.S. Navy facility has been in operation since 1941. The site is used to train naval personnel and maintain and refuel naval aircraft. Wastes generated by these activities, including lead-based acids, were deposited in three on-site landfills. Soil and groundwater contain high levels of VOCs, PCBs, and heavy metals. The contamination threatens over 100 municipal and private drinking water wells that service nearly 85,000 residents in Weymouth and Rockland, as well as large tracts of wetlands.

Response actions included removing PCB-contaminated soil and excavating jet fuel spills (1986). Tanks and cylinders of waste were removed in 1992. Site investigation studies are in progress and will eventually recommend a final remediation plan.

Sources: PNPLS-MA; Tufts Library, 46 Broadway Street, Weymouth, MA, 02188; Rockland Public Library, 366 Union Street, Rockland, MA, 02370.

Sullivan's Ledge
Bristol County

This former quarry operated as a New Bedford city landfill from the 1940s to the 1970s. It received a variety of municipal and industrial wastes, including capacitors, tires, demolition materials, and fuel oil. Tests in 1982 revealed that the soil and groundwater are contaminated with PCBs, VOCs, and PAHs. The contaminants have migrated to surrounding wetlands and streams and threaten the water supplies of about 100,000 people within 3 miles of the site.

Response actions included fencing (1984–1985). EPA studies have recommended groundwater extraction and treatment, capping, leachate collection, wetland dredging and restoration, and long-term groundwater monitoring. Construction of these remedies will begin in 1997.

Agreements with about 30 PRPs require the parties to finance the clean-up work.

Sources: PNPLS-MA; Wilkes Branch, New Bedford Public Library, 1911 Acushnet Avenue, New Bedford, MA, 02745.

W. R. Grace & Company
Middlesex County

Three chemical companies have operated at this site near Acton and Concord manufacturing sealants, latex, plastics, and resins. Wastewaters were discharged into unlined lagoons on the property. Testing in 1978 showed that two municipal wells in the area were contaminated with VOCs and heavy metals; soils contain cadmium, arsenic, and VOCs. The contamination has threatened the drinking water of nearby communities as well as Fort Pond Brook and the Assabet River.

Response investigations included removing tanks (1982–1983) and groundwater extraction and treatment (1985). Soil excavation and treatment began in 1994. Data from long-term monitoring programs will determine if additional remedies are necessary.

A 1980 Consent Decree with the viable PRP requires the party to finance the site investigation and remediation work.

Sources: PNPLS-MA; Acton Public Library, 486 Main Street, Acton, MA, 01720.

Wells G & H
Middlesex County

Testing of municipal wells G and H in the city of Woburn (population 36,000) in 1979 revealed that the water and soil are contaminated with VOCs, PAHs, PCBs, and pesticides. The wells were later shut down. Five nearby industrial facilities were subsequently found to be contributing to the contamination. In addition to Woburn's water supply, the contamination and surface runoff flows into the Aberjona River. River sediments contain anomalous levels of chromium, zinc, mercury, and arsenic.

Response actions included fencing and removing drums of waste. Site investigations by the EPA recommended a final remediation plan of soil excavation and vapor treatment, groundwater extraction and air-stripping, and capping (1993–1997). Studies regarding remediation of the Aberjona River are continuing.

Agreements have been signed with a number of PRPs that require the parties to finance the site investigations and remediation work.

Sources: PNPLS-MA; Thompson Public Library, 45 Pleasant Street, Woburn, MA, 01801.

MICHIGAN

Adam's Plating
Ingham County
A dry cleaning company was located on this site in Lansing in the 1950s. During its operation an underground storage tank leaked solvents into the ground. Adam's Plating has operated on the site since 1964. Wastewaters were discharged into the sewer system in the 1970s and are now held in underground tanks for treatment. Toxic water flooded a nearby basement in 1980, which prompted a state investigation. The soil is contaminated with arsenic, chromium, and trichloroethylene. About 2,000 people reside within 1 mile of the site.

An EPA site investigation concluded in 1993 that the contaminated soil should be excavated and removed, followed by monitoring. This remedy should be completed by 1997.

Sources: PNPLS-MI; Lansing Public Library, 401 South Capital Street, Lansing, MI, 48933.

Aircraft Components (D & L Sales)
Berrien County
Located in Benton Harbor, this site was a storage facility for radium-containing aircraft gauges from about 1940–1990. Radiation surveys in 1994–1995 discovered high beta/gamma radiation readings on the property. Radiation contamination has migrated to the surrounding properties, the Paw Paw River, and related wetlands. The contamination threatens the Paw Paw River ecosystem and the health of nearby residents.

The EPA and the State of Michigan initiated site investigations in 1996 to define the extent of the contamination.

Source: EPA Publication 9320.7-071, Volume 3, No. 1.

Albion-Sheridan Landfill
Calhoun County
This landfill site near Albion operated from 1966 to 1981. It received a variety of municipal and industrial wastes, including heavy metal sludges, oil, and grease. Testing in the 1980s revealed that the groundwater and soil contain high levels of arsenic, iron, barium, manganese, and VOCs. The contamination threatens water wells that serve about 14,000 people within 3 miles of the site as well as the Kalamazoo River.

Response actions included removing drums and fencing the site (1990). A detailed site investigation by the EPA (1991–1997) is in progress and will recommend a final remediation plan.

Four PRPs were issued a Universal Administrative Order in 1990 to perform the clean-up work.

Sources: PNPLS-MI; Albion Public Library, 501 South Superior Street, Albion, MI, 49224.

Allied Paper/Portage Creek/Kalamazoo River
Kalamazoo and Allegan Counties
Allied Paper Incorporated is one of several paper mills that recycled carbonless copy paper from 1957 to 1971, releasing PCBs into Portage Creek and the Kalamazoo River. Soil and groundwater on-site and the surface water, sediments, and aquatic life in a 3-mile stretch of Portage Creek and a 35-mile stretch of the Kalamazoo River contain toxic levels of PCBs. The contamination also threatens Lake Michigan.

Response actions included fencing contaminated soil areas and posting warnings along the affected waterways (1990–1991). Detailed site investigations are in progress (1992–1997) and will soon recommend a final remediation plan.

Three PRPs have signed an AOC to conduct the site studies and clean-up work.

Sources: PNPLS-MI; Kalamazoo Public Library, 315 South Rose Street, Kalamazoo, MI, 49007; Comstock Township Library, 6130 King Highway, Comstock, MI, 49041; Allegan Public Library, 180 South Sherwood, Allegan, MI, 49010.

American Anodco
Ionia County
American Anodco Incorporated has cleaned aluminum automobile parts at this site since 1962. Wastewaters containing heavy metals and spent

MICHIGAN

chemicals were discharged into seepage lagoons on-site. Testing has shown that the groundwater and soil contain high levels of phosphorus and heavy metals, including arsenic, copper, chromium, and lead. The contamination threatens the recreational Grand River and water wells within 3 miles of the site, which serve about 12,000 people.

Response actions included draining the lagoon and removing sludges (1987). Site investigations from 1987 to 1993 recommended only groundwater monitoring for 2 years to check arsenic levels in the groundwater. A final decision regarding deletion from the NPL is pending.

An AOC (1987) with American Anodco required the party to conduct the site investigations.

Sources: PNPLS-MI; Ionia Public Library, 126 East Main Street, Ionia, MI, 48846.

Anderson Development Company
Lenawee County
This company in Adrian manufactured organic chemicals at this site during the 1970s. Anderson produced the chemical MBOCA (Curene 442), used to cure polyurethanes and epoxy resins. Testing showed that this carcinogenic chemical was present in high amounts in soil, groundwater, and sediments

within a 2-mile radius of the plant. The contamination threatens the health and welfare of about 25,000 people who reside within 3 miles of the site. MBOCA was detected in preschool children living near the site.

Response actions included cleaning contaminated homes and connecting affected residents to clean water supplies (1981). In 1992 an EPA site investigation recommended low-temperature thermal desorption of the contaminated soil (1992–1993). Cleanup goals have been met, and monitoring is continuing.

Anderson Development Company, the PRP, signed an AOC with the EPA in 1986 to finance the site investigations and remediation work.

Sources: PNPLS-MI; Adrian City Library, 143 East Maumee Street, Adrian, MI, 49221.

Auto Ion Chemicals
Kalamazoo County
This company conducted chrome electroplating from 1963 to 1973. Liquid wastes and sludges were discharged in a lagoon and holding tanks on-site. Operations were suspended by the State of Michigan because of numerous pollution violations. Testing revealed that the soil and groundwater are contaminated with VOCs, PAHs, arsenic, cadmium, lead, nickel, chromium, chloride, and cyanide. The chemicals have penetrated the Kalamazoo River. The contamination threatens the water of about 3,000 residents within a mile of the site.

Response actions included excavating soils and contaminants in 1985. The site was also fenced. More contaminated soil was removed in 1993. Groundwater studies indicated pollution levels were low enough that only monitoring would be required.

The EPA identified 42 PRPs, who signed a Consent Decree (1990) that required them to finance the remediation work.

Sources: PNPLS-MI; Kalamazoo Public Library, 315 South Rose Street, Kalamazoo, MI, 49007.

Avenue E Groundwater Contamination
Grand Traverse County
Attention was first drawn to this site in Traverse City when residents complained about bad-tasting drinking water in 1980. Testing showed that the groundwater and soil are contaminated with VOCs, phthalates, and jet fuel residues. Subsequent investigations identified the nearby U.S. Coast Guard Air Station as the source of the contamination. Various leaks and spills of jet fuel, together with waste oil, solvents, and chemicals from an old dump, have been documented on the property. The contamination threatens the recreational Grand Traverse Bay and the water supply of about 16,000 people in the area.

Response actions included connecting affected residents to public water supplies. A groundwater extraction and treatment system was installed in 1985 and is still in operation. An ongoing soil bioremediation program was initiated in 1985.

The U.S. Coast Guard signed a Consent Order in 1985 that required it to finance the site investigations and remediation work.

Source: PNPLS-MI.

Barrels Inc.
Ingham County
This company recycled steel drums on the property from 1964 to 1981. Contaminated rinsewater from residual contents was dumped on the ground. Sampling by the state showed that the soil and groundwater are contaminated with VOCs, lead, zinc, chromium, and PCBs. The contamination threatens the drinking water of 133,000 residents in Lansing and Holt as well as the nearby recreational Grand River.

Response actions included removing contaminated soil, drums, and underground tanks in 1986. A detailed site investigation by the EPA is planned.

Source: PNPLS-MI.

Bendix Corporation
Berrien County
Bendix Corporation has manufactured automotive brake systems at this site since the 1960s. Waste solutions from electroplating, organic solvents, heavy metals, and process wastewater were held in seepage lagoons on the property. In 1982 a nearby private well was shut down because of high VOC levels. Tests showed that the groundwater contains excessive levels of VOCs. Private wells within 3 miles of the site supply about 4,500 people in the area.

Detailed site investigations (1989–1997) are nearing completion and will recommend a final remediation plan.

Bendix Corporation, the PRP, signed an AO in 1989 with the EPA to undertake the site investigation.

Sources: PNPLS-MI; Maud Preston Polenski Memorial Library, 500 Market Street, St. Joseph, MI, 49085.

Berlin and Farro
Genesee County
An industrial liquid waste incinerator operated on this site near Swartz Creek from 1971 to 980. Liquid wastes were stored in unlined ponds and underground tanks or discharged into an agricultural drain. Steel drums were crushed and buried on-site. Testing by the State of Michigan in the late 1970s showed

that the groundwater and soil contain VOCs, PCBs, and pesticides. The contamination has penetrated Slocum Drain and Swartz Creek and threatens the private water wells of the local rural population.

Response actions included removing drums and sludge in 1981. Liquid pesticides were pumped out of the underground tanks in 1982. Additional soil excavations and fencing were completed in 1982–1983. In 1991 the EPA recommended a final remediation plan of groundwater extraction and air-stripping and soil excavation, solidification, and capping. Construction should be under way by 1997.

Minimum settlements were negotiated between the EPA and small volume PRPs. Consent Decrees with the remaining PRPs require them to conduct the clean-up work.

Sources: PNPLS-MI; Perkins Library, 8095 Civic Drive, Swartz Creek, MI, 48473.

Bofors Nobel Inc.

Muskegon County

Industrial chemicals have been produced at this site since 1960. Various liquid wastes, including sludges, were kept in unlined lagoons on the property until the 1970s. One of the lagoons overflowed into Big Black Creek in 1975. Testing by the state has shown that the soil and groundwater contain methylene chloride, benzene, aniline, and benzidine (VOCs). About 7,000 people depend on water drawn from wells within 3 miles of the site.

Detailed site investigations recommended installing a groundwater extraction and treatment system (1995). A decision regarding soil treatment is expected by 1997.

Sources: PNPLS-MI; Muskegon County Library, 5384 Apple Avenue, Muskegon, MI, 49440.

Burrows Sanitation

Van Buren County

This site near Hartford was used as a disposal area for plating waste, sludges, coolants, and oils from 1970 to 1977. Lagoons occasionally flooded during rainstorms. Testing in the 1980s showed that the soil and groundwater contain heavy metals and VOCs. The contamination threatens the local water supplies of about 150 people as well as nearby recreational wetlands and the Paw Paw River.

Response actions included excavating contaminated sludges and soils and fencing (1984). EPA-recommended soil treatments and a groundwater pump-and-treat system were constructed from 1989 to 1993. The remedies have reduced contamination to acceptable levels, and groundwater monitoring is continuing.

An Administrative Order (1984) and Consent Decree (1989) required the PRPs to finance the site investigations and clean-up work.

Sources: PNPLS-MI; Hartford Public Library, 15 Franklin Street, Hartford, MI, 49057.

Butterworth 2 Landfill

Kent County

This 185-acre landfill site in Grand Rapids operated from the 1950s to the 1980s. It accepted a variety of municipal and industrial wastes, including plating sludges, paints, organic solvents, and liquid wastes in sealed drums. Testing in 1982 showed that the groundwater and soils are contaminated with VOCs, heavy metals (including lead), PCBs, pesticides, pyrene, and chrysene. The contamination threatens the nearby recreational Grand River, which sometimes serves as a water source for Grand Rapids, and its aquatic life.

Response actions included removing contaminated soil and constructing fencing (1986). In 1992 the EPA recommended a final remediation plan consisting of repairing the landfill cap and monitoring groundwater and river water. Construction is expected to be under way by 1997.

Five PRPs entered into Consent Decrees (1986) and Administrative Orders on Consent (1993) with the EPA to finance the site investigations and remediation work.

Sources: PNPLS-MI; Grand Rapids Public Library, 713 Bridge Street Northwest, Grand Rapids, MI, 49504; Grand Rapids Main Library, 60 Library Plaza Northeast, Grand Rapids, MI, 49503.

Cannelton Industries

Chippewa County

The main industry at this site along the St. Mary's River in Sault Ste. Marie was a tannery that operated from 1900 to 1958. Chemical and animal wastes were discharged into a wetland along the river or burned along the shore. Testing by the state in 1978 revealed that the soil, groundwater, river water, and sediments contained excessive amounts of heavy metals (chromium, lead, arsenic, mercury, and iron). About 1,200 people within 1 mile of the site use private wells. Other communities downstream use the recreational St. Mary's River as a water source. The contamination also threatens fish and wildlife habitats, including that of the threatened bald eagle.

Response actions included excavating contaminated areas and releasing gas build-ups (1988). Dikes were built along the river in 1989. Fencing was constructed in 1991. EPA studies from 1988 to 1992 recommended the removal of contaminated soil, debris, and sediments and disposal in a new landfill on the property. Construction is expected to be under way by 1997.

EPA-identified PRPs are participating in the site investigations and clean-up work.

Sources: PNPLS-MI; Gayliss Public Library, 541 Library Drive, Sault Sainte Marie, MI, 49783.

Carter Industrials

Wayne County

A scrap metal salvage yard operated at this site in Detroit from 1971 to 1986. Electrical transformers and capacitors were dismantled to recover copper components, releasing PCB-contaminated oils. Contaminated scrap was piled in the yard. Toxic oils were often spilled or leaked from drums and barrels. Testing between 1984 and 1986 by the state revealed that PCB, arsenic, cadmium, and lead contamination was present in the soil on-site, in neighboring backyards, and in the sewer lines that discharged into the Detroit River. About 35,000 people live within 1 mile of the site. The contaminants may also be present in fish in the Detroit River.

Response actions included excavating off-site contaminated soil (including residential yards) and treating the streets by high-pressure cleaning or repaving (1986). PCB-tainted soils, equipment, drums, and other debris were removed by the EPA from 1986 to 1989. Other emergency actions included covering waste piles and building runoff control systems. EPA studies from 1989 to 1991 recommended treating the contaminated soil with low-temperature desorption or incineration. Construction should begin by 1997.

An Administrative Order (1989) and Consent Decree (1993) require the PRPs to undertake the site investigations and remediation work.

Sources: PNPLS-MI; Detroit Public Library, 5201 Woodward Avenue, Detroit, MI, 48202.

Chem Central

Kent County

This company has processed and distributed industrial chemicals from this site in Wyoming since 1957. Liquid chemicals and hazardous wastes leaked from pipes and bulk storage tanks. Testing by the state in 1977 discovered high levels of VOCs, phthalates, and PCBs in the soils and groundwater. About 15,000 people live within 1 mile of the site. The contamination has penetrated Cole Drain and nearby Plaster Creek.

In response to a court order, the PRP installed a runoff control system and groundwater extraction and air-stripping system in 1984. Contaminated soils were also removed (1985). Detailed site investigations (1987–1991) recommended a soil vapor extraction system, expansion of the groundwater treatment system, and collecting free-product oil. These systems are expected to be operational by 1997.

Chem Central signed an AOC (1987) and UAO (1992) with the EPA that require the PRP to finance the site investigation and remediation work.

Sources: PNPLS-MI; Wyoming Public Library, 3350 Michael Street Southwest, Wyoming, MI, 49509.

Clare Water Supply

Clare County

Two production wells that serve the town of Clare were discovered to be contaminated with VOCs in 1982. Production was reduced from these wells and blended with clean water from other wells. Soil and groundwater contain bromoform, vinyl chloride, trichloroethane, and dichloroethane. Industrial operations in the well field area have used solvents and degreasers. About 4,000 people live within 3 miles of the site.

Response actions including installing air-strippers on the contaminated wells in 1991. In 1992 the EPA recommended a remediation plan of soil vapor extraction and groundwater extraction and treatment. Construction is expected to begin by 1997.

The EPA identified four companies as PRPs in 1985. An AOC (1985) and UAOs (1991, 1993) require the parties to finance the site investigations and remediation work.

Sources: PNPLS-MI; Garfield Library, 4th and McEwan Streets, Clare, MI, 48617.

Cliff Dump

Marquette County

This landfill near Marquette operated from 1954 to the 1960s. It received waste from a nearby charcoal manufacturing company, including wood tars. Testing has shown that the soil and groundwater are contaminated with VOCs, PAHs, and phenols. The contamination threatens the recreational Dead River and the drinking water supply of nearby homes.

Response actions included fencing the site (1984). Site investigations led the EPA to recommend excavation and biotreatment of contaminated soil, incinerating tar, restricting groundwater access, and groundwater monitoring.

The PRPs signed a Consent Order (1984) with the EPA to finance the site investigations and remediation work.

Sources: PNPLS-MI; Peter White Public Library, 217 North Front Street, Marquette, MI, 49855.

Duell and Gardner Landfill

Muskegon County

This landfill near Muskegon operated from the 1940s to 1975. It received a variety of municipal and industrial wastes, including laboratory wastes and sludges. Testing in the 1970s showed that the soil and groundwater contain VOCs (chloroform and carbon tetrachloride), aniline, PCBs, and crystal violet. About 1,200 people live within 2 miles of the site.

Response actions included removing drums and contaminated soil and fencing the site (1986). The State of Michigan and the EPA recommended a remediation plan of low-temperature thermal soil treatment, groundwater extraction and carbon adsorption treatment, and capping. Construction should be under way by 1997.

Sources: PNPLS-MI; Dalton Township Hall, 1616 East Riley Thompson Road, Muskegon, MI, 49445.

Electrovoice
Berrien County
This property near Buchanan has been a manufacturing site since the 1920s. Various companies have made insulating materials, transmissions, and electric sound reproduction equipment. Electroplating wastes were discharged into unlined ponds on the property. Paint and glue wastes were poured into a dry well. Testing by the state showed that the groundwater and soil contain VOCs, heavy metals, and cyanide. About 10,000 nearby residents depend on local groundwater sources, and recreational McCoy Creek is less than 3,000 feet from the site.

Site investigations by Electrovoice Corporation, the EPA, and the State of Michigan recommended a final remediation plan of soil vapor extraction, groundwater extraction and treatment, and soil excavation. Construction should be under way by 1997.

A Consent Order (1987) between the EPA and Electrovoice requires the company to conduct the site investigations.

Sources: PNPLS-MI; Buchanan Public Library, 117 West Front Street, Buchanan, MI, 49107.

Forest Waste Products
Genesee County
This landfill near Otisville operated from 1972 to 1978. It received a variety of wastes, including oils, sludges, paints, PCB-contaminated roofing material, acids, and resins. Testing has shown that the soil and groundwater contain VOCs, PCBs, phthalates, PAHs, chromium, and lead. The contamination threatens two drinking water aquifers, nearby wetlands, and recreational Butternut Creek.

Response actions included fencing the site (1984). Contaminated sludges, soils, and liquid wastes were removed from the property (1988–1990). The EPA recommended excavating and incinerating contaminated soil and installing a slurry wall, a leachate collection system, and a 30-year groundwater monitoring program.

The EPA has identified PRPs who are required to finance the site investigations and clean-up work through an AOC.

Sources: PNPLS-MI; Forest Township Library, 130 East Main Street, Otisville, MI, 48463.

G & H Landfill
Macomb County
This landfill operation between Utica and Rochester accepted municipal and industrial wastes, including large volumes of waste oils and paint thinners, from 1955 to 1974. Testing has shown that the soil, groundwater, and nearby pond sediments have high amounts of VOCs, phthalates, PAHs, PCBs, and heavy metals.

Fencing and runoff controls were installed from 1982 to 1986. In 1990 the EPA recommended capping; constructing a slurry wall; excavating, solidifying, and capping PCB-contaminated soil; installing a groundwater pump-and-treat system; and restoring wetlands. Construction should be completed by 1997.

Sources: PNPLS-MI; Shelby Township Library, 51680 Van Dyke Avenue, Utica, MI, 48316.

Grand Traverse Overall Supply Company
Leelanau County
This commercial laundering facility in Gerilickville operated from 1953 to the 1980s. Liquids wastes and spent solvents were discharged into holding ponds and a dry well on the property. Testing in the 1970s found that the soil and groundwater contained high levels of VOCs. At least ten nearby wells were also contaminated. About 1,200 residents live within 3 miles of the site; the recreational Cedar Lake, Cedar Lake Outlet, and Grand Traverse Bay are also within 3 miles.

Response actions included connecting affected residents to clean water supplies and drilling new wells. The lagoons were drained and contaminated soils removed in the late 1970s. An EPA study (1988–1992) showed that no further clean-up work is necessary. Groundwater monitoring is continuing, and the site will be deleted from the NPL.

Sources: PNPLS-MI; Traverse Area District Library, 322 Sixth Street, Traverse City, MI, 49684.

Gratiot County Landfill
Gratiot County
This landfill site near St. Louis operated during the 1970s and received a variety of waste, including chemicals from the Michigan Chemical Corporation. In 1977 the State of Michigan discovered that

the groundwater, soil, and sediments in nearby ponds contained high levels of polybrominated biphenyls (PBBs) and VOCs. City wells that supply about 5,000 residents are within 3 miles of the site; the recreational Pine River is less than 2 miles away.

Baseline studies by the State of Michigan (1977–1984) recommended controlling runoff, building a clay cap and slurry wall, and installing a groundwater extraction and evapo-transpiration treatment system. The slurry wall has not contained contaminated groundwater, and new studies are evaluating alternative treatments.

Velsicol Chemical Corporation (previously Michigan Chemical Company) reached a settlement with the State of Michigan in 1982 regarding financing remediation work at the site.

Sources: PNPLS-MI; City Hall, 129 West Emerson Street, Ithaca, MI, 48847.

H & K Sales

Ionia County

Aircraft parts, including radioactive luminescent dials and gauges, were stored on this site in Belding from the 1940s through the 1990s. Tests by the State of Michigan in 1994 revealed high levels of radon and gamma radiation in the warehouse area. The contamination is a threat to the Flat River and the health of nearby residents and on-site workers.

Response actions include site investigations that are expected to recommend an action plan by 1997–1998.

Source: EPA Publication 9320.7-071, Volume 3, No., 1.

H. Brown Company Incorporated

Kent County

This battery recycling facility in Walker has been in operation since 1961. Until 1978 batteries were opened, the lead recovered, and the acids dumped on the ground. Testing has shown that the soil and groundwater are contaminated with VOCs and heavy metals. About 3,000 people live within 3 miles of the facility. The contamination threatens nearby wetlands and the recreational Grand River.

Response actions included temporary capping of lead-contaminated soils and fencing (1991). An EPA investigation (1988–1992) recommended excavation and stabilization of contaminated soil, a clay cap, slurry wall, and monitoring of deeper groundwater. Construction is expected to be completed by 1997.

The PRPs have been noncompliant with a UAO, and the EPA has been funding the work at the site.

Sources: PNPLS-MI; Kent County Public Library, 1331 Walker Village Drive, Walker, MI, 49504.

Hedblum Industries

Iosco County

Automobile parts were manufactured on this site from 1958 to 1985 by a number of different companies. Various liquid wastes, including degreasers and cooling and rinse waters, were poured on the ground on the 10-acre site. Nearby wells were found to be contaminated in 1973. Inspections by the state revealed that the soil and groundwater are contaminated with VOCs (including trichloroethylene and vinyl chloride). The towns of Oscoda and Au Sable are within 3 miles of the site. The contamination also flows toward recreational wetlands and the Au Sable River.

Response actions included drilling new wells for affected residents or connecting them to city water supplies. In 1989 the EPA recommended a groundwater pump-and-treat system and groundwater monitoring. The groundwater treatment system became active in 1992 and will operate until at least 1998.

PRPs identified by the EPA have been active in carrying out the site studies and clean-up work.

Sources: PNPLS-MI; Oscoda Public Library, 110 South State Street, Oscoda, MI, 48750.

Hi-Mill Manufacturing Company

Oakland County

This company has made aluminum, brass, and copper tubing on the site since 1946. Metals are anodized and degreased. Spent acid bath solutions containing heavy metals were poured into a clay-lined lagoon, which often overflowed into an adjacent marshland. Testing in the 1970s showed that the groundwater and soils contain excessive levels of VOCs. About 13,000 people within 3 miles of the site use private drinking water wells.

Response actions included removing contaminated soil, sludge, and water from the lagoon. A new well was drilled to provide Hi-Mill employees with safe drinking water. Site investigations from 1989 to 1993 recommended only groundwater monitoring because the shallow groundwater is not used for domestic purposes.

Hi-Mill Manufacturing Company signed a Consent Order (1988) with the EPA that required it to conduct the site investigation.

Sources: PNPLS-MI; Highland Township Library, 205 West Livingston Street, Highland, MI, 48031.

Ionia City Landfill

Ionia County

This landfill in Ionia operated from the 1930s to 1969. It received a variety of municipal and industrial wastes, including sealed drums of industrial liquids and solvents. Testing showed that the groundwater and soil contain excessive levels of chromium, lead, barium, and VOCs. The contamination threatens the water supplies of about 6,000 people within 1 mile of the site, a municipal well field, and the recreational Grand River.

Response actions included removing drummed waste and installing fencing (1985). A temporary cover was also constructed. In 1989 the EPA recommended in-situ vitrification of soil, which should be operational by 1997. Groundwater remedies are pending.

Sources: PNPLS-MI; Hall-Fowler Memorial Library, 126 East Main Street, Ionia, MI, 48846.

J & L Landfill
Oakland County
This landfill near Rochester Hills operated from 1951 to 1980. It received a variety of municipal and industrial wastes, including dusts from emission control filters and alkaline slag. Testing has shown that the soil and groundwater contain excessive levels of heavy metals, such as manganese, chromium, and nickel. Many of the 1,500 people within 1 mile of the site use shallow water wells. The contamination also flows into the recreational Clinton River and the Rochester-Utica Recreational Area.

Response actions included connecting affected residents to the city water system. EPA studies from 1991 to 1994 called for a landfill cap. Groundwater studies are in progress.

Sources: PNPLS-MI; Rochester Hills Public Library, 500 Olde Towne Road, Rochester, MI, 48307.

K & L Avenue Landfill
Kalamazoo County
This landfill in Kalamazoo operated in the 1960s and 1970s. It received a variety of municipal and industrial wastes. It was closed in 1979 after nearby wells were determined to be polluted. Testing by the State of Michigan showed that the groundwater and soil contain VOCs, phenols, PCBs, and heavy metals. About 8,000 people live within 2 miles of the site. The contamination also threatens recreational Bonnie Castle Lake and Dustin Lake.

Response actions included providing bottled water and hook-ups to the city water system to affected residents (1980–1981). In 1990 the EPA recommended groundwater extraction and bioremediation treatment, fencing, a landfill cap, gas vents, and long-term groundwater monitoring. Construction will begin in 1996–1997.

Sources: PNPLS-MI; Oshtemo Library, 7265 West Main Street, Kalamazoo, MI, 49009.

Kayden Corporation
Muskegon County
Kayden Corporation has manufactured bearings at this site since 1941. Various liquid wastes, including sludges, were discharged into unlined ponds or Ruddiman Creek. Testing has shown that the groundwater and soil are contaminated with VOCs, cyanide, and heavy metals. About 700 people within 3 miles of the site use private water wells. The contamination threatens recreational Ruddiman Creek and Muskegon Lake.

Contaminated soil and sludge were removed in 1986. A groundwater extraction system was installed in 1988. Studies regarding soil contamination are in progress.

Kayden Corporation has financed the site investigations and remediation work.

Source: PNPLS-MI.

Kentwood Landfill
Kent County
This landfill near Kentwood operated from the 1950s to 1976. It received a variety of municipal and industrial wastes, including sludges. In the 1970s leachate from the landfill polluted nearby Plaster Creek. Testing showed that the groundwater and soil are contaminated with VOCs and heavy metals. About 100 people within 1 mile of the site obtain drinking water from private wells.

In 1991 the EPA recommended capping, gas venting, a leachate collection system, and extracting and treating groundwater. Construction was started in 1994–1995.

Kentwood and Kent County, the PRPs, signed an AOC (1985) and Consent Decree (1991) to undertake the site investigations and remediation work.

Sources: PNPLS-MI; Kent County Library, 4700 Kalamazoo Avenue Southeast, Kentwood, MI, 49508.

Kysor Industrial Corporation
Wexford County
Kysor Industrial Corporation, which builds temperature control systems for cars, has been operating on this industrial site in Cadillac since the 1970s. Waste sludges and solvents were discharged into unlined pits on the property until 1979. Testing found that the groundwater and soil are contaminated with VOCs, phenol, chromium, and other heavy metals. About 110,000 people live within 3 miles of the site, and the Cadillac municipal wells are less than 3,000 feet away. The contamination also threatens recreational Lake Cadillac and the Clam River.

In 1989 the EPA recommended a remediation plan of groundwater extraction and treatment by air-stripping and carbon adsorption and soil vapor extraction. Pilot studies were completed in 1995, and construction should be under way by 1997.

The EPA has identified PRPs who are participating in the site investigations and clean-up work.

Sources: PNPLS-MI; Cadillac-Westford Library, 411 South Lake Street, Cadillac, MI, 49601.

Liquid Disposal Inc.

Macomb County

This former gravel pit near Utica was used as a landfill from 1964 to 1982. It received a variety of municipal and industrial wastes, including paint thinners, sludges, oil, and grease. Testing in the 1980s showed that the groundwater and soil are contaminated with VOCs, heavy metals, PAHs, pesticides, and PCBs. The contamination threatens the local drinking water of about 4,000 people as well as the Clinton River and associated wetlands and the Shadbush Nature Study Area.

Response actions included removing contaminated soil, repairing dikes, and pumping out the lagoon (1982). The lagoons were capped in 1983. Removal of various debris and solid waste was carried out between 1985 and 1990. The EPA recommended a final remediation plan of soil solidification and capping, groundwater extraction and treatment with air-stripping and ion exchange, and a landfill cap. Construction began in 1992 and is still in progress.

Liquid Disposal Inc. was forced into bankruptcy in 1982. Federal funds are being used for the clean-up work.

Sources: PNPLS-MI; Shelby Township Library, 51680 Van Dyke Avenue, Utica, MI, 48087.

Lower Ecorse Creek Dump

Wayne County

This former wetland and dump site near Wyandotte operated from 1945 to 1955. It was filled in with a variety of industrial waste, including ferric ferrocyanide. Residents have since built homes on or near the area. Testing by the EPA in 1989 showed that the soil and groundwater are contaminated with ferric ferrocyanide and possibly cyanide gas.

Response actions included temporarily capping the site with topsoil and vegetation (1989–1990). Affected residents were permanently relocated. A detailed site investigation by the EPA (1994–1997) will recommend a final remediation plan.

Sources: PNPLS-MI; Bacon Memorial Public Library, 45 Vinewood, Wyandotte, MI, 54656.

Mason County Landfill

Mason County

This landfill site near Lundington operated from 1972 to 1978. It received a variety of municipal and industrial wastes, including liquids and sludges. Surface runoff and seeping leachate entered Iris Creek in 1980–1981. Testing in the 1980s showed that the groundwater and soil contain VOCs and heavy metals (sodium and lead). The contamination threatens wetlands, Babbin Pond, Iris Creek, and the private drinking wells of about 1,000 local residents.

In 1989 the EPA recommended a remediation plan of capping the landfill, fencing, deed restrictions, and long-term groundwater monitoring. The cap was completed in 1991. Groundwater monitoring has shown that the contaminants are present in nonhazardous levels, and no further action is required. Groundwater monitoring is continuing.

One of the PRPs, Acme Disposal, signed a Consent Order from the EPA in 1978. Mason County and Acme Disposal were later sued by owners of contaminated properties in 1981–1982. Mason County bought both properties.

Sources: PNPLS-MI; Ludington Public Library, 217 East Ludington Street, Ludington, MI, 49431.

McGraw Edison Corporation

Calhoun County

This plant in Albion owned by McGraw Edison Corporation operated from 1958 to 1980. It produced air conditioners and humidifiers, and sludges and oil wastes were discharged on the property. Testing in the 1980s showed that the groundwater and soil are contaminated with VOCS (trichloroethylene). Forty-five nearby private wells are also contaminated. About 11,000 people live with 3 miles of the site, and the recreational Kalamazoo River is less than 3,000 feet away.

Response actions included connecting affected residents to clean water supplies and removing contaminated soil. A groundwater extraction and air-stripping system was installed in 1984. Soils will be flushed so that the contaminants will enter the groundwater and be recovered. More than 4 billion gallons of groundwater have been treated. Groundwater monitoring will continue until 2030.

McGraw Edision signed a Consent Decree with the EPA in 1984. The new owner of the site, Cooper Industries, has taken over clean-up activities.

Source: PNPLS-MI.

Metamora Landfill

Lapeer County

This former gravel pit near Metamora operated from 1955 to 1980. It received a variety of municipal and industrial wastes, including about 35,000 sealed drums of toxic liquid waste. Testing has shown that the soil and groundwater contain VOCs, PCBs, and heavy metals (arsenic). The contamination threatens the water supplies of about 1,000 residents within 1 mile of the site.

Response actions included excavating and incinerating nearly 28,000 drums of waste (1986–1995). In 1990 an EPA study recommended groundwater extraction and treatment using air-stripping and chemical precipitation, a gas collection and flare system, and capping. Construction should be under way by 1997.

Sources: PNPLS-MI; Lapeer County Library, 4024 Oak Street, Metamora, MI, 48455; Margerite de Angeli Library, 921 West Nepessing Street, Lapeer, MI, 48446.

Michigan Disposal Service (Cork Street Landfill)

Kalamazoo County

This landfill site in industrial Kalamazoo has been in operation since 1925. It received a variety of unregulated, largely undocumented municipal and industrial wastes before the 1950s. It received fly ash and paper mill waste in the 1980s. Testing has shown that the soil and groundwater contain VOCs, arsenic, lead, and iron. The contamination threatens the recreational Davis Creek and the Kalamazoo River as well as private and municipal water wells that serve about 50,000 people.

Response actions included installing a clay cap on part of the landfill and a leachate collection system. Site investigations (1987–1991) recommended capping the entire site and installing a groundwater extraction and treatment system.

Consent Decrees (1987) with the city of Kalamazoo and Michigan Disposal Service require the parties to fund the site investigations and remediation work.

Sources: PNPLS-MI; Kalamazoo Public Library, 315 South Rose Street, Kalamazoo, MI, 49007.

Motor Wheel Incorporated

Ingham County

Located in Lansing, Motor Wheel Incorporated made wheels for the car industry from 1938 to 1978. Various wastes, including paints and solvents, were discharged on the ground or buried in drums. Testing in the 1980s showed that the groundwater and soils contain VOCs and PCP. Later sand and gravel mining on the property exposed buried waste. The contamination threatens the drinking water of nearby residents and businesses.

Response actions included removing buried tanks and drums (1982–1993). Site investigations (1987–1991) recommended a landfill cap and the installation of a groundwater extraction and treatment system. Construction should be completed by 1997.

Administrative Orders on Consent (1987, 1993) and a Consent Decree (1994) with three PRPs require the parties to finance the site investigations and clean-up work.

Sources: PNPLS-MI; Lansing Public Library, 401 South Capitol Avenue, Lansing, MI, 48933.

Muskegon Chemical Company

Muskegon County

This property near Whitehall has been a chemical manufacturing site since 1975. Materials were stored on-site in tanks and sumps that leaked chemicals, including heptane. Testing in 1981 revealed that the groundwater and soils contain high levels of VOCs, including xylene. The contamination threatens the private well water of about 6,500 residents within 3 miles of the site as well as recreational Mill Pond, Mill Pond Creek, and White Lake.

Response actions included creating a hydraulic barrier to keep the contamination plume in place. In 1993 the EPA recommended a final remediation plan of groundwater extraction and treatment by carbon adsorption. A soil remedy will be selected by 1997.

The PRPs have been active in studying and remediating the site.

Sources: PNPLS-MI; Whitehall City Library, 414 East Spring Street, Whitehall, MI, 49461.

North Bronson Industrial Area

Branch County

This waste disposal area near North Bronson operated from 1939 to 1981. It held various liquid industrial wastes, including electroplating sludges, in seepage lagoons. Groundwater and soil contain excessive levels of VOCs, heavy metals, cyanide, and PCBs. About 3,000 people within 3 miles of the site use the local groundwater.

Response actions included connecting affected residents to the city water supply system. An EPA investigation (1987–1993) will recommend a final remediation plan.

Sources: PNPLS-MI; Bronson Library, 207 North Matteson Street, Bronson, MI, 49028.

Northernaire Plating

Wexford County

This chromium- and nickel-plating company near Cadillac operated from 1971 to 1981. Liquid wastes were discharged into the city sewer system. In 1978 nearby private wells were found to be polluted with chromium. In the company's last years of operation, liquid wastes were improperly stored in tanks and drums on the unfenced site. The groundwater and soil are contaminated with cadmium, chromium, and VOCs. The contamination threatens private and municipal well water within 1 mile of the site.

Response actions included removing thousands of gallons of liquid cyanide and acid waste (1983). In 1985 the EPA recommended excavating contaminated soil and razing and decontaminating the buildings (1989–1991). Groundwater will be pumped and treated through an air-stripping and carbon adsorption system that will be operational by 1997.

Sources: PNPLS-MI; Cadillac-Westford Library, 411 South Lake Street, Cadillac, MI, 49601.

Novaco Industries
Monroe County
A tool and die manufacturing company operated on this site in Temperance in the 1970s and 1980s. Activities included the chrome plating of auto parts. Nearby wells were found to be contaminated with chromium in 1979. The groundwater and soil are contaminated with low levels of chromium.

EPA site investigations from 1983 to 1986 recommended a groundwater extraction and treatment system and long-term monitoring. About 40 million gallons of groundwater have been treated to date. Chromium levels in the groundwater are now considered safe, but monitoring will continue.

Novaco Industries participated in the site clean up.

Sources: PNPLS-MI; Monroe County Library, 8575 Jackman Road, Temperance, MI, 48182.

Organic Chemicals
Kent County
Companies refined, transported, and stored petroleum on this site in Grandville from the 1930s to 1978. From 1979 to 1991 solvents and pharmaceutical chemicals were processed. Wastewater was poured into a seepage basin from 1968 to 1980. Chemical spills on the property have also been documented by state officials. Testing showed that the groundwater and soil are contaminated with VOCs. About 9,000 people live within 3 miles of the site. The contamination threatens nearby recreational wetlands and ponds.

Response actions included removing sludges (1981). Drummed waste was removed in 1987. EPA studies recommended a groundwater extraction and carbon adsorption treatment system that should be operational by 1997. A soil remedy is still being considered.

The PRPs conducted site investigations in 1987.

Sources: PNPLS-MI; Grandville Public Library, 3141 Wilson Avenue, Grandville, MI, 49418.

Ossineke Groundwater Contamination
Alpena County
This area of groundwater contamination in Ossineke was caused by fuel spills and leaks from underground storage tanks and by wastewater from a dry cleaning shop. Testing in 1977 revealed that the groundwater in nearby wells contains VOCs, petroleum, and chlorinated hydrocarbons. About 1,200 people live within 3 miles of the site. The contamination threatens nearby wetlands and Devils River.

Response actions included drilling deeper replacement wells for affected residents. EPA studies between 1989 and 1991 determined that leaking underground storage tanks were the source of the contamination. The EPA has recommended transferring this NPL site to the federal Leaking Underground Storage Tank (LUST) Program.

Sources: PNPLS-MI; NBD Alpena Bank, 11686 U.S. Highway 23 South, Ossineke, MI, 49766.

Ott Chemical/Story Chemical/ Cordova Chemical Company
Muskegon County
Three former chemical manufacturing companies operated on this site near North Muskegon from the 1950s to 1985. A variety of agricultural and medical chemicals were produced. Chemical wastes were stored in drums or discharged into an unlined lagoon or Little Bear Creek. Testing in the 1970s revealed that the groundwater and soil contain VOCs, SVOCs, and pesticides. About 4,000 people within 3 miles of the site use the local groundwater. The contamination also threatens Little Bear Creek, Bear Creek, and Muskegon Lake.

Response actions included removing sludges, contaminated soil, and solid waste; razing buildings; and installing municipal water lines for affected residents (1978). The EPA recommended a final remediation plan of groundwater extraction and treatment that should be operational by 1997. Soil remedies are presently being considered.

Ott Chemical Company and Story Chemical Company conducted clean ups on the site in 1968 and 1977 by order of the State of Michigan. Cordova Chemical Company signed a Consent Order (1977) that requires participation in the remediation work.

Sources: PNPLS-MI; Walter Memorial Library, 1522 Ruddiman Avenue, North Muskegon, MI, 49445.

Packaging Corporation of America (PCA)
Manistee County
ABBCo (later PCA) began operating a pulp mill on this site near Filer City, originally a waste disposal area, in 1949. Various liquid wastes were discharged into Manistee Lake from 1949 to 1951 and later into

unlined seepage ponds, which were used until 1976. Testing has shown that the soil and groundwater contain phenols and heavy metals (lead and chromium). About 10,000 people live within 3 miles of the site. The contamination has polluted recreational Manistee Lake, which connects with Lake Michigan.

In 1993 the EPA recommended a remediation plan of groundwater monitoring only because 95 percent of the plume had already moved into Manistee Lake.

Sources: PNPLS-MI; Manistee County Library, 95 Maple Street, Manistee, MI, 49660.

Parsons Chemical Works
Eaton County
Located near Grand Ledge, this agricultural chemical processing plant operated from 1945 to 1979. Wastewater was discharged into drain fields or a sewer line that discharged into the Grand River. Testing in 1979–1980 by the state showed that the groundwater and soil contain excessive levels of dioxin, heavy metals (lead, mercury, and arsenic), DDT, dieldrin, and chlordane. The chemicals have been found in sediments from the Grand River. About 11,000 people depend on water wells within 3 miles of the site.

Response actions included fencing and excavating contaminated soils (1985–1994). The EPA is presently studying in-situ vitrification as a soil remedy. A pilot test is under way.

Sources: PNPLS-MI; Grand Ledge Public Library, 131 East Jefferson Street, Grand Ledge, MI, 48837.

Peerless Plating Company
Muskegon County
This electroplating shop in Muskegon operated from 1937 to 1983. Spent solvents and wastewaters were discharged into unlined seepage lagoons on the property. Testing by the State of Michigan in 1983 revealed that the soil and groundwater contain cadmium, chromium, VOCs, lead, and cyanide. About 1,500 people within 3 miles of the site use the local groundwater. The contamination also threatens Little Black Creek and Mona Lake.

Response actions included connecting affected residents to the municipal water supply (1983–1988). The site was decontaminated and solid and liquid wastes and debris removed (1983–1991). An EPA study (1989–1993) recommended soil stabilization and soil vapor extraction and the extraction and air-stripping of contaminated groundwater. Construction is in progress.

In 1983 the State of Michigan sued the PRP, which closed the shop.

Sources: PNPLS-MI; Norton Shores Library, 705 Seminole Road, Muskegon, MI, 49442.

Petoskey Municipal Well Field
Emmet County
This well field serving the city of Petoskey was found to contain VOCs. The contamination plume originates from the former Petoskey Manufacturing Company die casting plant, which operated from 1946 to the 1960s. Soil at the site contain heavy metals, VOCs, and pesticides. About 11,000 residents and a number of private wells are within 3 miles of the site.

Response actions included removing contaminated soil and partial capping (1982–1983). Site investigation studies (1987–1993) have recommended a groundwater extraction and air-stripping system.

Petoskey Manufacturing Company, the PRP, was operating under an AO (1984) and a CO (1987) before it went bankrupt in 1990. Clean-up activities are now federally funded.

Sources: PNPLS-MI; Petoskey Public Library, 451 East Mitchell Street, Petoskey, MI, 49770.

Rasmussen's Dump
Livingston County
Originally a gravel pit, this site near Hamburg accepted municipal and industrial wastes from the 1960s to 1972. Later gravel mining exhumed about 2,000 drums of liquid waste. Testing has shown that the soil and groundwater are contaminated with VOCs and PCBs. About 2,000 people within 3 miles of the site use local groundwater.

Response actions included excavating drums and other waste (1984–1985). Additional removals were conducted between 1985 and 1990. In 1991 the EPA recommended a remediation plan of groundwater extraction and treatment (chemical precipitation, bioremediation, air-stripping, and activated carbon), groundwater monitoring, capping, and fencing. Construction is in progress.

The EPA-identified PRPs signed an Administrative Order that requires the parties to finance the site investigation and remediation work.

Sources: PNPLS-MI; Hamburg Library, 7225 Stone Street, Hamburg, MI, 48139.

Rockwell International Corporation, Allegan
Allegan County
Two plants on this site in Allegan built parts for heavy equipment and machinery from the 1900s to 1991. Various wastewater, plating wastes, sludges, and oils were discharged into the Kalamazoo River or unlined settling ponds until 1971. Seepage from underground storage tanks entered the Kalamazoo River in 1976. Testing has shown that the soil and groundwater contain high levels of free-product oil, VOCs, PCBs, heavy metals (including lead, arsenic,

and mercury), cyanide, and PAHs. The river also contains these pollutants. About 7,000 people within 3 miles of the site use the shallow groundwater for drinking purposes.

Response actions included recovering free-product oil (1976–1991) and removing sludge (1978). Detailed site investigations to be completed by 1997 will recommend a final remediation plan.

The PRP, Rockwell International, signed a Consent Order with the EPA in 1988 to finance the site investigation and remediation work.

Sources: PNPLS-MI; Allegan Public Library, 331 Hubbard Street, Allegan, MI, 49010.

Rose Township Dump

Oakland County

This landfill site near Holly operated in the 1960s. It received a variety of municipal and industrial wastes, including paint sludges, solvents, and PCBs, that were buried on-site or discharged into unlined lagoons. Testing in the 1980s showed that the groundwater and soil contain heavy metals, PCBs, phthalates, DDT, and VOCs. About 5,000 people in the area use the local groundwater for domestic purposes. The contamination also threatens nearby wetlands and Buckhorn Lake.

Response actions included removing contaminated soil and fencing the site (1980). Drummed waste was removed in 1986. The EPA recommended a remediation plan of soil excavation and incineration, soil washing, groundwater extraction and treatment, and groundwater monitoring. The systems should be operational by 1997.

The EPA has identified 28 PRPs. About half of the parties signed Consent Decrees (1988) with the EPA that require them to finance the site investigations and remediation work.

Sources: PNPLS-MI; Holly Township Library, 1116 North Saginaw Street, Holly, MI, 48442.

Roto-Finish Company

Kalamazoo County

Roto-Finish Company made smooth-finished metallic products in Portage from 1950 to 1988. Prior to 1980, wastewater and spent solutions were discharged into unlined lagoons that frequently overflowed. Testing has shown that the groundwater and soil contain VOCs, MOCA, and heavy metals (including arsenic and lead). About 50,000 people live within 3 miles of the site. The contamination threatens Davis Creek, the recreational Kalamazoo River, and private and municipal wells that serve over 100,000 people,

Response actions included excavating lagoon sludges (1982). Site investigations (1991–1997) that will recommend a final remediation plan are nearing completion. An interim groundwater extraction and treatment system is being considered.

The PRPs are conducting the site investigations.

Sources: PNPLS-MI; Portage Public Library, 300 Library Lane, Portage, MI, 49002.

SCA Independent Landfill

Muskegon County

This former landfill near Muskegon Heights operated from 1965 to 1987. It received a variety of municipal and industrial wastes. Testing by the state in the 1980s revealed that the groundwater and soil contain VOCs (including benzene). About 10,000 people live in the area. The contamination threatens nearby wetlands and Black Creek, a trout stream.

SCA closed the site according to state regulations, including installing gas vents, capping, and monitoring groundwater (1985). Ongoing site investigations will recommend a final remediation plan in 1997.

Consent Agreements (1983, 1993) require the PRPs to finance the site investigations and remediation work.

Source: PNPLS-MI.

Shiawassee River

Livingston County

A 52-mile stretch of the Shiawassee River has been contaminated by aluminum manufacturing businesses that operated in the 1970s. Wastewater containing PCBs was either discharged into the river or held in unlined ponds on the properties. PCB concentrations have been identified in fish, sediments, and riverbank soil and on the industrial sites. The towns of Byron, Vernon, and Corunna are on the river downstream from these sites. About 1,400 people live near the manufacturing plants.

Response actions included excavating the lagoon and contaminated sediments. Portions of the Shiawassee River were dredged in 1982. A detailed site investigation by the EPA with final recommendations for clean up are expected by 1997.

Cast Forge Company, a PRP, conducted some of the early response actions.

Sources: PNPLS-MI; Howell Township Hall, 3525 Byron Road, Howell, MI, 48843.

South Macomb Disposal Authority

Macomb County

This former gravel pit near Mt. Clemens operated as a landfill from the 1960s to 1975. It received a variety of municipal and industrial wastes, including hazardous chemicals and heavy metals. Testing in 1983–1984 showed that the soil and groundwater contain VOCs, heavy metals, and nitrate. Nearby wells were also contaminated. The contamination threatens the water supplies of about 250 nearby residents as well as recreational McBride Drain and Clinton River.

Response actions included fencing the site and connecting affected residents to the municipal water supply system. Leachate controls were also installed. In 1991 the EPA recommended the extraction and treatment of groundwater through air-stripping, carbon adsorption, and oxidation/precipitation. Decisions regarding soil remedies and landfill excavations are pending.

Sources: PNPLS-MI; Macomb County Library, 16480 Hall Road, Mt. Clemens, MI, 48044.

Southwest Ottawa County Landfill
Ottawa County
This landfill site near Holland operated from 1968 to 1981. It received a variety of municipal and industrial wastes, including solvents, sludges, oil, and heavy metals. Testing in the 1980s showed that the groundwater and soils contain VOCs and heavy metals. A number of nearby residents use the local groundwater for domestic purposes. The contamination also threatens Lake Macatawa and Lake Michigan.

Response actions included connecting affected residents to the Holland municipal water supply system. A groundwater extraction and activated carbon adsorption treatment system became operational in 1987. Groundwater treatment and monitoring are continuing.

Sources: PNPLS-MI; Park Township Offices, 52 152nd Avenue, Holland, MI, 49424.

Sparta Landfill
Kent County
This former landfill near Sparta operated from the 1940s to 1972. It received a variety of municipal and industrial wastes, including sludges and liquid waste. Testing by the state in 1978 detected VOCs in the groundwater. A number of the 2,000 people within 1 mile of the site use the local groundwater for domestic purposes. The contamination also threatens the Rogue River, which supplies water to the city of Rockford.

Bottled water was supplied to affected residents, and in some cases deeper wells were built. Detailed site investigations (1993–1997) will recommend a final remediation plan.

PRPs identified by the EPA are conducting the site investigations.

Sources: PNPLS-MI; Sparta Township Library, 80 North Union, Sparta, MI, 49345.

Spartan Chemical Company
Kent County
Spartan Chemical Company has processed liquid industrial chemicals at this site in Wyoming since the 1950s. Prior to 1963 wastewater was discharged into holding ponds or on the ground. Various tests from 1975 to 1981 revealed that the groundwater is contaminated with VOCs. Nearby wells were also found to be contaminated. About 7,000 people live within 3 miles of the site, many of whom use the local groundwater for domestic purposes.

Response actions included removing leaking underground storage tanks (1986). Affected residents were connected to municipal water supplies. A groundwater extraction and air-stripping treatment system was also installed. A soil vapor extraction system will probably be installed by 1997.

Spartan Chemical Company signed a Consent Order with the EPA to finance the site investigations and clean-up work. The PRP later declared bankruptcy, and clean-up operations are being funded by the EPA and the State of Michigan.

Sources: PNPLS-MI; Wyoming Public Library, 3350 Michael Southwest, Wyoming, MI, 49509.

Spiegelberg Landfill
Livingston County
This former gravel pit near Hamburg was used as a landfill from 1966 to 1977. It received a variety of municipal and industrial wastes, including paint sludges. Testing in the 1980s revealed that the soil and groundwater contain lead and VOCs. About 18,000 people within 3 miles of the site use the local groundwater for drinking purposes.

Response actions included excavating contaminated soil and waste (1988–1989). EPA studies recommended a groundwater extraction and treatment system using air-stripping and carbon adsorption, which should be operational by 1997.

Sources: PNPLS-MI; Hamburg Library, 7225 Stone Street, Hamburg, MI, 48139.

Springfield Township Dump
Oakland County
This waste dump near Davisburg received a variety of chemical wastes, including drummed liquids and sludges, from 1966 to 1968. Testing revealed that the soil and groundwater contain high levels of VOCs, arsenic, lead, PCBs, pesticides, phthalates, barium, and cadmium. About 100 people within 1 mile of the site use private water wells.

Response actions included erecting a fence and removing drummed waste and contaminated soil (1975–1980). In 1990 the EPA recommended installation of an extraction and carbon adsorption treatment system for contaminated groundwater along with vacuum extraction, incineration, and solidification of contaminated soil. The groundwater system became operational in 1993.

Sources: PNPLS-MI; Springfield Township Hall, 650 Broadway Street, Davisburg, MI, 48019.

State Disposal Landfill
Kent County
This former landfill near Grand Rapids operated from 1966 to 1977. It received a variety of municipal and industrial wastes, including heavy metals. Testing in 1985 by the State of Michigan detected excessive levels of VOCs and heavy metals (barium and nickel) in the groundwater. About 13,000 people depend on private and municipal water wells within 3 miles of the site, some of which are contaminated.

Response actions included connecting affected residents to the municipal water supply system (1989–1990). Fencing was also erected, and contaminated wells were closed. Detailed site investigations (1991–1996) will recommend a final remediation plan.

The PRP, Waste Management of North America, is conducting the site investigations.

Sources: PNPLS-MI; Plainfield Branch Library, 2650 Five Mile Road Northeast, Grand Rapids, MI, 49505.

Sturgis Municipal Wells
St. Joseph County
Three of the four municipal wells that serve Sturgis were found to be contaminated with VOCs in 1982. The source of the VOCs may be nearby Ross Laboratories, whose production wells also contained VOCs. These were subsequently closed. The city of Sturgis has a population of about 10,000.

In 1991 detailed site investigations recommended groundwater extraction and air-stripping, soil vapor extraction, and removal of contaminated soil. The groundwater system became active in 1994 and will operate until 2001.

PRPs identified by the EPA are conducting the clean-up work.

Sources: PNPLS-MI; Sturgis City Library, 130 North Nottawa, Sturgis, MI, 49091.

Tar Lake
Antrim County
Iron was produced on this site near Mancelona by a number of companies. The last manufacturing operation ended in 1944. Wastes and sludges were discharged into an unlined lagoon on the property. A phenol-lead groundwater contamination plume nearly 3 miles in length originates from the site. Sludges contain iron, nickel, chromium, and copper. About 3,000 people who live within 3 miles of the site use the local groundwater for domestic purposes.

Response actions included site investigations. In 1992 the EPA recommended excavating contaminated tar, sludge, and soil and installing a groundwater containment system. Construction should be completed by 1997.

The Fifty-Sixth Century Antrim Iron Company, a PRP, signed an AO with the EPA (1986) to conduct the site investigations and clean-up work.

Sources: PNPLS-MI; Mancelona Public Library, 202 West State Street, Mancelona, MI, 49659.

Thermo-Chem
Muskegon County
This liquid waste disposal and recycling center near Muskegon operated from 1969 to 1980. Various liquid wastes and wastewater, including paints, antifreeze, sludges, and solvents, were discharged into pits and lagoons. Testing by the state revealed that the groundwater and soil contain VOCs. About 10,000 people in the local area use the local groundwater. The contamination also threatens Black Creek and Mona Lake.

Response actions included fencing (1989) and removing underground storage tanks (1991). EPA site investigations (1987–1991) recommended clearing the site of surface debris and structures, soil removal and incineration, in-situ soil vapor extraction, and groundwater extraction and filtration/air-stripping. Construction should be completed by 1998.

A group of PRPs signed a Consent Order (1987) with the EPA to fund the site investigations and remediation work.

Sources: PNPLS-MI; Hackley Public Library, 316 West Webster, Muskegon, MI, 49440.

Torch Lake
Houghton County
This lake in the Upper Peninsula near Houghton has been polluted by about 200 million tons of copper mine tailings that were dumped in the lake from the 1890s to 1969. Mine drainage water, sanitary waste, and explosives residues have also entered the lake. Testing has shown that the lake water and sediments contain high levels of copper and cupric ammonium carbonate, which have damaged fish populations. About 4,000 people live within a mile of Torch Lake and use local groundwater.

Response actions included removing drums of toxic waste and contaminated soils from the shoreline (1988–1990). Submerged tailings and slag piles are being covered with sediment and vegetation (1992–1997). Other site investigations are in progress.

Sources: PNPLS-MI; Portage Lake District Library, 105 Huron Street, Houghton, MI, 49931.

U.S. Aviex
Cass County
This company made automotive fluids on this site near Niles from the 1960s to 1978. A spill of di-

methyl ether in 1972 polluted nearby private wells. Testing from about 1975 to 1985 showed that groundwater and soil contain excessive levels of VOCs. The contamination threatens the water supplies for nearby homes, businesses, and schools.

Response actions included supplying bottled water to affected residents and in some cases digging new wells (1973–1987). Residents were later connected to the city water supply system. A groundwater extraction and air-stripping treatment system was installed in 1983. Soil flushing was conducted from 1991 to 1993.

After being sued by the State of Michigan in 1982, U.S. Aviex declared bankruptcy in 1986. Clean-up work has been funded by the EPA and the State of Michigan.

Sources: PNPLS-MI; Howard Township Hall, 1345 Barron Lake Road, Niles, MI, 49120.

Verona Well Field

Calhoun County

This well field serving the city of Battle Creek was found to be contaminated with VOCs in 1981. Various industries in the area have been identified as point sources. About 55,000 people depend on groundwater drawn from the well field.

Response actions included providing alternate water supplies to affected residents and drilling barrier wells to control the contamination plume (1983–1984). A groundwater extraction and air-stripper/carbon adsorption system was also installed. A soil vapor extraction system was built on the Thomas Solvent Company source area in 1993.

Thomas Solvent Company, a PRP, filed for bankruptcy in 1984. In 1992 the remaining PRPs were ordered to undertake the site investigations and remediation work.

Sources: PNPLS-MI; Willard Library, 7 West Van Buren Street, Battle Creek, MI, 49017.

Wash King Laundry

Lake County

A laundry operation was active on this site near Baldwin from 1962 to 1991. Laundry wastewater was discharged into unlined lagoons. Testing in 1973 revealed that the groundwater and soil contain perchloroethylene, phosphorus, sodium, chloride, lead, and arsenic. The contamination penetrated at least 30 nearby private wells.

Response actions included connecting affected residents to city water supplies (1984). In 1993 the EPA recommended excavating contaminated soil and installing a groundwater pump-and-treat system. Construction should be completed by 1997.

A 1978 legal action forced Wash King Laundry, the PRP, to abandon and clean up the site. The EPA is funding the work because the company declared bankruptcy in 1991.

Sources: PNPLS-MI; Pathfinder Community Library, 812 Michigan Avenue, Baldwin, MI, 49304.

Wurtsmith Air Force Base

Iosco County

This 5,200-acre U.S. Air Force base near Oscoda operated from 1923 to 1993. Military activities included gunnery practice, vehicle and aircraft maintenance, and refueling support. Various liquid wastes, including degreasers, de-icing solutions, fuels, paint by-products, and heavy metal sludges were discharged in unlined pits on the base. Testing in 1977 showed that the groundwater and soil contained VOCs, including vinyl chloride. The contamination threatens Van Etten Lake, Lake Huron, Allen Lake, and the water supply for the town of Oscoda.

Detailed site investigations that will recommend a final remediation plan are in progress.

Source: PNPLS-MI.

MINNESOTA

Agate Lake Scrap Yard

Cass County

Agate Lake Scrap Yard near Brainerd operated from 1952 to 1982. It recycled aluminum, copper, and lead by using two homemade smelters that ran on transformer oils and halogenated solvents. Spills and leaks were frequent. Ash piles were left on the property when the company went out of business in 1982. The groundwater and soil contain VOCs, PCBs, dioxins, furans, and heavy metals. The contamination threatens private water wells that serve about 1,200 residents and has entered recreational Agate Lake.

Response actions included removing transformers, drummed wastes, furnaces, and contaminated soil in 1983. In 1994 the EPA recommended a remedy of natural attenuation of groundwater and long-term groundwater monitoring.

The PRPs performed the site investigations according to a 1986 UAO with the State of Minnesota.

Sources: PNPLS-MN; Brainerd Public Library, 416 South 5th Street, Brainerd, MN, 56401.

MINNESOTA

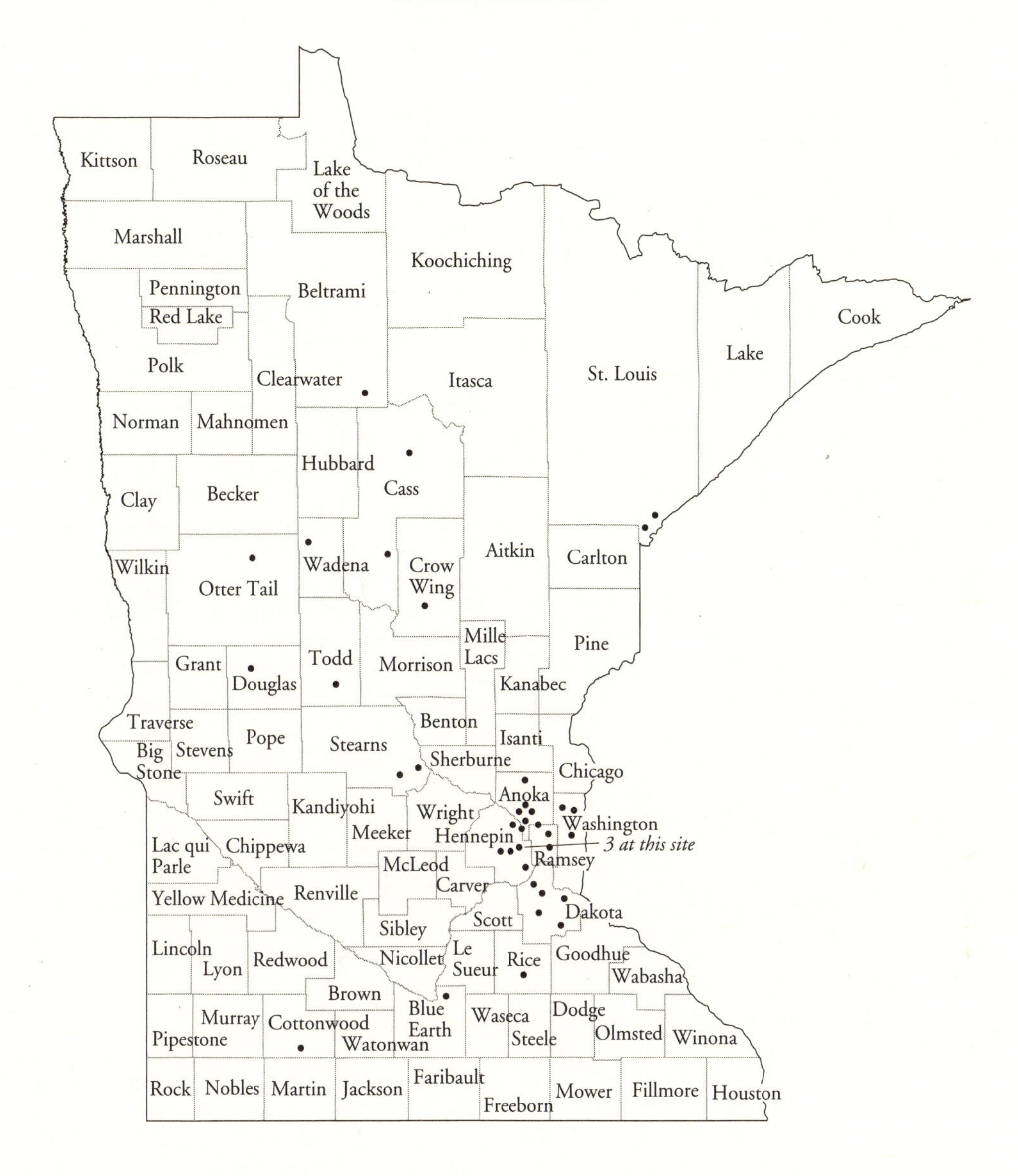

Arrowhead Refinery Company
St. Louis County
Cannery and waste oil recycling businesses operated on this site in Hermantown from the 1940s to 1977. Wastes generated during oil recycling included acidic heavy metal sludges that were discharged into lagoons and ditches. The pollutants flowed into neighboring wetlands and streams. Testing from 1979 to 1980 by the EPA revealed that the groundwater, surface water, and soil contain VOCs, PAHs, and heavy metals. The contamination threatens surrounding wetlands and private wells that supply about 3,000 residents in the area.

Response actions included runoff controls (1980) and fencing (1990). In 1986 the EPA recommended removing sludge and contaminated sediments, supplying alternative clean water supplies to affected residents, and installing a groundwater extraction and treatment system. The groundwater plant became operational in 1992.

EPA-identified PRPs have resisted a UAO (1990) from the EPA and subsequent efforts by the EPA to involve them in the remediation process.

Sources: PNPLS-MN; Duluth Public Library, 520 West Superior Street, Duluth, MN, 55802.

Boise Cascade/Onan
Anoka County
A wood treatment plant operated on this site in Fridley from 1921 to 1961. Timbers were soaked with creosote and PCP. The property has since been developed into new commercial and manufacturing facilities. Creosote contamination was discovered during construction. Testing by the EPA revealed high levels of organic compounds, creosote derivatives, and phenols in the groundwater and soil. The contamination threatens the water source for about 3,000 nearby residents and schools.

Response actions included excavating and removing contaminated soil and groundwater (1986) and fencing the site (1986). Groundwater and air monitoring is in progress.

Five PRPs negotiated settlements with the EPA. One company, Medtronic, signed a Consent Decree (1984) with the State of Minnesota to help fund the site studies.

Sources: PNPLS-MN; Minnesota Pollution Control Agency, 520 Lafayette Road, St. Paul, MN, 55155.

Burlington Northern, Brainerd Plant
Crow Wing County
Burlington Northern Railroad operated a railroad tie treatment facility on this site in Brainerd from 1907 to 1985. Wooden ties were soaked in creosote and fuel oil. Waste solutions and sludges were discharged on-site into lagoons. The groundwater and soil contain PAHs and heavy metals. The chemicals have migrated off-site and threaten private water wells and the Mississippi River.

Response actions included recovering free-product oil, excavating sludges and contaminated soil, and erecting fencing (1985–1987). Soil and groundwater studies are in progress.

Burlington Northern signed a Consent Decree in 1985 that requires the company to fund the remediation work.

Sources: PNPLS-MN; Brainerd Public Library, 206 North Seventh Street, Brainerd, MN, 56401.

Dakhue Sanitary Landfill
Dakota County
This landfill near Cannon Falls has been in operation since 1971. It has received municipal and industrial wastes, including VOC solutions, solvents, and sludges. Testing has indicated that the groundwater and soil contain VOCs, cadmium, and lead. The contamination threatens Pine Creek, the recreational Cannon River, and the water wells that serve about 700 residents,

Response actions included constructing surface runoff controls (1990). The site was capped and a leachate collection system installed in 1991–1992. In 1993 the EPA recommended natural attenuation of the contamination plume and groundwater monitoring.

A 1984 Consent Order requires the PRPs to conduct the site investigations.

Sources: PNPLS-MN; Cannon Falls Library, 306 West Mill Street, Cannon Falls, MN, 55009.

East Bethel Landfill
Anoka County
This landfill site in East Bethel has been in operation since 1969. It received a variety of municipal and industrial wastes, including solvents, inks, acids, paints, thinners, drummed hazardous wastes, transformers, and ether. Testing by the EPA revealed that the groundwater and soil contain high levels of VOCs, arsenic, barium, cadmium, mercury, lead, PAHs, and petrochemicals. The contamination threatens wetlands, Ned's Lake, and private wells that serve about 2,000 local residents. Volatized contaminants from the site could also be inhaled.

Site investigations carried out by the State of Minnesota were completed in 1993. A groundwater extraction and treatment system was installed in 1994. Other decisions are pending.

Sources: PNPLS-MN; East Bethel City Hall, 2241 21st Avenue, Northeast Cedar, MN, 55011.

FMC Corporation
Hennepin County
This tract of land near Fridley was used as a naval ordnance production complex and a waste disposal area from 1941 to 1969. Various waste products, including plating solutions, paint sludges, oils, ash, and solvents were discharged or buried on the site. Testing has shown that the groundwater and soil contain VOCs. The contamination threatens the Mississippi River and the drinking water source for about 500,000 people in the Minneapolis area.

Response actions included removing contaminated soil and installing a gas detection system (1983). A groundwater extraction and treatment system became operational in 1987.

Legal agreements with the State of Minnesota and the EPA require the PRPs to conduct the site investigations and remediation work.

Sources: PNPLS-MN; Minnesota Pollution Control Agency, 520 Lafayette Road, St. Paul, MN, 55155.

Freeway Landfill
Dakota County
This landfill near Burnsville, operating since 1971, has received a variety of commercial and municipal wastes, including battery casings and aluminum slag. Testing has shown that the groundwater and soil contain VOCs and heavy metals. The contamination threatens the Minnesota River and the municipal wells that supply the 36,000 residents of Burnsville.

Response actions included pumping and air-stripping contaminated groundwater from a nearby quarry (1986) and building a cap and gas control system (1995–1997).

PRPs have been identified and are expected to contribute to the remediation effort.

Sources: PNPLS-MN; Minnesota Pollution Control Agency, 520 Lafayette Road, St. Paul, MN, 55155.

General Mills/Henkel
Hennepin County
General Mills operated a research facility on this site in Minneapolis from 1930 to 1977. Various production wastes and spent solvents were discharged into pits on the property. Testing indicates that the groundwater and soil contain VOCs. The contamination is a threat to 5,000 people who live within 1 mile of the site.

Response actions included the installation of a groundwater extraction and air-stripping system in 1985. Treatment will continue until 2012.

The PRP, General Mills, has performed the remediation work.

Sources: PNPLS-MN; Minnesota Pollution Control Agency, 520 Lafayette Road, St. Paul, MN 55155.

Joslyn Manufacturing Company
Hennepin County
A wood treatment facility operated on this site in Brooklyn Center from the 1920s to1980. Wood timbers were treated with creosote, PCP, and copper-arsenate solutions. Wastewaters were discharged into a wetland by Twin Lakes. Sludges were also buried on the property. Testing has revealed that the groundwater and soil contain PCP, creosote, PAHs, and oil. The contamination threatens Shingle Creek, the Mississippi River, and recreational Twin Lakes. Nearby residents are connected to the city water supply.

Response actions included removing liquid wastes and sludge (1981–1982). The area was fenced in 1986. A groundwater extraction and treatment system and soil bioremediation program were initiated in 1989. Groundwater treatment will continue for 30 years.

A Consent Order (1985) requires Joslyn Manufacturing Company to remediate the site.

Sources: PNPLS-MN; Southdale Public Library, 7001 York Avenue South, Edina, MN, 55435.

Koch Refining Company
Dakota County
This company in Rosemount has refined oil on this site since 1969. End products include kerosene, jet fuel, gasoline, asphalt, butane, propane, and petroleum coke. Sampling by the State of Minnesota in the 1970s revealed that the groundwater and soil contain VOCs, PAHs, phenols, and lead. The chemicals have penetrated nearby private water wells. About 4,500 people depend on water drawn from wells within 3 miles of the site.

Response actions included supplying affected residents with clean water supplies. Groundwater and soil investigations are in progress. Recommendations are expected by 1997.

Koch Refining Company, the PRP, signed a Consent Agreement (1985) that requires the company to remediate the site.

Sources: PNPLS-MN; Minnesota Pollution Control Agency, 520 Lafayette Road, St. Paul, MN, 55155.

Koppers Coke
Ramsey County
Koppers Coke in St. Paul operated from 1911 to 1978 and produced coke, coal tars, and coal tar distillates. Various wastes, including VOC washes, were stored, buried, or discharged on the property. The site is now occupied by a business park. Groundwater and soil contain VOCs, heavy metals, PAHs, and phenols. The contamination threatens local irrigation and agricultural wells and recreational Como Park.

Response actions included removing liquid wastes and contaminated soils (1982). The site was also capped. Detailed site investigations were carried out from 1989 to 1994. Final recommendations included in-situ groundwater bioremediation, which became operational in 1996.

In 1978 Koppers signed an agreement that requires the company to shut down the plant and conduct the site investigations.

Source: PNPLS-MN.

Kummer Landfill
Beltrami County
This former landfill near Bemidji operated from 1971 to 1984. It received a variety of municipal and demolition wastes, including fly ash. Testing in 1982–1983 revealed that the groundwater and soil contain VOCs. The chemicals have penetrated nearby wells. The contamination threatens water wells that supply about 15,000 people as well as nearby wetlands and Lake Bemidji.

Response actions included connecting affected residents to clean water supplies (1991). Runoff controls, capping, fencing, and deed restrictions were completed by 1992. Groundwater monitoring is continuing. A groundwater extraction system is also planned.

A 1994 Consent Decree requires the PRPs to fund the site investigations and remediation work.

Sources: PNPLS-MN; Northern Township Town Hall, 445 Town Hall Road, Bemidji, MN, 56601.

Kurt Manufacturing Company
Anoka County
Kurt Manufacturing Company has been making computer parts in Fridley since 1960. VOC-containing solvents are used extensively and stored on-site. Testing in 1982 by the state revealed that the groundwater and soils contain VOCs. The contamination threatens the Mississippi River and the drinking water supplies of about 163,000 people.

Contaminated soil and metal shavings were removed from the property in 1984. In 1986 the State of Minnesota recommended a groundwater pump-and-treat system and long-term groundwater monitoring. Construction began in 1994.

A 1984 agreement with the State of Minnesota requires the PRP to fund the site investigations and clean-up work.

Sources: PNPLS-MN; Minnesota Pollution Control Agency, 520 Lafayette Road, St. Paul, MN, 55155.

Lagrand Landfill
Douglas County
This former landfill near Alexandria operated from 1974 to 1985. It accepted a variety of municipal and industrial wastes and had a history of violations. Diesel fuel was buried on-site in 1980. Subsequent testing has shown that the groundwater and soil contain VOCs. The contamination threatens the private water wells that supply about 1,200 residents as well as wetlands and streams.

Response actions included site investigations from 1987 to 1992. In 1992 the State of Minnesota recommended gas venting, sealing of contaminated wells, erosion controls, access restrictions, capping, and long-term groundwater and gas monitoring.

Sources: PNPLS-MN; Alexandria Public Library, Seventh and Fillmore Streets, Alexandria, MN, 56308.

Lake Elmo
Washington County
Groundwater and soil near the Lake Elmo Airport near Baytown were discovered to be contaminated with VOCs in 1987. Further investigations indicate that the main plume of contamination originates from the hangar at the airport. Trichloroethane was used regularly as a degreaser and paint-stripper. About 26,000 residents depend on the local groundwater for drinking water purposes. The contamination also threatens the St. Croix River.

Site investigations are in progress.

Source: PNPLS-MN.

Lehiller/Mankato Site
Blue Earth County
This 6,400-acre site near Mankato was used as a landfill from 1925 to 1950. Various wastes were accepted, including petroleum products and fuels. Testing in 1981 showed that the groundwater and soils contain VOCs and petroleum by-products. The contamination threatens the water supplies of about 30,000 residents as well as the Minnesota and Blue Earth Rivers.

Response actions included connecting affected residents to clean water supplies (1984–1985). In 1985 the EPA recommended groundwater extraction and air-stripping. The system will continue to treat groundwater until 1999.

Sources: PNPLS-MN; Minnesota Valley Regional Library, 100 East Main Street, Mankato, MN, 56002.

Long Prairie Groundwater Contamination
Todd County
Groundwater and soil under the town of Long Prairie were discovered to contain VOCs in 1983. Contaminated wells were shut down. Investigations indicate that the VOCs originated from a former dry cleaning business that dumped sludge and spent solvents in a perforated drain on the property. The contamination threatens the water supplies for about 3,000 residents.

The EPA recommended a final remediation plan of groundwater extraction and activated carbon treatment and contaminated soil venting. The systems should be operational by 1997. Affected residents were also connected to clean water supplies.

Sources: PNPLS-MN; City Hall, 42 Third Street North, Long Prairie, MN, 56347.

MacGillis and Gibbs Company
Ramsey County
This former wood treatment site in New Brighton has been in operation since the 1920s. Wooden timbers and poles were soaked with creosote and PCP-containing oils from 1950 to 1970. Since then the wood has been treated with chromated copper arsenate. Various wastes and sludges were discharged into lagoons and a pond on the site. Groundwater and soils contain PAHs, PCP, copper, chromium, arsenic, dioxins, and furans. The contamination threatens

wetlands, streams, and the drinking water supplies for thousands of nearby residents.

Response actions included removing drummed waste, excavating sludges and soils, and capping with clay. In 1992 the EPA recommended soil washing and incineration of contaminated wood debris, solidification of metal-contaminated soil, and groundwater extraction and treatment. The remedies are presently being designed.

The PRPs have participated in the site studies and remediation work.

Sources: PNPLS-MN; Ramsey County Library, 1941 West County Road, Arden Hills, MN, 55112.

N. L. Industries/Taracorp
Hennepin County
Since 1903 this industrial site in St. Louis Park has been the location of various businesses involved in metal refining, lead smelting, scrap material, and auto parts. Wastes and slag were stored on-site or discharged into process sewers. Testing has indicated that the groundwater and soil are acidic and contain sulfates, dissolved solids, and lead. The contamination threatens Minnehaha Creek, associated wetlands, the Mississippi River, and the private water wells of nearby residences.

Response activities included restricting access, removing contaminated soils, capping, revegetating, and long-term groundwater monitoring (1985–1988). Monitoring will be conducted for 30 years.

A Consent Order required N. L. Industries to conduct the site investigations and remediation work.

Sources: PNPLS-MN; St. Louis Park City Hall, 5005 Minnetonka Boulevard, St. Louis Park, MN, 55416.

Naval Industrial Reserve Ordnance
Anoka County
This U.S. naval facility in Fridley has built weapons systems since the 1940s. Various production wastes, including paints, solvents, oils, and heavy metals, were stored, leaked, discharged, or buried on the 83-acre property. Trichloroethylene was used as a degreaser until 1987. Testing has shown that the groundwater and soil contain VOCs. The contamination threatens the Mississippi River and the drinking water sources for about 30,000 residents.

Response actions included excavating contaminated soils (1983–1984). A site investigation was conducted from 1984 to 1988. A groundwater extraction and treatment system was installed in 1991 and will be in operation until 1999. A soil remedy is still being considered.

Sources: PNPLS-MN; Anoka Public Library, Fridley Branch, 410 Northeast Mississippi Street, Fridley, MN, 55432.

New Brighton/Arden Hills
Ramsey County
A 24-square-mile area of groundwater contamination north of Minneapolis–St. Paul was discovered through routine testing of municipal water wells in 1981. Contaminated wells in New Brighton were shut down and new wells constructed. The groundwater contains PCBs, chromium, arsenic, and VOCs, including benzene, toluene, and xylene. Soil in the area also contains PCBs. Further investigations showed that the source of the contamination is the Twin Cities U.S. Army Ammunition Plant.

Response actions included supplying clean water to affected residents and businesses (1983–1984), incinerating PCB-contaminated soils (1989), and building three groundwater extraction and carbon adsorption systems (1991–1995). Additional studies are in progress.

Sources: PNPLS-MN; New Brighton City Hall, 803 Fifth Avenue Northwest, New Brighton, MN, 55112.

Nutting Truck and Caster Company
Rice County
This manufacturing company produced tools and other equipment on this site in Faribault from 1959 to 1984. Sludges and other waste products were discharged into an unlined pit on the property. Subsequent testing revealed that the groundwater and soils contain VOCs and cadmium. The contamination threatens the drinking water supplies for the 16,000 residents of Faribault.

Response actions included excavating and capping the pit. A groundwater pump-and-treat system using passive aeration began operation in 1992.

Nutting Truck and Caster Company, the PRP, signed a Consent Order with the State of Minnesota to conduct the site investigations and remediation work.

Sources: PNPLS-MN; Minnesota Pollution Control Agency, 520 Lafayette Road, St. Paul, MN, 55155.

Oak Grove Sanitary Landfill
Anoka County
This former landfill near Cedar operated from the 1960s to 1984. It received a variety of municipal and industrial wastes, including acids, sludges, paints, pesticides, solvents, inks, and heavy metals. Sampling by state officials in 1984 revealed that nearby private wells contain high levels of VOCs and heavy metals. The groundwater and soil contain VOCs, phenols, phthalates, and heavy metals. The contamination threatens water wells that serve

about 10,000 people in the area as well as wetlands and the Rum River.

Response actions included fencing, capping, deed restrictions, and air and groundwater monitoring (1988—1992). Groundwater monitoring began in 1993.

PRPs have conducted site studies and remediation work according to two UAOs (1992) with the EPA.

Sources: PNPLS-MN; Oak Grove Board, 1990 Northwest Nightingale Street, Cedar, MN, 55011.

Oakdale Dump
Washington County
This former dump near Oakdale operated from the 1940s to 1961. Drummed chemical waste was buried in long trenches in the 1950s. VOCs were detected in private and municipal wells in the 1980s. The groundwater and soil contain VOCs, which threaten the local water supplies for about 2,500 residents.

Response actions included removing waste materials, establishing a monitoring well network, and installing a groundwater pump-and-treat system (1984–1985). The system continues to operate.

Minnesota Mining and Manufacturing, one of the PRPs, signed an agreement with the EPA that required the company to fund the remediation work.

Sources: PNPLS-MN; Minnesota Pollution Control Agency, 520 Lafayette Road, St. Paul, MN, 55155.

Perham Arsenic Site
Otter Tail County
Pesticides were mixed and formulated on this site in Perham from the 1930s to 1947. Unused pesticides were buried on the site when the operations closed. The site is now part of the county fairgrounds. Eleven employees were poisoned by arsenic when they drank from a well on the site in 1972. The groundwater and soil also contain arsenic. The contamination threatens the health of the 2,000 residents of Perham.

Response actions included capping (1982), excavating arsenic-contaminated soil, and recapping (1985). Affected residents and businesses were connected to the city water supply system. In 1994 the EPA recommended groundwater extraction, filtration, and aluminum adsorption to remove arsenic. The system should be operational by 1997.

Sources: PNPLS-MN; Perham Public Library, 100 Third Street Northeast, Perham, MN, 56573.

Pine Bend Landfill
Dakota County
This active landfill near inner Grove Heights has been in operation since 1972. It has received a variety of municipal and industrial wastes, including organic solvents, chlorides, halogens, and heavy metals. Testing by the EPA revealed that the groundwater and soil contain arsenic, VOCs, PAHs, and various chloride compounds. The chemicals have penetrated nearby private wells. The contamination threatens nearby farms and agricultural wells, the Mississippi River, and wells that supply about 16,000 residents.

Response actions included supplying affected residents with bottled water. A permanent clean water supply system should be completed by 1997. Groundwater studies are nearing completion and will recommend a final remediation plan.

The PRPs signed an agreement with the State of Minnesota in 1985 that requires the companies to conduct the site investigations.

Sources: PNPLS-MN; Dakota County Library System, 1340 Wescott Road, Eagen, MN, 55123.

Reilly Tar and Chemical
Hennepin County
This former coal tar distillation and wood treatment plant in St. Louis Park operated from 1917 to 1972. Various production wastes were discharged on the property into ponds or ditches that connected to an adjacent marsh. The property has since been developed as a residential neighborhood. Testing has shown that the groundwater and soil contain petrochemicals, VOCs, and creosote derivatives. The compounds have entered seven municipal wells, which have since been shut down. The contamination threatens the drinking water supplies for about 40,000 people as well as wetlands and nearby streams.

Response actions included the construction of a groundwater extraction and carbon filtration system that has been operational since 1985. Affected residents were connected to clean water supplies. Contaminated wetlands were filled in to control the spread of the contamination (1986). Deed restrictions were imposed in 1989. Other site studies are in progress.

Legal agreements (1984, 1986) between Reilly Tar and Chemical Corporation and the EPA require the company to fund the site investigations and remediation work.

Sources: PNPLS-MN; St. Louis Park Library, 3240 Library Lane, St. Louis Park, MN, 55426.

Ritari Post & Pole Company
Wadena County
This former wood treatment facility near Sebeka operated from 1959 to the 1970s. Wooden timbers were soaked with creosote and PCP and allowed to drip-dry. Sludges were also discharged on the ground. Nearby wells contain high levels of PCP. Testing has

shown that the groundwater and soils contain PCPs, phenols, and dioxin. The contamination threatens wetlands, the recreational Cat River, and wells that supply about 500 residents.

The State of Minnesota completed an investigation in 1994. Recommendations include bioremediation of PCP-contaminated soil, incineration of dioxin-contaminated soil, soil washing, and groundwater monitoring. The systems are currently being designed.

Sources: PNPLS-MN; Wadena Public Library, 304 First Street, Wadena, MN, 56482.

South Andover Site
Anoka County
Various small businesses have operated on this site in Minneapolis since the 1950s. Activities have included auto salvage, auto repair, and metal and oil recycling. Inks, solvents, paints, grease, and other chemical solutions were also stored on-site. Spills, leaks, and discharges have contaminated the groundwater and soils with VOCs, arsenic, PCBs, antimony. lead, and PAHs. The contamination threatens the water supplies for 13,000 residents as well as nearby marshes and wetlands.

Response actions included removing drummed waste (1981, 1986), fencing (1989), and removing old tires (1989). In 1988 the EPA recommended connecting affected residents to clean water supplies and groundwater monitoring (operational in 1994). Contaminated soils were biotreated or excavated and incinerated off-site (1992).

The EPA has identified 16 PRPs. A 1993 Consent Decree requires the companies to undertake the remedial actions.

Sources: PNPLS-MN; Andover City Hall, 1685 Crosstown Boulevard Northwest, Andover, MN, 55403.

St. Augusta Landfill
Stearns County
This former landfill near St. Augusta operated from 1955 to 1971. It received a variety of municipal and industrial wastes, including paints, sludges, ash, and solvents. The landfill was capped in 1983. Testing has shown that the groundwater and soil contain heavy metals, VOCs, atrazine, and phthalates. The contamination threatens Johnson Creek, the Mississippi River, and the water supply for about 2,500 people.

Site investigations begun in 1991 are nearing completion and will recommend a final remediation plan.

The EPA has identified about 40 PRPs.

Sources: PNPLS-MN; Great River Regional Library, 405 St. Germain Street, St. Cloud, MN, 56301.

St. Louis River Site
St. Louis County
This industrial site near Duluth has been contaminated by a steel mill and coking operation that were active from about 1915 to the 1950s. Activities included manufacturing coke, iron, wire, pig iron, and tar. Various wastes were stored or discharged on the property. The groundwater and soil contain PAHs, VOCs, tars, and PCBs. The air may also contain contaminated dust and VOCs. The contamination threatens the St. Louis River, Lake Superior, and the health of about 1,000 residents.

Response actions included razing buildings and cleaning the site. Tar seeps were excavated and burned in 1993. Other site investigations are nearing completion.

PRPs have participated in the site studies and response actions.

Sources: PNPLS-MN; Duluth Public Library, 520 West Superior Street, Duluth, MN, 55802.

St. Regis Paper Company
Cass County
This wood treatment facility near Cass Lake has been in operation since the 1950s. Lumber was treated with creosote and PCP. Waste solutions were discharged into disposal ponds on-site. The groundwater, surface water, and soil contain PAHs, phenols, heavy metals, dioxins, PCP, and arsenic. The contamination threatens Pike Bay, Cass Lake, and wells that serve about 1,000 local residents.

Response actions included excavating and storing contaminated soil (1986). A groundwater extraction and carbon adsorption circuit was installed in 1987.

Consent Orders (1985) require Champion International Company to fund the site studies and cleanup work.

Source: PNPLS-MN.

Twin Cities Air Force Reserve Base
Hennepin County
This U.S. Air Force property near the Minneapolis–St. Paul International Airport was in operation from 1944 to 1972. Various wastes, including paint sludge, fuel sludge, and heavy metals, were buried in a landfill on the property. The site has periodically been covered by floodwaters from the Minnesota River. The groundwater and soils contain heavy metals, petrochemicals, and VOCs. The contamination threatens wells that serve about 65,000 people as well as the Minnesota Valley National Wildlife Refuge and the Minnesota River.

Response actions included fencing, removing hazardous liquids, and excavating contaminated soil (1987). In 1992 a site investigation recommended deed restrictions, groundwater monitoring, and the natural attenuation of the contamination plume.

Sources: PNPLS-MN; Southdale Hennepin Library, 7001 York Avenue South, Edina, MN, 55435.

University of Minnesota
Rosemount Research Center
Dakota County
The University of Minnesota operated this disposal site, which received liquid wastes, batteries, PCB-bearing oils, acids, ammonia, ether, and other chemical laboratory wastes from 1968 to 1973. Testing has shown that the groundwater and soil contain heavy metals, VOCs, nitrates, pesticides, dioxins, and PCBs. The contamination threatens water wells that serve about 10,000 residents.

Response actions have included supplying affected residents with clean water supplies and building a groundwater extraction system. Lead-contaminated soil was removed in 1990; PCB-contaminated soil was incinerated on-site. Revegetation was completed in 1994.

A 1986 agreement with the State of Minnesota requires the University of Minnesota to conduct the investigation and removal actions.

Sources: PNPLS-MN; Minnesota Pollution Control Agency, 520 Lafayette Road, St. Paul, MN, 55155.

Waite Park Wells
Stearns County
Four city wells in the Waite Park area near St. Cloud were discovered to contain VOCs in 1984. Investigations revealed that the sources of the contamination were activities by Burlington Northern Railroad and the Electric Machinery Company, which generated a variety of wastes that were leaked, spilled, or discharged on the premises. Wastes included oils, grease, solvents, paints, PCBs, resins, coolant water, and heavy metals. The contamination threatens the Sauk and Mississippi Rivers and the health of the 4,000 residents of Waite Park.

Response actions included shutting down contaminated wells and later installing an air-stripper for these wells. A groundwater extraction and air-stripping system was installed on the site. Contaminated soils will be stabilized through gas and leachate collection systems and reburied with a liner. Construction will be completed in 1997.

Sources: PNPLS-MN; Waite Park Library, 612 North Third Street, Waite Park, MN, 56387.

Washington County Landfill
Washington County
This former landfill near Lake Elmo operated from 1969 to 1975. It received a variety of municipal and commercial wastes. Testing in 1981 revealed that the groundwater contains VOCs and heavy metals, which have entered nearby wells. The contamination threatens the drinking water supplies of nearby residents and agricultural wells.

Response actions included connecting affected residents to clean water supplies (1983–1984, 1992). A groundwater extraction and air-stripping system was installed in 1983. The landfill was also capped and fenced.

A Consent Order (1984) and UAO (1992) require the PRPs to perform site studies and remediation work.

Sources: PNPLS-MN; Washington County Library, 2150 Radio Drive, Woodbury, MN, 55125.

Waste Disposal Engineering
Anoka County
This former landfill near Andover operated from about 1963 to 1983. It received a variety of industrial and municipal wastes, including paints, oils, acids, sludges, and solvents. Testing has shown that the groundwater, surface water, and soil gas contain VOCs. The contamination threatens Coon Creek and the drinking water supplies of local residents.

A site investigation (1984–1987) recommended a groundwater treatment system using carbon adsorption, a soil cap, filling in a contaminated wetland and constructing a new wetland, and long-term groundwater monitoring. Construction activities were completed in 1996. Groundwater treatment is continuing.

Sources: PNPLS-MN; Andover City Hall, 1685 Crosstown Boulevard, Andover, MN, 55304.

Whittaker Corporation
Hennepin County
Operations on this site in Minneapolis from the 1940s to the 1960s included processing war materials (including antifreeze, oils, and fuels), industrial coatings, and resins. Raw materials were stored in drums or underground tanks. Waste solutions were discharged into a swamp on the premises. Testing in 1978 discovered that the groundwater and soil contain excessive levels of heavy metals and VOCs. The contamination threatens nearby water wells and the Mississippi River.

Response actions included excavating drums and contaminated soil (1985). A groundwater extraction and air-stripping system was also installed.

Sources: PNPLS-MN; Minnesota Pollution Control Agency, 520 Lafayette Road, St. Paul, MN, 55155.

Windom Dump
Cottonwood County
Various municipal and industrial wastes, including paint sludges, were buried in this former gravel pit near Windom from the 1930s to 1971. Testing has shown that the groundwater and soil contains VOCs, arsenic, cadmium, and chromium. The contamination is a threat to water wells that supply about 5,000 residents.

Response actions included fencing the site and installing monitoring wells. Erosion controls and a clay cap were also built. A groundwater extraction and treatment system was constructed in 1989.

Sources: PNPLS-MN; Windom Public Library, 904 Fourth Avenue, Windom, MN, 56101.

MISSISSIPPI

Chemfax
Harrison County
This chemical company near Gulfport has manufactured petroleum hydrocarbon resins for paraffin wax since 1955. Wastewater and cooling water were discharged into ditches and unlined ponds on the property. Air sampling by the EPA in 1990 revealed high levels of VOCs, including benzene and toluene. The contaminated air threatens the health of about 50,000 residents and workers within 4 miles of the site.

Site investigations are planned.

Source: PNPLS-MS.

Flowood Site
Rankin County
Manufacturing companies on this site have produced corrugated boxes, ceramic tiles, and stoneware cooking pots since the 1950s. Various liquid and solid wastes were buried or discharged into pits, ditches, and ponds. Testing in 1982–1983 revealed that the groundwater, soils, and sediments in Lake Marie and Neely Creek contain lead and other heavy metals. The contamination threatens nearby streams and wetlands as well as the drinking water supplies for the 950 residents of Flowood.

In 1988 the EPA recommended excavating and solidifying contaminated soils (1991–1993), deed restrictions, and groundwater monitoring. The EPA will likely propose this site for deletion from the NPL by 1997.

An AOC (1986) and CD (1989) require the PRPs to conduct the site investigations and clean-up work.

Sources: PNPLS-MS; Pearl Public Library, 3470 Highway 80, East Pearl, MS, 39208.

Newsom Brothers
Marion County
This site near Columbia was used as a sawmill and produced turpentine and resins from the 1930s to the 1980s. Wood preserving compounds using PCP and diesel oil were also manufactured. Until 1977 liquid wastes, including phenols, oil, and grease, were discharged directly into a nearby stream. Drummed waste was also buried on the property. Testing has shown that the groundwater, soil, and nearby stream sediments contain VOCs, PAHs, PCBs, oil, and heavy metals. The contamination threatens domestic and agricultural wells, nearby streams, and the Pearl River.

Response actions included removing drummed waste (1984, 1987, 1988). The EPA recommended excavating and incinerating tars, capping, constructing erosion controls, and long-term groundwater monitoring. This work was completed in 1994–1995. The excavations discovered more drummed waste, and additional studies are under way.

PRPs have participated in the site investigations and remediation work.

Sources: PNPLS-MS; South Mississippi Regional Library, 900 Broad Street, Columbia, MS, 39429.

Potter Company
Copiah County
Manufacturing operations have been conducted on this site in Wesson since 1953. Finished products have included electrical components, capacitors, and electromagnetic filters. Various waste solutions, including solvents, degreasers, and PCB-containing capacitor fluids, were discharged on the ground until 1986. Testing revealed that the soil and groundwater contain VOCs and PCBs. The chemicals have also been found on neighboring properties and in municipal wells. The contamination threatens the drinking water supplies of about 2,000 local residents.

Response actions included excavating and storing contaminated soil (1987–1989) closing contaminated wells (1989), drilling new wells (1989), and capping contaminated areas. A detailed site investigation is planned.

The response actions were conducted by the Potter Company.

Source: PNPLS-MS.

MISSISSIPPI

Texas Eastern Kosciusko Compressor Station
Attala County
This compressor station for the Texas Eastern Pipeline System near Kosciusko was built in the 1950s. PCB-containing fluids were used in the high-speed turbine engines until 1979. Various PCB-contaminated wastes and solutions released during operations and maintenance were discharged on-site. In 1987 testing by the EPA revealed that the soil and groundwater contain PCBs. PCBs were also detected in nearby streams, sediments, and fish populations. The contamination threatens recreational Little Conehoma Creek, Conehoma Creek, and Yockanookany River. Drinking water supplies are not threatened.

A detailed site investigation is planned.

Source: PNPLS-MS.

MISSOURI

Bee Cee Manufacturing Company
Dunklin County
This company manufactured aluminum windows and doors at this site in Malden from 1964 to 1983. Contaminated wastewater was discharged on the ground. Testing in 1981 showed that the groundwater and soil contain anomalous levels of chromium and aluminum. Nearby agricultural wells are also contaminated. The pollutants threaten the domestic water sources of about 20,000 residents, including the city of Malden.

Response actions included removing the contaminated soil (1992). Site investigations in progress will recommend a final remediation plan by 1997.

Sources: PNPLS-MO; Malden Public Library, 113 North Madison Street, Malden, MO, 63863.

Big River Mine Tailings
St. Francois County
Six hundred acres of mine tailings from nearby lead and zinc mining were piled on this site near Desloge from 1929 to 1958. Rains and flooding have caused tailings material to slide into recreational Big River. Testing in 1981–1982 showed that the tailings piles are rich in lead, cadmium, and zinc and have contaminated the river water, sediments, and aquatic life in Big River. Particulate heavy metals in the air also threaten local residents. About 24,000 people live within 4 miles of the site.

Response actions included stabilizing the tailings piles, controlling runoff, and capping. Ongoing site investigations will recommend a final remediation plan by 1997.

An AOC (1994) between the State of Missouri and Doe Run Mining Company requires the PRP to finance the site investigations and clean-up work.

Sources: PNPLS-MO; Desloge Public Library, 209 North Desloge Drive, Desloge, MO, 50613.

Conservation Chemical Company
Jackson County
Conservation Chemical Company operated this chemical storage and processing facility in Kansas City from 1960 to 1980. Waste solutions and sludges, including cyanide and metal hydroxides, were discharged in basins and lagoons on the property. Pesticides, arsenic, and phosphorus were also processed and stored. Testing in the early 1980s showed that the soil and groundwater contain heavy metals, cyanide, phenols, dioxins, PCBs, and VOCs. The contamination has penetrated the Missouri River and Little Blue River. The contamination plume has moved off-site and threatens the water supplies of about 150 nearby residents and the city of Independence.

In 1987 the EPA recommended removal of debris, a landfill cap, groundwater extraction and treatment, and long-term groundwater monitoring. The groundwater system will operate from 1990 to 2020.

Lawsuits brought by the EPA against the PRPs (1982–1985) resulted in agreements that require the PRPs to finance certain site investigations and clean-up work.

Sources: PNPLS-MO; Mid-Continent Public Library, 317 West Highway 24, Independence, MO, 64050.

Ellisville Site
St. Louis County
A waste oil recycling company operated on this site near Ellisville from the 1960s to the 1970s. Various waste oils, oil by-products, and other chemical wastes were stored in drums or buried in pits on the property. Testing has shown that the soil and groundwater are contaminated with dioxin and VOCs. The contamination threatens Caulks Creek, the Missouri River, and the drinking water supplies for about 5,000 people within 4 miles of the site,.

Response actions included excavating and removing drummed waste, fencing, constructing runoff controls, and removing contaminated soil (1981–1984). In 1991 the EPA recommended additional soil excavations and off-site incineration. Soil treatment is in progress.

Sources: PNPLS-MO; EPA Information Trailer, I-44, Lewis Exit, Times Beach, MO, 63025.

MISSOURI

Fulbright Landfill

Greene County

Two former landfills on this site near Springfield operated from 1962 to 1974. They received a variety of municipal and industrial wastes, including cyanide, paint sludges, metal-plating solutions, pesticides, and waste oil. Testing revealed that the groundwater and soil contain VOCs and heavy metals, including chromium and cyanide. The contamination threatens the water source for about 10,000 nearby residents as well as agricultural wells, the recreational Lake McDaniel, and Little Sac River.

Response actions included removing solid waste and drums of waste (1990–1992). Groundwater and surface water monitoring was started in 1992 and is continuing. Soil and groundwater treatments may be considered later.

Consent Orders (1986) and Consent Decrees (1990) between the EPA and three PRPs require the companies to finance the site investigations and remediation work.

Sources: PNPLS-MO; Greene County Library, 397 East Central, Springfield, MO, 65801.

Kem-Pest Laboratories

Cape Girardeau County

This pesticide-herbicide laboratory near Cape Girardeau operated from 1965 to 1975. Waste solutions were discharged into a storage lagoon on-site. The lagoon was closed in 1981 by filling it with clay. Testing in 1984 and 1989 showed that the shallow groundwater and soil contain VOCs and the pesticides heptachlor, chlordane, endrin, aldrin, and dieldrin. The contamination threatens the drinking water of about 63,000 residents as well as the Mississippi River.

Response actions included site investigations from 1984 to 1989. Contaminated soil was removed from 1988 to 1993. The EPA ruled that no treatment is necessary for the groundwater, although it will be monitored.

The EPA has identified PRPs who are funding the site investigations and remediation work.

Sources: PNPLS-MO; Cape Girardeau Public Library, 711 North Clark Street, Cape Girardeau, MO, 63701.

Lake City Army Ammunition Plant
Jackson County
This 4,000-acre U.S. Army base near Independence has stored and manufactured ammunition and weapons since 1941. Various hazardous wastes, including oils, solvents, explosives, and heavy metals, were buried on the site. Inspections have shown that the groundwater and soil contain high levels of VOCs, lead, chromium, arsenic, and explosives derivatives. The contamination threatens the drinking water of about 200 local residents as well as the Little Blue and Missouri Rivers.

Response actions included installing air-strippers on water wells on the base. Site investigations from 1987 to 1992 recommended groundwater monitoring and possibly soil treatment. A final decision is pending.

Sources: PNPLS-MO; Mid-Continent Public Library, Blue Springs, MO; Lake City Ammunition Plant, Independence, MO, 64050.

Lee Chemical
Clay County
Lee Chemical Company packaged chemicals on this site near Liberty from 1966 to 1974. Drummed chemicals were left on the site when the company went out of business. Testing in 1980 showed that the groundwater and soil are contaminated with trichloroethylene. City wells also contained TCE. The contamination threatens agricultural wells, Shoal Creek, the Missouri River, and the water supplies for about 25,000 people.

Drummed waste chemicals were removed in 1977. Contaminated wells were shut down, and the buildings on the site demolished (1982–1983). New city wells were drilled in 1982. In 1990 the EPA recommended soil flushing and a groundwater extraction and treatment system. Construction was completed in 1994.

The city of Liberty bought the property from Lee Chemical and is the PRP. A 1992 Consent Order requires the city to finance the site investigations and remediation work.

Sources: PNPLS-MO; Liberty Public Library, 1000 South Kent Street, Liberty, MO, 64048.

Minker/Stout/Romaine Creek
Jefferson County
These sites were built up with dirt fill taken from a former horse arena near Imperial that was sprayed with dioxin-contaminated oil in the 1970s. Some of the fill has subsequently eroded into Romaine Creek. About 500 people live within 1 mile of the site, and the dioxin has penetrated Romaine Creek.

Response actions included excavating contaminated soil from residences in the Minker, Sullins, and Cashel neighborhoods. Eleven families were permanently relocated. The EPA has recommended the off-site incineration of the stored contaminated soil and creek sediments. This should be completed by 1997.

A 1990 Consent Decree requires the PRPs to undertake the remediation plan.

Sources: PNPLS-MO; EPA Information Trailer, I-44, Lewis Exit, Times Beach, MO, 63025.

Missouri Electric Works
Cape Girardeau County
Electric motors and transformers were serviced and reconditioned on this site in Cape Girardeau from 1954 to 1992. Waste transformer oil and other solutions were often spilled, leaked, or discharged on the ground. In 1987 the EPA identified PCBs, VOCs, and chlorinated hydrocarbons in the soil and groundwater. The contamination threatens agricultural wells, Cape La Croix Creek, the Mississippi River, and the drinking water sources for about 40,000 residents.

Response actions included constructing runoff controls. In 1990 the EPA recommended the incineration of contaminated soil, a groundwater extraction and air-stripping/carbon adsorption system, and long-term groundwater monitoring.

The EPA has identified over 100 PRPs who have funded the site investigations. A 1991 Consent Decree with 15 PRPs requires these companies to undertake the remediation work.

Sources: PNPLS-MO; Cape Girardeau Public Library, 711 North Clark Street, Cape Girardeau, MO, 63701.

Oronogo-Duenweg Mining Belt
Jasper County
This area of past mining activity near Joplin is part of the Tri-State Mining District. Lead and zinc were mined from 1848 to the 1960s. At least 10 million tons of waste rock and mine tailings are stockpiled on the 64,000-acre site. Smelting operations have contaminated a large area of soil through airborne transmission of heavy metal dust. Testing from 1977 to 1994 has shown that the soil and groundwater contain excessive levels of lead, zinc, and cadmium. The

contamination threatens the water source for about 1,500 people within 3 miles of the site.

Response actions included connecting affected residents to clean water supplies. Contaminated soils were removed from a number of residences and businesses. High-lead soils in residences near the old smelter will be excavated in 1997. Site investigations are expected to recommend a final remediation plan by 1997.

An AOC (1991) and UAO (1994) with PRPs require the companies to undertake site investigations and clean-up actions.

Sources: PNPLS-MO; Webb City Public Library, 101 South Liberty Street, Webb City, MO, 64870; Joplin Public Library, 300 Main Street, Joplin, MO, 64801.

Quality Plating

Scott County

This metal-plating company near Sikeston operated from 1978 to 1983. Wastewater and spent acidic and metal-plating solutions were discharged into an unlined lagoon on the property. Testing by the state revealed that the groundwater and soils contain heavy metals such as lead, chromium, and silver. The local groundwater is used by about 130 residents within 1 mile of the site.

Response actions included removing contaminated soil in 1992. EPA investigations from 1991 to 1995 recommended a final remediation plan of groundwater extraction and treatment.

Quality Plating had a history of effluent violations prior to going out of business. The clean up is being funded by the State of Missouri and the EPA.

Sources: PNPLS-MO; Sikeston Public Library, 221 North Kings Highway, Sikeston, MO, 63801.

Shenandoah Stables

Lincoln County

This horse arena near Moscow Mills became contaminated in 1971 when the grounds were sprayed with dioxin-contaminated oil for dust control. The toxic soil was removed and used as fill material on a highway construction project and in a wetland. Testing in 1982 showed that the soil in the arena and wetland is contaminated with dioxin. The contamination threatens about 100 people who live within 1 mile of the site.

Response actions included closing the arena in 1988. In 1988 the EPA recommended excavating and storing the contaminated soil in a secured area on-site. This was completed in 1990. The material will be transported to a thermal treatment plant in Times Beach.

The PRPs have funded the site investigations and remediation work.

Sources: PNPLS-MO; Moscow Mills City Hall, 500 Highway MM, Moscow Mills, MO, 63362.

Solid State Circuits

Greene County

Solid State Circuits Incorporated manufactured circuit boards on this site in Republic. VOCs found in a municipal well led to the discovery of additional contamination on the Solid State Circuits property. Groundwater and soil contain VOCs, including tetrachloroethylene and chloroform. Plating wastes and solvents were stored in a sump that was later buried. The contamination threatens recreational Schuyler Creek and the drinking water of about 6,000 residents.

Response actions included fencing, removing contaminated soil, and installing monitoring wells (1984–1985). Detailed site investigations by the EPA recommended groundwater extraction and air-stripping and long-term groundwater monitoring. The system became operational in 1994 and will run until 2034.

A Consent Decree (1991) with Solid State Circuits requires the company to fund the remediation work.

Sources: PNPLS-MO; Greene County Library, 393 East Central, Springfield, MO, 65801.

St. Louis Airport

St. Louis County

This site near St. Louis consists of three areas that were used to store radioactive waste from 1947 to the 1960s. The site was later developed and includes part of the St. Louis airport. Testing has shown high radon-222 concentrations in the air. Soil and groundwater contain uranium, thorium, and radium. The contamination threatens the health of about 36,000 people within 3 miles of the site.

Response actions included fencing and the excavation and storage of contaminated soil (1984–1986). Additional soil was stockpiled in 1995. Detailed site investigations (1990–1996) will recommend a final remediation plan in 1997–1998.

Sources: PNPLS-MO; St. Louis Public Library, 1301 Olive Street, St. Louis, MO, 63102.

Syntex Facility

Lawrence County

Operations on this site near Verona consisted of manufacturing vitamins, animal feed, hexachlorophene, and dioxin. Liquid wastes containing the by-product dioxin were discharged into pits and lagoons on the property. Testing has shown that the groundwater and soil contain low levels of dioxin. The contamination has penetrated the Spring River. About 700 people live within 3 miles of the site.

Response actions included excavating and incinerating the contaminated soil off-site, backfilling, and building a clay cap. Equipment on the property will be decontaminated. In 1993 it was recommended only that the groundwater be monitored for 2 years.

Syntex Corporation signed a Consent Order (1982) and Consent Agreement (1983) that require the company to fund the clean-up work.

Sources: PNPLS-MO; Varon Elementary School, 1011 Ella, Verona, MO, 65769.

Times Beach Site
St. Louis County
The road system in the town of Times Beach was sprayed with dioxin-contaminated waste oil in the 1970s. The Meramec River flooded, further spreading the contamination. All residents and businesses were permanently relocated in 1983. Roadways and soils are contaminated with dioxin, as are fish and sediments in the Meramec River. By 1986 the town was deserted and became the property of the State of Missouri. About 4,000 people live within 2 miles of the former town site.

Response actions included relocating all the town's residents and excavating and storing contaminated soil. Fencing and road blockades restrict the site, which is guarded by 24-hour security staff. A temporary incineration unit should be in operation by 1997 and will process contaminated soil from Times Beach and other NPL sites in Missouri.

A 1990 Consent Decree with PRPs requires the companies to fund the remediation work and site restoration.

Sources: PNPLS-MO; EPA Information Trailer, I-44, Lewis Exit, Times Beach, MO, 63025.

Valley Park
St. Louis County
This area of groundwater contamination in Valley Park was first noticed in 1982. City and private wells were found to contain various VOCs, including tetrachloroethylene. Probable point sources are the various industries in the town. Groundwater and soil in these areas are contaminated with VOCs. About 3,000 people use the local groundwater for drinking purposes.

The city of Valley Park installed air-strippers at its water treatment plant (1986). In 1989 the city was connected to the county water supply system. VOC-contaminated soil was removed from some businesses in 1990. Soil vapor extraction, groundwater extraction and air-stripping, and groundwater monitoring was selected for the Wainwright Industries property. These remedies should be operational by 1997. Other studies are continuing.

A number of PRPs have been identified by the EPA. Some have conducted site investigations and remediation on their properties.

Sources: PNPLS-MO; Valley Park Library, 320 Benton Street, Valley Park, MO, 63088.

Weldon Spring Ordnance Works
St. Charles County
This former U.S. Army base near St. Charles operated from 1941 to 1944 as an explosives manufacturing plant. It is now used as an Army Reserve training area. Various explosive-laden wastewaters were stored in lagoons on the property. In 1987 lead, explosives (TNT and DNT) were detected in the soil and groundwater. The contamination threatens about 75,000 people within 3 miles of the site who depend on local sources for drinking water. The Missouri and Mississippi Rivers are also at risk.

Response actions included decontaminating buildings and removing solid waste from the area. Site investigations (1987–1997) are nearing completion and will recommend a final remediation plan.

Sources: PNPLS-MO; Weldon Training Area, 7301 Highway 94 South, St. Charles, MO, 63304.

Weldon Spring Quarry
St. Charles County
This abandoned quarry property near St. Charles was used as a landfill and explosives manufacturing plant by the U.S. Army from 1941 to 1944. Sulfonate-laden wastewaters were frequently discharged onto the ground through leaks and spills. Explosives waste materials were deposited in the quarry. From 1957 to 1965 the plant processed uranium and thorium concentrates, and solid and liquid wastes were buried in open pits. Testing has shown that the groundwater and soil contain TNT, DNT, heavy metals, radionuclides, PAHs, and PCBs. The contamination threatens the Missouri and Mississippi Rivers and the local drinking water supplies of about 75,000 people.

Response actions included decontaminating buildings, removing solid waste and debris, and building runoff controls (1987–1994). Detailed site investigations (1986–1996) will recommend a final remediation plan.

Sources: PNPLS-MO; Spencer Creek Library, 425 Spencer Road, St. Peters, MO, 63376.

Westlake Landfill
St. Louis County
This former quarry near Bridgeton operated as a landfill from the 1960s to 1970s. It received a variety of municipal, industrial, and hazardous wastes, including uranium ore residues from processing operations. Testing by the Nuclear Regulatory Commission in 1981–1982 showed that the soil and groundwater contain radioactive compounds. Nearby properties are also radioactive. The contamination threatens the health of nearby residents.

Detailed site investigations (1994–1996) will recommend a final remediation plan.

Source: PNPLS-MO.

Wheeling Disposal Service Landfill
Andrew County
This former landfill near Amazonia operated from the 1970s to 1986. It received a variety of municipal and industrial wastes, including heavy metals, paint, solvents, pesticides, and leather tanning sludge. Testing revealed that the groundwater and soil contain VOCs, arsenic, chromium, nickel, and lead. Contaminated leachate is seeping from the property. The contamination threatens the local drinking water source for about 5,000 residents.

Detailed site studies recommended capping and long-term groundwater monitoring. Additional remedies will be considered if natural attenuation of the groundwater does not occur.

PRPs signed a Consent Decree (1991) that requires the companies to conduct the site investigations and remediation work.

Sources: PNPLS-MO; Rolling Hills Library, 514 West Main Street, Savannah, MO, 64458.

MONTANA

Anaconda Company Smelter
Deer Lodge County
This mining smelter complex operated for about 100 years. Smelting waste has contaminated about 6,000 acres in the Deer Lodge Valley. On the site are huge volumes of tailings, furnace slags, and flue dusts. Testing has shown that the air, soil, and groundwater contain arsenic, cadmium, lead, copper, zinc, and beryllium. The nearby community of Mill Creek had the highest contamination levels of any town in the area. In addition to streams, wildlife habitats, and wetlands, the contamination threatens the drinking water supply and air quality within the 6,000-acre area.

Response actions included relocating the residents of Mill Creek and destroying the town (1986–1987). The most toxic soil, waste piles, buried beryllium waste, flue dust, and other smelter waste were either removed, stabilized, or capped from 1991 to 1993. Site investigations (1991–1996) will recommend soil and groundwater treatment plans in 1997–1998.

ARCO, the PRP, has signed Administrative Orders on Consent (1984, 1986, 1988, 1990) and a Consent Decree (1988) to undertake the site investigations and remediation work.

Sources: PNPLS-MT; Mansfield Library, University of Montana, Missoula, MT, 59812; Hearst Free Library, 401 Main Street, Anaconda, MT, 59711; U.S. EPA Records Center, Drawer 10096, 301 South Park, Helena, MT, 59626.

Burlington Northern
Park County
This industrial site in the city of Livingston has been in operation since 1883. Main activities included servicing and maintaining train cars and manufacturing train parts. Liquid wastes discharged on-site in holding ponds, sumps, or pits included chlorinated cleaners, petroleum by-products, oils, and degreasers. Fuels frequently leaked from storage tanks or were spilled. Testing revealed that the soil and groundwater are contaminated with VOCs and diesel fuel. Pancreatic cancer rates in the Livingston area are twice the national average. Municipal and private wells have been shut down because of VOC content. The contamination plume threatens the drinking water and health of 7,000 Livingston residents as well as the Yellowstone River.

Response actions included removing underground tanks, sludges, and contaminated soils and supplying alternative water supplies (1988–1994). Soil vapor extraction wells were also installed. Ongoing site investigations should be completed by 1997 and will recommend a final remediation plan.

Sources: PNPLS-MT; Livingston Public Library, 228 Callendar, Livingston, MT, 59047; Montana State Library, Capital Complex, State Repository Program, Helena, MT, 59626; Mansfield Library, University of Montana, Missoula, MT, 59812; Hearst Free Library, 401 Main Street, Anaconda, MT, 59711; U.S. EPA Records Center, Drawer 10096, 301 South Park, Helena, MT, 59626.

MONTANA

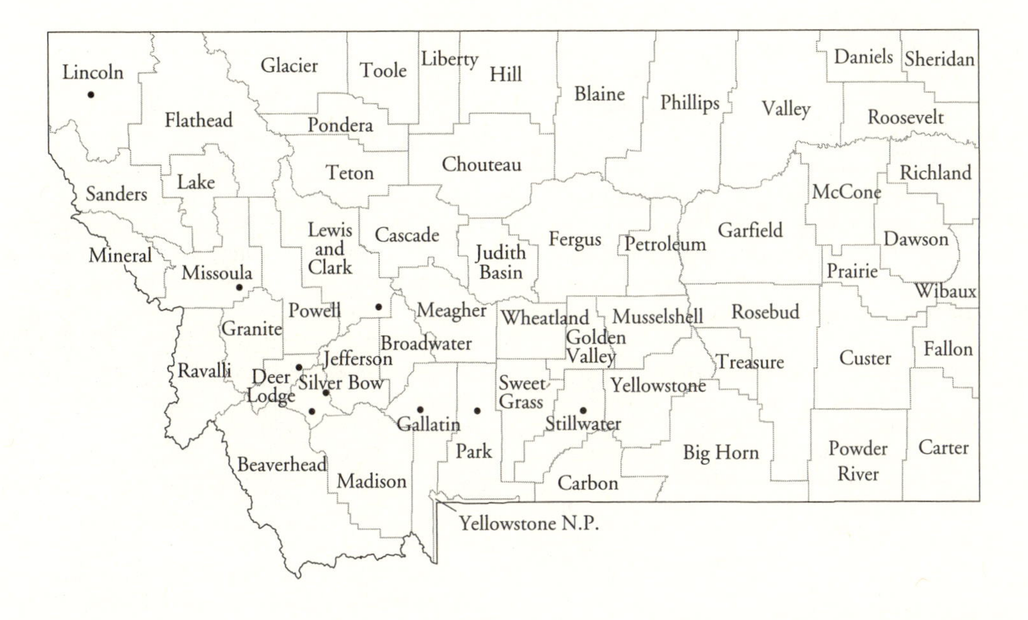

East Helena Site
Lewis and Clark County
ASARCO Mining Company has operated a lead smelter near the town of East Helena for over 100 years. Testing has shown that the air, groundwater, and soil contain arsenic, cadmium, lead, and zinc. Although the groundwater is not used for drinking, it is used for agricultural purposes. Wind has deposited the heavy metals in soil in East Helena and surrounding farmland.

Response actions included removing contaminated soil from playgrounds, schools, homes, and streets (1991–1996). About 400 residential yards need to be replaced. Former holding ponds will be excavated and the contaminated soil smelted (1990–1999). Ongoing site investigations will address groundwater and soil remediation.

Administrative Orders of Consent (1984, 1998, 1991) require ASARCO to fund the site investigations and remediation work.

Sources: PNPLS-MT; Mansfield Library, University of Montana, Missoula, MT, 59812; Hearst Free Library, 401 Main Street, Anaconda, MT, 59711; U.S. EPA Records Center, Drawer 10096, 301 South Park, Helena, MT, 59626.

Idaho Pole Company
Gallatin County
This wood treatment plant near Bozeman began operations in 1946. Wood timbers and utility poles were originally treated with creosote and later with PCP. Spills, leaks, and surface discharge of wastewater were documented during early operations. Testing in 1978 showed that the soil and groundwater contain PCP, PAHs, dioxins, and VOCs. The contamination has penetrated Rocky Creek. About 1,300 people use groundwater pumped from wells within 3 miles of the site.

Response actions included runoff controls and monitoring wells (1978–1983). Sludges have been removed. In 1992 the State of Montana recommended a remediation plan of extraction and biodegradation of groundwater, soil excavation and off-site biological treatment, and long-term groundwater monitoring. Construction should be under way by 1997.

The PRPs are participating in the site investigations and remediation work.

Sources: PNPLS-MT; Mansfield Library, University of Montana, Missoula, MT, 59812; Hearst Free Library, 401 Main Street, Anaconda, MT, 59711; U.S. EPA Records Center, Drawer 10096, 301 South Park, Helena, MT, 59626.

Libby Groundwater Contamination Site
Lincoln County
This site in Libby is the location of a Champion International Corporation lumber and plywood mill that operated from 1946 to 1969. Waste solutions

and tank sludges were occasionally buried on-site. Sampling by the EPA in 1979 revealed that the soil and groundwater contain excessive levels of PCP, PAHs, heavy metals, and dioxins. The contamination threatens Flower Creek, Libby Creek, the Kootenai River, and the drinking water supply of about 11,000 residents of Libby.

Response actions included distributing city water to affected residents (1985). In 1988 the EPA recommended excavation and biotreatment of contaminated soil, capping, bioremediation of extracted groundwater, and long-term monitoring. All construction has been completed, and systems are in operation.

Champion International is the main PRP and has agreed to remediate the site through a 1985 AOC and 1989 CD.

Sources: PNPLS-MT; Lincoln County Offices, 418 Main Avenue, Libby, MT, 59923; Mansfield Library, University of Montana, Missoula, MT, 59812; Hearst Free Library, 401 Main Street, Anaconda, MT, 59711; U.S. EPA Records Center, Drawer 10096, 301 South Park, Helena, MT, 59626.

Milltown Reservoir Sediments

Missoula County

This site consists of 120 million cubic feet of contaminated sediments behind a dam on the Clark Fork River near Milltown. The dam was built in 1906 and trapped the sediments polluted by mining and smelting operations upstream in the Upper Clark Fork Valley. Testing in 1981 found Milltown's water wells to contain high levels of heavy metals, including arsenic and manganese. The sediments in the river also contain copper, zinc, and cadmium. The contamination threatens drinking water supplies, recreational activities on the river, and wildlife habitats. About 100 people live in Milltown.

Response actions included connecting all residents to alternate clean water supplies (1984) and drilling a new municipal well (1985). Detailed site investigations (1989–1997) that will select groundwater, soil, and river sediment treatment plans are nearing completion.

The PRP signed an AOC that requires the company to finance the site investigations and clean-up work.

Sources: PNPLS-MT; Mansfield Library, University of Montana, Missoula, MT, 59812; Hearst Free Library, 401 Main Street, Anaconda, MT, 59711; U.S. EPA Records Center, Drawer 10096, 301 South Park, Helena, MT, 59626.

Montana Pole and Treating Company

Silver Bow County

This former wood treatment plant in Butte operated from 1946 to 1983. Wooden timbers and poles were treated with PCP. Waste solutions were discharged into Silver Bow Creek. Treated poles were stored on the site in open barns. Various PCP-contaminated wastes and oils and contaminated equipment were left on the property. Testing has shown that the groundwater and soil contain high levels of PCP, dioxins, furans, VOCs, and heavy metals. The contamination threatens Silver Bow Creek and the water supplies of local residents.

Response actions included excavating contaminated soils, fencing, installing an interim groundwater pump-and-treat system, and recovering free-product oil (1988–1992). In 1993 the EPA recommended a plan of excavation and bioremediation of contaminated soil and groundwater extraction and bioremediation. Construction should be completed by 1997–1998.

Three PRPs have been identified, including ARCO and Burlington Northern Rail. The EPA sued ARCO for previous costs in 1991, a case that has yet to be resolved.

Sources: PNPLS-MT; Montana School of Mines and Technology, Butte, MT, 59701; Mansfield Library, University of Montana, Missoula, MT, 59812; Hearst Free Library, 401 Main Street, Anaconda, MT, 59711; U.S. EPA Records Center, Drawer 10096, 301 South Park, Helena, MT, 59626.

Mouat Industries

Stillwater County

Mouat Industries operated this chrome ore processing facility near Columbus from 1957 to 1963. Waste solutions were apparently discharged on the property. Chromium oxides were visible on the surface in 1976. Testing revealed that the groundwater and soil contain excessive levels of hexavalent chromium. The contamination threatens the local drinking water supply for about 300 residents, agricultural water sources, and the Yellowstone River.

Response actions included fencing and removing contaminated soil (1990–1994). The EPA will recommend a final remediation plan when ongoing site investigations are completed in 1997.

PRPs identified by the EPA conducted the initial soil removal actions.

Sources: PNPLS-MT; Mansfield Library, University of Montana, Missoula, MT, 59812; Hearst Free Library, 401 Main Street, Anaconda, MT, 59711; U.S. EPA Records Center, Drawer 10096, 301 South Park, Helena, MT, 59626.

Silver Bow Creek/Butte

Silver Bow County

This area of contamination covers 140 miles of riverbank habitat along Silver Bow Creek and the Clark

Fork River. Over 100 years of mining, milling, and smelting activity have polluted the river. Tailings piles along the rivers have washed downstream, and fish kills have been documented. Air, groundwater, soil, river water, and river sediments along these routes contain excessive amounts of heavy metals, including lead and mercury.

Response actions included excavating and stabilizing mine dumps, cleaning up contaminated yards, hauling away solid waste and debris, pumping and treating acid mine drainage, excavating and treating holding pond soils, building runoff controls, chemical fixation of submerged tailings, and interim groundwater extraction and treatment programs (1988–1996). A number of studies that are addressing other aspects of the clean up are in progress.

A series of AOCs and UAOs has required PRPs, including Anaconda and ARCO, to conduct the site investigations and remediation work.

Sources: PNPLS-MT; Mansfield Library, University of Montana, Missoula, MT, 59812; Hearst Free Library, 401 Main Street, Anaconda, MT, 59711; U.S. EPA Records Center, Drawer 10096, 301 South Park, Helena, MT, 59626.

NEBRASKA

Bruno Association
Butler County
This area of groundwater contamination in Bruno was detected in 1986. Former grain storage operations on the site had applied VOCs to the grains as fumigants. Two municipal wells, groundwater, and local soils contain VOCs (including chloroform and carbon tetrachloride). The contamination threatens the drinking water supplies for the town of Bruno.

Affected residents were supplied with alternative water supplies (1989–1990) until new city wells could be installed. A site investigation is planned.

The EPA has identified four PRPs.
Source: PNPLS-NE.

Cleburn Street Well
Hall County
This municipal well for the city of Grand Island was found to be contaminated with tetrachloroethylene in 1986 and was later shut down. Further studies showed that a former solvent company and three laundry operations have contaminated the groundwater and soil with VOCs. The contamination threatens the private wells serving about 1,200 residents and over 300 agricultural wells used to irrigate farmland.

Response actions included the installation of a groundwater extraction system to contain the contamination plume. Detailed site investigations from 1988 to 1996 will recommend final groundwater and soil clean-up strategies.

At least four PRPs have been identified.
Sources: PNPLS-NE; Edith Abbott Memorial Library, 211 North Washington Street, Grand Island, NE, 68801.

Cornhusker Army Ammunition Plant
Hall County
This U.S. Army ammunition plant near Grand Island manufactured ammunition and fertilizers in the 1940s. Waste explosives were detonated in a burn pit; other solid and liquid wastes were buried in landfills on the property. Testing has shown that the groundwater and soil are contaminated with explosives byproducts, lead, chromium, and cadmium. The contamination plume extends at least 3 miles beyond the base and has penetrated over 500 private wells. About 30,000 people live within 4 miles of the site, and numerous agricultural wells are used for irrigation and watering livestock.

Response actions included providing affected residents with bottled water and city water supply hookups (1986). Contaminated soils were removed in 1987–1988. In 1994 the construction of an interim groundwater extraction and treatment system was started. Other studies are in progress.
Sources: PNPLS-NE; Grand Island Public Library, 212 North Washington Street, Grand Island, NE, 68802.

Hastings Groundwater Contamination
Adams and Clay Counties
A number of municipal wells in the city of Hastings were determined to contain VOCs in 1983. Further investigation discovered various industrial or military point sources, including a landfill and ammunitions dump. Operations at the industrial sites included the use of toxic grain fumigants and chlorinated solvents. Testing has delineated a wide area of groundwater and soil contaminated with VOCs, heavy metals, and PAHs. The contamination threatens the drinking water of about 25,000 residents, many agricultural wells, and recreational waterways.

Response actions have included soil vapor extraction, soil excavation, soil incineration, groundwater extraction and treatment, capping, gas monitoring, and groundwater monitoring (1989–1996). Due to the complexity of the plume and the variety of point sources, a number of investigations are still in progress.

NEBRASKA

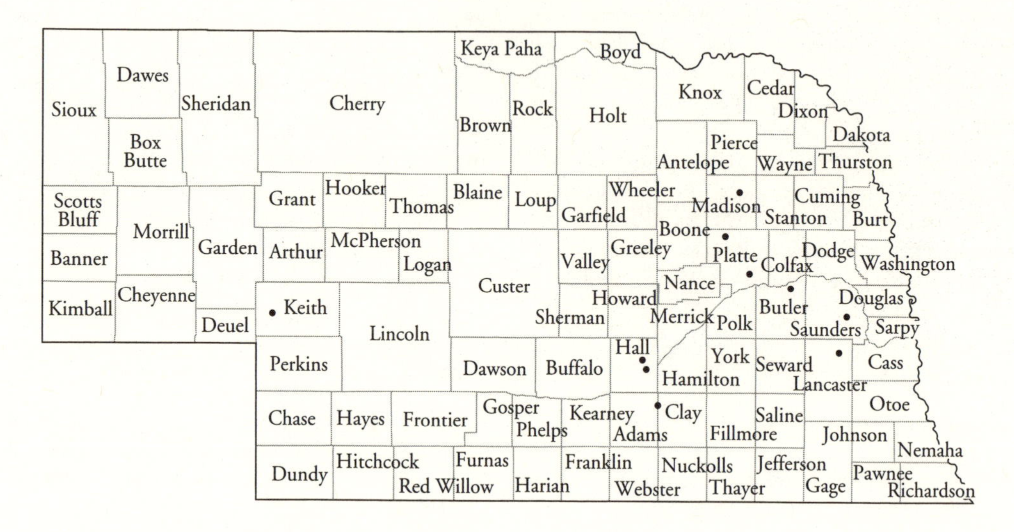

Numerous legal agreements between the State of Nebraska, the Department of Defense, the city of Hastings, and a number of PRPs require the PRPs to fund the site investigations and clean-up work on their own sites.

Sources: PNPLS-NE; Hastings Public Library, Fourth and Denver Streets, Hastings, NE, 68901.

Lindsay Manufacturing Company
Platte County
The Lindsay Company is a galvanizing plant in the town of Lindsay. Waste acidic solutions and heavy metals were discharged into an unlined pond on the property. Testing in 1983 showed that the groundwater and soils contain heavy metals (zinc, chromium, and lead) and VOCs. The contamination threatens numerous agricultural wells and the water supplies of about 3,000 local residents.

Response actions included closing the pond in 1983. Interim groundwater pump-and-treat systems were installed in 1984 and 1989. A site study completed in 1990 recommended a final remediation plan of continuing with groundwater extraction and treatment, soil vapor extraction, and groundwater monitoring. These systems should be operational by 1997.

The PRP, Lindsay Manufacturing Company, has been involved in site restoration since 1983. A 1992 Consent Decree required the company to fund the remediation work.

Sources: PNPLS-NE; Columbus Public Library, 2504 14th Street, Columbus, NE, 68801.

Nebraska Army Ornance Plant
Saunders County
This 17,000-acre military base near Mead manufactured bombs, ammonium nitrate, and other explosives from 1942 to 1956. Three Atlas missile silos were also located on the property. The base has since been subdivided and sold to different individuals and businesses. Testing has shown that the groundwater and soil contain VOCs, PCBs, and a variety of explosives by-products and chemicals. The contamination threatens agricultural wells used for irrigation and watering livestock and private water wells that serve about 500 nearby residents.

Response actions included connecting affected residents to clean water supplies or carbon filtration systems. Various contaminated soil removals were conducted between 1991 and 1995. Ongoing site investigations to be completed in 1997 will recommend a final remediation plan for the remaining soil and contaminated groundwater.

Sources: PNPLS-NE; Ashland Public Library, 207 North 15th Street, Ashland, NE, 68003.

Ogallala Groundwater Contamination
Keith County
Nearly half of the municipal wells in Ogallala were discovered to contain excessive amounts of VOCs in 1987. Further investigation identified two industrial point sources nearby that have manufactured electrical components since the 1960s. The soil and groundwater at these sites contain VOCs, which have spread into surrounding wells. About 5,000 people depend on private and public water wells within 4

miles of the site. The local groundwater is also used for agricultural purposes.

Detailed site investigations were initiated in 1994. A final remediation plan is pending.

PRPs identified by the EPA include TRW Incorporated, American Shizuki Corporation, and Ogallala Electronics.

Sources: PNPLS-NE; Ogallala Public Library, Ogallala, NE, 69153.

Sherwood Medical Corporation
Madison County
This company has manufactured disposable medical supplies at this site in Norfolk since 1962. For 5 years drains in the building discharged into Sherwood Lake. From 1968 to 1974 liquid wastes were discharged into an underground storage tank and leach fields. Sampling by the EPA and State of Nebraska from 1987 to 1989 revealed that local wells contained excessive levels of VOCs. Later tests determined that Sherwood Medical Corporation was the point source. The contamination threatens Sherwood Lake and wells that serve about 6,000 nearby residents.

Response actions included supplying affected residents with bottled water or water filtered through a carbon treatment system. New wells were drilled and the septic tank and settling basin removed. In 1993 the EPA recommended low-temperature thermal treatment of contaminated soil, long-term groundwater monitoring, and complete removal of the underground tanks and pipelines. Construction is expected to begin in 1996–1997.

Sherwood Medical Corporation signed AOCs (1989, 1991) that require the company to undertake the site investigations and remediation work.

Sources: PNPLS-NE; Norfolk Public Library, Norfolk, NE, 69153.

Tenth Street Site
Platte County
This area of groundwater contamination in Columbus was identified in the 1980s. Sampling indicated that the groundwater, soil, and municipal well water contained VOCs. Nearby point sources include a former scrap metal yard and dry cleaning operations. The local groundwater serves the domestic needs of about 20,000 residents.

An EPA site investigation was completed in 1994 and will recommend a final remediation plan.

Sources: PNPLS-NE; Columbus Public Library, 2504 Fourteenth Street, Columbus, NE, 68601.

Waverly Groundwater Contamination Area
Lancaster County
This area of groundwater contamination in the city of Waverly was first detected in 1982. Three municipal wells were found to contain heavy metals, VOCs, nitrates, and sulfates. Further work identified the source area as being a former U.S. Department of Agriculture grain processing plant that operated from 1955 to 1965. Grain was treated with carbon tetrachloride and carbon disulfide (fumigants). Soils contain VOCs. The contamination has polluted nearby wells and threatens the drinking water supply for about 2,000 local residents as well as Salt Creek. The local groundwater is also used for agricultural purposes.

Response actions included shutting down the contaminated wells and drilling new wells. Soil vapor extraction and groundwater extraction/air-stripping systems were installed in 1989. An additional groundwater extraction well was added in 1993. The systems will operate until about 2000.

The USDA is the PRP and has undertaken the site investigations and remediation work.

Sources: PNPLS-NE; City Hall, 10350 North 141st Street, Waverly, NE, 68462.

NEVADA

Carson River Mercury Site
Lyon, Storey, and Churchill Counties
This stretch of the Carson River from Carson City to the Lahontan Reservoir is contaminated from the mercury that was used to process gold and silver from former mining operations near Virginia City. Mercury is found in elevated concentrations in river sediments, soil, and old tailings piles. Sediments in the Stillwater Wildlife Refuge are also contaminated. In addition to affecting fish and wildlife, the mercury threatens the drinking water supplies of about 2,500 local residents.

Response actions included removing tailings piles. An EPA site investigation is under way and should be completed by 1997, including recommendations for a final remediation plan.

Sources: PNPLS-NV; Ormsby Public Library, 900 North Roop Street, Carson City, NV, 89701.

NEVADA

Humboldt
Elko
Washoe
Pershing
Lander
Eureka
Churchill
Storey
White Pine
Lyon
Ormsby
Douglas
Mineral
Nye
Esmeralda
Lincoln
Clark

Auburn Road Landfill

Rockingham County

Located in the town of Londonderry, this landfill is a 200-acre site that received a variety of municipal and industrial wastes from the 1960s to 1980, including spent chemical solutions, tires, and demolition debris. EPA testing in 1986 showed that the soil and groundwater are contaminated with arsenic, VOCs, PCBs, and pesticides. The groundwater migrates downgrade toward 500 homes with private wells within 1 mile of the site. Surface runoff and contaminated groundwater also threaten Cohas Brook and the Merrimack River.

Response actions included connecting affected residents to the city water system (1987). Seven families were temporarily relocated when the EPA excavated hazardous waste from the site in 1987. Fencing and additional waste removal were completed in 1988. In 1989 the EPA recommended a treatment plan of groundwater extraction and treatment using chemical precipitation, air-stripping, and carbon adsorption. Additional studies are in progress to evaluate the arsenic contamination. The property has been capped and revegetated (1995–1996).

Sources: PNPLS-NH; Leach Public Library, 276 Mammoth Road, Londonderry, NH, 03053.

Beede Waste Oil

Rockingham County

This former waste oil recycling and storage center in Plaistow operated from 1926 to 1994. Multiple companies have used this site to process used oils for asphalt production and road dusting. Oils were stored in a lagoon and underground tanks until about 1991. Many of the reprocessed oils contained PCBs or heavy metals. Buried drums of VOC waste were discovered on the property in 1992. Testing has revealed that the groundwater and soils contain VOCs, PCBs, and heavy metals, including copper, zinc, and arsenic, which have penetrated nearby wells and wetlands. The contamination is a threat to local drinking water supplies, Kelley Brook, and associated wetlands.

Response actions included removing drummed waste from the site in 1992. Affected residents have also been connected to clean water supplies. This property was proposed as an NPL site in 1996; a final decision is expected in 1997–1998. More site studies are planned. An interim remedy will be to install a system to recover oil from the groundwater, which should be operational by 1997.

Source: EPA Publication 9320.7-071.

Coakley Landfill

Rockingham County

This site received a variety of municipal and industrial wastes, including incinerator residues, from the Portsmouth area from 1972 to 1985. The soil and groundwater are contaminated with VOCs, phenol, methyl ethyl ketones, and heavy metals (arsenic, chromium, and lead). The contamination has penetrated the water wells of a number of private homes in the area; small businesses, motels, and restaurants are also nearby.

Response actions included connecting affected residents to the municipal water system and groundwater monitoring (1989). The State of New Hampshire, in studies from 1986 to 1990, recommended consolidating contaminated soil and sediment, fencing, capping, recovering and burning landfill gas, installing a groundwater pump-and-treat system, and long-term monitoring. The final design of the systems is expected by 1997, followed by 2 years of construction.

About 60 PRPs have been identified by the EPA. To date 30 parties, through a 1992 Consent Order, have agreed to finance the remediation plan.

Sources: PNPLS-NH; North Hampton Public Library, 235 Atlantic Avenue, North Hampton, NH, 03862.

Dover Municipal Landfill

Stratford County

Located on the edge of the town of Dover, this landfill operated from 1960 to 1980. A variety of municipal and industrial wastes were accepted, including organic solvents, construction debris, wastewater treatment plant sludge, and leather tanning by-products. In 1977 it was determined that organic solvents, VOCs, organic compounds, and heavy metals (arsenic, chromium, and lead) were penetrating the groundwater. Air releases of VOCs were also documented. In addition to about 50 homes and a few business facilities, two important municipal water wells are within 1 mile of the site. Seepage from the site has also contaminated the recreational Cocheco River and neighboring wetlands less than 500 feet away.

Response actions included connecting affected residents to city water (1981). Studies by the State of New Hampshire, the EPA, and the PRPs (1984–1991) recommended capping, leachate and runoff control systems, and groundwater extraction and treatment. The remediation plan should be operational by 1997.

A number of PRPs, through an Administrative Agreement and Consent Decree (1992), have agreed to finance the site investigations and remediation work.

Sources: PNPLS-NH; Dover Public Library, Carnegie Building, 73 Locust Street, Dover, NH, 03820.

Fletcher's Paint Works

Hillsborough County

Fletcher's Paint Works manufactured paints and stains in this commercial/residential area near down-

NEW HAMPSHIRE

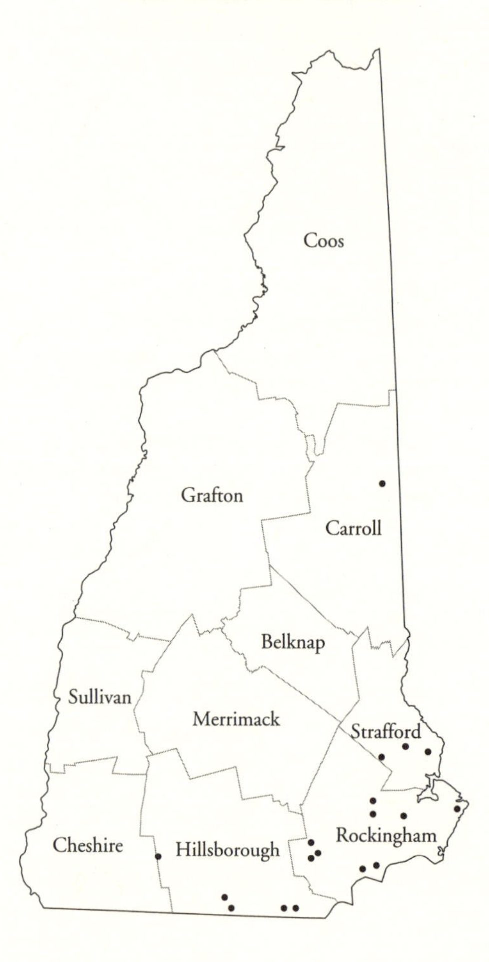

town Milford from 1949 to 1991. Paint chemicals (including naphtha, resins, and mineral spirits) were stored in drums behind the plant or in underground storage tanks. Improper handling and storage and frequent spills and leaks contaminated the soil and groundwater with PCBs, VOCs (benzene and toluene), and heavy metals (nickel and lead). About 12,000 residents, as well as private and commercial water wells, are located within 3 miles of the site. The nearby recreational Souhegan River may also be contaminated.

Response actions included removing drummed waste (1988), temporarily capping contaminated soil areas, and fencing the site (1991). The buildings and on-site materials were removed in 1993. The yards of three nearby homes were detoxified in 1995. A site investigation and feasibility study is nearing completion and will recommend a final remediation plan by 1997.

Sources: PNPLS-NH; Wadleigh Memorial Library, 49 Nashua Street, Milford, NH, 03055.

Kearsarge Metallurgical Corporation
Carroll County
Located near Conway, Kearsarge Metallurgical Corporation manufactured stainless castings at this site from 1964 to 1982. Wastes were disposed of or stored on-site, including a 20-foot-tall pile of rusting drums and metal scrap. The soil and groundwater are contaminated with VOCs (including trichloroethane), chromium, nickel, and copper. About 3,000 people in the area use the local groundwater for drinking. The site is adjacent to Pequawket Pond and threatens more distant municipal well fields.

Response actions included removing drummed waste and installing monitoring wells. The waste pile, a septic tank and its contents, and contaminated soil were removed in 1992. A groundwater pump-and-treat system using a clarifier (for heavy metals) and air-stripping was installed in 1993 and will operate for at least 10 years. Long-term monitoring will continue.

Litigation against the PRP (1983) is pending.

Sources: PNPLS-NH; Conway Public Library, Main Street, Conway, NH, 03813.

Keefe Environmental Services

Rockingham County

This site near the town of Epping operated as a chemical waste storage facility from 1978 to 1981. Various wastes such as paint sludges, heavy metals, acids, oils, and solvents were stored in aboveground tanks and a lined lagoon. Soil and groundwater are contaminated with VOCs (trichloroethane and benzene). The groundwater is used for drinking water by about 2,000 people within 3 miles of the site. The contamination threatens the recreational Piscassic River and its wetland areas.

Response actions included pumping out the lagoon (1981) and removing toxic liquids and storage tanks (1983–1984). The lagoon was drained and excavated in 1984. Site investigations by the State of New Hampshire and the EPA recommended a groundwater treatment system using air-stripping and carbon adsorption. The system became operational in 1993. Groundwater monitoring is continuing.

The owner declared bankruptcy in 1981. About 120 PRPs settled with the EPA through a 1986 Consent Agreement. The EPA sued the nonsettling parties in 1989, and a UAO (1990) and CD (1992) require them to finance the remediation work.

Sources: PNPLS-NH; Harvey-Mitchel Memorial Library, 52 Main Street, Epping, NH, 03042.

Mottolo Pig Farm

Rockingham County

This abandoned pig farm near the town of Raymond was a disposal site for chemical manufacturing waste, including toluene, xylene, and methyl ethyl ketone, from 1975 to 1979. Tests showed that the groundwater and soil are contaminated with VOCs and arsenic. The pollutants have migrated into tributaries of the Exeter River, the drinking water source for the towns of Exeter, Hampton, and Stratham. About 1,600 people within 3 miles of the site depend on the local groundwater for drinking purposes.

Response actions included removing drummed waste (1980) and contaminated soil (1981–1982) and improving erosion controls. Site investigations recommended surface runoff and leachate collection, temporary caps, vacuum extraction, and groundwater monitoring. The collection system was installed in 1992. The groundwater plume is expected to attenuate as the surface systems reduce the volume of contaminants. Fencing was installed in 1992. The vacuum extraction system should be operational by 1997.

K. J. Quinn Company, one of several PRPs, signed an AO (1988) with the EPA that required it to finance the site investigation study.

Sources: PNPLS-NH; Dudley-Tucker Library, 6 Epping Street, Raymond, NH, 03077.

New Hampshire Plating Company

Hillsborough County

Located near Merrimack, this electroplating facility operated from 1961 to 1985. Waste plating solutions were stored on-site in unlined lagoons. Soil and groundwater are contaminated with VOCs, cyanide, and heavy metals (including cadmium). About 40,000 people depend on water from wells within 4 miles of the site. The contaminated groundwater flows toward the recreational Merrimack River and Horseshoe Pond, less than 1,000 feet away.

Response actions included chemically neutralizing the lagoon (1987). Contaminated sludge and soil was consolidated on-site and the site fenced in 1990. Buildings were razed and a temporary cap installed in 1994. The site investigation is expected to be completed by 1997.

New Hampshire Plating Company was sued by the EPA and the State of New Hampshire in 1982 for violating state and federal regulations. The court ordered the company to finance site studies and become compliant, which forced its bankruptcy.

Sources: PNPLS-NH; Merrimack Public Library, 470 Daniel Webster Highway, Merrimack, NH, 03054.

Ottata & Goss/Kingston

Rockingham County

This site near Kingston operated as a drum refurbishing plant from the 1950s to 1979. Drums were washed in a caustic solution that was discharged in a dry well near the South Brook waterway. Toxic runoff and seepage killed fish in South Brook and Country Pond in the 1970s. Sludges and incinerator residues were also processed on-site. Tests showed that the soil and groundwater contain high levels of VOCs, PCBs, acidic compounds, and heavy metals. About 5,000 people live within 3 miles of the site, which is also near the recreational Powwow River and Country Pond. PCBs have been detected in South Brook and nearby wetlands.

Response actions included fencing and removing drummed waste, contaminated soil, and flammable sludge (1980). Contaminated soils were excavated and treated by low temperature thermal aeration

(1989). Groundwater and additional soil treatment systems are expected to be operational by 1997.

A civil action suit by the Justice Department in 1980 found the PRPs liable for the contamination. A group of PRPs settled with the EPA and contributed to the site studies. The remaining parties, through a CD (1993) with the EPA, will finance the rest of the remediation work.

Sources: PNPLS-NH; Kingston Public Library, Main Street, Kingston, NH, 03848.

Pease Air Force Base
Rockingham County
Pease Air Force Base, located near Portsmouth and Newington, was operational from the 1950s to 1991. Various wastes such as fuel, oils, solvents, and lubricants from aircraft maintenance activities were disposed of on-site. The soil and groundwater are contaminated with VOCs (trichloroethylene), PAHs, pesticides, fuel derivatives, and heavy metals.

Response actions included removing drummed waste and contaminated soil and installing groundwater pump-and-treat systems. A soil vapor extraction system is expected to be operational by 1997. Other contaminated soils will be excavated and removed, followed by backfilling and capping. Some source-specific investigations are still under way. Additional groundwater extraction and treatment systems should be installed by 1997.

The U.S. Air Force will finance the site investigations and clean-up work.

Sources: PNPLS-NH; Pease Air Force Base, 61 International Drive, Pease Air Force Base, NH, 03803.

Savage Municipal Water Supply
Hillsborough County
This well field is located west of Milford and supplied about half of the town's water from 1961 to 1983. A state sampling program in 1983 found the soil and groundwater to be contaminated with VOCs (tetrachloroethylene, trichloroethylene, and vinyl chloride) and heavy metals (chromium, nickel, and lead). The well was later closed. Agricultural and industrial sites nearby have been identified as sources of contamination.

Response actions included supplying bottled water and city water hook-ups to affected residents (1993). Site studies concluded in 1991 that the best treatment would be extraction and treatment of the groundwater, along with natural attenuation of the contamination plume. Construction is expected to be completed in 1997–1998.

Hitchiner Manufacturing Company, a wire and cable company, bought the well site from the city of Milford. A CO (1987) and two CDs (1994) with the PRPs require them to finance the site studies and remediation work.

Sources: PNPLS-NH; Wadleigh Memorial Library, 21 Nashua Street, Milford, NH, 03055.

Somersworth Sanitary Landfill
Strafford County
This landfill site near the town of Somersworth operated from the mid-1930s to 1981. A variety of municipal and industrial wastes were received, including sludges, acids, dyes, pharmaceutical wastes, potash, and solvents. Testing in 1981 revealed that the soil and groundwater contain high levels of arsenic, chromium, lead, PAHs, VOCs, and various inorganic compounds. The immediate vicinity is residential, and part of the site has been developed into a park. The contamination threatens private wells in the area. Nearby streams carry contamination into the Salmon Falls River, the water source for Somersworth and Berwick, Maine.

Response actions included a site investigation and feasibility study. Conclusions in 1994 called for constructing a groundwater chemical treatment wall, capping, and possibly installing groundwater pump-and-treat systems.

Twenty PRPs signed a Consent Decree with the EPA (1995) that requires them to fund the site investigation studies and remediation work.

Sources: PNPLS-NH; Somersworth Public Library, 27 Main Street, Somersworth, NH, 03878.

South Municipal Water Supply Well
Hillsborough County
This city well served as the main source of water for Peterborough (which has about 5,000 residents) from 1952 to 1982. The state shut down the well in 1982 after discovering that the water contained VOCs (chloroform, benzene, and toluene) and PCBs. The source was identified as the New Hampshire Ball Bearings facility, less than 1,500 feet away. Wastewater and solvents were periodically dumped on the property from 1946 to the 1970s. About 100 homes are located within 1 mile of the New Hampshire Ball Bearings site. Nearby wetlands and the recreational Contoocook River are also threatened.

A clean-up plan consisting of a groundwater pump-and-treat system, soil vacuum extraction, and excavation and removal of PCB-laden soils was recommended in 1989. Construction of these remedies was completed by 1994.

A CD (1989) and UAO (1990) between the EPA and New Hampshire Ball Bearings Company required the company to fund the remediation work.

Sources: PNPLS-NH; Peterborough Town Library, Maine and Concord Streets, Peterborough, NH, 03458.

Sylvester
Hillsborough County
Originally a sand pit, this site was used as an illegal hazardous waste dump from the 1960s to 1979. A variety of waste, including sludges and liquid chemicals, were buried or discharged on-site. The soil and groundwater are contaminated with VOCs and heavy metals. About 1,000 residents and five private wells are located within 2,000 feet of the site. The landfill is beside Lyle Reed Brook, which enters the Nashuya River, another drinking water source.

Response actions included removing drums of liquid waste (1979) and fencing (1980). A temporary groundwater recirculation system was installed in 1981–1982. A slurry wall surrounds the site, which was also capped. Treatment through groundwater extraction and air-stripping and soil vacuum extraction ended in 1996. Monitoring will continue for at least 5 more years.

The EPA has signed several CDs with PRPs to recover costs for the site investigation studies and remediation work.

Sources: PNPLS-NH; New Hampshire Department of Environmental Protection Office, 6 Hoyden Drive, Concord, NH, 03301.

Tibbetts Road
Strafford County
This site was used from 1944 to 1958 as a staging area for drums of liquid waste, including kerosene, oil, grease, antifreeze, and other solvents. Most of the drums rusted, leaking toxic solutions into the ground. Tests in 1984 showed that the soil and groundwater have high levels of VOCs, solvents, arsenic, PCBs, and dioxin. About 2,100 people within 3 miles of the site depend on the local groundwater for drinking.

Response actions included removing the leaking drums (1984). Contaminated soils were removed from 1985 to 1988. Affected residents were connected to clean water supplies in 1987. EPA studies in the early 1990s recommended vacuum extraction and groundwater pumping and treatment, which will be installed in 1997–1998.

The EPA has identified a number of PRPs. The most viable, Ford Motor Company, signed a CD in 1994 that requires the company to finance the remediation work, including constructing a new water supply system for affected residents.

Sources: PNPLS-NH; Barrington Public Library, Star Route, Barrington, NH, 03825.

Tinkham Garage
Rockingham County
An automotive service garage operated at this site in Londonderry from 1978 to 1979. Oily wastewater from washing activities was discharged on-site. Soil and groundwater contain high levels of VOCs. Contamination in nearby wells was detected in 1983. About 500 people live on or adjacent to the area of contamination.

Response actions included supplying city water to affected residents (1983). Site investigations from 1987 to 1994 recommended soil vacuum extraction, soil excavation, groundwater pumping and treatment, and groundwater monitoring. Soil remediation was completed in 1995.

Twenty-three PRPs agreed to finance the site investigation and remediation work.

Sources: PNPLS-NH; Leach Library, 276 Mammoth Road, Londonderry, NH, 03053.

Town Garage/Radio Beacon
Rockingham County
Tests on residential and commercial wells in this area of Londonderry in 1984 showed that some were contaminated with VOCs, including dichloroethane. Subsequent investigations by the State of New Hampshire and the EPA have failed to find the source of the contamination. About 7,500 people are supplied by wells within 3 miles of the Radio Beacon site.

Response actions included connecting affected residents to the city water supply in the late 1980s. An EPA study in 1992 recommended natural attenuation of the contamination plume and continued groundwater monitoring.

Sources: PNPLS-NH; Leach Library, 276 Mammoth Road, Londonderry, NH, 03053.

A. O. Plymer
Sussex County
This manufacturing company has produced resins, paper coatings, and polymers at this site near Sparta since the 1960s. Solvents and waste solutions were spilled, leaked, or discharged on the property. Testing in the late 1970s revealed that nearby wells contained VOCs. Additional tests showed that the soil and groundwater contain VOCs and phenols. The contamination threatens private and public water wells that serve about 1,000 people as well as wetlands and the Wallkill River.

Response actions included removing drummed waste and contaminated soil (1980, 1981). Monitoring wells were installed in 1982. Site investigations were conducted from 1986 to 1991. The EPA recommended soil vapor extraction and a groundwater extraction and carbon adsorption treatment system. Construction should be under way by 1997.

Sources: PNPLS-NJ; Sparta Public Library, 22 Woodport Road, Sparta, NJ, 07871.

American Cyanamid
Somerset County
This chemical manufacturing company has produced pharmaceuticals, dyes, elastics, and pigments at this site in Bound Brook since the 1930s. Various wastes such as VOC solvents and sludges were discharged into lagoons on the property. Testing has shown that the groundwater and soil contain VOCs, heavy metals, and cyanide. The contamination threatens the Raritan River and the water supplies for about 14,000 people.

Response actions included the installation of a groundwater treatment plant that will operate until 1998. Contaminated soils were removed from 1992 to 1995. Additional studies are in progress.

AOCs (1981, 1988) require the PRPs to conduct the site investigations and clean-up work.

Sources: PNPLS-NJ; Somerset County Public Library, North Bridge Street, Bridgewater, NJ, 08807.

Asbestos Dump
Morris County
Operations from 1927 to the 1950s at a former asbestos products manufacturing company in Millington left behind asbestos waste piles along the Passaic River. The groundwater and soil contain heavy metals, VOCs, phthalates, asbestos, and phenols. The contamination threatens about 700 nearby residents. Asbestos fibers from exposed asbestos waste at the surface could be carried in the air. The site also drains into the Passaic River and the Great Swamp National Wildlife Refuge.

Response actions included constructing runoff controls along the Passaic River (1983), fencing, and capping exposed asbestos waste (1990). In 1988 the EPA recommended additional capping and runoff controls, stabilizing asbestos waste, in-site soil stabilization, and long-term monitoring. These remedies are currently being designed. Other studies are in progress.

PRPs funded the site investigations and clean-up work until they declared bankruptcy in 1990.

Sources: PNPLS-NJ; Passaic Township Library, 91 Central Avenue, Stirling, NJ, 07980.

Bog Creek Farm
Monmouth County
This former landfill near Farmingdale operated in the 1970s. It received a variety of waste, including sludges, organic solvents, heavy metals, and paints. Testing has shown that the groundwater and soil contain VOCs, phthalates, pesticides, and heavy metals. The compounds have penetrated nearby ponds and streams, including Squankum Brook. The contamination threatens the private wells in the area, which supply about 1,500 people.

Response actions including pumping out liquid wastes (1984). In 1988 the EPA recommended erosion controls, excavating contaminated soils, capping, fencing, and monitoring. Contaminated sediments from Squankum Brook were incinerated in 1990. A groundwater pump-and-treat system will be operated from 1995 to 2005.

Sources: PNPLS-NJ; Howell Township Public Library, Preventorium Road, Howell, NJ, 07731.

Brick Township Landfill
Ocean City
This former landfill operated from the 1940s until 1979. It received a variety of municipal and industrial wastes, including solids, sludges, and chemical wastes. Testing in 1987 showed that the groundwater and soil contain excessive levels of cadmium, VOCs, and pesticides. The contamination threatens water wells within 3 miles of the site, which serve about 60,000 residents.

Response actions included cleaning up surface debris and installing gas vents (1982). A site investigation conducted from 1990 to 1992 recommended capping, fencing, gas venting, erosion controls, and groundwater monitoring.

Consent Orders (1982, 1985) require the PRPs to conduct the site investigations and clean-up activities.

Sources: PNPLS-NJ; Brick Township Public Library, 401 Chambers Bridge Road, Brick Town, NJ, 08723.

NEW JERSEY

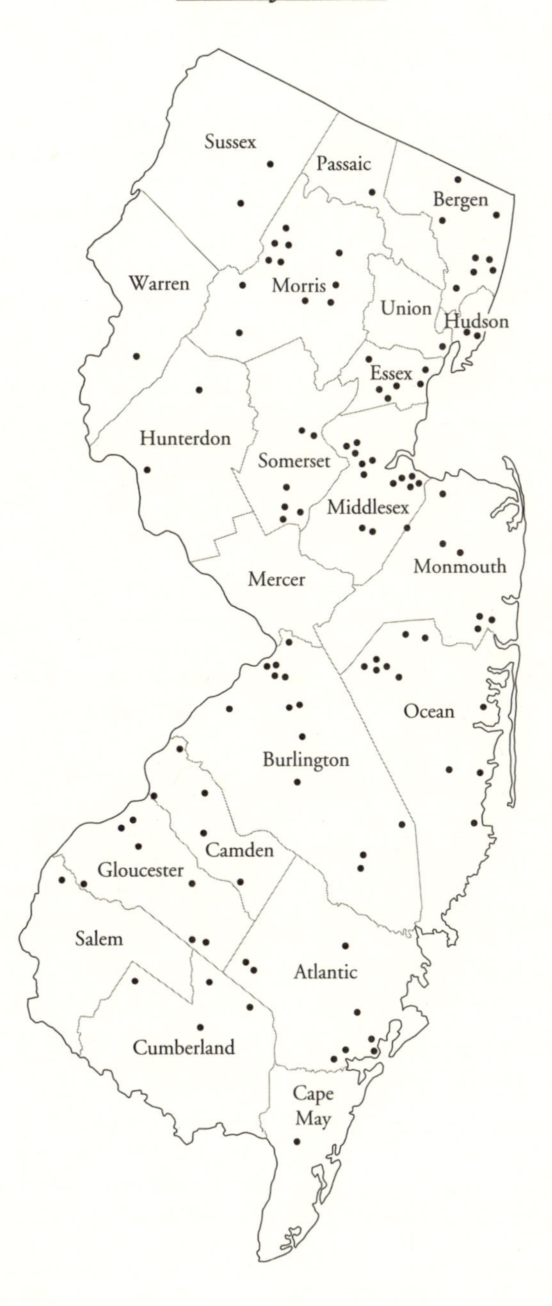

Bridgeport Rental & Oil Services
Gloucester County
A waste oil recycling plant operated on this site near Bridgeport in the 1960s and 1970s. Waste solutions and oils were discharged into an unlined lagoon or storage tanks on the property. Some sludges and oils were contaminated with PCBs. Runoff from the lagoons has been documented. Testing has shown that the groundwater and soil contain PCBs, VOCs, lead, cadmium, chromium, and barium. The contamination threatens the local drinking water supplies of about 1,000 people. Wetlands, Cedar Swamp, Little Timber Creek, and the Delaware River are also threatened.

Response actions included improving erosion controls (1981), connecting affected residents to water filtration units (1982), and removing contaminated soil and sludge (1984, 1990). In 1984 the EPA recommended removing contaminated sludges, soil, and storage tanks; groundwater excavation and treatment; and constructing a permanent water supply line (1987–1988). Contaminated soil and sludge is presently being incinerated on the site. The design of the groundwater remedy is under way.

The PRPs have participated in the remediation efforts according to a 1982 Consent Decree with the EPA and Department of Justice.

Sources: PNPLS-NJ; Logan Township Library, Center Square Road and School Lane, Swedesboro, NJ, 08085.

Brook Industrial Park

Somerset County

This multiple industrial area in Bound Brook has produced chemical products, pesticides, and electroplated metals since 1971. Various wastes including chemicals, organic solvents, heavy metal sludges, dioxins, and other pesticides were leaked, spilled, or deliberately discharged on the property or into the Raritan River. Testing has shown that the groundwater and soil on the site contain VOCs, dioxin, heavy metals, pesticides, and lead. The contamination threatens the recreational Raritan River and nearby private and public wells that supply water to about 600,000 people.

Response actions included capping contaminated soils (1983) and fencing (1990). A detailed site investigation (1989–1994) recommended excavating and treating contaminated soil, razing contaminated buildings, and extracting and treating groundwater.

Fines have been paid by Jame Fine Chemicals, a PRP, as a result of two Consent Orders (1980).

Sources: PNPLS-NJ; Bound Brook Public Library, 402 East High Street, Bound Brook, NJ, 08805.

Burnt Fly Bog

Monmouth County

This wetland near Marlboro served as a dumping site in the 1950s and 1960s. Various industrial wastes, including waste oil, sludges, and drummed solvents from nearby recycling and salvage operations, were spread in the bog. Testing has shown that the soil, sediments, surface water, and groundwater contain PCBs and heavy metals, including lead. The contamination threatens nearby private and public wells, wetlands, the New Jersey Pine Barrens resource area, and Deep Run.

Response actions included fencing and stabilizing the lagoon and sludge areas (1982). In 1983 the EPA recommended excavating contaminated soils and sludge, recontouring, and groundwater monitoring. The removal actions were completed in 1994. Other studies are in progress.

Sources: PNPLS-NJ; Montmouth County Library, 1 Library Court, Marlboro, NJ, 07746.

Caldwell Trucking Company

Essex County

This former waste disposal site near Fairfield operated from the 1950s to 1973. It received a variety of municipal and industrial wastes, including septic sludges. Testing has indicated that the groundwater and soil contain VOCs, metals, PAHs, and PCBs. The contamination has penetrated over 100 water wells in the area as well as Deepavaal Brook, which flows into the Passaic River. The contamination threatens the private drinking water supplies of about 2,000 people within a few miles of the area.

Response actions included fencing and capping the lagoon sites (1990). In 1986 the EPA recommended thermal treatment of contaminated soils, groundwater extraction and air-stripping, providing alternate water supplies to affected residents, and groundwater monitoring. Soil treatment was conducted from 1993 to 1996. The groundwater system has yet to be built.

Sources: PNPLS-NJ; Anthony Pio Costa Memorial Library, 261 Hollywood Avenue, Fairfield, NJ, 07004.

Chemical Control

Union County

A hazardous waste processing facility operated on this site in Elizabeth from 1970 to 1978. Chemicals, including acids, arsenic, cyanide, flammable solvents, radioactive compounds, PCBs, gases, pesticides, and biological agents, were stored, treated, or disposed of on the premises. Testing has shown that the soil and groundwater contain VOCs, pesticides, acids, and heavy metals. Sediments, fish, and shellfish in the Elizabeth River also contain these chemicals. The contamination threatens the Elizabeth River and the drinking water source for about 15,000 residents.

Response actions included removing surface waste and constructing surface runoff controls. (1990). In 1987 the EPA recommended the solidification and capping of contaminated soils (completed in 1994). More erosion controls were also built. Long-term monitoring is in progress.

In 1990 the EPA signed a Consent Decree with nearly 200 PRPs that requires the companies to perform the remediation work.

Sources: PNPLS-NJ; Director's Office, Department of Health, Welfare and Housing, 50 Winfield Scott Plaza, Room G5, Elizabeth, NJ, 07021.

Chemical Insecticide Corporation
Middlesex County
Various chemicals, including pesticides, were processed on this site near Edison from 1954 to 1970. Complaints from neighbors prompted an investigation that revealed that the soil and groundwater contain arsenic and pesticides such as DDT, dioxin, and lindane. Surface waters tainted with arsenic and dinoseb, a herbicide, flow into Mill Brook. The contamination is a threat to the water supplies for about 80,000 people in the area.

Response actions included fencing and installing surface runoff controls. A landfill cap was installed from 1993 to 1994. Groundwater studies are nearing completion.

Chemical Insecticide Corporation, the PRP, declared bankruptcy in 1970. Federal funds have paid for the clean-up work to date.

Sources: PNPLS-NJ; Edison Public Library, 340 Plainfield Avenue, Edison, NJ, 08817.

Chemical Leaman Tank Lines
Gloucester County
This tank-washing facility near Bridgeport has been in operation since 1961. From 1961 to 1975 contaminated wastewaters were discharged into unlined lagoons, Cedar Swamp, and Moss Branch Creek. Testing in 1981 discovered neighboring wells that contain VOCs. The groundwater and soils contain VOCs (including benzene and toluene), arsenic, chromium, zinc, and phthalates. The contamination threatens the water wells that supply about 4,000 local residents.

Response actions included connecting affected residents to new water supply lines or providing carbon treatment units (1987, 1993). In 1989 the EPA recommended groundwater extraction and treatment through chemical precipitation and air-stripping. Construction on this system is expected to be under way by 1997. Studies regarding soil and wetland remediation are in progress.

The PRP, Chemical Leaman Tank Lines, has assisted in the remediation work according to three Consent Orders (1985, 1989, 1991) with the EPA.

Sources: PNPLS-NJ: Clerk's Office, Logan Township Municipal Building, 73 Main Street, Bridgeport, NJ, 08014.

Chemsol Incorporated
Middlesex County
This solvent recycling facility in Piscataway operated from the 1950s to 1964. Leaks, spills, and discharges on the site have contaminated groundwater and soil with VOCs, pesticides, PCBs, and heavy metals. Purchased by a real estate company in 1978, the property is surrounded by residential or commercial lots. The contamination has penetrated nearby wells and threatens the water supply systems that serve a wider radius.

PCB-contaminated soils were removed in 1988. Uncovered solid wastes were removed in 1990–1991. An interim groundwater extraction and treatment system became operational in 1994. Additional groundwater and soil remediation studies are in progress.

A 1992 UAO with four PRPs requires the companies to design and construct the selected remedies.

Sources: PNPLS-NJ; Kennedge Public Library, 500 Hoes Lane, Piscataway, NJ, 08854.

Ciba-Geigy Corporation
Ocean County
This 1,400-acre manufacturing site in Dover Township produced dyes, pigments, epoxy, and resins from 1951 to 1990. Various solid and liquid wastes, including lime sludges, drummed waste, calcium sulfate, and demolition debris, were buried or discharged into landfills or lagoons on the property. Testing has indicated that the groundwater and soil contain VOCs and heavy metals. The groundwater flows into Toms River and nearby wetlands. The contamination also threatens municipal, industrial, and private water wells that serve about 65,000 people, an elementary school, and Winding River Park along the Toms River.

Response actions included removing drummed waste (1980). A groundwater pump-and-treat system was installed in 1991. An off-site groundwater treatment plant and river and groundwater monitoring programs should be operational by 1997.

The PRP signed a Consent Decree with the EPA in 1993 that requires the company to reimburse past clean-up costs and undertake the remediation work.

Sources: PNPLS-NJ; Ocean County Public Library, 101 Washington Street, Toms River, NJ, 08753.

Cinnaminson Contamination
Burlington County
A 400-acre area of groundwater in Cinnaminson was discovered to be contaminated with VOCs (chloroform, benzene, and vinyl chloride) in the 1980s. The VOCs are derived from nearby landfill operations that took in municipal and industrial wastes from the 1950s to 1980. The contamination threatens the public and private water wells within 2 miles of the site, which serve about 20,000 residents.

Response actions included detailed site investigations. In 1990 the EPA recommended a groundwater extraction and treatment system. Construction should be under way by 1997. The clay cap will also be reexamined.

A UAO between the PRP and the EPA requires the company to undertake the remediation work.

Sources: PNPLS-NJ; Municipal Building, 1621 Riverton Road, Cinnaminson, NJ, 08077.

Combe Fill, North Landfill

Morris County

This former landfill site in Mt. Olive Township operated from 1966 to 1978. It received a variety of municipal and industrial wastes, including heavy metal and sewage sludges. Testing in 1979 discovered that the groundwater and soil contained VOCs, phthalates, zinc, and cyanide, which had also penetrated nearby drinking wells. Air samples contained methane and VOCs. Surface runoff flows into nearby creeks and wetlands. The contamination threatens local water wells that supply about 10,000 residents.

Response actions included site investigations. In 1986 the EPA recommended grading, capping, installing leachate and surface runoff collection systems, gas venting, and fencing. Construction has been completed and groundwater and air are being monitored.

The landfill owner declared bankruptcy in 1981, and the remediation work has been paid for with federal funds.

Sources: PNPLS-NJ; Morris County Free Library, 30 East Hanover Avenue, Whippany, NJ, 07891.

Combe Fill, South Landfill

Morris County

This former landfill in Chester and Washington Townships operated from the 1940s to 1981. It received a variety of municipal and industrial wastes, including sewage sludge, chemicals, and oils. Testing has shown that the groundwater and soil contain VOCs and the air contains various gases and VOCs. Surface runoff has entered Trout Brook. The contamination threatens the water wells that serve about 500 people in the area. Local groundwater is also pumped for agricultural purposes.

Response actions included connecting affected residents to clean water supplies and fencing the site. In 1986 the EPA recommended capping the landfill, gas venting, runoff controls, and installing a groundwater pump-and-treat system. Construction should be completed by 1997.

The owner went bankrupt in 1981.

Sources: PNPLS-NJ; Washington Township Library, 146 Schooleys Mt. Road, Long Valley, NJ, 07853.

Cosden Chemical Coatings

Burlington County

Companies on this site in Beverly manufactured industrial coatings from the 1940s to 1989. Various solvents and waste solutions were stored on-site in drums and tanks, which were left behind when the owner abandoned the site in 1985. Testing by the EPA in 1988 discovered that the soil and groundwater contain VOCs, PCBs, and heavy metals (chromium, lead, zinc, and copper). The contamination threatens public and private wells in the area, which supply about 70,000 residents, and the nearby recreational Delaware River.

Response actions included removing drummed waste and erecting fencing (1989). In 1992 the EPA recommended stabilizing and capping contaminated soil, razing buildings on the site, and constructing a groundwater pump-and-treat system. These activities are in progress.

The PRP failed to undertake clean-up work after it declared bankruptcy in 1985.

Sources: PNPLS-NJ; City Hall, 446 Broad Street, Beverly, NJ, 08010.

CPS and Madison Industries

Middlesex County

Zinc compounds for pharmaceuticals and various organic chemicals used in lubricants and oil field chemicals have been processed on this site in Old Bridge Township since the 1960s. Related wastes, including spent halogenated solvents, have contaminated the soil and groundwater through spills, leaks, or discharges. Contaminants included VOCs, zinc, cadmium, copper, and lead, which have also been detected in recreational Prickett's Pond.

Response actions included the installation of an interim groundwater extraction and treatment system in 1991. Detailed site investigations are in progress.

A 1988 Administrative Order requires the two PRPs, CPS and Madison Industries, to conduct the site studies.

Sources: PNPLS-NJ; Old Bridge Public Library, 516 Cottrell Road, Old Bridge, NJ, 08857.

Curcio Scrap Metal

Bergen County

This scrap metal recycling facility in Saddle Brook Township has been in operation since the 1960s. In 1982 PCB-containing oils from transformers were spilled on the premises. Later spills of oil and hydraulic fluid also occurred. Testing has shown that the groundwater and soil contain PCBs, heavy metals, and VOCs. Nearby surface waters are also contaminated. The contamination threatens the nearby Schroeders Brook and the local wells that supply about 30,000 residents with drinking water.

Contaminated soil was removed and incinerated in 1991 and 1994. Groundwater studies are in progress.

Various legal agreements require the PRPs to participate in the site investigations and clean-up work.

Sources: PNPLS-NJ; Saddle Brook Free Public Library, 340 Mayhill Street, Saddle Brook, NJ, 07662.

Dayco Corporation

Morris County

This vinyl wall covering plant in Wharton Borough operated in the 1970s. Various solid and liquid wastes were held in lagoons beside the Rockaway River. The groundwater and soil contain VOCs, phthalates, and PCBs, which have entered the Rockaway River. The contamination threatens the Rockaway River and associated wetlands and the local drinking water supplies for about 27,000 people.

Response actions included removing contaminated soil from the property in 1982. Detailed site investigations were conducted from 1986 to 1994. Free-product organic liquids are presently being separated from the groundwater. A groundwater extraction and treatment system will also be constructed.

According to a 1986 Administrative Order, the PRP will conduct the site investigations and remediation work.

Sources: PNPLS-NJ; Wharton Public Library, 1519 South Main Street, Wharton, NJ, 07885.

De Rewal Chemical Company

Hunterdon County

This manufacturer of textile preservatives and agricultural fungicides operated on the site from 1970 to 1973. Other chemicals such as metals, acids, and fertilizers were stored or processed. A major acidic chromium spill was reported in 1973. Testing has shown that the groundwater and soil contain VOCs, cadmium, chromium, copper, lead, and PAHs. The contamination threatens the drinking water supply for about 3,000 local residents as well as nearby streams and wetlands.

The EPA recommended extraction and thermal treatment of soil, solidification and capping of soil, deed restrictions, and long-term monitoring. A contaminated residential well was connected to a filtration system in 1990. Other studies are in progress.

Source: PNPLS-NJ.

Delilah Road

Atlantic County

This former gravel pit operation in Egg Harbor Township was used as a landfill from 1974 to 1980. It accepted various municipal, construction, and industrial wastes, including sludges, VOCs, heavy metals, and flammable solvents. In 1982 the EPA determined that the groundwater and soil contain VOCs and heavy metals, which were also detected in nearby private wells, Jarrets Run, and Absecon Creek. The contamination threatens nearby wetlands and the well systems that supply about 3,600 residents.

A site investigation was completed in 1990. The EPA recommended a final remediation plan of capping, surface runoff controls, gas venting, fencing, deed restrictions, and long-term groundwater monitoring. Construction of these remedies is under way.

Sources: PNPLS-NJ; Atlantic County Library, Swift Drive, Bargaintown, NJ, 08232.

Denzer & Schafer

Ocean County

Silver is recycled from photographic materials on this 5-acre site in Bayville. From 1974 to 1981 stripping solutions were discharged into septic tanks on the property. Shredded film was also burned. Testing has indicated that the groundwater and soil contain arsenic, chromium, lead, mercury, silver, and VOCs. The contamination threatens Potters Creek, Barnegat Bay, and private and public wells that supply about 26,000 residents.

Response actions included a site investigation from 1987 to 1993. The EPA recommended connecting affected residents to city water hook-ups, removing underground storage tanks, and removing shredded film waste. Clean-up work is in progress.

Sources: PNPLS-NJ; Ocean County Library, 101 Washington Street, Toms River, NJ, 08753.

Diamond Alkali Company

Essex County

The Diamond Alakali Company manufactured pesticides and other chemicals from the 1940s to 1969 on this long-time industrial site beside the Passaic River. Sampling in 1983 revealed that the groundwater and soil contain dioxin, pesticides, and VOCs. The contamination threatens anyone who may come into contact with the contaminated soil. Contaminated groundwater flows into the Passaic River.

Response actions included fencing, capping, and removing dioxin-contaminated soil and debris from neighboring properties (1983). In 1987 the EPA recommended building a slurry wall and a groundwater extraction and treatment system. Buildings on the site will also be demolished. Activities are under way. Studies regarding the impact on the Passaic River are in progress.

Various legal agreements between 1984 and 1993 require 5 PRPs to fund the site investigations and clean-up work.

Sources: PNPLS-NJ; Newark Public Library, 5 Washington Street, Newark, NJ, 07101.

D'Imperio Property
Atlantic County
This former landfill site in Hamilton Township operated as an illegal dump in the 1970s. It received a variety of wastes, including drummed VOC solvents and heavy metal sludges. Testing has shown that the soil and groundwater contain VOCs, acids, and heavy metals. The contamination threatens the private and public wells within 3 miles of the site, which supply about 6,000 residents, and the Pineland Reserve, Babcock Swamp, and Babcock Creek.

Response actions included fencing (1982). In 1987 the EPA removed buried drums of waste and contaminated soil. A groundwater pump-and-treat system will be operational in 1997. A landfill cap will also be considered.

A 1993 AOC requires 14 PRPs to conduct the groundwater studies and remediation work.

Source: PNPLS-NJ.

Dover Municipal Well
Morris County
This municipal well in the town of Dover was discovered to contain VOCs (halogenated solvents) in 1980. The well was taken out of service. A point source has yet to be identified. About 32,000 people reside within 3 miles of the site.

A site investigation by the EPA recommended the extraction and treatment of the contaminated groundwater and long-term groundwater monitoring. Construction is in progress. The EPA is continuing to try to identify the source of the VOCs.

Sources: PNPLS-NJ; Dover Free Library, 32 East Clinton Street, Dover, NJ, 07801.

Ellis Property
Burlington County
Various drummed wastes were stored on this former dairy farm in Evesham and Medford Townships in the 1970s. About 300 drums of toxic waste were discovered on the property, many of them rusted and leaking. Sampling revealed that the groundwater and soils contain VOCs, arsenic, lead, chromium, and PCBs. The contamination threatens Sharps Run and associated wetlands and local water wells that supply about 3,000 residents.

Response actions included removing drummed waste and contaminated soil in 1983. Buildings were destroyed 4 years later, and acidic soil was neutralized by mixing it with lime. All the drums had been removed by 1990. An investigation from 1987 to 1992 recommended the extraction and treatment of contaminated groundwater and soil. Construction is expected to be under way by 1997.

Sources: PNPLS-NJ; Evesham Library, Route 70, Marlton, NJ, 08053.

Evor Phillips Leasing
Middlesex County
A waste haulage, treatment, and incineration facility operated on this site near Sayreville and Perth Amboy in the 1970s. Contaminated wastewaters were discharged into unlined lagoons, and drummed chemical waste was buried. Testing by the EPA revealed that the soils and groundwater contain VOCs, heavy metals, and phthalates. The contamination threatens the drinking water supplies for the residents of Perth Amboy and Sayreville as well as nearby recreational streams and wildlife areas.

Response actions included removing drummed waste (1983). In 1992 the EPA recommended removing all the drummed waste and installing an interim groundwater pump-and-treat system while wider-range groundwater testing continues off-site.

An AOC with EPA-identified PRPs requires the companies to fund the clean-up work.

Sources: PNPLS-NJ; Old Bridge Public Library, 516 Cottrell Road, Old Bridge, NJ, 08857.

Ewan Property
Burlington County
This site in Shamong Township operated as an industrial waste disposal center in the 1970s. Up to 8,000 drums of chemical waste were buried in trenches or dumped on the ground. Testing has revealed that the groundwater and soil contain VOCs, chromium, aluminum, and other heavy metals. The contamination threatens the water supplies for about 400 nearby residents, agricultural wells, streams, and the New Jersey Pine Barrens, a sensitive wildlife habitat.

Response actions included fencing (1988). In 1988–1989 the EPA recommended the removal and off-site incineration of drummed waste, solvent extraction and washing of contaminated soil, and groundwater extraction and treatment. Construction is expected to be under way by 1997.

The EPA has identified 30 PRPs, 17 of which will participate in the remediation work.

Sources: PNPLS-NJ; Shamong Township Clerk, 60 Willow Grove Road, Vincetown, NJ, 08088.

FAA Technical Center
Atlantic County
This 5,000-acre site near Atlantic City began operations in 1942 as a U.S. Navy base and later became a research center for the Federal Aviation Administration. Atlantic City International Airport and a New Jersey Air National Guard station are also on the

property. Various military and FAA activities have contaminated the soil and groundwater with VOCs, PCBs, PAHs, phthalates, free-product petroleum, heavy metals, and pesticides. The contamination threatens Doughty's Mill Stream and the drinking water supplies for Atlantic City's 40,000 residents and 125,000 annual visitors.

Response actions have included removing contaminated soil (1989–1990), installing interim groundwater extraction and air-stripping systems (1992), removing free-product jet fuel, and soil venting (1994). Other studies in progress are nearing completion and will recommend remediation plans.

Sources: PNPLS-NJ; Atlantic County Library, 2 South Farragut Avenue, May's Landing, NJ, 08330.

Fair Lawn Well Field
Bergen County
Three city wells in the town of Fair Lawn were discovered to contain VOCs in 1978. Subsequent studies indicated that the groundwater plume originated from the nearby Fisher Scientific Company and Sandvik Incorporated. Other industries are also in the area. The contamination threatens the water supplies for the 32,000 residents of Fair Lawn.

Response actions included removing contaminated soil (1984) and installing air-strippers on the contaminated wells (1987). An EPA site investigation (1992–1996) will make final recommendations for remediating the site.

A 1984 AO with Sandvik and a 1986 AO with Fisher Scientific require the companies to undertake the clean-up efforts.

Sources: PNPLS-NJ; Pine Public Library, 10-01 Fair Lawn Avenue, Fair Lawn, NJ, 07410.

Florence Landfill
Burlington County
This landfill in Florence and Mansfield Townships operated from 1973 to 1981. It received a variety of municipal and industrial wastes, including phthalates, heavy metals, and vinyl chlorides. Testing has revealed that the groundwater and soil contain VOCs, arsenic, chromium, lead, and PAHs. The contamination threatens residential wells that serve about 4,500 people, agricultural wells, Assiscunk Creek, and the Delaware River.

Response actions included capping (1982) and installing a leachate collection system. In 1986 the EPA recommended upgrading the cap, building slurry walls, gas venting, removing contaminated lagoon sediments, and long-term groundwater monitoring. Systems should be operational by 1997.

PRPs refused to comply with a 1979 Consent Order until they were sued by the State of New Jersey.

Sources: PNPLS-NJ; Burlington County Library, West Woodlane Road, Mt. Holly, NJ, 08060.

Fort Dix Landfill Site
Burlington County
A landfill operated from 1950 to 1984 on the U.S. Army Fort Dix property near Pemberton. Various wastes, including sludges, paints, thinners, pesticides, solvents, and construction debris, were buried in trenches. Chemical spills have also been recorded. Testing has shown that the groundwater, surface water, and soils contain VOCs, manganese, lead, and cadmium. The contamination threatens the water wells that supply about 13,000 residents as well as Cannon Run, Rancocas Creek, and associated wetlands.

Response actions included capping, fencing, and long-term groundwater monitoring. Site investigations are in progress.

Sources: PNPLS-NJ; Fort Dix Environmental Office, Building 5512, Texas Avenue, Fort Dix, NJ, 08640.

Franklin Burn
Gloucester County
Illegal copper recycling activities were conducted on this site in Franklin Township from the 1960s to 1988. Copper scrap, including wires, capacitors, and transformers, was burned on the ground to remove paint and insulation. Testing in 1989 revealed that the numerous burn piles contained dioxins, furans, PCBs, heavy metals, and pesticides. The contamination is a threat to the groundwater drinking supplies for the area.

Response actions included EPA removal of the burn piles and fencing the site (1989). Contaminated ash and soil were stockpiled and temporarily capped (1992–1993). Additional studies are nearing completion. A final remediation plan is expected by 1997–1998.

PRPs have been identified but do not have the funds to undertake the site investigations.

Source: EPA Publication 9320.7-071, Volume 3, No. 1.

Fried Industries
Middlesex County
Floor finishing products, detergents, adhesives, algaecides, and antifreeze have been manufactured on this property in East Brunswick Township. The groundwater and soil have been contaminated by various wastes that were stored, discharged, or buried on-site. Pollutants include VOCs, copper, and arsenic. The contamination threatens Bob Brook, Mill Pond, Lawrence Brook, the Raritan River, and the water supplies for about 25,000 residents.

Response actions included removing liquid wastes (1985), connecting affected residents to clean water supplies, fencing (1989), and removing drummed and solid waste (1989–1992). In 1994 the EPA recommended excavating contaminated soil and installing a groundwater extraction and carbon treatment system. Construction is expected to be under way by 1997.

A 1985 Consent Decree with the PRP required the company to abandon operations.

Sources: PNPLS-NJ; East Brunswick Library, 2 Jean Walling Civic Center Drive, East Brunswick, NJ, 08816.

Garden State Cleaners Company
Atlantic County

This dry cleaning establishment in Minotola has operated since 1966. Steam and wastewaters from the facility have contaminated the groundwater and soil with VOCs (tetrachloroethylene and trichloroethylene). The contamination has penetrated nearby residential wells and the city water system and threatens the water supplies for an overall population of about 10,000. The local groundwater is also used for agricultural purposes.

The EPA recommended a remediation plan of soil vapor extraction, groundwater extraction and treatment through air-stripping/carbon adsorption, and long-term monitoring. The soil treatment system should be operational by 1997.

Sources: PNPLS-NJ; Buena Borough Municipal Building, 616 Central Avenue, Minotola, NJ, 08341.

Gems Landfill
Camden County

This former landfill site in Gloucester Township operated from the 1950s to 1980. It received a variety of wastes, including asbestos, solvents, and sludges. Testing has shown that the groundwater, surface water, and soil contain VOCs, heavy metals, and pesticides (DDE, DDD, and DDT). The contamination threatens Holly Run and Briar Lake as well as the water supplies for about 10,000 people.

Response actions included removing debris and fencing the site (1983). The EPA recommended capping, installing a leachate treatment system and surface runoff controls, connecting affected residents to clean water supplies, and long-term monitoring. All work was completed by 1995.

A 1988 Administrative Order required 131 PRPs to conduct the remediation work.

Sources: PNPLS-NJ; Gloucester Library, 1650 Blackwood Road, Blackwood, NJ, 08012.

Glen Ridge Radium Site
Essex County

This site consists of 306 residential properties in Glen Ridge that were built on or near radioactive waste materials generated by a radium-processing company from 1900 to 1920. The radioactive waste was used as fill or mixed with cement for sidewalks and basement foundations. The soil is contaminated with radium and its by-products, including radon gas and gamma radiation.

Response actions included the installation of interim radon ventilation systems and gamma ray shielding in private homes (1983). Additional homes were protected using subslab depressurization systems (1990–1991). The removal of all contaminated materials, including soils, from homes and yards is expected to be completed by 1998. Groundwater studies are in progress.

Sources: PNPLS-NJ; Glen Ridge Public Library, 240 Ridgewood Avenue, Glen Ridge, NJ, 07028.

Global Sanitary Landfill
Middlesex County

This former landfill near Sayreville operated from the 1960s to 1984. It received a variety of municipal and industrial wastes, including paints, thinners, and solvents. A portion of the landfill collapsed during heavy rains in 1984 and contaminated Cheesequake Creek Tidal Marsh. Testing revealed that the groundwater and soil contain VOCs, including methylene chloride. The contamination threatens private and public wells that serve about 90,000 residents, nearby tidal wetlands, and the commercial and recreational Raritan Bay.

Response actions included a site investigation from 1990 to 1991. The EPA recommended capping, erosion controls, and a leachate collection system. Construction should be completed by 1997.

The State of New Jersey received negotiated settlements from PRPs.

Sources: PNPLS-NJ; Sayreville Public Library, 1050 Washington Road, Parlin, NJ, 08859.

Goose Farm
Ocean County

Companies that made polysulfide rubber and rocket fuel propellants used this site in Plumsted Township as a hazardous waste dump from the 1940s to the 1970s. Manufacturing wastes were buried in drums or discharged on the property. Testing in 1980 revealed that the groundwater and soil contain VOCs, SVOCs, PCBs, and heavy metals. The contamination threatens the water supplies for nearby residents, Lahaway Creek, and the recreational Delaware River. Surface water and groundwater are also used for agricultural purposes.

Response actions included the installation of a groundwater pump-and-treat system and the removal of drummed waste (1980–1981). In 1985 the EPA recommended groundwater treatment, soil flushing, and soil excavation. PCB-contaminated soils were removed in 1988–1989. The groundwater treatment system will run from 1993 to 1999.

The PRP, Morton-Thiokol, agreed to conduct the remediation work according to a 1988 Consent Decree with the EPA and State of New Jersey.

Sources: PNPLS-NJ; Ocean County Library, 101 Washington Street, Toms River, NJ, 98753.

Helen Kramer Landfill

Gloucester County

This former gravel pit in Mantua Township operated as a landfill from 1963 to 1981. It received a variety of municipal and industrial wastes, including sludges, oils, solvents, plastics, acids, pesticides, heavy metals, and paints. Testing has shown that the groundwater and soils contain VOCs, arsenic, cadmium, and lead. Various gases are also released into the air. The contamination threatens Edwards Run, Mantua Creek, the Delaware River, and the local groundwater that serves the water needs of about 10,000 residents. Surface water is also used for irrigation purposes.

In 1985 the EPA recommended capping, gas collection, construction of a slurry wall and a groundwater/leachate collection trench, and treatment through air-stripping and carbon adsorption. All remedies have been completed or are in operation.

The EPA is suing PRPs for recovery of past cleanup costs.

Sources: PNPLS-NJ; Cloucester County Library, 200 Holly Dell Drive, Sewell, NJ, 08080.

Hercules Gibbstown Plant

Gloucester County

Chemicals have been manufactured on this site in Gibbstown since the 1940s. DuPont Corporation and Hercules Incorporated discharged solvents, acids, oils, and other liquid wastes into unlined lagoons from the 1940s to 1974. The groundwater and soil contain VOCs, phenols, lead, and other heavy metals. The contamination threatens Cronmell Creek, the recreational Delaware River, and water wells that serve about 13,000 people,

Detailed site investigations that are nearing completion will make recommendations for a final remediation plan in 1997–1998.

Hercules Incorporated, one of the PRPs, is undertaking the site investigations according to a 1986 AOC.

Sources: PNPLS-NJ; Greenwich Township Municipal Building, Broad and Walnut Streets, Gibbstown, NJ, 08027.

Higgins Disposal

Somerset County

This former landfill in Franklin Township operated without a permit from the 1950s to 1986. It received a variety of municipal and industrial wastes, including spent solvents and PCB-tainted oils. In 1986 the State of New Jersey discovered VOCs and PCBs on adjacent property. Further testing indicated that the groundwater and soil contain VOCs, PCBs, and heavy metals. The contamination threatens private and municipal wells within 3 miles of the site, which service about 11,000 residents, the Delaware-Raritan Canal, Millstone River, and Raritan Bay.

Response actions included removing contaminated soil in 1992. Drummed waste was excavated and fencing installed in 1993. Recommendations from ongoing site investigations are expected by 1997.

Source: PNPLS-NJ.

Higgins Farm

Somerset County

Drummed industrial waste was buried on this cattle farm in Franklin Township in the 1970s. The illegal site was discovered when VOCs were found in a neighbor's well in 1985. Removal actions punctured some of the drums, and liquid wastes were released into the ground. Further testing showed that the groundwater and soil contain VOCs, SVOCs, heavy metals, dioxins, and pesticides. The contamination threatens the health and drinking water supplies of nearby residents.

Response actions included excavating contaminated soil and debris (1987, 1992) and supplying carbon filters to affected residents (1989). In 1992 the EPA recommended soil gas surveys, groundwater extraction and treatment, and long-term monitoring. The groundwater system became operational in 1996. Permanent clean water supplies were installed for affected residents in 1994.

PRPs have refused to participate in the site studies or remediation work.

Sources: PNPLS-NJ; Franklin Township Public Library, 485 DeMott Lane, Somerset, NJ, 08872.

Hopkins Farm

Ocean County

Thiokol Chemical Company used this farm site in Plumsted Township to dump various wastes from 1962 to 1965. Wastes included pesticides, VOCs, and heavy metal sludges. Testing has shown that the groundwater and soil contain VOCs. The contamination threatens the water supplies for the town of New Egypt and Fort Dix Military Reservation.

Recommendations from a detailed site investigation by the State of New Jersey (1987–1995) are pending.

Sources: PNPLS-NJ; Ocean County Library, 101 Washington Street, Toms River, NJ, 08753.

Horseshoe Road
Middlesex County
Chemicals were manufactured and processed by multiple companies on this 9-acre parcel near Sayreville from the 1960s to 1987. Production wastes such as contaminated wastewaters, sludges, pesticides, and heavy metals (including mercury) were stored, buried, or abandoned on the property. Testing has shown that the soil and groundwater contain VOCs (PCBs), SVOCs, heavy metals, and pesticides. The contamination threatens water wells within 4 miles of the site, which service about 14,000 people, as well as wetlands and streams.

Response actions included removing drummed waste (1980–1981) and contaminated soil (1987). A detailed site investigation is planned.

The State of New Jersey has identified 18 PRPs; they have refused to participate in the site clean up, despite legal actions by the state.

Source: PNPLS-NJ.

Imperial Oil/Champion Chemicals
Monmouth County
This oil processing and storage facility near Morganville blends and reprocesses waste oil, which is held in about 60 on-site storage tanks. Some processing wastes were discharged into a nearby stream. Groundwater and soil contain VOCs, PCBs, PAHs, petrochemicals, phthalates, and heavy metals. The contamination threatens local drinking water supplies, waterways, wetlands, and Lake Lefferts.

Response actions included removing contaminated soil (1991) and free-product petrochemicals (1993–1995). Drummed waste was also removed. Fencing was erected in 1991. In 1990 the EPA recommended restoring wetlands, which should be completed by 1997–1998. A groundwater extraction and precipitation/carbon adsorption system is being installed. Other studies are in progress.

Sources: PNPLS-NJ; Marlboro Township Hall, 1979 Township Drive, Marlboro, NJ, 07746.

Industrial Latex Corporation
Bergen County
Chemical adhesives, latex, and rubber compounds were manufactured on this site near Wallington from 1951 to 1983. Production chemicals included acetone, heptane, hexane, methyl ethyl ketone, methylene chloride, and PCBs. A state inspection in 1981 found about 250 leaking drums of chemical waste. Waste solutions had also been dumped in an on-site septic system. An additional 1,600 leaking drums of VOC-type solvents were discovered in 1983. The groundwater and soil contain PCBs and VOCs. The contamination threatens the water supplies of about 15,000 residents.

Response actions included removing contaminated liquids, drummed waste, and tanks (1986) and installing fencing (1994). Recommendations from a 1989–1992 site study by the EPA included the excavation and low-temperature thermal desorption of contaminated soil, razing of structures, and monitoring. A groundwater study should be completed by 1997.

Sources: PNPLS-NJ; John F. Kennedy Memorial Library, 92 Hathaway Street, Wallington, NJ, 07057.

Jackson Township Landfill
Ocean County
This former titanium mining pit in Jackson Township was used as a landfill from 1972 to 1980. It received a variety of municipal and industrial wastes, including spent solvents and acids. Nearby wells were discovered to contain VOCs in 1977. The soil and groundwater contain VOCs and heavy metals. The contamination threatens the drinking water supplies for about 4,000 residents.

Response actions included connecting over 130 affected homes to clean water supplies. A site investigation from 1988 to 1994 determined that no cleanup actions were required.

Jackson Township signed a 1988 Consent Order to conduct the site investigation.

Sources: PNPLS-NJ; Ocean County Library, 101 Washington Street, Toms River, NJ, 08753.

Jis Landfill
Middlesex County
This former landfill and waste transfer station in South Brunswick operated from 1956 to 1980. It received various municipal and industrial wastes, including battery casings, sludges, paints, solvents, and pesticides. The groundwater and soil contain VOCs, aldrin, and heavy metals. The chemicals have penetrated nine nearby residential wells and threaten the drinking water supplies for local residents and nearby waterways.

Response actions included connecting affected residents to clean water supplies (1992). A site investigation is in progress.

Legal agreements with 23 PRPs (1984) require the companies to fund the site investigations and remediation work.

Sources: PNPLS-NJ; South Brunswick Public Library, 110 Kingston Lane, Monmouth Junction, NJ, 08852.

Kauffman & Minteer
Burlington County
From 1960 to 1980 contaminated wastewaters from washing out tanker trucks were discharged into unlined lagoons on this site. Contaminants include or-

ganic chemicals, plastics, resins, oils, and alcohol. The lagoons flooded and flowed off-site in 1984. Testing has shown that the soil and groundwater contain VOCs and heavy metals. The contamination threatens water wells that supply about 1,000 local residents.

The site was fenced in 1990. The EPA drained the lagoon in 1991. A site investigation (1989–1996) will recommend a final remediation plan.

The PRP has conducted the initial response actions according to a 1983 Consent Order with the EPA.

Sources: PNPLS-NJ; Springfield Township Municipal Building, Jobstown Road, Jacksonville, NJ, 08041.

Kin-Buc Landfill
Middlesex County
This former landfill in Edison Township operated from the 1940s to 1976. It received a variety of municipal and industrial wastes, including oils, sludges, PCB-containing fluids, and chlorinated VOC solvents. Testing has revealed that the groundwater and soil contain PCBs, heavy metals, and VOCs. The chemicals have penetrated nearby Edmonds Creek, Rum Creek, the Raritan River, and their aquatic life. The contamination threatens the drinking water supplies of about 4,000 residents.

Response actions included removing oil waste and leachate (1980–1991) and drummed waste (1984). Capping and slurry walls were installed from 1993 to 1996. Contaminated wetland sediments were removed from 1994 to 1996. Other studies are in progress.

PRPs have participated in the site studies and clean-up work.

Sources: PNPLS-NJ; Edison Township Library, 340 Plainfield Avenue, Edison, NJ, 08817.

King of Prussia
Camden County
King of Prussia Technical Corporation dumped treated industrial waste at this site near the New Jersey Pine Barrens from 1970 to 1973. At least 15 million gallons of toxic solutions were discharged into unlined lagoons on the site. Testing has shown that the groundwater and soil contain heavy metals (copper, nickel, and chromium), phthalates, and VOCs. The contamination threatens Winslow Wildlife Area, the Great Egg Harbor River, and water wells that serve about 4,000 residents.

Response actions included fencing (1988) and removing drummed waste and contaminated soil (1989). Additional removals were conducted in 1990 and 1991. Soil washing was completed from 1993 to 1994. A groundwater treatment plant became operational in 1996.

A 1991 AOC requires five PRPs to fund the site investigations and clean-up work.

Sources: PNPLS-NJ; Camden City Library, 418 Federal Street, Camden, NJ, 08101.

Landfill and Development Company
Burlington County
This former gravel pit served as a landfill operation in the 1960s and 1970s. It received a variety of municipal, industrial, and commercial wastes, including sewage sludge. Testing in 1973 revealed that the groundwater and soil contain VOCs and heavy metals. The contamination threatens Rancocas Creek, Smithville Canal, recreational Smithville Lake, and the water supplies of about 100 local residents.

Response actions included connecting threatened residents to city water supplies. The landfill was also capped. Detailed site investigations that will recommend a final remediation plan are in progress.

Sources: PNPLS-NJ; Burlington County Library, West Woodlane Road, Mt. Holly, NJ, 08060.

Lang Property
Burlington County
This illegal landfill in Pemberton Township accepted various industrial waste in the 1970s, including tires and spent solvents. Testing by the state revealed that the soil and groundwater contain VOCs and heavy metals. The contamination threatens Rancocas Creek, Pineland National Reserve, and private and public water wells that supply about 11,000 residents.

In 1986 the EPA recommended a final remediation plan of excavating contaminated soil and debris, fencing, installing a groundwater pump-and-treat system, and long-term monitoring. Soil clean up was completed in 1988. The groundwater treatment system became operational in 1992 and will be completed by 1997.

Sources: PNPLS-NJ; Community Library, 348 Lakehurst Road, Brownsmill, NJ, 08015.

Lipari Landfill
Gloucester County
This former landfill in Mantua Township operated from 1958 to 1971. It received a variety of industrial and municipal wastes, including paints, solvents, thinners, filter residues, resins, and formaldehyde. The groundwater, surface water, and soil contain VOCs, heavy metals, phthalates, and phenols. Some of these compounds have also been detected in air emissions. The contamination threatens orchards, Chestnut Creek, Rabbit Run Creek, Alcyon Lake, the Delaware River, related wetlands, and water wells that supply about 11,000 residents.

Response actions included installing fencing (1982–1985), surface runoff controls (1984), and a groundwater pump-and-treat system (1992). A system for the removal of contaminated sediments from streams and wetlands is presently being designed. About 10 million gallons of groundwater have been treated since 1992.

Consent Decrees (1982, 1993, 1994) require five PRPs to fund the site investigations and remediation work.

Sources: PNPLS-NJ; McCowan Library, 15 Pitman Avenue, Pitman, NJ, 08071.

Lodi Municipal Well
Bergen County
Groundwater in the borough of Lodi was discovered to contain VOCs, uranium, and radium-226 in the mid-1980s. The contaminated wells were shut down in 1987, but the EPA has not been able to identify the source of the chemicals. The contamination threatens the drinking water supply for about 25,000 residents.

Response actions including connecting affected residents to clean water supplies (late 1980s). An EPA investigation completed in 1993 indicated that the contamination was naturally occurring. The EPA decided a remedial action was not required.

Sources: PNPLS-NJ; Lodi Memorial Library, 1 Memorial Drive, Lodi, NJ, 07655.

Lone Pine Landfill
Monmouth County
Located in Freehold Township, this former landfill operated from 1959 to 1979. It received a variety of municipal and industrial wastes, including sewage sludges and VOC-solvents. Leachate has polluted the Manasquan River with VOCs, cadmium, and lead. The soil and groundwater contain VOCs and heavy metals, including arsenic. The contamination threatens local drinking water supplies, wetlands, a reservoir, and the Manasquan River.

Response actions included installing a slurry wall, capping, and methane gas venting (1991–1993). A groundwater extraction and treatment plant was installed in 1994 and is operational.

EPA-identified PRPs have conducted the remediation work according to Consent Decrees (1992, 1993) with the EPA.

Sources: PNPLS-NJ; Monmouth County Public Library, 25 Breod Street, Freehold, NJ, 07728.

Mannheim Avenue Dump
Atlantic County
This former landfill in Galloway Township operated in the 1960s and 1970s. It received a variety of wastes, including degreasing sludge and lead-contaminated glazes from a local china manufacturer. Testing in the mid-1980s revealed that the soil and groundwater contain VOCs (trichloroethylene) and heavy metals (lead). The contamination threatens private and public wells that serve about 1,500 residents as well as nearby streams.

Response actions included removing asphaltic sludges and erecting fencing (1985). Monitoring wells were drilled in 1986. Contaminated soil was excavated in 1989. In 1990 the EPA recommended groundwater extraction and air-stripping, long-term monitoring, and installing carbon filtration units for affected residents. The groundwater treatment system became operational in 1994.

Lenox Incorporated, the main PRP, is required to conduct the clean-up work according to a UAO (1987), AOC (1988), and CD (1991) with the EPA.

Sources: PNPLS-NJ; Galloway Township Public Library, 30 West Jim Leeds Road, Pomona, NJ, 08240.

Maywood Chemical Company
Bergen County
Maywood Chemical Company processed radioactive thorium ore on this site in Maywood, Lodi, and Rochelle Park Boroughs from 1916 to the 1960s. Various wastes, including radioactive materials, lithium, oils, detergents, and alkaloids, were discharged into surface impoundments. Some materials were used as fill for local construction projects. Testing has shown that the groundwater and soil contain radioactive materials, VOCs, SVOCs, heavy metals, and pesticides. The contamination threatens the Saddle River and water wells that serve about 30,000 people.

Response actions included removing contaminated soil and debris (1984–1986). Twenty-five residential properties have been decontaminated since 1986. A final remediation plan will be recommended by the EPA after reviewing data from recently completed, long-term site studies.

The PRPs are participating in the clean-up work according to a 1987 AOC and 1991 UAO.

Sources: PNPLS-NJ; Maywood Public Library, 459 Maywood Avenue, Maywood, NJ, 07607.

Metaltec Aerosystems
Sussex County
Various metal products were manufactured on this site in Franklin Borough from 1965 to 1980. Various waste products were stored or discharged on the site. In 1980 state officials detected high levels of VOCs and heavy metals in the groundwater and soil. The chemicals also penetrated nearby private and municipal wells, which were shut down. The contamination

threatens the Wildcat Brook, the Wallkill River, and water supplies of about 5,000 residents.

Over the period 1986 to 1989 the EPA recommended excavating and treating contaminated soil, connecting affected residents to clean water supplies, and groundwater extraction and treatment through air-stripping and carbon adsorption. Soil treatments should be completed by 1997. The groundwater treatment plant is expected to be operational by 1997–1998.

Sources: PNPLS-NJ; Sussex County Library, 101 Main Street, Newton, NJ, 07860.

Monitor Devices
Monmouth County
This company manufactured circuit boards on this site in Lakewood Industrial Park from 1977 to 1981. Copper, tin, lead, nickel, gold, cyanide, sulfuric acid, fluoboric acid, and acetone were used in the manufacturing process. Wastes were stored, leaked, spilled, or discharged into lagoons on the property. Testing revealed that the groundwater and soil contain excessive levels of heavy metals and VOCs. Lawsuits by the State of New Jersey and Monmouth County prompted the company to declare bankruptcy in 1988. The contamination threatens private and public wells within 3 miles of the site, which serve about 26,000 people.

Response actions included site investigations by the State of New Jersey (1986–1991) and the EPA (1992–present). The EPA will recommend a final remediation plan after all the data have been assessed.

Sources: PNPLS-NJ; Wall Township Library, Highway 35, Sea Girt, NJ, 08750.

Monroe Township Landfill
Middlesex County
This former landfill near Monroe operated from 1955 to 1978. It received a variety of municipal and industrial wastes, including spent VOC solvents, hydrochloric acid, demolition debris, and sludges. Leachate was discovered in neighboring properties and drinking wells in 1978–1979. Testing showed that the groundwater and soil contain VOCs, phenols, and heavy metals. The contamination threatens water wells that supply about 12,000 residents, agricultural wells, and nearby streams.

Response actions included connecting affected residents to clean water supplies (1979–1980). A dike and leachate collection system was also installed (1984) and the site capped. Although the site was deleted from the NPL in 1994, long-term monitoring is continuing.

A 1979 Consent Order and 1986 AOC require the PRPs to remediate the site.

Sources: PNPLS-NJ; Jamesburgh Public Library, 229 Gatzmer Road, Jamesburgh, NJ, 08831.

Montclair/West Orange Radium Site
Essex County
This 120-acre residential area in Montclair and West Orange Townships was built on radium-contaminated fill from a nearby radium processing facility. The waste was also mixed into cement that was used for sidewalks and foundations. About 440 homes contain excessive levels of radon gas and gamma radiation.

Response actions included installing temporary radon gas venting systems (1983) and gamma radiation shields (1983–1990). In 1989 the EPA recommended removing contaminated soil and long-term monitoring. An estimated 41,000 cubic yards of contaminated soil and concrete will be removed by 1998.

Sources: PNPLS-NJ; Montclair Public Library, 50 South Fullerton Avenue, Montclair, NJ, 07042.

Montgomery Township Housing Development
Somerset County
An area of groundwater near the Montgomery Township Housing Development was discovered to be contaminated with VOCs in 1978. Testing showed that a number of private wells contained VOCs. A point source has not been identified. The contamination threatens the local drinking water supplies, Beden Brook, and the Millstone River.

Response actions included connecting affected residents to the city water supply (1981). Site investigations (1984–1988) recommended the installation of a groundwater extraction and air-stripping system. Construction is in progress.

Sources: PNPLS-NJ; Somerset County Library, North Bridge Street, Bridgewater, NJ, 08807.

Myers Property
Hunterdon County
Pesticides were formulated and processed on this site in Franklin Township in the 1940s. Various wastes, including drummed chemicals and asbestos, were discovered on the site. The groundwater and soil contain excessive levels of VOCs, pesticides (including DDT), SVOCs, dioxins, and heavy metals. The contamination threatens Cakepoulin Creek, the Raritan River, and wells that serve about 1,000 local residents.

Response actions included removing contaminated soil and drummed waste (1984) and fencing (1987–1988). In 1989 the EPA recommended excavating contaminated soil, soil washing, groundwater extraction and treatment, and groundwater monitoring. Construction will begin in 1998.

The PRPs signed a Consent Decree (1991) with the EPA to design and implement the remediation work.

Sources: PNPLS-NJ; Hunterdon County Library, Route 12, Flemington, NJ, 08822.

N. L. Industries

Salem County

This former lead smelter near Pedricktown operated from 1972 to 1984. Lead from automotive batteries was also recycled. Spent acids, slags, and waste rubber materials were buried or discharged in a landfill on the property. Testing has shown that the groundwater and soil contain heavy metals, including lead. The contamination threatens the recreational Delaware River and wells that supply about 4,000 residents.

Response actions included fencing and capping (1989–1990) and removing surface waste, slag piles, and steel drums. Slag was also treated and removed in 1992. Contaminated stream sediments were excavated in 1993. In 1994 the EPA recommended a groundwater pump-and-treat system and excavation of lead-contaminated soils. Additional studies are in progress.

N. L. Industries, the main PRP, has participated in the site clean up through AOCs (1982, 1983) and a Consent Order (1986). An additional 31 PRPs signed a UO in 1992 that requires the companies to participate in the remedial actions.

Sources: PNPLS-NJ; Penns Grove Library, South Broad Street, Penns Grove, NJ, 08069.

Nascolite Corporation

Cumberland County

Nascolite Corporation manufactured polymethyl methacrylate sheets (plexiglass) at this site near Millville from 1953 to 1980. Various liquid wastes and wastewater were stored in underground tanks that leaked contaminants into the soil and groundwater. Pollutants include polymethyl methacrylate, asbestos, VOCs, heavy metals, and phthalates. The contamination threatens water wells that supply about 20,000 residents and agricultural wells.

Response actions included removing contaminated soil (1981), capping, and, fencing (1987). Affected residents were connected to permanent clean water supplies. Construction of a groundwater treatment system is expected to be under way by 1997. Contaminated soils will be solidified and reburied.

PRPs identified by the EPA have participated in the clean-up work.

Sources: PNPLS-NJ; Millville Public Library, 210 Buck Street, Millville, NJ, 08332.

Naval Air Engineering Center

Ocean County

This 7,400-acre U.S. Navy military base near Lakehurst has been in operation since the 1920s. Various activities have included training, testing, and weapons systems research and development. Related wastes were discharged or buried in trenches, pits, lagoons, and wetlands on the base. Testing has shown that the groundwater and soil contain heavy metals, VOCs, phenols, petroleum, and pesticides. The contamination threatens Manapaqua Brook, Toms River, Obhanan Creek, Harris Creek, North Ruckels Creek, and water wells that supply about 10,000 residents.

Response actions included installing four groundwater pump-and-treat systems (1990, 1993, 1994, 1995), removing contaminated soil (1990–1992), and soil vapor extraction (1994–1995). Other studies are in progress.

Sources: PNPLS-NJ.; Ocean County Library, 101 Washington Street, Toms River, NJ, 08753.

Naval Weapons Station Earle

Monmouth County

This 11,100-acre U.S. Navy weapons station in Colts Neck has stored and renovated munitions since the 1940s. Various wastes from these activities, including paint, lead bullets, solvents, sludges, and heavy metals were discharged or buried on-site. The groundwater and soil contain heavy metals and PCBs. The contamination threatens private and public wells that serve about 2,000 residents, agricultural wells, and the Swimming, Manasquan, and Shark Rivers.

Response actions included a detailed site investigation (1989–1996). Recommendations are pending.

Sources: PNPLS-NJ; Colts Neck Library, 15 Heyers Mill Pond, Colts Neck, NJ, 07722.

Pepe Field

Morris County

This former landfill site in the town of Boonton operated from the 1920s to 1950. It received a variety of waste, including oils, soap products, and industrial solvents. The site was later converted into a recreational area. Testing has shown that the groundwater and soil contain heavy metals and VOCs. The contamination threatens the Rockaway River and the water supplies of local residents.

Response actions included a site investigation from 1985 to 1989. In 1989 the EPA recommended capping, installing a gas collection system, upgrading the leachate control system, and groundwater monitoring. Construction should be under way by 1997.

Sources: PNPLS-NJ; Boonton Holmes Library, 621 Main Street, Boonton Town, NJ, 07005.

Picatinny Arsenal
Morris County
The U.S. Army has operated an ammunition arsenal on this 6,500-acre property in Rockaway Township for over 100 years. It currently designs and manufactures new explosives and propellants. Associated waste materials, including a variety of sludges, were buried or discharged in over 100 landfills on the base. The groundwater and soil contain heavy metals, VOCs, PCBs, pesticides, dioxins, and nitrates. The chemicals have been detected in nearby streams and private wells. The contamination threatens the local drinking water supplies for base personnel and residents, Lake Denmark, and Picatinny Lake.

Response actions included removing buried drums and connecting affected residents to clean water supplies (1992–1994). In 1989 the EPA recommended extracting and treating groundwater with air-strippers, which became operational in 1992. Additional studies regarding soil treatments are in progress.

Sources: PNPLS-NJ; Office of Environmental Affairs, Picatinny Arsenal, NJ, 07806.

Pijak Farm
Ocean County
This farm near New Egypt was used as a dump from 1963 to 1970. It received a variety of waste, including research chemicals and drummed solutions and solvents. Liquid wastes were simply poured on the ground. In 1980 state officials discovered that the groundwater and soil contained VOCs, phthalates, DDT, and PCBs. The contamination threatens recreational Crosswicks Creek and water wells that serve about 1,500 people.

Response actions included removing drummed waste, excavating contaminated soil, and installing a groundwater treatment system and erosion controls (1990–1994). Groundwater monitoring is in progress.

A 1985 AOC requires the PRPs to conduct the site investigations and remediation work.

Sources: PNPLS-NJ; Ocean County Library, 101 Washington Street, Toms River, NJ, 08753.

PJP Landfill
Hudson County
This former landfill in Jersey City operated from 1968 to the 1970s. It received a variety of chemical and industrial wastes. The site also had a history of underground fires. Testing has shown that the groundwater and soil contain heavy metals, phenols, pesticides, and VOCs. The contamination threatens the local drinking water supplies and the Hackensack River. About 12,000 people live within 1 mile of the site.

Response actions included the installation of a gas venting system in 1986. A site investigation was recently completed by the State of New Jersey. Recommendations are pending.

Sources: PNPLS-NJ; Jersey City Public Library, 678 Newark Avenue, Jersey City, NJ, 07306.

Pohatcong Valley
Warren County
Groundwater supplying private and public wells in Washington and Franklin Townships was discovered to contain VOCs in 1978–1979. Contaminated wells were either closed or fitted with carbon filtration units. A point source has not been identified. At least 79 contaminated wells had been identified by 1985. About 12,000 residents depend on the local groundwater for drinking purposes.

Response actions included connecting affected residents to clean water supplies. An ongoing groundwater study should be completed by 1997.

Sources: PNPLS-NJ; Washington Borough Municipal Building, 100 Belvidere Avenue, Washington, NJ, 07882.

Pomona Oaks Residential Wells
Atlantic County
Water wells supplying about 200 homes in rural Galloway Township were discovered to contain VOCs (benzene and xylene) in 1982. Contaminated wells were closed and affected residents connected to the town of Absecon's water supply. A point source has yet to be identified. Nearly 10,000 residents live within 3 miles of the site.

Response actions included supplying affected residents with bottled water (1985) until a permanent water supply system was established in 1989. An investigation (1986–1990) showed that the VOC levels had fallen to acceptable levels. Although groundwater monitoring will continue, the EPA will delete this site from the NPL.

Sources: PNPLS-NJ; Galloway Township Public Library, 30 West Jim Leeds Road, Pomona, NJ, 08240.

Price Landfill No. One
Atlantic County
This former gravel pit in Egg Harbor Township operated as a landfill from 1971 to 1976. Various wastes included drummed liquids, sludges, oil, grease, septic sewage, and heavy metals. Testing in the 1980s revealed that the groundwater and soil contain VOCs and heavy metals. The contamination threatened Atlantic City's water wells, which were replaced with

more distant wells (1983–1985). The contamination now threatens Absecon Creek and the drinking water supplies for about 1,000 residents.

Response actions included supplying affected residents with clean water supplies in 1981. The EPA recommended fencing, groundwater extraction and treatment, leachate collection, capping, and long-term groundwater monitoring. Construction of the groundwater treatment system will be under way by 1997.

About 50 PRPs have contributed funds to the clean up through negotiated settlements with the EPA (1982–1990).

Sources: PNPLS-NJ; Atlantic City Public Library, 1 North Tennessee Avenue, Atlantic City, NJ, 08401.

Radiation Technology

Morris County

Activities related to radiation sterilization, manufacture of architectural products, and hardwood floor finishing have been carried out on this site near Lake Denmark since 1970. Various chemical wastes and solutions were improperly stored or disposed of on the site. Nearby water wells were discovered to contain VOCs in 1981. Testing has shown that the groundwater, surface water, and soil contain VOCs. The contamination threatens the domestic groundwater supplies for about 20,000 residents, Lake Denmark, and nearby streams.

Response actions included closing the contaminated wells (1981) and removing tanks, drums, and contaminated soil (1993). In 1994 the State of New Jersey recommended groundwater extraction and air-stripping. The system is currently being designed.

Radiation Technology, the PRP, signed a Consent Order (1983) and AOC (1987) that require the company to fund the site investigations and remediation work.

Sources: PNPLS-NJ; Rockaway Township Public Library, 61 Mt. Hope Road, Rockaway Township, NJ, 07866.

Reich Farm

Ocean County

This illegal dump site in Dover Township was used by a waste hauling operation in 1971. Various wastes, including organic solvents, sludges, plastics, and resins, were buried there in trenches. Testing showed that the groundwater and soil contain VOCs, which contaminated nearby wells. About 106,000 residents use the local groundwater for domestic purposes. The contamination also threatens Toms River.

Response actions included removing about 5,000 drums of waste and contaminated soil (1972–1974). Nearly 150 wells were closed in 1974. Affected residents were connected to the nearby municipal water supply. The EPA recommended groundwater extraction and treatment through air-stripping and carbon adsorption, soil excavations, and backfilling. Construction of these remedies is under way.

A number of PRPs have been identified by the EPA. A 1990 Consent Decree with Union Carbide requires the company to conduct the remediation work.

Sources: PNPLS-NJ; Ocean County Library, 101 Washington Street, Toms River, NJ, 08753.

Renora Incorporated

Middlesex County

This waste transfer station near Bonhamtown operated from 1978 to 1982. Various hazardous materials, including waste oils, were stored, processed, or abandoned on the site. Improper handling, spills, and leakage contaminated the groundwater and soil with heavy metals, VOCs, PAHs, and PCBs. The contamination threatens the groundwater supplies for the 3,000 residents of Bonhamtown in Edison Township.

Response actions included removing drummed waste and contaminated soil during the 1980s. In 1987 the EPA recommended excavating and treating contaminated soil by bioremediation and groundwater extraction and treatment. PCB-contaminated soil was removed from the site in 1990. Construction of the soil and groundwater remedies should be under way by 1997.

Sources: PNPLS-NJ; Edison Main Library, 340 Plainsfield Drive, Edison, NJ, 08817.

Rockaway Borough Well Field

Morris County

A 2-square-mile area of groundwater in Rockaway Borough was discovered to contain high amounts of VOCs in 1980. Contaminated municipal wells were subsequently shut down. Further investigations identified at least six point sources for the contamination. The local groundwater serves the domestic needs of about 11,000 residents.

Response actions included connecting affected residents to clean water supplies in 1981. An EPA investigation in 1985 identified PRPs and recommended groundwater extraction and treatment through air-stripping/chemical precipitation. Additional studies are in progress.

The EPA has identified six PRPs. A 1994 Consent Decree with Thiokol Corporation requires the company to undertake the remediation work and to pay $800,000 toward past clean-up costs.

Sources: PNPLS-NJ; Rockaway Burough Municipal Comples, 1 East Main Street, Rockaway Township, NJ, 07866.

Rockaway Township Wells
Morris County
This 2-square-mile area of VOC-contaminated groundwater in Rockaway Township was discovered in 1979–1980. Further investigations identified gas stations, freight yards, and industrial sites as point sources. VOCs were derived from chlorinated solvents and fuel components. The contamination threatens Meadow Brook and the groundwater that serves the domestic needs of about 13,000 local residents.

Response actions included constructing an air-stripping unit and conducting a site investigation. Contaminated wells were outfitted with an air-stripping and carbon adsorption system in 1980. In 1993 the EPA recommended groundwater extraction and upgrading the present air-stripping unit. Construction is in progress.

Sources: PNPLS-NJ; Rockaway Township Public Library, 61 Mt. Hope Road, Rockaway Township, NJ, 07866.

Rocky Hill Municipal Well
Somerset County
This 2-acre area of VOC-contaminated groundwater in Rocky Hill Borough was discovered in 1978. Affected municipal wells were closed between 1978 and 1982 until an air-stripper could be installed to clean the groundwater. In 1988 a septic tank system was identified as the probable point source. The contamination threatens the water supplies for about 8,000 residents.

Response actions included installing an air-stripper and conducting a site investigation. In 1988 the EPA recommended groundwater extraction and air-stripping, sealing contaminated wells, connecting affected residents to clean water supplies, and groundwater monitoring. Construction is under way.

Sources: PNPLS-NJ; Somerset County Library, North Bridge Street, Bridgewater, NJ, 08807.

Roebling Steel Company
Burlington County
Roebling Steel Company has manufactured steel and wire and processed polymers, insulation, and refrigeration units on this site in Florence since the 1920s. Waste products were stored, buried, or discharged on the premises, including sludge lagoons and slag heaps. Testing has shown that the groundwater and soil contain asbestos, PCBs, heavy metals, and PAHs. The chemicals have been found on adjacent properties, including a playground. The contamination threatens the recreational Delaware River, associated wetlands, and water wells within 3 miles of the site, which serve about 13,000 residents.

Response actions included removing surface debris, containers, drums, baghouse dust, and various hazardous liquid wastes such as cyanide, sulfuric acid, and phosphoric acid. In 1990 the EPA recommended additional removals of debris, drums, and contaminated soil (completed in 1991). Slag piles were remediated in 1994–1996. Other studies are in progress.

The EPA reached a settlement in 1992 with the PRP.

Sources: PNPLS-NJ; Florence Township Public Library, 1350 Hornberger Avenue, Roebling, NJ, 08554.

Sayreville Landfill
Middlesex County
This former landfill near Sayreville operated from 1970 to 1977. It received a variety of municipal and industrial wastes, including hazardous chemicals. Testing has shown that the groundwater and soil contain heavy metals, phenols, VOCs, PAHs, pesticides, and chloroform. The contamination threatens the South River, associated wetlands, and the drinking water supplies of about 70,000 residents,

Response actions included excavating drummed waste in 1982. In 1990 the EPA recommended the removal and incineration of drummed waste, capping, and installing runoff controls and a methane gas collection system. Construction is in progress.

AOCs (1986, 1991) require the PRPs to conduct the site investigations and clean-up work.

Sources: PNPLS-NJ; Sayreville Free Public Library, 1050 Washington Road, Parlin, NJ, 08859.

Scientific Chemical Processing
Bergen County
This former waste processing facility near Carlstadt operated from the 1970s to 1980. Various hazardous wastes were stored in tanks and drums on the property. Illegal dumping also occurred. Testing revealed that the groundwater and soils contain VOCs, PCBs, PAHs, heavy metals, petrochemicals, and dieldrin. The chemicals have penetrated Peach Island Creek. The contamination threatens the health of about 15,000 nearby residents, industrial water supplies, Peach Island Creek, and tidal wetlands.

Response actions included removing tanks (1985–1986). A site investigation (1985–1990) recommended constructing a slurry wall and an interim groundwater extraction and treatment system, which became operational in 1992. Other studies are in progress.

Executives at Scientific Chemical Processing were convicted in federal court of illegal disposal activities. An AOC (1983), UAO (1985), and UAO (1990) require about 140 PRPs to participate in the remediation work.

Sources: PNPLS-NJ; Dermody Public Library, 420 Hackensack Street, Carlstadt, NJ, 07072.

Sharkey Landfill
Morris County
Sharkey Landfill operated near East Hanover from 1945 to 1975. It received a variety of municipal and industrial wastes, including VOCs and heavy metal sludges. Testing has shown that the groundwater and soil contain VOCs, lead, cadmium, and chromium. Leachate has penetrated the Whippany and Rockaway Rivers. The contamination threatens drinking water supplies for the towns of Montville and East Hanover, the Whippany and Rockaway Rivers, and associated wetlands.

In 1986 the EPA recommended capping, gas venting, groundwater extraction and treatment, surface runoff controls, fencing, and long-term monitoring. Construction will begin shortly.

Sources: PNPLS-NJ; Morris County Public Library, 30 East Hanover Avenue, Whippany, NJ, 07981.

Shieldalloy
Gloucester County
Chromium alloy and other metal products are manufactured on this property in the Borough of Newfield. In the 1970s waste products were discharged into unlined lagoons, which contaminated the groundwater and soil with VOCs and chromium. Hudson's Branch is also polluted. The contamination threatens Hudson's Branch, the Maurice River, and water wells that supply about 60,000 people.

Response actions included an interim groundwater pump-and-treat system that was installed in 1979. A new system that uses electrochemical treatment was built in 1988. Other studies are in progress.

The PRP, Shieldalloy Corporation, signed AOCs (1984, 1988) that require the company to undertake the site investigations and remediation work.

Sources: PNPLS-NJ; Newfield Library, Catawba Avenue, Newfield, NJ, 08344.

South Brunswick Landfill
Middlesex County
This former landfill operated from the 1950s to 1978. It received a variety of wastes, including hazardous chemicals and pesticides. In 1980 testing revealed that the groundwater, soil, surface water, and stream sediments contained VOCs and heavy metals, including iron. The contamination threatens the water supplies of nearby residents as well as a nearby stream.

Response actions included the installation of a leachate collection system and slurry wall, capping, and gas ventilation system. The site will be monitored until at least 2016.

The PRPs have conducted the site investigations and remediation work according to an AOC.

Sources: PNPLS-NJ; South Brunswick Public Library, 110 Kingston Lane, Monmouth Junction, NJ, 08852.

South Jersey Clothing Company
Atlantic County
Military uniforms were assembled on this site in Minotola in the 1970s. The finished clothing was dry-cleaned using trichloroethylene, and the wastewaters were discharged on the ground. Sampling in 1981 revealed that the groundwater and soil contain VOCs. The contamination threatens wells that serve about 9,000 residents, agricultural wells, and streams.

Response actions included excavating contaminated soils (1981) and installing an interim groundwater pump-and-treat system (1985). The EPA recommended soil vapor extraction, groundwater extraction and air-stripping, and long-term monitoring. Construction began in 1996.

The PRP, South Jersey Clothing Company, does not have the financial resources to undertake the clean-up work. Federal funds will be used to remediate the site.

Sources: PNPLS-NJ; Buena Borough Municipal Building, 616 Central Avenue, Minotola, NJ, 08341.

Spence Farm
Ocean County
This former farm near New Egypt was used as a hazardous waste landfill from the 1950s to the 1970s. Wastes included drummed solvents and laboratory wastes that leaked from rusted containers. The groundwater and soil contain VOCs, phthalates, phenols, heavy metals, PAHs, and PCBs. The contamination threatens water wells that serve about 2,000 local residents, agricultural wells, and Crosswicks Creek.

Response actions included the installation of monitoring wells (1982). In 1984 the EPA recommended removing surface and buried debris and waste, excavating contaminated soil, and groundwater monitoring. Removal actions were completed by 1990. PCB-contaminated soil was discovered during the excavations and is being removed. Groundwater monitoring indicated that the water is uncontaminated.

The PRP signed a 1985 AO to remediate the site.

Sources: PNPLS-NJ; Ocean County Library, 101 Washington Street, Toms River, NJ, 08753.

Swope Oil & Chemical Company
Camden County
Chemical compounds, oils, paints, and solvents were reprocessed on this site in Pennsauken Township from 1965 to 1979. Waste materials were discharged into a sludge lagoon on the premises. The groundwater and soil contain PCBs, VOCs, phthalates, and heavy metals. The contamination threatens private and public wells within 3 miles of the site, which serve about 20,000 residents.

Response actions included removing tanks and sludge, capping, and erecting fencing (1984). In 1985 the EPA recommended removing additional tanks and buildings, capping, and soil excavation. A soil vapor extraction system became operational in 1996. Groundwater will be monitored for 5 years to determine if contamination is decreasing.

The PRPs are conducting the clean-up work.

Sources: PNPLS-NJ; Clerk's Office, Township of Pennsauken, 5605 North Crescent Blvd., Pennsauken, NJ, 08110.

Syncon Resins
Hudson County
Alkyd resin carriers for the paint and varnish industries were manufactured on this industrial site near Newark. Various wastes were stored in tanks, drums, and unlined lagoons. The groundwater and soil contain VOCs, heavy metals (lead and nickel), PCBs, DDT, and aldrin. The contamination threatens the Passaic and Hackensack Rivers and the groundwater drinking supplies for about 10,000 residents within 3 miles of the site.

Response actions included removing drummed waste and surface debris (1982–1984) and erecting fencing (1990). In 1986 the EPA recommended decontaminating the buildings, pumping out storage tanks, capping, soil flushing, and a groundwater extraction and treatment system. Other studies are in progress.

Sources: PNPLS-NJ; Kearny Public Library, 318 Kearny Avenue, Kearny, NJ, 07302.

Tabernacle Drum Dump
Burlington County
Drummed solvents, paints, and sludges were dumped on this property near Bozarthtown in 1976–1977. Severe leakage from the rusting containers was noted in 1984. Follow-up sampling by the State of New Jersey revealed that the soil and groundwater contain VOCs and heavy metals (chromium, cyanide, and lead). The contamination threatens private wells in the area that serve about 500 people, agricultural wells, and nearby streams.

Response actions included removing drums and contaminated soils in 1984. The EPA recommended a final remediation plan of groundwater extraction and air-stripping, which became operational in 1994 and will run to 1999.

A UO (1984) and CD (1989) require two PRPs to perform the remediation work.

Sources: PNPLS-NJ; Tabernacle Township Building, 163 Carranza Road, Tabernacle Township, NJ, 08088.

Universal Oil Products
Bergen County
Chemicals were produced on this 75-acre site near East Rutherford from 1932 to 1979. Various wastewaters, solvents, and chemicals were discharged into unlined lagoons on the property. Testing has shown that the groundwater and soil contain VOCs, PCBs, lead, and PAHs. The contamination threatens wells that serve about 37,000 local residents, the municipal wells for Wallington Township that are within 3 miles of the site, and the Hackensack River Basin and associated wetlands.

Response actions included removing contaminated solutions and soil from the lagoons in 1990. A recently completed site investigation recommends thermal desorption of contaminated soil, capping, and leachate collection. Construction should begin by 1997.

An AOC with the PRP requires the company to undertake the site investigations and remediation work.

Sources: PNPLS-NJ; East Rutherford Memorial Library, 143 Boiling Spring Avenue, East Rutherford, NJ, 07073.

Upper Deerfield Township Landfill
Cumberland County
This former landfill in upper Deerfield Township operated from 1938 to 1983. It received a variety of municipal and industrial wastes, including heavy metals and various sludges. Testing by state officials in 1983 revealed that the groundwater and soil contain low levels of VOCs and heavy metals, including mercury. The contamination has penetrated nearby water wells and threatens the drinking water supplies for about 7,000 residents.

Response actions included supplying affected residents with clean water supplies (1983–1986). A site investigation was completed from 1987 to 1991. In 1991 the EPA ruled that no remediation actions were necessary. A comprehensive monitoring plan for the air and groundwater is being developed.

The EPA has failed to identify any viable PRPs.

Sources: PNPLS-NJ; Cumberland County Library, 800 East Commerce Street, Bridgeton, NJ, 08302.

U.S. Radium Corporation
Essex County
A radium extraction and processing company operated on this site in the city of Orange from 1917 to

1926. Radioactive waste was left on the property or used as fill on construction projects in the area. About 110 properties have been identified as containing radon gas, radium-226, or other radioactive materials. About 50,000 people live within 1 mile of the manufacturing site.

Response actions included the installation of radon gas venting systems and gamma radiation shielding in affected homes and buildings. The manufacturing site was fenced in 1989. The excavation of radium-contaminated material from about 120 residential and business properties is under way and should be completed in 1998. Other studies are in progress.

Source: PNPLS-NJ.

Ventron

Bergen County

A chemical processing plant operated on this site in Wood-Ridge Borough from 1929 to 1974. Chemical wastes were buried on the property. A food distribution center now occupies the site. Testing has shown that the groundwater, soil, and nearby stream sediments contain heavy metals (mercury and zinc) and PCBs. Aquatic life in nearby wetlands and Berry's Creek contain elevated levels of PCBs. The contamination threatens the drinking water supplies of about 12,000 nearby residents as well as streams and associated wetlands.

Response actions included removing mercury-contaminated soils on ten nearby properties in 1990. Site investigations are in progress.

Sources: PNPLS-NJ; Carlstadt Public Library, 420 Hackensack Street, Carlstadt, NJ, 07062.

Vineland Chemical Company

Cumberland County

Vineland Chemical Company has manufactured arsenic-based herbicides on this site in Vineland since 1950. Various wastes were contained in unlined lagoons, and arsenic-tainted solids were stored in unprotected piles on the site. Testing showed that the groundwater and soil contain heavy metals (arsenic, cadmium, antimony, and lead) and VOCs. Workers on the property have elevated arsenic concentrations in their bloodstreams. The contamination threatens Maurice River, recreational Union Lake, and wells that supply drinking water to about 60,000 residents.

Response actions included fencing contaminated areas and removing hazardous chemicals (1992–1993). Soil flushing to treat the arsenic-contaminated soil should begin by 1997. Construction of a groundwater extraction and treatment system is expected to be under way by 1997. Once operational, the system will be active for 13 years. Clean-up actions for the contaminated river and lake sediments are also being planned.

A 1994 Consent Decree requires the PRP to contribute toward the site investigations and clean-up work.

Sources: PNPLS-NJ; Vineland Public Library, 1058 East Landis Avenue, Vineland, NJ, 08360.

Vineland State School

Cumberland County

Unregulated discharge, burial, or incineration of wastes occurred from the 1950s to the 1980s at the Vineland State School in Vineland. Wastes included pesticides, transformer oils, gasoline, asbestos, and various medical and maintenance chemicals from general operations. The groundwater and soil contain DDT, dieldrin, VOCs, phthalates, PCBs, and heavy metals (lead and mercury). The contamination threatens the local groundwater supplies that serve the domestic needs of about 14,000 residents within 3 miles of the site.

Response actions included connecting affected residents to clean water supplies, removing contaminated soil, capping, and fencing (1988). Groundwater monitoring to ensure that the initial response actions are cleaning the groundwater is in progress.

Sources: PNPLS-NJ; Vineland Public Library, 1058 East Landis Avenue, Vineland, NJ, 08360.

W. R. Grace

Passaic County

Thorium and rare earth elements were processed from monazite ore on this site in Wayne Township. Various waste products, including radioactive materials, were stored or buried on the site. Surface runoff contaminated adjacent properties. Soil and creek sediments contain radioactive thorium and related decay products. The contamination threatens the health and drinking water supplies of about 50,000 nearby residents.

Response actions included excavating contaminated soil (1985–1993). An ongoing site investigation should be completed by 1997 and will make final clean-up recommendations.

Source: PNPLS-NJ.

Waldick Aerospace Devices

Monmouth County

This former metal fabrication and electroplating facility near Girt operated from 1979 to 1984. Production wastes included degreasers, plating solutions, and various rinses. Wastewater was discharged on the ground. Testing has shown that the groundwater and soil contain VOCs and heavy metals, including cadmium and chromium. About 40,000 people live within 3 miles of the area. The contamination also threatens agricultural wells in the area.

Response actions included removing contaminated soil and drummed waste (1983–1985). An EPA investigation was completed in 1986–1987 and recommended razing structures, excavation and low-temperature thermal treatment of VOC-contaminated soil, and the off-site removal of heavy metal–contaminated soil. A groundwater extraction and air-stripping system is also planned.

Court actions led to the conviction and bankruptcy of the PRPs in 1984. Clean up is being conducted with federal funds.

Sources: PNPLS-NJ; Wall Township Library, 2700 Allaire Road, Wall, NJ, 07719.

Welsbach and General Gas Mantle (Camden Radiation)

Gloucester County

Gas mantles were manufactured on this site near Camden and Gloucester. The neighborhood is now mixed residential/commercial, and testing in 1981 revealed high gamma radiation levels in the area. The contamination threatens the health of on-site workers and residents.

Response actions included the purchase of highly radioactive properties by the State of New Jersey, temporary capping, and the relocation of affected persons. Other site investigations are in progress.

Source: EPA Publication 9320.7-071, Volume 3, No. 1.

White Chemical Corporation

Essex County

The White Chemical Corporation manufactured acid chlorides and flame-resistant compounds on this site in Newark from 1983 to 1990. Various hazardous wastes were improperly stored or handled and frequently leaked or spilled. Investigations have shown that a catastrophic event could release the toxic substances at levels that would endanger the health of about 60,000 residents and workers within 3 miles of the site.

Response actions included the emergency removal of drummed waste in 1990. Subsequent removal actions occurred from 1991 to 1993. A detailed site investigation to address the possibilities of soil and groundwater contamination should be under way by 1997.

PRPs have participated in the removal actions.

Source: PNPLS-NJ.

Williams Property

Cape May County

In 1979 several hundred drums of liquid wastes were emptied on the ground at this location in Swainton. The solutions contaminated nearby water wells. Testing has shown that the soil and groundwater contain phthalates, heavy metals, and VOCs. The contamination threatens public and private wells within 3 miles of the site, which supply water to about 5,000 residents. The local groundwater is also pumped for agricultural purposes.

Response actions included removing contaminated sludge and soil in 1980. Contaminated wells were closed in 1985, and affected residents were connected to clean water supplies. In 1987 the EPA recommended groundwater extraction and treatment and revegetation. Soil treatments were completed in 1991. Beginning in 1996, the groundwater will be treated by biological, carbon adsorption, hydrogen peroxide, and ultraviolet techniques.

After a prolonged legal battle (1984–1993), the PRP, Wheaton Industries, signed a 1993 Consent Decree with the EPA that requires the company to reimburse past clean-up costs.

Sources: PNPLS-NJ; Cape May Public Library, Mechanic Street, Cape May, NJ, 08210.

Wilson Farm

Ocean County

Located in Plumsted Township, this former farm operated as a dump site for Thiokol Chemical Company in the 1960s to 1970s. Wastes included sludges, inorganic chemicals, and solvents. Testing has shown that the groundwater, soil, and nearby stream sediments contain VOCs, pesticides, and heavy metals. The contamination threatens Borden's Run Creek, Collier's Mill Lake, a wildlife refuge, and private water wells that supply about 3,000 residents.

Response actions included removing drummed waste, sludges, and contaminated soil in 1980. A site investigation conducted from 1987 to 1992 reported no additional actions were warranted. Groundwater monitoring is continuing to ensure clean-up goals are met.

Sources: PNPLS-NJ; Ocean County Library, 101 Washington Street, Toms River, NJ, 08753.

Woodland Route 532 Dump

Burlington County

This illegal chemical waste dump in Woodland Township operated from 1956 to the 1960s, receiving drummed chemical waste and various sludges. Testing has shown that the groundwater and soil contain VOCs, SVOCs, heavy metals, and pesticides. The contamination threatens the New Jersey Pinelands, Goodwater Run, a cranberry bog, and wells that serve about 1,000 nearby residents.

Response actions included erecting fencing in 1986. Studies by the State of New Jersey and the EPA recommended the excavation and removal of contaminated soil and a groundwater extraction and treatment system using air-stripping, chemical precipitation, and biological treatment. Soil remedies were completed in 1991. The groundwater system will be operational by 1997.

AOCs (1990, 1991) require the PRPs to conduct the site investigations and remediation work.

Sources: PNPLS-NJ; Woodland Township Municipal Building, Main Street, Chatsworth, NJ, 08019.

Woodland Route 72 Dump

Burlington County

This illegal industrial dump near Manahawkin operated from the 1950s to the 1960s. It received drummed wastes, sludges, and radioactive material. Testing discovered that the groundwater and soil contain VOCs, SVOCs, heavy metals, pesticides, PCBs, acids, DDT, and gamma radiation from thorium, radium, and uranium. The contamination threatens Pinelands National Reserve, Pope Branch and associated wetlands, blueberry and cranberry fields, and the local drinking water supplies for about 1,000 residents.

Response actions included fencing in 1986. Site investigations by the State of New Jersey and the EPA recommended removing contaminated soil and installing a groundwater extraction and treatment system using air-stripping, chemical precipitation, and bioremediation. The soil removal was completed in 1991. The groundwater system is expected to be operational by 1997.

AOCs (1990, 1991) require the PRPs to conduct the site investigations and remediation work.

Sources: PNPLS-NJ; Woodland Township Municipal Building, Main Street, Chatsworth, NJ, 08019.

NEW MEXICO

AT & SF

Bernalillo County

The Atchison, Topeka, and Santa Fe (AT & SF) Railway Company operated this wood treatment facility in the city of Albuquerque from 1907 to 1972. Lumber for railroad ties and bridge timbers was treated with creosote and oil. Waste solutions and sludges were discharged into an unlined pit on the site. An inspection by the EPA revealed that the soil and groundwater are contaminated with arsenic, barium, lead, benzo(a)pyrene, naphthalene, acenaphthylene, anthracene, phenanthrene, dibenzofuran, and xylene (mostly constituents from creosote). The contaminated aquifer is the water source for about 200 private and city wells within 4 miles of the site, which supply about 44,000 people.

A site investigation study was completed in 1996. A final remediation plan has yet to be selected.

AT & SF signed an AO with the EPA in 1994 to finance the site investigation and feasibility study.

Sources: PNPLS-NM; Albuquerque Public Library, 501 Copper Avenue Northwest, Albuquerque, NM, 87102.

AT & SF (Clovis)

Curry County

Located near the town of Clovis, this Atchison, Topeka, and Santa Fe Railroad property operated as a refueling and train washing station from the mid-1950s to the 1970s. Contamination in on-site water wells was discovered in the late 1970s. The soil and surface water in a shallow lake are contaminated with boron, fluoride, chloride, and petroleum hydrocarbons. About 31,000 people, some of whom depend on private water wells, live within 3 miles of the site.

Response actions included fencing the site and building a dike around the lake to stop runoff contamination (1988–1992). The lake has been evaporated and the sediments removed and biodegraded on-site. The detoxified soil will later be capped. Groundwater monitoring is continuing.

AT & SF is the single PRP and is remediating the site through an AO with the EPA (1983).

Source: PNPLS-NM; Clovis-Carver Public Library, Fourth and Mitchell Streets, Clovis, NM, 88108.

Cal West Metals

Socorro County

From 1979 to 1981 Cal West Metals recycled batteries and smelted the recovered lead at this plant a few thousand feet north of the town of Lemitar. EPA tests from 1984 to 1989 showed that the soil and pond sediments on-site contained high levels of lead, threatening the local groundwater. Municipal and private wells within 3 miles of the site supply water to about 1,000 residents.

Response actions included an EPA site investigation (1990–1992). The remediation plan (1995–1996) calls for groundwater monitoring, excavation and treatment of contaminated soil and sediments, and reburial of the sediments on-site.

The EPA has identified three PRPs, one of whom has financial resources that can be applied to the clean-up cost.

Sources: PNPLS-NM; Socorro Public Library, 404 Park Street Southwest, Socorro, New Mexico, 87801.

NEW MEXICO

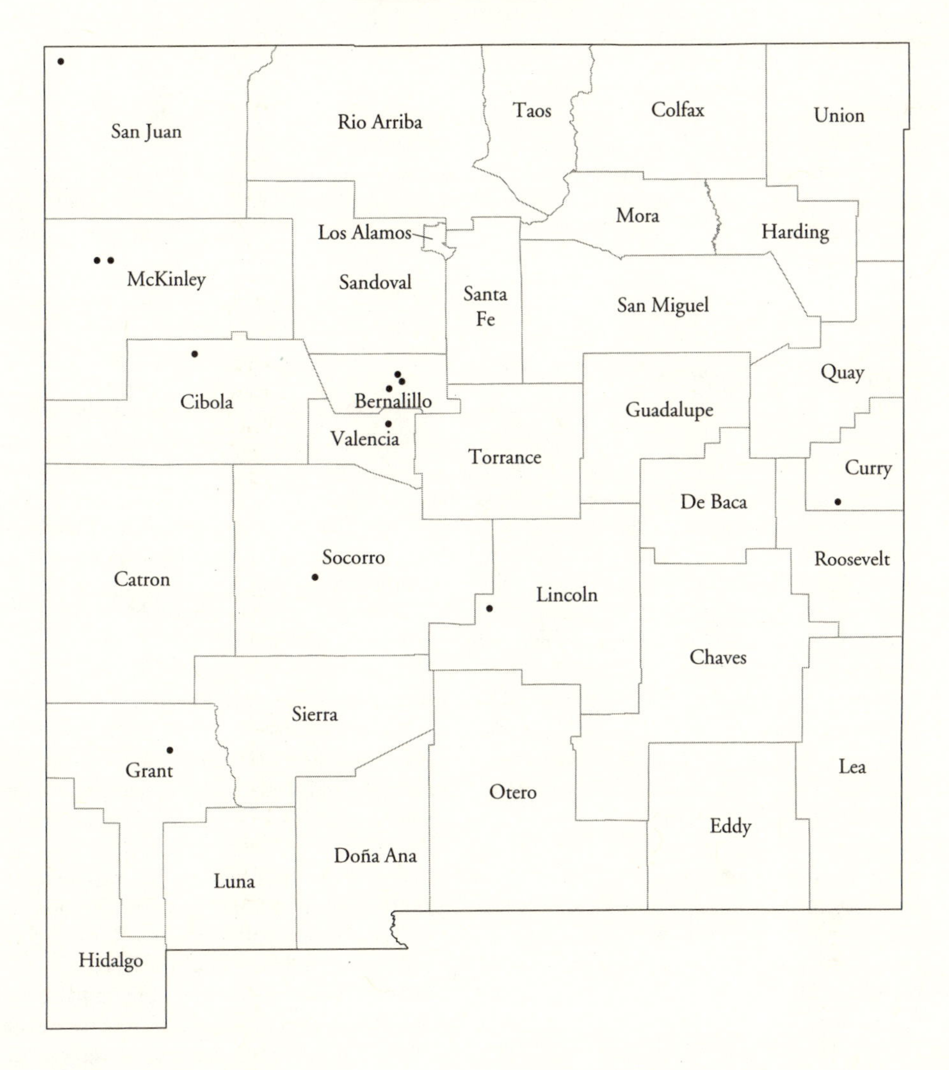

Cimarron Mining Corporation
Lincoln County
This site near Carrizozo recovered gold from transported ore using cyanide leaching. It operated from 1979 to 1982, and Cimarron Mining Company filed for bankruptcy the following year. In 1984 the EPA found that the groundwater and soil were contaminated with cyanide and heavy metals (lead). About 30 municipal drinking wells within 3 miles of the site supply drinking water to 1,000 residents and irrigate farmland.

Response actions included fencing the site (1987) and removing high-hazard waste material (1991–1992). The final remediation plan calls for the extraction and transport of contaminated groundwater to a publicly owned treatment plant (1991–1996). The solidification, stabilization, and disposal of contaminated soils and tailings on-site should be under way by 1997.

Sources: PNPLS-NM; Carrizozo City Hall, 100 Fifth Street, Carrizozo, NM, 88301.

Cleveland Mill
Grant County
Lead, zinc, and copper ore was mined on this site about 5 miles northeast of Silver City from 1910 to 1919. A site examination in 1985 revealed uncovered and unlined piles of sulfidic tailings from the mill operations, which have washed into Little Walnut Creek. The creek water is acidified and contaminated with heavy metals, including lead, zinc, arsenic, cadmium, and beryllium. A number of private wells

within 3 miles of the site supply water to about 1,200 nearby residents.

Response actions included an EPA-funded site investigation (1990–1993). The recommended remediation plan consists of excavation and removal of contaminated soil and sediments and revegetation, which will probably begin in 1997.

A 1995 Consent Decree was signed between the EPA and three PRPs to finance the clean-up plan.

Sources: PNPLS-NM; Silver City Library, Silver City, NM, 88061.

Homestake Mining Company
Cibola County
This mining and processing facility is located about 6 miles north of Milan. Built in 1958, the mill is still processing uranium. Groundwater contamination was first detected in 1961. About 250 acres of tailings piles and two large tailings ponds have contaminated soil and shallow groundwater with radium, selenium, uranium, and radon. The contamination threatens the water supplies of four subdivisions within 2 miles of the site.

Response actions included connecting affected residents to alternate water supplies (1985–1989). Radon covers will be installed over the tailings piles. The injection of clean water downgradient from the contamination plume has driven it back about a mile and contained it on-site. Over a trillion gallons of contaminated water have been extracted.

Homestake Mining Company, the PRP, has financed the remediation efforts.

Sources: PNPLS-NM; New Mexico State University, Grants Library, 1500 Third Street, Grants, NM, 87020.

Lee Acres Landfill
San Juan County
This federal land was leased by San Juan County as a landfill with waste trenches and unlined lagoons from 1962 to 1985. Toxic vapors caused local residents to become ill in 1985. Chemical tests showed the soil and shallow groundwater to be contaminated with 1,1-dichloroethane, 1,1,1-trichloroethane, 1,2-dichloroethylene, benzene, and manganese. About 600 people live within 3 miles of the site, and the contamination threatens the private wells in the area.

Response actions included fencing the site and filling in the liquid waste lagoons (1985). Affected residents were connected to clean water supplies in 1987. A site investigation and feasibility study completed between 1992 and 1996 recommended removing toxic sludge and sediments, capping part of the landfill, and groundwater monitoring. Natural attenuation of the contaminated groundwater plume is also expected.

Three PRPs will finance the cost of the site investigation and clean-up plan.

Sources: PNPLS-NM; Farmington Public Library, 100 West Broadway Street, Farmington, NM, 87401.

Pagano Salvage
Valencia County
A metal salvage company operated at this location near Los Lunas. EPA testing in 1987 identified pesticides and PCBs in soil and debris from waste oils. The contamination threatens the drinking water supplies of about 11,000 people within 3 miles of the site. Local groundwater is also used for crop irrigation.

Response actions included removing the worst soil contamination (1987–1989). Additional soil and debris were excavated in 1990, and the site was deleted from the NPL list in 1992. Monitoring is continuing.

Five PRPs identified by the EPA refused to do the work but later settled with the EPA through a Consent Decree.

Sources: PNPLS-NM; Los Lunas Public Library, 460 Main Street, Los Lunas, NM, 87031.

Prewitt Abandoned Refinery
McKinley County
This abandoned oil refinery near Prewitt included waste pits and tanks. The soil and groundwater are polluted with lead, benzene, toluene, ethylbenzene, asbestos, PAHs, xylene, and petroleum sludges. Nearby private wells have been contaminated. About 1,600 people use local groundwater drawn from public and municipal wells within 3 miles of the site.

Response actions included fencing and installing carbon filtration systems on contaminated wells or providing clean water hook-ups to affected residents (1989–1990). A site investigation completed in 1992 recommended groundwater extraction and treatment, excavation and disposal of asbestos and lead-bearing soils, landfarming hydrocarbon-soaked soil, and soil vapor extraction. Construction of these remedies is under way.

PRPs (ARCO, Rexene, and El Paso Natural Gas), through an agreement with the EPA (1993), are required to finance the site investigation and clean up.

Sources: PNPLS-NM; Prewitt Fire House, Highway 66, Prewitt, NM, 87045.

Rinchem Company
Bernalillo County
Located in an industrial part of Albuquerque, Rinchem transports, stores, processes, and recycles hazardous waste. Tests have shown that the soil and groundwater on-site are contaminated with indus-

trial solvents such as trichloroethylene, trichloroethane, acetone, and methylisobutylketone.

Response actions have included the installation of groundwater monitoring wells. A final remediation plan will be recommended when the site investigation is completed.

Seven PRPs have been identified, and Rinchem is participating in the site investigation.

Sources: PNPLS-NM; Albuquerque Public Library, 501 Copper Avenue, Northwest, Albuquerque, NM, 87102.

South Valley
Bernalillo County
Companies occupying this site in an industrial section of Albuquerque handle and store organic chemicals and petroleum fuels. Spills, leaks, and improper handling have led to the contamination of soil and groundwater with halocarbons (trichloroethane), benzene, ethylbenzene, toluene, and xylene, which resulted in the closure of two municipal wells in 1981. The chemicals have penetrated all three aquifers that supply water to the city of Albuquerque. Chlorinated solvent leaks also caused the shut down of about 20 private wells in the area. The greatest threat is to a residential area immediately north of the site.

Response actions included digging out soils contaminated with chlorinated solvents and oil (1988). Affected wells were closed and plugged, and shallow groundwater pump-and-treat systems were installed in 1992. Deeper groundwater extraction wells were drilled in 1995 and became operational in 1996. Other systems yet to be installed will recover free-product oil and soil vapors.

Twelve PRPs have been identified by the EPA. Nine of these companies are contributing to the cost of the site investigation and clean-up operations.

Sources: PNPLS-NM; Albuquerque Public Library, 501 Copper Avenue, Northwest, Albuquerque, NM, 87102.

United Nuclear Corporation
McKinley County
This site contains about 100 acres of radioactive tailings and an abandoned mill that processed uranium ore from 1977 to 1982. A 1979 spill resulted in the contamination of Rio Puerco with 100 million gallons of radioactive slurry. EPA studies from 1986 to 1988 revealed that the soil and shallow groundwater are contaminated with heavy metals, nitrates, sulfates, radium, thorium, aluminum, ammonia, and iron. Four water wells are within 4 miles of the site; a Navajo reservation is immediately north of the site.

A site investigation completed in 1988 recommended groundwater extraction and treatment systems to remediate the contamination. The groundwater systems were installed in 1989, and extracted water is being evaporated. The mill was decommissioned and the tailings site covered (1993). A radon cover over the tailings will be in place by 1997.

The EPA issued a UAO to United Nuclear Corporation to implement and finance the clean-up plan.

Sources: PNPLS-NM; Gallup Public Library, 115 West Hill Avenue, Gallup, NM, 87301.

NEW YORK

Action Anodizing and Plating
Suffolk County
Metal electroplating has been performed on this site near Nassau since 1968. Activities include cadmium plating, chromate coating, metal dyeing, degreasing, and anodizing electronic aluminum parts with sulfuric acid. Various wastewaters were discharged into pits on the property until 1980. The groundwater and soil contain high levels of heavy metals. The contamination threatens Amityville Creek, Great South Bay, related wetlands, and public drinking water wells within 3 miles of the site, which serve over one million people.

Response actions included a site investigation from 1989 to 1992. The EPA concluded that contamination levels were low enough to warrant a "no action" ruling. Monitoring is continuing.

Sources: PNPLS-NY; Babylon Town Hall, 200 East Sunrise Highway, Lindenhurst, NY, 11757.

American Thermostat Company
Greene County
Thermostats were manufactured on this site in South Cairo from 1954 to 1985. Illegal waste disposal activities, including dumping chemical wastes and sludges into the septic system or on the grounds of the 8-acre property, were documented by the State of New York in 1981. Subsequent tests showed that the groundwater and soil contain high levels of VOCs, which have also penetrated nearby streams that flow into Catskill Creek. Private wells are also contaminated. The contamination threatens the local drinking water supplies of about 5,000 residents and recreational Catskill Creek and associated wetlands.

Response actions included connecting affected residents to clean water supplies. A groundwater extraction system and three air-stripping towers were built in the mid-1980s and have treated about 10 million gallons of water to date. A soil

NEW YORK

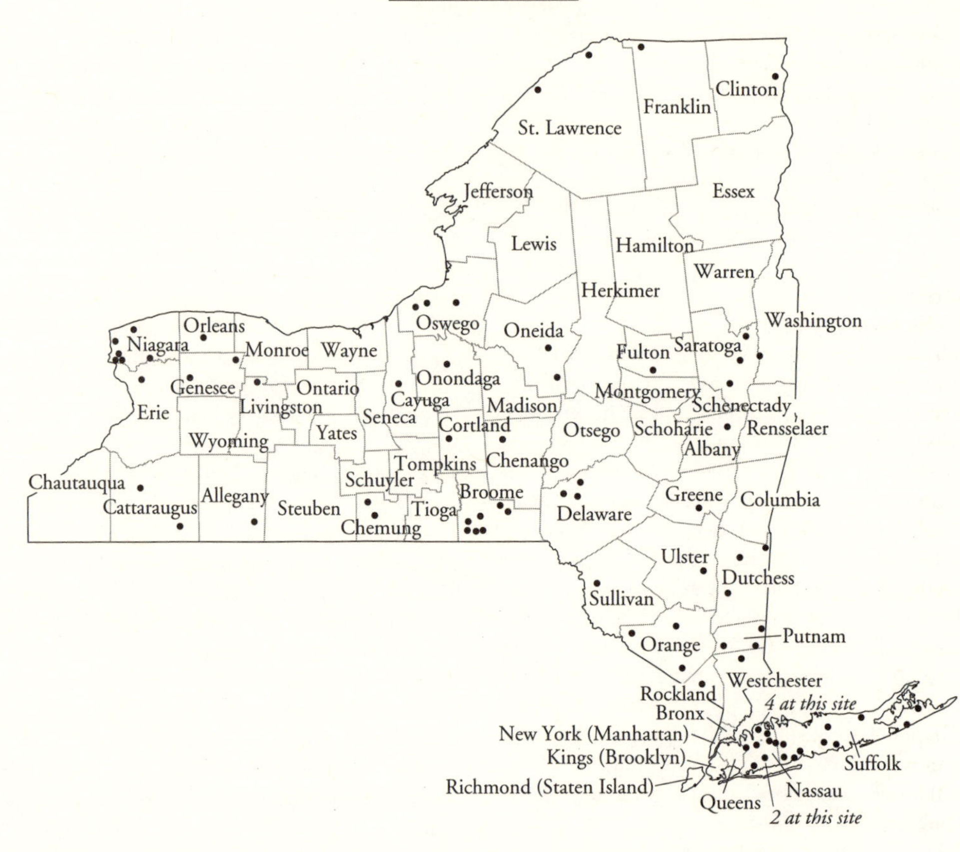

vacuum extraction system was installed in 1989. Permanent clean water supplies were constructed for affected residents in 1991–1992. Some contaminated soil will also undergo low-temperature thermal treatments.

The PRP conducted the initial response actions before the company declared bankruptcy in 1985. Subsequent work has been carried out with federal funds.

Sources: PNPLS-NY; Catskill Town Office, 439 Main Street, Catskill, NY, 12414.

Anchor Chemicals
Nassau County
Various chemicals were produced on this site near Hicksville from 1964 to 1984. Chlorinated organic solvents and other chemical liquids were stored in underground tanks that leaked. Testing in 1982 showed that the groundwater and soil contain excessive levels of VOCs. The contamination threatens private and public wells within 3 miles of the area, which serve nearly 100,000 residents. Groundwater is also used for agricultural needs.

Response actions included fencing the site. An investigation (1989–1997) that will recommend a final remediation plan is nearing completion.

The PRPs are undertaking the site investigation according to a 1989 AOC.

Sources: PNPLS-NY; Hicksville Library, 169 Jerusalem Avenue, Hicksville, NY, 11801.

Applied Environmental Services
Nassau County
This hazardous waste recycling facility near Glenwood Landing operated from 1980 to 1983. Fuels and liquid chemicals were stored in aboveground tanks on the property. Improper handling and storage resulted in numerous spills and leaks. The groundwater and soil contain VOCs, petrochemicals, and PCBs. The contamination threatens the local drinking water supplies for about 20,000 residents as well as streams, recreational fishing and swimming areas, Mott Cove, and Hempstead Harbor.

Response actions included digging a trench to capture leaking leachate and removing drummed chemical waste (1984–1985). A site investigation was conducted from 1987 to 1991. The EPA recommended soil vapor extraction and groundwater extraction and bioremediation. Construction should be completed in 1997.

An AOC (1987) and Consent Decree (1992) require the PRPs to conduct the site investigations and remediation work.

Sources: PNPLS-NY; Sea Cliff Village Library, Central Avenue, Sea Cliff, NY, 11579.

Batavia Landfill
Genesee County
This former landfill near Batavia operated from the 1960s to 1980. It received a variety of municipal and industrial wastes, including oils, spent solvents, and heavy metal sludges. Leachate from the landfill has entered nearby Galloway Swamp, a protected wetland. Testing has shown that the groundwater and soil contain heavy metals. phenols, and VOCs. The contamination threatens the local drinking water supplies for about 7,000 people and nearby streams and wetlands.

Response actions included removing drummed waste and contaminated soil (1990–1991). In 1993 the EPA recommended connecting affected residents to clean water supplies. A recently completed site investigation will recommend a final remediation plan.

Administrative Orders of Consent (1984, 1990, 1993) require seven PRPs to conduct the site investigations and remediation work.

Sources: PNPLS-NY; Richmond Library, 19 Ross Street, Batavia, NY, 14202.

Brewster Well Field
Putnam County
Groundwater in the well field supplying the town of Brewster was found to be contaminated with VOCs in 1978. Further investigations revealed that a dry cleaning operation was the likely point source. Various dry cleaning wastes and solutions had been discharged into a dry well on the property since the 1950s. Testing showed that the groundwater and soil contain VOCs, which have also penetrated the East Branch of the Croton River, part of New York City's water supply system. The contamination threatens the local drinking water sources for about 2,500 residents and nearby streams and wetlands.

Response actions included the city of Brewster's installation of an air-stripper to clean the contaminated water. In 1986 the EPA recommended expanding the extraction and treatment system, which has now been completed. Contaminated soils, sediments, and sludge were excavated and removed between 1989 and 1991.

Sources: PNPLS-NY; Brewster Public Library, 79 Main Street, Brewster, NY, 10509.

Brookhaven National Laboratory
Suffolk County
This 5,265-acre site near Upton has been managed by the U.S. Army, the Atomic Energy Commission, and the U.S. Department of Energy. Various activities have included nuclear and biomedical research and development, including the construction of particle accelerators and nuclear reactors. Solid and liquid wastes, including radioactive slurries, sewage sludge, animal carcasses, medicinal waste, and laboratory debris, were buried or discharged on the property. The groundwater and soil in 28 areas contain VOCs, radioactive materials, PCBs, PAHs, and heavy metals. The contamination threatens the Peconic River, associated wetlands, and wells that supply drinking water to about 16,000 residents.

Response actions included removing contaminated soil (1992–1993). Underground storage tanks have also been removed. A number of site studies are nearing completion and will recommend a final remediation plan.

Sources: PNPLS-NY; Brookhaven National Laboratory Library, Building 477A, Upton, NY, 11973.

Byron Barrel and Drum
Genesee County
This former gravel pit near Batavia was illegally used as a hazardous waste landfill in the 1970s. Hundreds of drums of solid and liquid chemical wastes were opened, drained, and covered with a thin layer of dirt. Other drums had been buried or abandoned on-site. Testing revealed that the groundwater and soil contain VOCs, heavy metals, and PCBs. The contamination threatens Oak Orchard Creek, associated wetlands, and private and public wells within 3 miles of the area, which serve about 3,000 residents.

Response actions included removing drummed waste and contaminated soil in 1984 and 1990. In 1989 the EPA recommended soil flushing and groundwater extraction and treatment through air-stripping and carbon adsorption. Construction is expected to be under way by 1997.

Administrative Orders (1984, 1989) require the PRPs to undertake the response actions and clean-up work.

Sources: PNPLS-NY; Gillam-Grant Library, 6966 West Bergen Road, Bergen, NY, 14416.

Carroll and Dubies
Orange County
This former waste disposal site near Port Jervis operated in the 1970s and 1980s. Various industrial liq-

uid wastes were held in seven lagoons on the property. Piles of solid waste were left exposed when the operation closed in 1989. Testing has shown that the groundwater and soil contain VOCs, phthalates, and heavy metals. The contamination threatens the local drinking water supplies for about 3,500 residents, Cold Creek, and the Neversink River.

Response actions have included site investigations that will soon recommend a final remediation plan.

A 1990 AOC requires two PRPs to conduct the site investigations.

Source: PNPLS-NY.

Circuitron Corporation
Suffolk County
Circuit boards were produced on this site near Farmingdale from 1981 to 1986. Various chemical wastes from plating and scrubbing activities were held in tanks or discharged into drain systems and leaching pools. Hundreds of drums of dangerous chemicals were abandoned on the site when the facility was closed in 1986. Testing has shown that the groundwater and soil contain heavy metals and VOCs. The contamination threatens private and public wells that serve about 115,000 people.

Response actions included removing drummed waste and tanks and contaminated soil (1987–1988). Surface debris was removed and fencing erected in 1992. The EPA recommended soil vapor extraction, excavating lagoon sediments, and decontaminating the structures. The buildings were later demolished. These activities should be completed by 1997. A groundwater extraction and air-stripping system will also be installed.

Sources: PNPLS-NY; Farmingdale Public Library, Conklin Street, Farmingdale, NY, 11735.

Claremont Polychemical
Nassau County
Pigments for plastics and inks were manufactured on this site in Old Bethpage from 1966 to 1980. Liquid wastes and sludges were held in tanks that leaked or discharged into holding basins. Leaks, spills, seepage, and direct discharge contaminated the groundwater and soil with VOCs, copper, zinc, arsenic, chromium, lead, and manganese. Thousands of leaking tanks and drums were abandoned on the property when the company declared bankruptcy in 1980. The contamination threatens public and private wells within 3 miles of the site, which serve about 50,000 residents.

Response actions included removing drummed liquid wastes and storage tanks and erecting fencing (1988–1991). In 1990 the EPA recommended groundwater extraction and air-stripping, low-temperature thermal treatment of contaminated soil, and decontamination and dusting of buildings. These remedies will be completed by 1997.

Sources: PNPLS-NY; Plainview Public Library, 999 Old Country Road, Plainview, NY, 11803..

Colesville Municipal Landfill
Broome County
This former landfill near Colesville operated from 1965 to the 1980s. It received a variety of wastes, including solvents, dyes, heavy metals, and sludges. In 1984 state officials detected VOCs in the groundwater, soil, nearby stream sediments, and private wells. The contamination threatens the recreational Susquehanna River and private water wells that supply about 2,000 residents.

Response actions included connecting affected residents to clean water supplies. In 1991 the EPA recommended capping, a leachate collection system, groundwater extraction and air-stripping, and a permanent clean water supply system. Construction should be completed by 1997.

The PRPs have conducted the site investigations and remediation work according to a 1987 Consent Order.

Sources: PNPLS-NY; Colesville Town Hall, Walton Street, Colesville, NY, 13787.

Conklin Dumps
Broome County
This former landfill near Conklin operated in the 1960s. It received a variety of municipal and industrial wastes, including sludges and heavy metals. Leachate and surface runoff flow into Carlin Creek. The groundwater and soil contain heavy metals and VOCs. The contamination threatens nearby streams, wetlands, Carlin Creek, the Susquehanna River, and the local drinking water supplies for about 2,000 residents.

Response actions included a site investigation that was completed in 1991. The EPA recommended capping, leachate collection and treatment, and waste consolidation and reburial. All construction activities should be completed by 1997.

The PRP signed a 1987 Consent Order that requires it to conduct the site investigation and remediation work.

Sources: PNPLS-NY; Conklin Town Hall, 1271 Conklin Road, Conklin, NY, 13748.

Cortese Landfill
Sullivan County
This former landfill near Tusten operated from 1970 to1981. It received a variety of municipal and industrial wastes, including solvents, thinners, sludges, and

oils. Testing in the mid-1970s showed that the groundwater, surface water, and soil contain VOCs and heavy metals. The contamination threatens private and municipal wells in the area and the recreational Delaware River, which is designated a Wild and Scenic River.

Response actions included a series of site investigations. In 1994 the EPA recommended removing about 8,000 buried drums of waste, capping, erosion controls, and a groundwater extraction and treatment system. Construction will begin in 1997–1998.

Consent Orders (1985, 1990) require the PRPs to conduct the site investigations.

Source: PNPLS-NY.

Cross County Landfill
Putnam County
This former landfill near Patterson operated from 1963 to 1974. It accepted a variety of municipal and industrial wastes, including solvents, drummed waste, and sludges. Testing has shown that the groundwater and soil contain VOCs and PCBs, which have been detected in the adjacent Great Swamp and Muddy Brook. The contamination threatens local drinking water supplies and nearby streams, wetlands, and associated wildlife habitats.

Response actions included removing exposed drummed waste in 1993–1994. The site was proposed to the NPL in June 1996; a final decision is expected in 1997–1998. Additional sampling and site investigations are planned.

Source: EPA Publication 9320.7-071.

Endicott Village
Broome County
A 100-acre area of groundwater that supplies Endicott Village was discovered in 1981 to be contaminated with VOCs. The contamination can be linked to the Endicott Landfill. Surface runoff flows into Nanticoke Creek and the Susquehanna River. The contamination threatens nearby waterways and wetlands and the drinking water supplies for about 25,000 residents.

Response actions included the installation of an air-stripper to treat the contaminated groundwater in 1984. Fencing was also erected. In 1987 the EPA recommended expanding the groundwater treatment system (completed in 1992) and long-term groundwater monitoring. Gas venting and a landfill cap will also be installed.

A Consent Order (1988) and Consent Decrees (1991, 1993) require the PRPs to conduct the site studies and remediation work.

Sources: PNPLS-NY; Clerk's Office, Municipal Building, 1009 East Main Street, Endicott, NY, 13760.

Facet Enterprises
Chemung County
Metal bicycle, automobile, and weapons parts were fabricated on this site in Elmira Heights from 1926 to 1978. Various liquid and solid wastes, including sludges, solvents, oils, and cyanide salts, were buried or discharged on the 39-acre site. Testing has shown that the groundwater and soil contain VOCs, petrocarbons, heavy metals, and PCBs, which have penetrated nearby wells. The contamination threatens nearby creeks and wetlands and the drinking water supplies for about 10,000 people.

Response actions included excavating about 500 buried drums and contaminated soil in 1992. Site investigations were conducted from 1983 to 1992. The EPA recommended excavating the remaining contaminated soil and installing a groundwater extraction and treatment system. Construction is under way.

A Consent Order (1983), Administrative Order (1986), and Consent Decree (1993) require the PRP to conduct the site investigations and remediation work.

Source: PNPLS-NY.

FMC Corporation (Dublin Road)
Orleans County
This former industrial landfill near Ridgeway operated from the 1930s to 1968. FMC Corporation buried coal ash, laboratory wastes, spent chemicals, and pesticide residues on the 30-acre property. Testing has shown that the groundwater and soil contain pesticides and heavy metals, including lead, mercury, and arsenic. The contamination threatens private water wells in the area, the New York State Barge Canal, and Jeddo Creek.

Response actions included a site investigation that was completed in 1991. In 1993 the State of New York recommended excavating and treating contaminated soil and groundwater extraction and treatment. Construction should be completed by 1997.

A Consent Order requires the PRPs to conduct the site investigations and remediation work.

Sources: PNPLS-NY; Lee-Whedon Memorial Library, 620 West Avenue, Medine, NY, 14103.

Forest Glen Mobile Homes
Niagara County
A mobile home park was constructed on this former landfill near Niagara Falls, which received chemical wastes. Testing showed that the soil and groundwater contain phenolic resins and PAHs. The contamination threatens the health of the local residents as well as local water supplies and East Gill Creek.

Response actions included temporary capping and fencing. In 1989 the EPA recommended the permanent relocation of the on-site residents. A detailed site investigation that will recommend a final remediation plan is nearing completion. To date 53 families have been relocated.

Sources: PNPLS-NY; EPA Public Information Office, 345 3rd Street, Suite 350, Niagara Falls, NY, 14303.

Fulton Terminals

Oswego County

Roofing materials and asphalt were manufactured or processed on this site near Fulton from 1936 to 1960. Various solid and liquid wastes were stored or discharged on the property. Hazardous wastes, including PCB-contaminated oils, were stored on the premises from 1972 to 1977. The groundwater, soil, and nearby stream sediments contain heavy metals and VOCs. The contamination threatens the local drinking water supplies and recreational Oswego River.

Response actions included fencing and removing storage tanks and contaminated soil (1986). Surface runoff controls were built in 1990. In 1989 the EPA recommended low-temperature thermal soil treatment and groundwater extraction and treatment using carbon adsorption. Construction is in progress.

A Consent Order (1986) and Consent Decrees (1990, 1991) require the PRPs to conduct the removal activities and remediation work.

Sources: PNPLS-NY; Fulton Public Library, 160 South 1st Street, Fulton, NY, 13069.

GCL Tie & Treating Company

Delaware County

This former wood treatment facility near Sidney operated from the 1940s to 1986. Lumber and railroad ties were soaked in creosote. Testing by the EPA indicated that the groundwater and soil contain creosote derivatives such as anthracene, chrysene, and benzo(a)pyrene. About 20,000 gallons of creosote wastes and sludges were abandoned on the property in 1986. The contamination threatens nearby streams, the Susquehanna River, and the local drinking water supplies for about 5,000 people.

Response actions included fencing and removing wastes, tanks, and contaminated soil (1991). Site studies that will recommend a final remediation plan are nearing completion.

Source: PNPLS-NY.

GE Moreau

Saratoga County

Solid and liquid wastes generated by General Electric Company were placed in a pit on this property near South Glens Falls from 1958 to 1968. Wastes included PCBs, VOC-containing solvents, oils, and sludges. Testing in the 1980s showed that the groundwater and soil contain VOCs and PCBs. The contamination threatens the local drinking water supplies for about 15,000 residents, Reardon Brook, Fort Edward Reservoir, and associated wetland habitats.

Response actions included connecting affected residents to clean water supplies, building a slurry wall, capping, excavating contaminated soil, and installing a groundwater treatment system (1981–1985). In 1985 the EPA recommended the continuance of these interim response actions and the solidification and reburial of PCB-contaminated soil. The groundwater treatment system was expanded in 1994. Monitoring is continuing.

General Electric has participated in the response actions, site studies, and remediation work.

Sources: PNPLS-NY; Crandel Library, City Park, Glens Falls, NY, 12801.

General Motors, Central Foundry Division

St. Lawrence County

This General Motors die-casting factory near Massena has been in operation since 1958, building mostly aluminum cylinder heads. Various waste products included PCB-containing hydraulic fluids and PCB-laden sludges, which were discharged into on-site lagoons. Testing has shown that the groundwater and soil contain PCBs and VOCs. The contamination threatens the local drinking water supplies, the St. Lawrence River, the Raquette River, and the St. Regis Mohawk tribal lands.

Response actions included the installation of a temporary cap in 1987. An EPA study recommended the excavation and treatment of contaminated soils and river sediments and groundwater extraction and treatment. Construction is expected to begin by 1997.

A Consent Order (1985) and UAOs (1992, 1992) require General Motors to undertake the site investigations and remediation work.

Sources: PNPLS-NY; Massena Public Library, 14 Glenn Street, Massena, NY, 13602.

Genzale Plating

Nassau County

Metal electroplating has been conducted on this site near Franklin Square on Long Island since 1915. Heavy metal–bearing solutions and sludges were discharged into three leaching pools at the rear of the one-half-acre site. Testing has shown that the groundwater and soil contain heavy metals, including nickel, cadmium, chromium, and iron. The contamination threatens the drinking water supplies for about 55,000 residents.

Response actions included removing sludges in 1982. A site investigation (1988–1991) recommended soil vacuum extraction, soil excavation, and an interim groundwater extraction and air-stripping system. Construction of these remedies began in 1994.

The PRP conducted the initial removal actions in 1982.

Sources: PNPLS-NY; Franklin Square Public Library, 19 Lincoln Road, Franklin Square, NY, 10110.

Goldisc Recordings
Suffolk County
Phonograph records were manufactured on this 34-acre site in Holbrook from 1968 to 1983. Various liquid wastes, including metal-plating sludges, solvents, and hydraulic fluids and oils were held in storage tanks, lagoons, or an on-site dump. Liquid wastes also entered a nearby wetland. Testing by Suffolk County revealed that the groundwater and soil contain chromium, nickel, lead, zinc, copper, and VOCs. The contamination threatens nearby wetlands and private and public wells that serve about 75,000 people.

A site investigation that will recommend a final remediation plan is nearing completion.

A number of legal actions by the EPA against the PRP require the company to undertake the site investigations.

Source: PNPLS-NY.

Griffiss Air Force Base
Oneida County
This U.S. Air Force base near Rome has been in operation since 1943. Various research and development activities have generated solid and liquid wastes, including lead from batteries, spent acids, solvents, and sludges. These wastes were held in storage tanks or discharged into lagoons and landfills on the 3,900-acre site. Testing has indicated that the groundwater and soil contain VOCs, PCBs, heavy metals, and ethylene glycol, some of which have entered private wells, agricultural wells, and Six Mile Creek. The contamination is a threat to local drinking water supplies, Six Mile Creek, and associated wetlands.

Response actions included removing underground tanks and contaminated soil (1985–1989). Affected residents have been connected to clean water supplies (1990–1991). The EPA will issue a final remediation plan after a site investigation is completed in 1997.

Source: PNPLS-NY.

Haviland Complex
Dutchess County
The water supplies for this neighborhood in Hyde Park were discovered to contain VOCs in 1981. The point source was discovered to be a local carwash and laundromat whose sewage and septic systems had failed. The contamination threatens Fall Kill Creek, associated wetlands, and the drinking water supplies for about 25,000 people.

Response actions have included the installation of carbon filtration units in affected homes. In 1987 the EPA recommended the installation of a groundwater extraction and treatment system. Additional studies have led to the EPA's reevaluation of its 1987 decision.

Sources: PNPLS-NY; Hyde Park Town Hall, Albany Port Road, Hyde Park, NY, 12538.

Hertel Landfill
Ulster County
This former landfill near Plattekill operated from 1963 to 1977. It received a variety of municipal and industrial wastes, including heavy metal sludges. Testing by state officials indicated that the groundwater and soil contain VOCs, arsenic, chromium, and manganese. The contamination is a threat to Black Creek, associated wetlands, and the local drinking water supplies for about 1,800 people.

An EPA site investigation was completed in 1991. The EPA recommended capping and the extraction and treatment of contaminated groundwater through chemical precipitation, filtration, and ultraviolet oxidation. Construction should be under way by 1997.

UAOs were sent to six PRPs in 1992. Only one, Ford Motor Company, has complied with the order to undertake the remediation of the site.

Sources: PNPLS-NY; Plattekill Public Library, Route 32, Modena, NY, 12548.

Hooker Chemical
Nassau County
Plastics, polyvinyl chloride, latex, and esters have been manufactured on this site in Hicksville since 1945. Various liquid wastes were discharged into sand sumps, lagoons, and dry wells on the property until about 1980. PCBs have also been spilled. The contamination threatens about 30 wells that serve about 60,000 nearby residents.

Response actions included a detailed site investigation that was completed in 1992. The EPA recommended the excavation and removal of PCB-contaminated soils. This work was completed in 1993. A decision regarding groundwater treatment is pending.

A 1988 Consent Order and 1991 UAO require the PRPs to conduct the site investigations and remediation work.

Sources: PNPLS-NY; Hicksville Public Library, 169 Jerusalem Avenue, Hicksville, NY, 11801.

Hooker Chemical, S-Area

Niagara County

This landfill near Niagara Falls was used by Occidental Chemical Corporation from 1947 to 1989. Various solid and liquid chemical wastes were buried on the 8-acre site. Liquid wastes were held in lagoons until 1989. Testing revealed that the groundwater and soil contain various chemicals, including VOCs, which have entered the Niagara River. The contaminants threaten the Niagara River and the drinking water supplies for about 75,000 residents.

Response actions included a site investigation that recommended a slurry wall, leachate collection, groundwater collection and carbon adsorption treatment, capping, and long-term monitoring. Much of the work has been completed or is in progress. The final cap will be installed by 1998. The city of Niagara Falls is building a new water treatment plant that will be finished in 1997.

The EPA sued the PRPs in 1979. A series of legal agreements between 1984 and 1991 require the PRPs to conduct the site investigations and remediation work.

Sources: PNPLS-NY; New York State Department of Environmental Conservation, 600 Delaware Avenue, Buffalo, NY, 14202.

Hooker-Hyde Park

Niagara County

Hazardous chemicals and other wastes were buried in this 15-acre landfill near Niagara Falls from 1953 to 1975. The groundwater and soil contain VOCs and dioxins, which have entered Bloody Run Creek and the Niagara River gorge. The contamination threatens the Niagara River and the health and drinking water supplies of local residents.

Response actions included groundwater extraction and treatment and runoff controls (1985). A leachate collection system was completed in 1990. A free-product liquid separation system became operational in 1994. Sediments from Bloody Run Creek were excavated in 1993. The landfill was capped in 1994–1995.

The PRP, Occidental Chemical Corporation, signed a 1981 Consent Decree that requires the company to undertake the clean up of the site.

Sources: PNPLS-NY; New York State Department of Environmental Conservation, 600 Delaware Avenue, Buffalo, NY, 14202.

Hooker-102nd Street

Niagara County

This former landfill near Niagara Falls operated from about 1943 to 1971. It received a variety of industrial wastes, including organic solvents, phosphates, sludges, fly ash, lindane, and hexachlorocyclohexane from nearby chemical companies. Part of the Love Canal Emergency Declaration Area, this landfill continues to discharge leachate into the Niagara River. The groundwater and soil contain VOCs, SVOCs, pesticides, phenols, chlorophenols, dioxins, furans, and heavy metals, including mercury. The contamination threatens the Niagara River and the health and drinking water supplies of local residents.

Response actions included capping, erecting fencing, and building a barrier to protect the Niagara River (1972). Numerous investigations were completed in 1990. The EPA recommended a remediation plan consisting of capping, soil solidification and consolidation, a slurry wall, dredging and incinerating contaminated river sediments, groundwater recovery and treatment, restricted access, and long-term monitoring. The design plans are presently being reviewed.

The PRPs were sued by the EPA in 1979. An Administrative Order (1991) between the EPA and two PRPs, Occidental Chemical Corporation and Olin Chemical Corporation, requires the companies to undertake the clean-up action.

Sources: PNPLS-NY; New York State Department of Environmental Conservation, 600 Delaware Avenue, Buffalo, NY, 14202.

Hudson River PCBs

Rensselaer, Washington, and Saratoga Counties

This 40-mile stretch of the Hudson River between the towns of Fort Edward and Troy has been contaminated with PCBs from wastes discharged into the river by General Electric Corporation's capacitor manufacturing plants. The contamination threatens the health of residents and ecosystems along the river and has resulted in fishing bans. The town of Waterford draws its drinking water from this portion of the Hudson River. Other communities further downstream, including Poughkeepsie, Rhinebeck, Highland, and Port Ewen, also use the Hudson River for drinking water.

Response actions included dredging contaminated sediments from the river in 1977–1978. Interim measures have included capping remnant shoreline contamination and stabilizing riverbanks. Other studies by the EPA are nearing completion.

The PRP, General Electric, is participating in the site studies, response actions, and remediation work.

Sources: PNPLS-NY; New York State Department of Environmental Conservation, 50 Wolf Road, Room 409, Albany, NY, 12233.

Islip Landfill

Suffolk County

This former landfill in Islip operated from 1963 to 1990. It received a variety of municipal and industrial wastes, including VOCs and other solvents. Testing in the 1980s revealed that the groundwater and soils contain VOCs, which have penetrated nearby wells. The contamination threatens the private and public water wells that supply about 85,000 people in the area.

Response actions included connecting affected residents to clean water supplies (1981, 1992). A site study was completed in 1992. The EPA recommended capping and groundwater extraction and treatment. The groundwater treatment system should be operational by 1997 and will operate for about 10 years.

Various legal actions require the PRPs to conduct the site investigations and remediation work.

Sources: PNPLS-NY; Central Islip Public Library, 33 Hawthorne Avenue, Central Islip, NY, 11722.

Johnstown City Landfill

Fulton County

This former landfill near Johnstown operated from 1947 to the 1980s. It received a variety of municipal and industrial wastes, including heavy metal sludges, sewage sludge, and materials from tanneries and textile mills. Testing has shown that the groundwater and soil contain heavy metals, VOCs, SVOCs, cyanide, and ammonia-nitrogen, which have been released as seeping leachate on surrounding properties, including Matthew Creek, where fish kills have been documented. The contamination threatens Matthew Creek, associated wetlands, and the local water supplies of about 30,000 people.

A site investigation was completed in 1992. In 1993 the EPA recommended capping and connecting affected residents to clean water systems.

The EPA has identified 14 PRPs. In 1988 the city of Johnstown agreed to conduct the site investigation.

Sources: PNPLS-NY; Johnstown Public Library, 38 South Market Street, Johnstown, NY, 12095.

Jones Chemicals

Livingston County

Chemicals have been manufactured and processed on this site in Caledonia since 1942. Products have included chlorine, sodium hypochlorite, ammonium hydroxide, mineral acids, and caustic soda. Improper handling and storage, leaks, and spills have polluted the immediate environment. Testing in 1986 revealed that the groundwater and soil contain VOCs, which have penetrated nearby wells and Caledonia's water supply system. The contamination threatens recreational wetlands and the drinking water supplies for about 6,000 local residents.

Response actions included removing underground tanks in 1985. Caledonia installed an air-stripper to clean the municipal water supplies. An air-stripper on the property was pilot-tested in 1994–1995. A site investigation (1991–1996) will recommend a final remediation plan.

A 1991 AOC requires the PRPs to undertake the site investigations and remediation work.

Source: PNPLS-NY.

Jones Sanitation

Dutchess County

This landfill in Hyde Park has been active since 1956. It has received a variety of municipal, septic, and industrial wastes, including oils, acids, solvents, plating sludges, pigments, phenols, and VOCs. Testing revealed that the groundwater, surface water, and soil contain oil, grease, VOCs. and heavy metals (chromium, lead, and mercury). The contamination threatens Maritje Kill, associated wetlands, and water wells that serve about 10,000 local residents.

Response actions included a site investigation (1991–1996) that will recommend a final remediation plan.

A 1991 AOC requires the PRP to conduct the site investigation.

Sources: PNPLS-NY; Hyde Park Town Hall, 627 Albany Post Road, Hyde Park, NY, 12538.

Katonah Municipal Well

Westchester County

This municipal supply well for the town of Katonah was discovered to contain VOCs in 1978. Subsequent investigations indicated that the likely point sources were four dry cleaning companies that disposed of wastewaters in septic systems. The groundwater and soil contain VOCs, pesticides, PAHs, PCBs, and heavy metals. The contamination threatens the local drinking water supply for about 6,000 Bedford residents as well as the Muscoot Reservoir, which is a source of drinking water for New York City.

In 1987 the EPA recommended drilling a new well, plugging the closed well, and installing an air-stripper to treat the contaminated groundwater. Construction was completed in 1993. Groundwater monitoring is in progress.

A Consent Order (1988), UAO (1988), and Consent Decree (1989) require the PRPs to undertake the site investigations and clean-up work.

Sources: PNPLS-NY; Bedford Hills Free Library, 26 Main Street, Bedford Hills, NY, 10507.

Kentucky Avenue Well Field

Chemung County

Trichloroethylene was discovered in Elmira's Kentucky Avenue Well Field in 1980. Private wells were also contaminated. Further studies indicated a Westinghouse Corporation plant was the source of the VOCs. Testing indicated that the groundwater and soil contain VOCs and heavy metals, which have penetrated nearby lakes and streams, including Newtown Creek. The contamination threatens streams, wetland habitats, and the water supplies for about 60,000 Elmira residents.

Response actions included connecting affected residents to clean water supplies (1985, 1986, 1989). In 1990 the EPA recommended groundwater extraction and air-stripping, which started in 1994.

An Administrative Order requires the PRP, Westinghouse, to conduct the site investigation and remediation work.

Sources: PNPLS-NY; Horseheads Town Hall, 150 Wygant Road, Horseheads, NY, 14845.

Li Tungsten Corporation

Nassau County

Tungsten ore was processed on this site in Glen Cove from the 1940s to 1984. Ammonium paratungstate, metal tungsten powder, and tungsten-carbide powder were also produced. Various solid and liquid wastes were stored, leaked, spilled, or discharged on the 26-acre site. Sampling in 1988 revealed that the groundwater and soil contain heavy metals, chlorides, sulfates, PCBs, cyanide, acids, and radioactive material. The contamination threatens wells within 4 miles of the site, which serve about 55,000 people.

Response actions included removing tanks, drums, and liquid wastes in 1988. New monitoring wells were also installed. An EPA site investigation (1992–1997) is nearing completion and will recommend a final remediation plan.

The EPA has identified a number of PRPs. One, Glen Cove Development Company, agreed to a 1990 AOC that requires it to conduct the initial clean-up actions.

Source: PNPLS-NY.

Liberty Industrial Finishing

Nassau County

Former manufacturing operations on this site in Farmingdale produced various metal parts from the 1930s to the 1980s. Activities included electroplating and painting. Various liquids wastes and sludges were held in storage tanks or lagoons on the 30-acre property. Testing showed that the groundwater and soil contain heavy metals, VOCs, and PCBs. The contamination threatens recreational Massapequa Creek, associated wetland habitats, and the drinking water supplies for about 100,000 residents.

Response actions included removing contaminated soil and sludge in 1978 and 1987. Storage tanks and PCB-contaminated soil were removed in 1994. A site investigation (1991–1994) will recommend a final remediation plan. Other studies are also in progress.

Various legal agreements require the PRPs to undertake the site investigations and remediation work.

Sources: PNPLS-NY; Southern Farmingdale Library, Merritt Road, Farmingdale, NY, 11735.

Little Valley

Cattaraugus County

Groundwater, including a number of private wells, in the Little Valley area near the city of Salamanca was discovered in 1982 to contain trichloroethene. Five potential point sources have been identified to date. The contamination threatens local drinking water supplies. Groundwater monitoring is under way to identify the source of the contamination.

The EPA will recommend clean-up actions after additional studies have been completed.

Source: EPA Publication 9320.7-071, Volume 3, No. 1.

Love Canal

Niagara County

This former landfill site near Niagara Falls operated from 1942 to 1952. It received a variety of hazardous wastes, including dioxins and pesticides. The landfill was covered, and the area became extensively developed with schools and homes. Complaints from residents about odors and residues in the water led to inspections in the 1970s that revealed that the groundwater and soil contain VOCs, dioxins, PAHs, pesticides, and heavy metals. The chemicals penetrated nearby streams, including Black Creek, Bergholtz Creek, Cayuga Creek, and the Niagara River. The contamination threatens streams and the health and drinking water of about 85,000 nearby residents.

Response actions included the installation of an interim leachate collection system, capping, and fencing (1978–1981). Contaminated homes and a school were demolished in the 1980s. A 40-acre cap was installed in 1985. Contaminated sediments were excavated from creeks and culverts and have been stockpiled for incineration (1985–1989). Other incineration treatments are planned. Contaminated soil was removed from a schoolyard in 1991–1992. Nearly $2.5 million dollars have been spent acquiring or cleaning up contaminated homes and businesses.

The PRP, Occidental Chemical Corporation, signed a Consent Decree in 1989 that requires the company to participate in the remedial actions.

Sources: PNPLS-NY; New York State Department of Environmental Conservation, Public Information Office, 9820 Colvin Boulevard, Niagara Falls, NY, 14304.

Ludlow Sand and Gravel
Oneida County
This former gravel pit near Paris was used as a disposal site from the 1960s to 1988. It received a variety of municipal and industrial wastes, including dyes, oils, sludges, sewage, PCBs, and animal carcasses. Testing revealed that the groundwater and soil contain VOCs, heavy metals, PCBs, and phenols. The contamination threatens the local drinking water supplies for thousands of residents as well as nearby streams, wetlands, and the Mohawk River.

In 1988 the EPA recommended consolidating and reburying contaminated soil, capping, leachate collection, lowering the water table, groundwater extraction and treatment, fencing, deed restrictions, and long-term groundwater monitoring. Other groundwater studies are in progress.

Sources: PNPLS-NY; Utica Public Library, 303 Genessee Street, Utica, NY, 13501.

Malta Rocket Fuel Area
Saratoga County
This former U.S. military facility near Malta and Stillwater operated from 1945 to 1984. It tested rockets and fuels and conducted other research for the Department of Energy and NASA. Most of the facility was sold to private interests in 1984. The various activities on the site have contaminated the soil and groundwater with VOCs, PCBs, and heavy metals. The contamination threatens the drinking water supplies for about 10,000 nearby residents.

Response actions included a site investigation (1989–1996) that will recommend a final remediation plan. A warning system was installed that will detect contamination migrating toward public water wells.

A 1989 Unilateral Order requires the PRP to fund the site investigations.

Source: PNPLS-NY.

Marathon Battery
Putnam County
Nickel-cadmium batteries were produced on this site near Cold Spring from 1952 to 1979. Liquid wastes were discharged into the city sewer system or the Hudson River until about 1965. Testing by state officials has shown that the groundwater and soil contain heavy metals and VOCs. The contamination threatens local drinking water supplies, recreational East Foundry Cove, the Hudson River, and associated wetlands.

Response actions included dredging contaminated sediments from Foundry Cove in 1972–1973. The buildings were secured and fenced in 1989. After a detailed site investigation, the EPA recommended dredging and consolidating sediments from East Foundry Cove Marsh, reconstructing the wetlands, diverting surface runoff, and long-term environmental monitoring. Buildings were decontaminated in 1993.The clean-up work should be completed by 1997.

A series of legal agreements between 1972 and 1993 required the PRPs to fund the site studies and remediation work.

Sources: PNPLS-NY; Cold Spring Town Hall, 234 Main Street, Cold Spring, NY, 10516.

Mattiace Petrochemical Company
Nassau County
This former chemical distribution facility in Glen Cove operated from the 1960s to 1987. Drums were also cleaned and rinsed on the site. Various wastes and chemicals were buried on the site or discharged into leaching pools. Testing revealed that the groundwater and soil contain VOCs. The contamination threatens the drinking water supplies of the local residents, Garvies Point Preserve, Glen Cove Creek, Hempstead Harbor, Long Island Sound, and associated wetlands.

Response actions included removing liquid wastes still stored on the site in 1988. About 400 buried drums were excavated in 1991–1992. In 1991 the EPA recommended soil vapor extraction, soil excavation, tank removals, and groundwater extraction and treatment. The clean-up work should be completed by 1997.

Sources: PNPLS-NY; Glen Cove Public Library, Glen Cove Avenue, Glen Cove, NY, 11542.

Mercury Refining
Albany County
Mercury from old batteries has been recycled on this site in Albany since 1956. Battery wastes were stored in drums. Improper handling and storage, leaks, spills, and discharges have contaminated the groundwater, surface water, and soil with mercury, zinc, nickel, arsenic, and PCBs. The chemicals have also entered nearby streams. Mercury vapors are also considered a risk to human health. The contamination threatens Patroons Creek, the Hudson River, associated wetlands, and the drinking water supplies for about 100,000 people.

Response actions included excavating contaminated soil and debris (1985), regrading, and capping. Equipment and the furnace on-site have been upgraded to be in compliance with state and federal regulations. Long-term environmental monitoring is in progress.

Consent Orders (1985, 1989, 1993) require the PRPs to conduct the remediation work.

Sources: PNPLS-NY; New York State Department of Environmental Conservation, 2176 Guilderland Avenue, Schenectady, NY, 12306.

Nepera Chemical Company
Orange County
Nepera Chemical Company placed pharmaceutical and other industrial chemical wastes in pits and lagoons on this property near Maybrook from 1953 to 1968. Testing showed that the groundwater and soil contain VOCs, heavy metals, phthalates, and pyridines. The contamination threatens Beaverdam Brook and the local drinking water supplies for about 10,000 residents.

Response actions included filling the lagoons and fencing the site in 1974. An ongoing site investigation is nearing completion and will recommend a final remediation plan.

The PRPs are conducting the site investigation.

Sources: PNPLS-NY; Harriman Village Hall, 1 Church Street, Harriman, NY, 10926.

Niagara County Refuse
Niagara County
This former landfill near Wheatfield operated from 1969 to 1976. It received a variety of municipal and chemical wastes, including VOCs, pesticides, and sludges. Testing has found that the groundwater, soil, surface water, and nearby stream sediments contain VOCs, SVOCs, pesticides, and heavy metals. The contamination threatens Black Creek, the Niagara River, associated wetlands, and the local drinking water supplies for about 10,000 residents.

A site investigation was conducted between 1987 and 1993. The EPA recommended regrading, capping, leachate collection and treatment, gas venting, deed restrictions, and long-term environmental monitoring.

The PRPs conducted the site investigation.

Sources: PNPLS-NY; New York State Department of Environmental Conservation, 600 Delaware Avenue, Buffalo, NY, 14202.

Niagara Mohawk Power Company
Saratoga County
Originally a coal gasification plant from 1853 to the 1940s, this site near Saratoga Springs is now an electric substation and vehicle repair shop. PCB-containing transformers are also stored on the premises. Coal tar wastes from the gasification process have contaminated the local environment. Testing has shown that the groundwater and soil contain PAHs, VOCs, DDT, and petrochemicals. The contamination threatens Loughberry Lake, Village Brook, Spring Run, and the drinking water supplies for about 11,000 residents.

A site investigation was completed from 1989 to 1992. The EPA will recommend a final remediation plan shortly.

The PRP signed a Consent Order (1989) with the EPA that requires it to conduct the site investigations.

Sources: PNPLS-NY; Saratoga Springs Public Library, 320 Broadway, Saratoga Springs, NY, 12866.

North Sea Landfill
Suffolk County
This active landfill near Southampton has been in operation since 1963. It has received a variety of municipal, septic, and industrial wastes, including sludges and solvents. Testing in 1986 showed that the groundwater and soil contain VOCs and heavy metals. Cadmium was detected in nearby Fish Cove. The contamination threatens local drinking water supplies and the recreational coves and wetlands of Little Peconic Bay and Fish Cove.

Response actions included connecting affected residents to clean water supplies (1981). The EPA recommended proper closure of the inactive sludge lagoons on the site. Clean up was completed in 1994. Groundwater monitoring is in progress.

The PRP conducted the site studies and remedial actions according to a 1987 Administrative Order and 1991 Consent Decree.

Sources: PNPLS-NY; Southampton College Library, Montauk Highway, Southampton, NY, 11968.

Old Bethpage Landfill
Nassau County
This former landfill near Oyster Bay operated from 1957 to 1986. It received a variety of municipal and industrial wastes, including drummed organic chemicals, heavy metal sludges, and incinerator residue. The groundwater and soils contain heavy metals and VOCs. Methane and VOCs have also been detected in air samples. The contamination threatens the drinking water supplies for about 10,000 local residents and the town of Farmingdale.

Response actions included the installation of a gas collection system (1982) and a partial cap. In 1988 the EPA recommended groundwater extraction and air-stripping, upgrading the gas and leachate collection systems, and long-term monitoring. Complete capping and the groundwater treatment system were completed in 1992.

Consent Decrees (1984, 1988) require the PRPs to conduct the site investigations and clean-up work.

Sources: PNPLS-NY; Plainview Public Library, 999 Old Country Road, Plainview, NY, 11803.

Olean Well Field

Cattaraugus County

More than a l-square-mile area of groundwater used by the city of Olean was discovered to contain VOCs in 1981. Over 50 municipal and private wells are contaminated. The VOCs are thought to be derived from local industrial operations and municipal and industrial dumps. Groundwater and soil at the business sites contain VOCs. The contamination threatens Olean Creek, the Allegheny River, and the drinking water supplies for about 20,000 residents.

Response actions included installing carbon filters in affected homes (1983–1985). Some contaminated soil was removed in 1990. The EPA recommended the installation of air-strippers to clean the groundwater, connecting affected residents to permanent clean water supplies, deed restrictions, and monitoring. The air-strippers became operational in 1993. Studies regarding surface clean ups at the point sources are nearing completion.

UAOs (1984, 1986) and AOs (1986, 1989, 1991) require six PRPs to conduct the site investigations and remediation work.

Source: PNPLS-NY.

Onondaga Lake

Onondaga County

Historically industrial activities discharged wastewaters directly into this lake near Syracuse, causing extensive environmental damage. The most harmful point sources have been chemical and plastic producers Allied Signal, Linden Chemicals and Plastics, and the Hanlin Group. Various liquid wastes from chlorine, soda ash, sodium hydroxide, and potassium hydroxide production, including mercury-contaminated process waters and VOC-tainted acid washwaters, were dumped into the lake for decades. Public fishing was banned in 1970. Testing has shown that the lake water contains mercury and bottom sediments contain PCBs, pesticides, creosotes, heavy metals (including mercury), PAHs, and VOCs. The contamination threatens the Seneca River, Onondaga Lake, and the health and drinking water supplies of local residents.

Response actions included undertaking a site investigation in 1993 that should be completed in 5 years. A final remediation plan will then be issued.

Allied Signal is funding the site investigation according to a 1992 Consent Order.

Sources: PNPLS-NY; Oleans Public Library, 2nd and Lauren Streets, Oleans, NY, 14760.

Pasley Solvents and Chemicals

Nassau County

Chemicals were stored and distributed at this site near Hempstead from 1969 to 1982. Improper handling and storage, leaks, and spills have contaminated the soil and groundwater with VOCs, chlorinated organic solvents, and petrochemicals. The contamination threatens public and private wells within 3 miles of the site, which serve over 100,000 people.

Response actions included a site study that was completed in 1992. The EPA recommended soil flushing, soil vacuum extraction, and groundwater extraction and air-stripping.

The EPA has identified PRPs that claim they cannot pay for any of the site studies or remediation work. The EPA is presently reviewing their financial viability.

Sources: PNPLS-NY; Nassau Public Library, 900 Jerusalem Avenue, Uniondale, NY, 11553.

Pfohl Brothers Landfill

Erie County

Located near the Buffalo International Airport, this former landfill operated from 1932 to 1971. It received a variety of municipal and industrial wastes, including paint and electroplating sludges, inks, dyes, solvents, and petroleum products. Testing has shown that the surface water, groundwater, and soil contain VOCs, phenols, dioxins, PCBs, and heavy metals. The contamination threatens local drinking water supplies, Aero Lake, Ellicott Creek, and associated wetlands.

A detailed site investigation is planned.

Source: PNPLS-NY.

Plattsburgh Air Force Base

Clinton County

This U.S. Air Force base near Plattsburgh has been in operation since 1955. Activities included supporting a tactical wing of the Strategic Air Command, aircraft maintenance, painting, and munitions storage and processing. Related solid and liquid wastes were buried in landfills on the 3,440-acre site. Testing has shown that the groundwater and soil contain VOCs, DDT, petrochemicals, chlorinated solvents, and free-product petroleum. The contamination threatens the Saranac and Salmon Rivers, Lake Champlain, and the local drinking water supplies for about 2,000 people.

Response actions included removing contaminated soils, tanks, and liquid wastes in 1992–1993.

Landfills were capped from 1992 to 1995. A series of other studies will be completed by 1998.

Sources: PNPLS-NY; Plattsburgh Public Library, 15 Oak Street, Plattsburgh, NY, 12901.

Pollution Abatement Services

Oswego County

A chemical waste incinerator operated on this site near Oswego from 1970 to 1977. Waste oils and other hydrocarbons were stored in lagoons or large tanks, many of which leaked. Spills and overflows from the lagoons drained into Wine Creek. Testing showed that the groundwater and soil contain heavy metals, VOCs, and PCBs. The contamination threatens Wine Creek, Lake Ontario, and private and public wells that serve about 30,000 people.

Response actions included building erosion controls (1976) and treating water from the lagoons (1977). Drummed wastes were removed and fencing erected in 1980–1981. An interim groundwater extraction and treatment system began operation in 1987. The EPA also recommended a slurry wall, capping, leachate collection, an expanded groundwater treatment system, and long-term monitoring. Most of these remedies were completed in 1986.

AOCs (1990, 1991) required the PRPs to conduct off-site investigations and related removal actions.

Sources: PNPLS-NY; Oswego City Hall, West Oneida Street, Oswego, NY, 13126.

Port Washington Landfill

Nassau County

This former gravel mine near North Hempstead operated as a landfill from the 1960s to 1983. It received a variety of municipal and industrial wastes, including sludges. Testing has shown that the soil, air, and groundwater contain VOCs. The contamination threatens the drinking water supplies for thousands of residents, Hempstead Harbor, and associated wetlands.

Response actions included the installation of a gas venting system in 1990. In 1989 the EPA recommended closing the landfill, capping, and upgrading the gas collection system. Capping was completed in 1994–1995. A groundwater extraction and air-stripping system is also planned.

A 1990 Consent Decree and Administrative Order require the PRPs to conduct the response actions.

Sources: PNPLS-NY; Port Washington Public Library, 245 Main Street, Port Washington, NY, 11050.

Preferred Plating Corporation

Suffolk County

Metal electroplating was conducted on this site near Babylon from 1951 to 1976. Various liquid wastes and sludges were held in on-site leaching pits. Heavy metals and VOC-bearing degreasers were commonly used to produce metal parts. Testing has shown that the groundwater, soil, and soil gas contain cadmium, lead, nickel, chromium, VOCs, and cyanide. The contamination threatens the drinking water supplies of about 17,000 residents.

In 1989 the EPA recommended groundwater extraction and treatment, soil excavation, and off-site treatment. The soil program was completed in 1994. Construction of the groundwater treatment plant is in progress.

A 1990 AOC and 1993 UAO require the PRPs to undertake some of the remediation work.

Sources: PNPLS-NY; West Babylong Library, 211 Route 109, West Babylon, NY, 11704.

Ramapo Landfill

Rockland County

This former landfill near Hillburn operated from 1972 to the 1980s. It received a variety of municipal and industrial wastes, including cosmetic, pharmaceutical, and paint sludges. Testing has shown that the groundwater and soil contain VOCs, heavy metals (mercury, lead, and chromium), SVOCs, and phenols. The contamination threatens the Ramapo River and private and public water wells that serve about 200,000 people.

Response actions included a detailed site investigation. In 1992 the EPA recommended capping, groundwater extraction and treatment, and environmental monitoring. The construction should be completed by 1997.

AOCs (1985, 1988) require the PRPs to conduct the site investigations and remediation work.

Sources: PNPLS-NY; Suffern Free Library, Maple and Washington Avenues, Suffern, NY, 10901.

Richardson Hill Landfill

Delaware County

This former landfill near Sidney Center operated during the 1960s. It received a variety of municipal, industrial, and hazardous wastes, including oils, sludges, fuel, PCBs, and VOCs. Sampling in the 1980s revealed that the soil and groundwater contain VOCs, hydrocarbons, and PCBs. The contamination threatens Herrick Hollow Creek, Cannonsville Reservoir, and the local drinking water supplies for about 1,200 residents.

Response actions included connecting affected residents to clean water supplies and installing a temporary cap. A site investigation started in 1987 is nearing completion and will recommend a final remediation plan.

Consent Orders (1987, 1993) require the PRPs to conduct the site investigations and remediation work.

Sources: PNPLS-NY; Sidney Public Library, Main Street, Sidney, NY, 13838.

Robintech/National Pipe Company

Broome County

Polyvinyl chloride pipe has been produced from inert resins on this site near Vestal since 1966. Plastic-coated cable is also manufactured. Improper storage, leaks, spillage, and waste disposal have contaminated the groundwater and soil with VOCs. Contaminated runoff has entered recreational Skate Estate. The contamination threatens Skate Estate, nearby streams, and the local drinking water for about 7,500 people.

The EPA completed an investigation in 1991, recommending groundwater extraction and air-stripping. Installation should be under way by 1997.

Legal agreements (1987, 1992) require the PRPs to conduct the site investigation and remediation work.

Sources: PNPLS-NY; Vestal Public Library, 320 Vestal Parkway East, Vestal, NY, 13850.

Rosen Brothers Scrap Yard

Cortland County

Wire screens and nails were manufactured on this property in Cortland in the 1960s. Industrial wastes were suspected of being buried on the 20-acre site. From 1970 to the 1980s the property was used as a scrap yard and dump. Cars and drummed waste were crushed, the liquid contents flowing onto the ground. Liquid wastes, including oils, were discharged into an unlined pit. Various municipal and industrial wastes were also buried. The owners were cited for waste handling violations in 1972, 1984, and 1985. Testing in 1986 confirmed that the soil and groundwater contain VOCs and heavy metals. The contamination threatens Perplexity Creek, the Tioughnioga River, Cortland City High School, and local water wells that supply about 25,000 people.

Response actions included removing drummed waste and contaminated soil (1987–1989) and fencing the site (1989). A site investigation was conducted from 1990 to 1993. An EPA recommendation is still pending.

Administrative and Unilateral Orders (1988, 1990) require the PRPs to conduct the site investigations and remediation work.

Sources: PNPLS-NY; Cortland Free Library, 32 Church Street, Cortland, NY, 13045.

Rowe Industries

Suffolk County

Rowe Industries manufactured electric motors and transformers on this property in Sag Harbor from the 1950s to 1980. Wastewaters and spent solvents were discharged on the surface or into drains that flowed into wetlands. Electronic devices are now built on the premises by Sag Harbor Industries. Testing revealed that the groundwater and soil contain VOCs, which have entered at least 46 private wells. The contamination threatens a wetland and the local drinking water supplies for about 6,000 people

Response actions included connecting affected residents to clean water supplies in 1985. In 1992 the EPA recommended the extraction and treatment of the contaminated groundwater. Installation should be under way by 1997.

A 1993 Consent Decree requires the PRPs to undertake the remediation work.

Sources: PNPLS-NY; John Jermain Memorial Library, Main Street, Sag Harbor, NY, 11963.

Sarney Farm

Dutchess County

This illegal landfill near Amenia operated from 1965 to 1969. It received a variety of municipal and industrial wastes, including sludges and solvents. Testing has shown that the groundwater and soil contain VOCs, including toluene, dichloroethane, and vinyl chloride, which have also penetrated nearby Cleaver Swamp. The contamination threatens the local drinking water wells for about 3,000 people, surface water used for agricultural purposes, and wetlands.

Response actions included the installation of a leachate collection and treatment system in 1987. In 1990 the EPA recommended excavating and removing drummed waste and low-temperature thermal treatment of contaminated soil. About 2,000 containers of waste have been removed to date. Soil treatment should be under way by 1997.

Sources: PNPLS-NY; Amenia Town Hall, Mechanic Street, Amenia, NY, 12501.

Sealand Restoration

St. Lawrence County

A waste disposal facility was operated on this former dairy farm near Lisbon in the 1970s. Industrial wastes included sludges and petroleum by-products that were buried or spread on the property. Testing has shown that the groundwater, surface water, and soil contain VOCs, heavy metals, aluminum, iron, zinc, lead, manganese, PCBs, phenols, and pesticides, which have entered nearby private wells. The contamination threatens local drinking water supplies and nearby streams and wetlands.

Response actions included removing above-ground wastes and contaminated soil (1984–1990) and connecting affected residents to clean water supplies (1993). A detailed site investigation is nearing completion.

An AOC (1993) required the PRPs to connect affected residents to clean water supplies.

Sources: PNPLS-NY; Lisbon Town Hall, 62 Main Street, Lisbon, NY, 13658.

Seneca Army Depot
Seneca County
This U.S. Army munitions and explosives storage and disposal depot near Romulus has been in operation since 1941. Related wastes, including detonated explosives and incinerator ash, were buried in a landfill on the 10,587-acre site. The groundwater and soil contain VOCs and heavy metals. The contamination threatens the local drinking water supplies for about 1,200 people and nearby wetlands.

Response actions included a site investigation that should be completed by 1997.

Source: PNPLS-NY.

Sidney Landfill
Delaware County
This former landfill near Sidney operated from 1964 to 1972. It received a variety of municipal and industrial wastes, including solvents, sludges, and PCB-bearing oils. Testing has shown that the groundwater and soil contain VOCs, PCBs, and heavy metals. The contamination threatens wetlands, springs, Herrick Hollow Creek, and drinking water wells that supply about 2,000 residents.

Response actions included posting warning signs around the perimeter of the site. An EPA investigation (1990–1997) is nearing completion and will recommend a final remediation plan.

Sources: PNPLS-NY; Sidney Public Library, Main Street, Sidney, NY, 13838.

Sinclair Refinery
Allegany County
An oil refinery operated on this site near Wellsville from about 1900 to 1958. Various wastes, including filters, sludges, pesticides, heavy metals, and fly ash, were buried in landfills on the property for 30 years. Erosion is beginning to undercut the landfill, which is located on the west bank of the Genesee River. Testing has shown that the groundwater and soil contain VOCs, SVOCs, and heavy metals. The contamination threatens the local drinking water supplies and the Genesee River.

Response actions included removing drummed waste and diverting the Genesee River away from the landfill (1983). Wellsville's municipal water supply intake was moved upstream from the landfill in 1985. Tanks and buildings were removed in 1993. The EPA recommended removing drummed waste, capping, channelizing the river, and fencing (completed 1991–1994). Contaminated soils were consolidated in 1993, and a groundwater extraction and treatment system is being constructed.

PRPs have participated in the site studies and remediation work.

Sources: PNPLS-NY; Howe Library, 155 N. Main Street, Wellsville, NY, 14895.

SMS Instruments
Suffolk County
Military aircraft components were refurbished on this site in Deer Park from 1971 to 1983. Various liquid wastes, including degreasers and jet fuel, were discharged into a leaching pool or underground storage tanks. Improper handling and storage, leaks, spills, and seepage have contaminated the groundwater and soil with VOCs and heavy metals. The contamination threatens Sampawams Creek, Guggenheim Lake, Belmont Lake State Park, and local drinking water supplies for about 125,000 people.

Response actions included pumping out the leaching pool and removing underground storage tanks (1983, 1988). In 1989 the EPA recommended groundwater extraction and air-stripping and soil vapor extraction. The soil program was concluded in 1993.The groundwater treatment system became operational in 1995 and will run to 1999.

Sources: PNPLS-NY; Deer Park Public Library, 44 Lake Avenue, Deer Park, NY, 11729.

Solvent Savers
Chenango County
Chemical wastes were reprocessed on this site in Lincklaen from about 1967 to 1974. Industrial solvents and degreasers were recycled. Resulting wastes, including sludges, were buried on the 13-acre property. Testing has shown that the groundwater, surface water, and soil contain VOCs, PCBs, and heavy metals. The contamination threatens local drinking water supplies, Mud Creek, and surface water used for agricultural needs.

Response actions included removing drummed waste and contaminated soil (1989, 1990, 1991, 1994). In 1990 the EPA recommended groundwater extraction and treatment by chemical precipitation and air-stripping, low-temperature thermal extraction or incineration of contaminated soil, and soil flushing and vapor extraction. These remedies should be operational by 1997.

AOCs (1989, 1991) require the PRPs to undertake the remediation work.

Sources: PNPLS-NY; Pond's Store, 567 Star Route, DeRuyter, NY, 13052.

Syosset Landfill

Nassau County

This former landfill near Oyster Bay operated from about 1933 to 1975. It received a variety of waste, including septic and industrial sludges, spent solvents, and degreasing agents. In 1974 local wells were discovered to be contaminated with VOCs. Landfill gases were also detected in a nearby school. Further testing showed that the groundwater and soil contain VOCs, heavy metals, and PCBs. The contamination threatens wells that supply about 60,000 people as well as businesses, schools, and a hospital.

Response actions included the installation of a gas ventilation system in 1982–1983. In 1990 the EPA recommended capping, expanding the gas collection system, fencing, deed restrictions, and long-term environmental monitoring. The remedies are presently being designed.

The PRPs have been active in conducting the site studies and response actions.

Sources: PNPLS-NY; Syosset Public Library, 225 South Oyster Bay Road, Syosset, NY, 11791.

Tri-Cities Barrel Company

Broome County

Steel storage drums have been reconditioned on this site in Fenton since 1955. Rinsewaters contaminated from barrel residues and caustic washing agents were discharged into unlined lagoons until about 1980. Tests have shown that the groundwater and soil contain PCBs, chlordane, VOCs, and heavy metals. The contamination threatens local drinking water supplies for about 4,000 residents, surface water used for agricultural purposes, and Osborne Creek.

A detailed site investigation (1992–1997) is nearing completion and will recommend a final remediation plan.

Source: PNPLS-NY.

Tronic Plating

Suffolk County

Electroplating and metal coating of electronic parts were conducted on this site in Farmingdale from 1968 to 1984. Various liquid wastes were stored in tanks, leaching pools, or discharged into a storm drain. Sampling by state officials in 1980 revealed that the groundwater and soils contain cyanide and heavy metals, which could also be carried in the air. The contamination threatens private and public wells in the area that serve about 17,000 people.

Response actions included removing liquid wastes from the property in 1993. A site investigation completed in 1992 recommended no action because contaminant levels were not high enough to threaten human health. Contaminated soils and sediments were removed in 1993.

The PRPs have conducted the site studies and removal actions according to an Administrative Order on Consent.

Sources: PNPLS-NY; Farmingdale Public Library, Main and Conklin Streets, Farmingdale, NY, 11735.

Vestal Water Supply Well 4-2

Broome County

This municipal water supply well for the town of Vestal was discovered to be contaminated with VOCs in 1980. The point source is a nearby chemical processing facility. Testing has shown that the groundwater contains VOCs, including trichloroethylene. The contamination threatens the drinking water sources for about 20,000 residents, wetlands, streams, and the Susquehanna River.

Response actions included closing the well in 1980. In 1989 an air-stripper/carbon filtration unit was installed to remove the VOCs. Soil studies are in progress.

Settlement agreements were signed between the State of New York and three PRPs.

Sources: PNPLS-NY; Vestal Town Hall, 605 Vestal Parkway, Vestal, NY, 13850.

Vestal Water Supply Well 1-1

Broome County

This municipal water supply well for the town of Vestal was discovered to be contaminated with VOCs in 1980. Further investigations identified businesses in a nearby industrial park as being point sources for the contamination. Testing has shown that the groundwater and soil contain VOCs and heavy metals. The contamination threatens nearby wetlands, Choconut Creek, the Susquehanna River, and the drinking water supplies for about 20,000 residents in Vestal.

Response actions included closing the well in 1980. A site investigation recommended air-stripping the contaminated groundwater in Well 1-1. An air-stripping unit was installed in conjunction with a new well in 1994, and soil vapor extraction is also under way. Groundwater monitoring is in progress.

Sources: PNPLS-NY; Vestal Town Hall, 605 Vestal Parkway, Vestal, NY, 13850.

Volney Municipal Landfill

Oswego County

This former landfill near Volney operated from 1969 to 1983. It received a variety of municipal and industrial wastes, including drummed chemicals, spent solvents, and sludges. Testing in the 1980s confirmed that the groundwater and soil contain heavy metals and VOCs. The contamination threatens the local drinking water supplies for about 1,000 people and nearby recreational streams and wetlands.

Response actions included capping, gas venting, and leachate collection (1979–1985). In 1987 the EPA recommended additional capping on the side slopes, expanding the leachate collection system, building slurry walls, and constructing a leachate treatment system. Construction should be under way by 1997.

AOCs (1979, 1990, 1993) require the PRPs to undertake the site studies and remediation work.

Sources: PNPLS-NY; Fulton Public Library, 160 South 1st Street, Fulton, NY, 13850.

Warwick Landfill
Orange County
This former landfill near Warwick operated from the 1950s to 1979, when it was closed by the state because of federal violations. It received a variety of municipal and industrial wastes, including spent solvents, acids, and sludges. Testing revealed that the groundwater and soil contain VOCs, SVOCs, heavy metals, and phenols. The contamination threatens streams, wetlands, recreational Greenwood Lake, and water wells that supply about 2,300 residents. Response actions included a first-phase site investigation that was completed in 1991. The EPA recommended interim capping until other studies currently in progress are completed.

Unilateral Administrative Orders (1992, 1993) require eight PRPs to conduct remediation work.

Sources: PNPLS-NY; Warwick Town Hall, 60 Main Street, Warwick, NY, 10990.

York Oil Company
Franklin County
Waste oils were recycled at this facility near Moira from 1962 to 1975. Metal storage tanks and crankcase oils, including PCB-containing oils, were processed on the site in 1980–1981. Wastes and oils were held in storage tanks or settling lagoons. The PCB-contaminated oils were sold as fuel oil or used to dust local roads. The lagoons overflowed with spring rains onto adjacent property, including wetlands. Testing revealed that the groundwater and soil contain heavy metals, VOCs, PCBs, and phenols. The contamination threatens wetlands, streams, associated wildlife habitats, and the private drinking water wells that serve about 2,000 residents.

Response actions included draining the lagoons and removing contaminated soils from the property and adjacent wetlands (1980). Other removal actions were conducted in 1983. The EPA stabilized leaking storage tanks in 1992. In 1988 the EPA recommended the excavation, solidification, and reburial of contaminated soil; removing contaminated oils in storage tanks; and extracting and treating contaminated groundwater. These systems are presently being designed. Off-site contamination is still being studied.

A Consent Decree (1990) and Consent Order (1992) require the PRPs to undertake the design and construction of the remediation program.

Sources: PNPLS-NY; Moira Town Hall, North Lawrence Road, Moira, NY, 12957.

NORTH CAROLINA

ABC One Hour Cleaners
Onslow County
This dry cleaning facility in Jacksonville has operated since 1954. Tetrachloroethylene was used as a solvent. Spent solvents and other wastes were buried or discharged on the site until about 1965. Nearby wells were closed because of high levels of VOCs. Further tests showed that the groundwater and soil on the site contain VOCs. The contamination threatens public wells within 3 miles of the site, which supply drinking water to about 40,000 residents.

In 1993 the EPA recommended a groundwater extraction and air-stripping system that is presently being designed. Soil vapor extraction was conducted from 1994 to 1995.

Sources: PNPLS-NC; Onslow County Library, 58 Doris Avenue East, Jacksonville, NC, 28540.

Aberdeen Pesticide Dumps
Moore County
Chemicals were manufactured in this area near Aberdeen from the 1930s to the 1960s. Various wastes, including pesticides, were illegally buried on the site. Complaints from local residents in 1984 led to an investigation by the state that discovered the dumpsites. Nearby wells were shut down because of high levels of lindane. Testing revealed that the groundwater and soil contain pesticides, including DDT, toxaphene, and benzene hexachloride. The contamination threatens streams, recreational Page's

NORTH CAROLINA

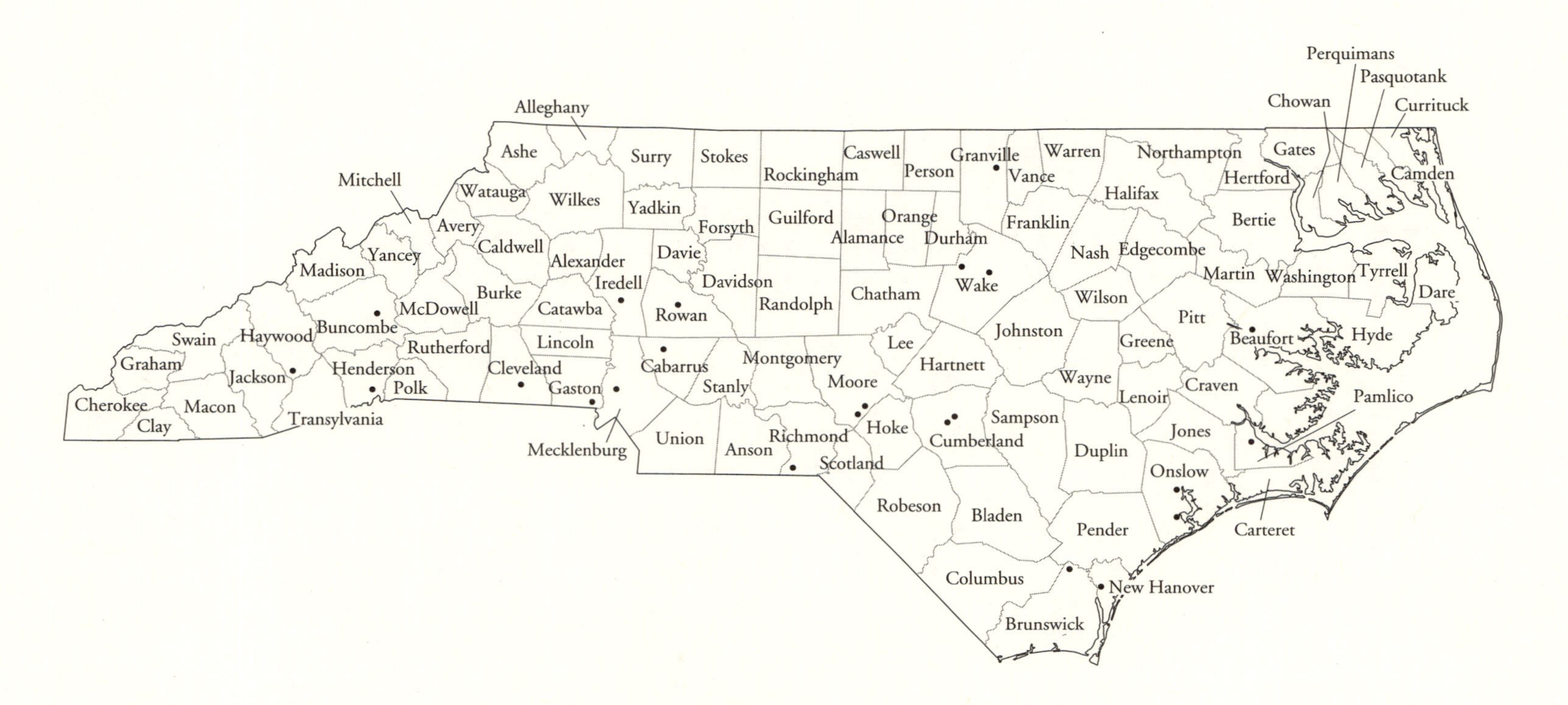

Lake, and public and private wells that supply about 6,000 people.

Response actions included removing containers of pesticides and contaminated soil (1985–1989). A site study was conducted from 1987 to 1992. The EPA recommended the removal and thermal treatment of contaminated soil, razing structures, and groundwater extraction and treatment. These systems will be operational in 1996–1997. Other studies are in progress.

UAOs (1990, 1993), AOCs (1994), and UAOs (1994) require about 25 PRPs to undertake the clean-up activities. The EPA is also suing the PRPs for the recovery of past clean-up costs.

Sources: PNPLS-NC; Aberdeen Town Hall, 115 North Poplar Street, Aberdeen, NC, 28315.

Benfield Industries
Haywood County
This organic and inorganic chemical processing center near Hazelwood operated from 1976 to 1986. Testing in 1985 showed that the soil contains high levels of PAHs. The contamination is a threat to local surface water and groundwater, water wells that serve about 2,000 people, Browning Branch, and Richland Creek.

A site investigation was conducted from 1990 to 1992. The EPA recommended fencing, landfarming of contaminated soil, the extraction and treatment of groundwater, and environmental monitoring. Construction is expected to be under way by 1997.

Source: Hazelwood Town Hall, 101 West Georgia Avenue, Hazelwood, NC, 28738.

By-Pass 601
Cabarrus County
Groundwater in this area in the town of Concord was found to contain lead. Multiple sources are suspected, including a scrap metal recycling facility that recovered lead from old batteries and dumped acid on the premises. Ten other sources are being investigated by the state. Testing has shown that the groundwater, soil, and surface water contain lead and chromium. The contamination threatens private wells that serve about 3,000 residents.

Response actions included the removal of battery debris and contaminated soil from the recycling facility (1982–1983). In 1993 the EPA recommended the excavation, solidification, and capping of contaminated soil. The work is expected to be completed by 1997.

Over 100 PRPs have signed a Consent Decree that requires the companies undertake the site investigations and remediation work.

Sources: PNPLS-NC; Cannon Memorial Library, 27 Union Street, North Concord, NC, 28025.

Camp Lejeune
Onslow County
The U.S. Marines established this base near Jacksonville in 1941. Various activities have generated solid and liquid wastes that were buried in 77 landfills on the 170-square-mile site. Testing has shown that the groundwater and soil contain pesticides, VOCs, fuels, petrochemicals, solvents, paint strippers, PAHs, heavy metals, aldrin, and PCBs. Contaminated wells on the base have been shut down. The contamination threatens the recreational New River and private and public wells that supply about 14,000 residents.

Response actions have included installing interim groundwater pump-and-treat systems and removing contaminated soil. A number of site investigations are nearing completion.

Sources: PNPLS-NC; Onslow County Library, 58 Doris Avenue East, Jacksonville, NC, 28540.

Cape Fear Wood Preserving
Cumberland County
Wood was treated with creosote and chromated copper-arsenate solutions on this site near Fayetteville from 1953 to 1983. Various wastes were held in an unlined lagoon. Treated lumber was stored on site. Testing has shown that the groundwater and soil contains heavy metals, VOCs, and PAHs. The contamination threatens nearby streams and wetlands and private and public wells that supply about 17,000 residents.

Response actions included removing sludge, creosote, tanks, and contaminated sediments (1985–1988). In 1989 the EPA recommended soil washing, soil bioremediation, soil solidification and reburial, and groundwater extraction and treatment. This program will be under way by 1997.

The PRPs conducted some of the initial response actions.

Sources: PNPLS-NC; Cumberland County Public Library, 300 Maiden Lane, Fayetteville, NC, 28301.

Carolina Transformer Company
Cumberland County
Transformers were recycled on this site near East Fayetteville in the 1960s and 1970s. Testing by the EPA in 1978 revealed that the groundwater, soil, and some nearby wells were contaminated with VOCs and PCBs. Ruptured capacitors were also found leaking PCB-bearing oil onto the ground. The contamination threatens the local drinking water supplies for about 3,000 people.

Response actions included removing contaminated soil in 1984. Affected residents were also connected to clean water supplies. The leaking capacitors were removed in 1990. A site investigation was completed in 1991. The EPA recommended soil solvent extraction, removal of solid waste and debris, and groundwater extraction and treatment. Construction will be under way by 1997.

The PRP rejected an AOC from the EPA to clean up the site and is being sued by the EPA for past clean-up costs and damages.

Sources: PNPLS-NC; Cumberland County Public Library, 300 Maiden Lane, Fayetteville, NC, 28301.

Celanese Corporation
Cleveland County
Polyester and filament yarn have been produced on this site in Shelby since 1960. Various chemical wastes were discharged on the surface until a wastewater plant was built about 1966. Other wastes were stored in tanks or drums. Testing has shown that the groundwater, soil, and nearby stream sediments contain VOCs, SVOCs (phthalates), and heavy metals. The contamination threatens nearby wells that serve about 1,000 residents, agricultural wells, and Buffalo Creek.

Response actions included a site investigation. In 1988 the EPA recommended the extraction and air-stripping of groundwater. The system became operational in 1993 and will run to 2013. Contaminated soil and sludge were excavated and solidified or incinerated in 1993.

Sources: PNPLS-NC; Cleveland County Public Library, 104 Howie Drive, Shelby, NC, 28151.

Charles Macon Lagoon
Richmond County
A former hazardous waste storage facility near Cordova operated from the 1970s to 1981. Liquid wastes, oil, and sludges were held in lagoons or drums on the 16-acre property. Lagoons occasionally overflowed. In 1985 the EPA determined that the groundwater and soil contain VOCs and heavy metals (barium and chromium). The contamination threatens streams, wetlands, the Pee Dee River, and private wells in the area that serve about 1,200 people.

Response actions included removing drummed waste and excavating the lagoons in 1983. The EPA recommended soil vapor extraction, groundwater extraction and air-stripping, and the removal of remaining wastes, tanks, and drums. The treatment systems will be operational by 1997.

A Consent Decree (1989) and UAO (1992) require the PRPs to refund past clean-up costs and carry out the remediation plan.

Sources: PNPLS-NC; Leath Memorial Library, 412 East Franklin Street, Rockingham, NC, 28379.

Chemtronics
Buncombe County
This active industrial plant near Swannanoa has produced explosives, fuels, and pharmaceutical chemicals since 1952. Various wastes, including acids and solvents, were buried in pits and trenches until about 1980. Testing has shown that the groundwater and soil contain VOCs (bromoform), TNT, and heavy metals. The contamination threatens local drinking water supplies, Bee Tree Creek, and the Pisgah National Forest.

Response actions included removing drummed waste in 1985. The EPA recommended capping, a gas collection system, a groundwater extraction/air-stripping circuit, and long-term environmental monitoring. The groundwater treatment system became operational in 1993 and will run for 30 years.

AOCs (1985, 1989) require three PRPs to conduct the site investigations and remediation work.

Sources: PNPLS-NC; Martha Ellison Library, Warren Wilson College, 701 Warren Wilson Road, Swannanoa, NC, 28778.

Cherry Point Marine Corp Station
Craven County
This U.S. Marine Corp air station near Havelock began manufacturing and repairing aircraft in 1942. Various wastes, including heavy metal–plating sludges, cyanide, solvents, paint strippers, petroleum, oils, and PCBs, were buried on the 11,500-acre site until 1982. Testing in 1986 showed that the groundwater and soil contain heavy metals and VOCs. Sediments from Slocum Creek contain PCBs and arsenic. The contamination threatens local drinking water supplies, Slocum Creek, and the recreational Neuse River.

A detailed site investigation is planned.

Source: PNPLS-NC.

FCX, Statesville Plant
Iredell County
Agricultural chemicals were blended and processed on this site in Statesville from 1940 to 1986. Illegal burial and spills of pesticides on the premises have contaminated the soil and groundwater with DDT, chlordane, VOCs, lindane, and halogenated organic solvents. The contamination threatens private and public wells within 3 miles of the site, which supply about 12,500 people.

Response actions included fencing (1989) and a site study (1989–1993). The EPA recommended a groundwater extraction and treatment system that should be operational by 1997. Contaminated soil will be thermally treated. Other studies are in progress.

The PRP filed for bankruptcy in 1985.

Sources: PNPLS-NC; Iredell County Library, 135 East Water Street, Statesville, NC, 28677.

FCX, Washington Plant

Beaufort County

Agricultural chemicals were processed on this site near Washington from 1945 to the 1980s. Various chemical and pesticide wastes were buried in trenches on the 6-acre property. Testing in 1985 showed that the soil and groundwater contain pesticides (DDT and chlordane), mercury, and VOCs. The contamination threatens nearby wetlands and local water supplies that serve about 3,000 residents.

Response actions included removing and thermally treating contaminated soil and fencing (1988–1992). A site investigation was completed in 1993. The EPA recommended a groundwater pump-and-treat system that should be operational in 1997.

The PRP filed for bankruptcy in 1985.

Sources: PNPLS-NC; Brown Library, 122 Van Norden Street, Washington, NC, 27889.

Geigy Chemical Corporation, Aberdeen Plant

Moore County

Various chemical companies have produced solid and liquid pesticides on this property near Aberdeen since 1947. Waste solutions and chemicals were held in lagoons or storage tanks. State officials discovered that neighboring wells contained pesticides in 1985. Testing in 1987 revealed that the groundwater, surface water, and soil contain toxaphene, DDT, and lindane. The contamination threatens recreational Aberdeen Creek and water wells that serve about 8,000 residents.

Response actions included removing contaminated soil and debris (1989, 1991). In 1992 the EPA recommended off-site incineration of contaminated soil and the construction of a groundwater treatment system. Construction will begin in 1997.

An AO (1988) and Consent Decree (1993) require three PRPs to conduct the site examination and remediation work.

Sources: PNPLS-NC; Aberdeen Town Hall, 115 North Poplar Street, Aberdeen, NC, 28315.

General Electric/Shepherd Farm

Henderson County

Luminaire lighting systems have been manufactured on this site in East Flat Rock since 1955. Related production wastes, including sludges and heavy metals, were buried on the nearby Shepherd Farm from 1957 to 1970. Testing in the 1980s showed that the groundwater and soil contain heavy metals, VOCs, and PCBs, which have also entered nearby wells. The contamination threatens the local water supplies for about 5,000 residents.

Response actions included connecting affected residents to clean water supplies. A detailed site investigation is nearing completion and will recommend a final remediation plan.

Sources: PNPLS-NC; Henderson County Library, 301 North Washington Street, Hendersonville, NC, 28792.

Jadco Hughes

Gaston County

A solvent recycling center operated on this site in Belmont from 1971 to 1975. Waste residues were stored in about 20,000 drums. Contaminated soil and debris were buried on the property. Testing has shown that the groundwater and soil contain PCBs, heavy metals, and VOCs. The contamination threatens the Catawba River and the local water supplies for about 5,000 residents.

Response actions included the removal of contaminated sediments in 1990. A site investigation (1986–1990) recommended soil flushing, soil vapor extraction, and groundwater extraction and treatment. Construction should be under way by 1997.

Legal agreements require the PRP to conduct the site investigations and remediation work.

Sources: PNPLS-NC; Gaston County Library, 111 Central Avenue, Belmont, NC, 28012.

JFD Electronics

Granville County

Television antennas were produced on this site in Oxford from 1962 to 1979. Wastewaters and heavy metal sludges were discharged into a lagoon on the 13-acre property. Spent solvents were also stored in leaking underground tanks. Tests have shown that the soil, groundwater, and surface water contain VOCs. VOCs have also been detected in sediments from recreational Fishing Creek. The contamination threatens the private drinking water sources for about 3,000 local residents.

Response actions included removing underground tanks, spent solvents, sludges, and contaminated soils in 1987–1988. An investigation was concluded in 1992 that recommended excavating additional contaminated soil and sludge and installing a groundwater extraction and air-stripping system.

A 1993 Consent Decree requires the PRPs to conduct the site studies and remediation work.

Sources: PNPLS-NC; Thorton Library, Main and Spring Streets, Oxford, NC, 27565.

Koppers Company, Morrisville Plant
Wake County
A sawmill and glue-laminated wood factory have operated on this site near Morrisville since the 1950s. Wood was also treated with PCP. Wastewaters were held in lagoons on the property until 1977. Testing has shown that the soil, groundwater, and surface water contain PCP, dioxins, and furans. The contamination threatens the local drinking water supplies for about 2,500 residents as well as Crabtree Creek, Koppers Pond, and Medlin's Pond.

Contaminated soil was removed in 1980 and 1986. Affected residents were connected to city water supplies in 1989. A site investigation conducted from 1989 to 1992 recommended the off-site incineration of contaminated soil and a groundwater pump-and-treat system.

Three AOs require the PRPs to conduct the site investigations and clean-up work.

Sources: PNPLS-NC; Cary Public Library, 310 South Academy Street, Cary, NC, 27511.

Martin-Marietta
Mecklenburg County
Chemical dyes have been produced on this site near Charlotte since 1936. Organic solvents were discovered in nearby wells in 1980. Various wastes, including distillate tars, were buried on the property or held in settling ponds. Testing has shown that the groundwater and soil contain VOCs. The contamination threatens the Catawba River and the drinking water supplies of about 10,000 residents.

Response actions consisted of excavating contaminated wastes and soils. In 1987 the EPA recommended groundwater extraction and treatment, capping, and soil excavation and treatment. Soil vacuum extraction is also being considered. The groundwater treatment system is in operation and will run until 1999.

Sources: PNPLS-NC; Mt. Holly Public Library, 245 West Catawba Avenue, Mt. Holly, NC, 28120.

National Starch and Chemical Company
Rowan County
Chemicals for the textile industry have been manufactured on this site near Salisbury since the 1970s. Various wastes were buried on the property or held in unlined lagoons. Testing has shown that the groundwater, soil, and surface water contain VOCs and heavy metals. These chemicals have also been detected in nearby streams. The contamination threatens streams and wetlands and the local drinking water supplies for about 8,000 residents.

Response actions included a site investigation. A groundwater extraction and air-stripping treatment system began operation in 1993 and will run for 30 years. Another extraction system is being designed that will operate for over 100 years.

The PRPs have participated in the site studies and remediation work.

Sources: PNPLS-NC; Rowan Public Library, 201 West Fisher Street, Salisbury, NC, 28144.

New Hanover County Airport
New Hanover County
Fire-fighting training occurred on this property near Wilmington from 1968 to 1979. Various fuels were burned in a pit and extinguished by training crews. Testing by the state in 1986 revealed that the groundwater and soil contain heavy metals, VOCs, and petrochemicals. The contamination threatens nearby streams and wetlands and wells that serve about 6,500 residents.

Response actions included removing sludge, tanks, and contaminated soil in 1990. In 1992 the EPA recommended a groundwater extraction and air-stripping system that should be operational by 1997.

A 1990 Consent Order, a Consent Decree (1993), and UAO (1994) require the PRPs to conduct the site investigations and remediation work.

Sources: PNPLS-NC; New Hanover Public Library, 201 Chestnut Street, Wilmington, NC, 28401.

North Carolina State University
Wake County
An agricultural research area that is part of North Carolina State University contains buried solvents, pesticides, heavy metals, radioactive material, and acids. The drummed waste was buried in trenches from 1969 to 1980. Testing has shown that the groundwater and soils contain high amounts of VOCs and heavy metals. The contamination threatens local private wells.

A detailed site investigation (1987–1995) will recommend a final remediation plan.

Sources: PNPLS-NC; Regional Public Library, Cameron Village Shopping Center, Raleigh, NC.

Potter's Septic Tank Service
Brunswick County
A local septic tank service buried related wastes, oil, and sludge on this property near Riegelwood in the 1970s. Tests by the state discovered contaminated wells that were later shut down. The groundwater and soil contain VOCs, petrochemicals, heavy metals, chloroform, and phenols. The contamination threatens Chinnis Branch and Rattlesnake Branch Creek as well as private wells that supply about 2,000 residents with drinking water,.

Response actions included removing oil, sludge, and contaminated soil from the property and from Rattlesnake Branch Creek. (1983–1984). In 1993 the EPA recommended groundwater extraction and treatment through chemical precipitation, flocculation, and air-stripping in addition to low-temperature thermal desorption of contaminated soil. These systems went into operation in 1994–1995.

Sources: PNPLS-NC; Columbus County Library, Route 2, Highway 87, Riegelwood, NC, 28456.

NORTH DAKOTA

Arsenic Trioxide Site
Richland, Ransom, and Sargent Counties
This 500-square-mile site is the result of multiple applications of arsenic-based pesticides to kill grasshopper plagues in the 1930s and 1940s. The problem was first detected when drinking water wells in Lidgerwood were found to contain high levels of arsenic. Testing has shown that the groundwater and soil contain arsenic. The contamination threatens many private and public drinking wells, including those for the towns of Lidgerwood, Wyndmere, and Milnor. The groundwater is also used for agricultural needs.

Response actions included capping certain areas and providing affected residents with clean water supplies (1986–1989). The EPA helped the communities upgrade their water treatment systems in 1990–1991. Drinking water is now high quality and colorless. The EPA will delete this site from the NPL by the year 2000.

Sources: PNPLS-ND; Lidgerwood Public Library, 15 Wiley Avenue North, Lidgerwood, ND, 58053.

Minot Landfill
Ward County
This municipal landfill near Minot operated from 1962 to 1971. It also accepted industrial waste, including oils, battery parts, lime sludges, and military construction debris. Testing has shown that the groundwater and soil contain VOCs, inorganic compounds, and pesticides. Gas emissions from the landfill contain VOCs (benzene and vinyl chloride). Surface runoff flows into the Souris River. The contamination threatens the Souris River and the local drinking water supplies for the 35,000 residents of Minot.

Response actions included fencing the site and building surface runoff controls (1990). A site investigation (1990–1993) recommended upgrading the cap, installing a leachate extraction system, gas venting, and groundwater monitoring. Construction should be completed by 1997.

Sources: PNPLS-ND; Minot Public Library, 516 2nd Avenue Southwest, Minot, ND, 58701.

NORTH DAKOTA

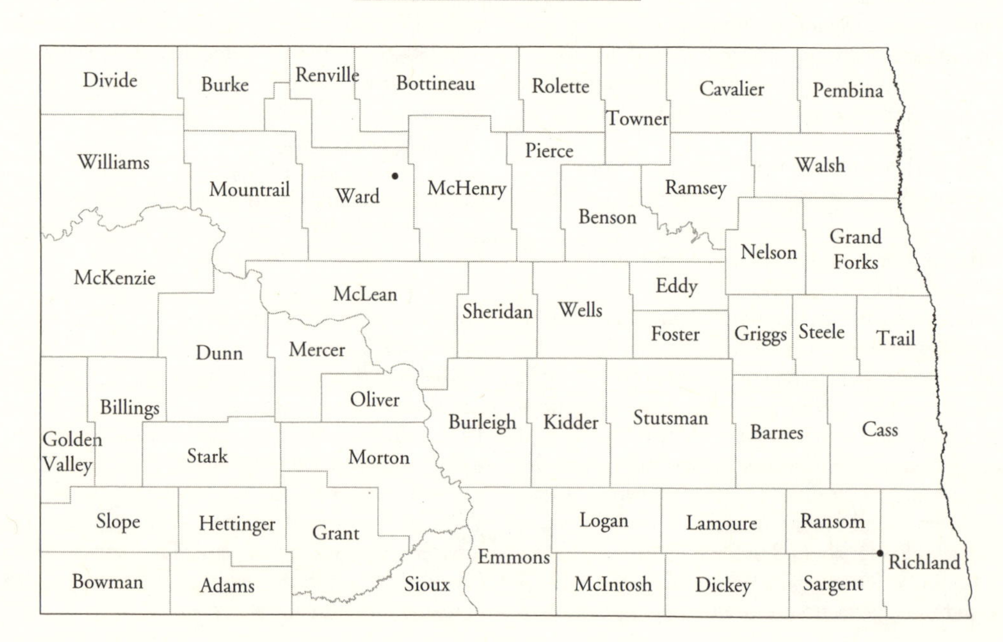

Air Force Plant 85
Franklin County
This U.S. Air Force aircraft production and maintenance facility near Columbus has been in operation since 1941. Various production wastes, including coal pile leachate, sulfuric acid, transformer oils, solvents, paint strippers, cyanide sludge, fuels, magnesium chips, PAHs, and heavy metal sludges, were stored, spilled, leaked, discharged, or buried on the 420-acre site. Sampling has shown that the soil, groundwater, surface water, and stream sediments contain chromium, copper, nickel, zinc, VOCs, mercury, PAHs, PCBs, acetone, and cyanide. The contamination threatens Mason's Run, recreational Big Walnut Creek, and the health and drinking water supplies of local residents and on-site employees.

Response actions have consisted of six ongoing site investigations. Once completed and reviewed, a final remediation plan will be issued.

Source: PNPLS-OH.

Allied Chemical and Ironton Coke
Lawrence County
This industrial site along the Ohio River near Ironton manufactured coke from 1917 to 1982 and tar from 1945 to the present. Various chemical wastes and wastewater were discharged into Ice Creek or unlined lagoons. Testing has shown that the groundwater, soil, and creek sediments contain VOCs, phenols, PAHs, cyanide, and heavy metals, including arsenic. The contamination threatens Ice Creek, the recreational Ohio River, and local wells that supply drinking water to about 16,000 residents.

Response actions included razing the coke plant. The EPA recommended a slurry wall, capping, groundwater extraction and treatment, alternate water supplies for affected residents, deed restrictions, on-site incineration of coke wastes, and soil bioremediation. Construction of these remedies should be under way by 1997.

A UAO (1989) requires the PRPs to contribute toward the site studies and remediation work.

Sources: PNPLS-OH; Briggs Lawrence County Public Library, 321 South Fourth Street, Ironton, OH, 45638.

Alsco Anaconda
Tuscarawas County
Aluminum has been refined and processed at this site owned by ARCO Chemical Company in Gnadenhutten since 1965. Various wastes were discharged into unlined lagoons and pits on the property, including heavy metal sludges. The lagoons occasionally overflowed into the Tuscarawas River. Testing has shown that the groundwater and soil contain cyanide, chromium, other heavy metals, and PCBs. The contamination threatens the recreational Tuscarawas River and local wells that supply about 3,000 residents.

Response actions included site investigations. In 1989 the EPA recommended excavating sludges and contaminated soil, removing buried drums, and long-term monitoring of surface water and groundwater.

Various agreements (1987, 1989, and 1993) require the PRPs to conduct the site investigations and remediation work.

Sources: PNPLS-OH; Gnadenhutten Public Library, 160 North Walnut Street, Gnadenhutten, OH, 44629.

Arcanum Iron and Metal
Darke County
Lead batteries were recycled on this property near Arcanum from the 1960s to 1982. Battery acids and lead oxide sludges were discharged on the surface or buried. A fish kill occurred in nearby Sycamore and Painter Creeks in 1964. In the 1970s the soil and groundwater were discovered to contain lead, antimony, and arsenic. The contamination threatens public and private wells within 1 mile of the site, including those serving the town of Arcanum.

Response actions included fencing in 1984. In 1986 the EPA recommended excavating contaminated soil and battery waste, demolishing structures, deed restrictions, treatability studies, and groundwater monitoring. The soil plan was modified to include capping and should be under way by 1997.

A 1979 Consent Decree with the PRP required the company to clean up the site. The PRP did not fully comply before it went out of business in 1982.

Sources: PNPLS-OH; Arcanum Public Library, 101 North Street, Arcanum, OH, 45304.

Big D Campground
Ashtabula County
This former gravel pit in Kingsville Township was used as a landfill from 1964 to 1976. It accepted a variety of industrial and municipal wastes, including chemical solutions, sludges, solvents, and drummed waste. Sampling in 1978 revealed that the groundwater, soil, and leachate contain VOCs and heavy metals (barium, chromium, lead, and nickel). The pollutants have penetrated Conneaut Creek. The contamination threatens the water supplies of about 4,000 nearby residents.

Response actions included capping and constructing surface runoff controls in 1983. In 1989 the EPA recommended removing drummed waste and contaminated soil, revegetation, fencing, deed restrictions, groundwater extraction and air-stripping, and long-term monitoring. Construction of the remedies

OHIO

has been completed, and the groundwater plant has been in operation since 1994.

The PRP, Olin Chemical Corporation, has been an active and willing participant in the site studies and clean-up work.

Sources: PNPLS-OH; Kingsville Township Library, 6006 Academy Avenue, Kingsville, OH, 44048.

Bowers Landfill
Pickaway County
Originally a gravel pit, this site near Circleville was used as a landfill from 1958 to 1968. It received a variety of domestic and industrial wastes, including chemical solvents. Testing in 1980 discovered that the groundwater, soil, and stream sediments contain barium, manganese, VOCs, phthalates, PCBs, PAHs, petrochemicals, pesticides, arsenic, and lead. The chemicals have entered nearby streams that flow into the Scioto River. The contamination threatens the health of about 2,000 people who live within 2 miles of the site.

A site investigation was conducted by the EPA from 1983 to 1989. In 1989 the EPA recommended removing surface debris, erosion controls, capping, fencing, groundwater monitoring, and creating a new wetland. These remedies were completed by 1993. Monitoring is continuing.

Sources: PNPLS-OH; Pickaway County Library, 165 East Main Street, Circleville, OH, 43113.

Buckeye Reclamation

Belmont County

This former landfill near St. Clairsville operated from the 1960s to 1979. It received a variety of wastes, including coal mine spoils, asbestos, and industrial sludges. The groundwater and soil contain VOCs and heavy metals that have penetrated King's Run, Little McMahon Creek, and neighboring wells. The contamination threatens the local drinking water supplies for about 120 residents.

A site investigation conducted from 1985 to 1991 recommended capping, surface runoff controls, groundwater collection and treatment using constructed wetlands, deed restrictions, and fencing. Construction should be under way by 1997.

Legal agreements (1985, 1992) with 20 PRPs require the companies to conduct the site investigations and clean-up work.

Sources: PNPLS-OH; St. Clairsville Public Library, 108 West Main Street, St. Clairsville, OH, 43950.

Chem-Dyne

Butler County

This industrial waste recycling and processing facility near Hamilton operated in the 1970s. Various chemicals, including antifreeze, VOC solvents, and PCBs, were leaked or spilled from storage drums and tanks. Spills or discharges were linked to five fish kills in the Great Miami River from 1976 to 1979. Groundwater and soil contain VOCs, heavy metals, pesticides, and PCBs that have penetrated nearby Ford Canal and the Great Miami River. The contamination threatens the local drinking water supplies for hundreds of residents as well as the recreational Great Miami River.

Drummed waste and contaminated soil were removed in 1980 and 1982 by the EPA. The site was also fenced. In 1985 the EPA recommended groundwater extraction and air-stripping, razing structures, removing contaminated soil, and capping. The groundwater treatment system became operational in 1988 and will continue for at least 10 years.

A Consent Decree (1985) between the EPA and nearly 200 PRPs requires the companies to conduct the remediation work.

Sources: PNPLS-OH; Hamilton Municipal Building, 20 High Street, Hamilton, OH, 45011.

Chemical And Minerals

Cuyahoga County

Located near Cleveland, this recycling facility operated from 1979 to 1980. Various wastes, including solvents, paints, tars, grease, and resins, were stored on-site in barrels and tanks. A fire in 1980 closed operations. Testing revealed that the soil was contaminated with VOCs. The contamination threatens the local groundwater and the Cuyahoga River.

Response actions included removing contained waste in 1981. The buildings were razed and contaminated soil removed in 1982.

A 1987 Consent Decree required PRPs to pay for nearly 90 percent of the EPA's clean-up costs.

Source: PNPLS-OH.

Coshocton Landfill

Coshocton County

This former landfill and coal mining pit operated in the early 1900s and from the 1950s to 1979. It received a variety of municipal and industrial wastes, including sludges. In 1977 the landfill caught fire and burned for 3 days. Leachate has been reported on surrounding properties. The groundwater and soil contain VOCs, heavy metals, PCP, and phenols. The contamination threatens local water wells and could be a threat to the 14,000 residents of Coshocton.

Response actions included removing drummed waste in 1985. In 1988 the EPA recommended capping, deed restrictions, fencing, and a gas venting system. Construction will be completed by 1997.

A 1989 UAO required the PRP to undertake the clean up of the site.

Sources: PNPLS-OH; Coshocton Public Library, 655 Main Street, Coshocton, OH, 43812.

Diamond Shamrock Corporation

Lake County

Chemicals, including acids, caustic soda, chlorine, chromic acid, and potassium dichromate, were manufactured on this site in Painesville from 1912 to 1972. Waste solutions were discharged into lagoons or landfills on the 500-acre property. In 1981 the EPA found that the soil and groundwater contained heavy metals, cyanide, VOCs, and PCBs. The contamination threatens various drinking water sources, including Lake Erie and the recreational Grand River. Several endangered species inhabit the Grand River ecosystem.

Response actions included razing all the buildings and removing PCB-contaminated waste. A detailed site investigation is planned.

Source: PNPLS-OH.

Dover Chemical Corporation

Tuscarawas County

This chemical processing plant near Dover has been in operation since 1949. Chemical products have included chlorinated products, paraffins, lubricants, flame retardants, phosphites, furans, and plasticizers. The groundwater and soil contain VOCs and polychlorinated compounds. The contamination threatens Sugar Creek and wells that supply about 28,000 residents.

Response actions included removing chemical wastes and contaminated soils from 1981 to 1986. Additional removals were conducted in 1991. A groundwater extraction and treatment system became operational in 1992. A detailed site investigation will be completed by 1997.

AOCs (1988, 1991) require the Dover Chemical Company to remediate the site. Dover has conducted all the response actions to date.

Sources: PNPLS-OH; Dover Public Library, 525 North Walnut, Dover, OH, 44622.

E. H. Schilling Landfill
Lawrence County
This former landfill near Hanging Rock operated from 1969 to 1980, receiving a variety of municipal and industrial wastes. Leachate has entered adjacent properties. The groundwater and soil contain heavy metals (nickel and arsenic), VOCs, and PAHs. Nickel has also been detected in the air. The contamination threatens Winkler Run, the Ohio River, and private and public drinking water wells in the area that serve about 2,000 residents.

In 1989 the EPA recommended capping, stabilizing contaminated soil, fencing, constructing a slurry wall and runoff controls, and treating leachate and liquid waste in an on-site treatment plant. The plant became operational in 1994 and will run to 1997. Groundwater monitoring is in progress.

Sources: PNPLS-OH; Briggs Lawrence County Library, 321 South Fourth Street, Ironton, OH, 45638.

Feed Materials Production Center
Hamilton and Butler Counties
Metallic uranium was produced on this site near Harrison from 1952 to 1989 by the U.S. Department of Energy. The uranium was used for nuclear weapon reactors. Various production wastes, including oils, solvents, and radioactive materials, were stored, buried, or discharged on the 1,450-acre site. The groundwater and soil contain uranium, radium, technetium, chromium, and VOCs. Radon gas has been detected in the air. The chemicals have penetrated nearby streams. The contamination threatens the recreational Great Miami River and the drinking water supplies for workers on the site (now a feed materials production center) and nearby residents.

Response actions included removing buildings and contaminated soils, installing a groundwater pump-and-treat system, and capping. Most site investigations have been recently completed and will recommend remediation plans.

The Department of Energy will conduct the site studies and clean-up work according to a series of Consent Decrees (1988, 1990, 1991).

Sources: PNPLS-OH; Public Environmental Information Center, 10845 Hamilton-Cleues Highway, Harrison, OH, 45030.

Fields Brook
Ashtabula County
Fields Brook flows through an industrial part of Ashtabula. Creek sediments contain PCBs, VOCs, PAHs, heavy metals, and phthalates derived from the industries in the area. The stream flows through a residential neighborhood and into the Ashtabula River, which empties into Lake Erie. The contamination threatens recreational users of the Ashtabula River and Lake Erie and the Lake Erie water supply for the city of Ashtabula.

The EPA recommended a remediation plan consisting of the excavation and thermal treatment of contaminated sediments and long-term monitoring. Other studies that are evaluating possible remedial actions for the Ashtabula River and affected wetlands are in progress.

Six PRPs identified by the EPA are participating in the site studies and clean-up work.

Sources: PNPLS-OH; Ashtabula District Library, 335 West 44th Street, Ashtabula, OH, 44004.

Fultz Landfill
Guernsey County
This former strip mine near Byesville was used as a landfill from the 1950s to 1985. It received a variety of municipal and industrial wastes, including drummed chemical solutions, heavy metals, and sludges. Leachate was observed seeping from the landfill in 1980 and has penetrated nearby ponds and waterways. The groundwater and soil contain heavy metals, VOCs, phenols, and phthalates. The contamination threatens ponds, Wills Creek, and the drinking water supplies for local residents.

In 1991 the EPA recommended fencing, leachate collection, capping, and a groundwater extraction and treatment system. Construction should be under way by 1997.

Sources: PNPLS-OH; Guernsey County Public Library, Byesville Branch, 100 Glass Avenue, Byesville, OH, 44632.

Industrial Excess Landfill
Stark County
A former gravel pit, this site near Hartville was used as a landfill from 1966 to 1980. It received a large quantity of waste from rubber companies in the

area. Testing in the mid-1980s showed that the groundwater and soil contain VOCs, heavy metals, cyanide, pesticides, and phthalates. VOCs have also been detected in the air. The contamination threatens nearby streams and wells that serve about 1,500 local residents.

Response actions included the installation of a methane venting system (1985–1989). Drummed wastes were also removed. Affected residents were connected to clean water supplies in 1987. In 1989 the EPA recommended capping, a groundwater pump-and-treat system, fencing, deed restrictions, and long-term monitoring. Construction activities should be under way by 1997.

Sources: PNPLS-OH; Hartville Library, 411 East Maple Street, Hartville, OH, 44632.

Laskin/Poplar Oil Company
Ashtabula County
Greenhouse and waste oil recovery operations from the 1890s to the 1970s on this site near Jefferson have contaminated the groundwater and soil. Waste oils from both activities were stored in tanks, pits, and lagoons on the property. Testing has shown that the groundwater and soil contain phenols, PAHs, VOCs, pesticides, PCBs, dioxins, and heavy metals. The contamination threatens local drinking water sources, Cemetery Creek, Mill Creek, and the Grand River.

Response actions included treating and discharging contaminated water and removing contaminated wastewater, oil, sludge, and drummed waste (1980–1986). Fencing was repaired in 1987. In 1989 the EPA recommended the excavation and thermal treatment of contaminated soil, capping, erosion controls, and monitoring groundwater and surface water. Clean-up activities were completed in 1993; groundwater monitoring is continuing.

Sources: PNPLS-OH; Ashtabula Public Library, 355 West 44th Street, Ashtabula, OH, 44004.

Miami County Incinerator
Miami County
This landfill site near Troy operated from 1968 to 1978. Various municipal and industrial wastes were received, including fly ash, sludges, solvents, and oils. Testing in the 1980s showed that the groundwater and soil contain VOCs, heavy metals, PCBs, PAHs, dioxins, and pesticides. The contamination has penetrated nearby streams and threatens Eldean Creek, the Great Miami River, and private and public wells that serve about 20,000 people.

Response actions included supplying affected residents with clean water supplies (1986–1989). A clay cap was built in 1989. The EPA recommended removing contaminated soil, soil vapor extraction, fencing, and groundwater extraction and treatment. Construction will be under way by 1997.

PRPs have been active in the site studies and remediation work.

Sources: PNPLS-OH; Miami Public Library, 419 West Main Street, Troy, OH, 45373.

Mound Plant
Montgomery County
This 306-acre U.S. Department of Energy complex near Miamisburg has developed nuclear weapons and technology since 1948. Various wastes, including solvents, paints, explosives, and radionuclides, were buried or discharged on the property. Testing has shown that the groundwater and soil contain VOCs, radionuclides, and plutonium. The contamination threatens nearby streams, the Great Miami River, and private and public water wells that serve about 17,000 people.

Site investigations started in 1990 will be completed between 1997 and 2007.

Sources: PNPLS-OH; CERCLA Reading Room, Miamisburg Senior Adult Center, 305 East Central Avenue, Miamisburg, OH, 45342.

Nease Chemical
Columbiana County
Pesticides, fire retardants, and cleaning and pharmaceutical compounds were manufactured on this site near Salem from 1961 to 1975. Investigations in 1982 revealed that drums of waste had been buried on the site. Testing showed that the groundwater, soil, and nearby stream sediments contain VOCs and mirex, a pesticide. High levels of mirex have been found in blood in local residents and workers as well as in fish and dairy cows. The contamination threatens local drinking water supplies, agricultural wells, Feeder Creek, Little Beaver Creek, and the Ohio River.

Response actions included removing buried wastes and contaminated sediments and sludges. A leachate collection system and interim groundwater treatment system were also installed (1993). An EPA investigation (1989–1997) is nearing completion and will recommend a final remediation plan.

PRPs have been active in the removal actions and site studies.

Sources: PNPLS-OH; Salem Public Library, 821 East State Street, Salem, OH, 44460; Lepper Library, 303 East Lincoln Way, Lisbon, OH, 44432.

New Lyme Landfill
Ashtabula County
This former landfill near New Lyme operated from 1969 to 1978. It received a variety of municipal and

industrial wastes, including construction debris, sludges, acids, oils, resins, and chlorinated solvents. Testing by the state revealed that the groundwater and soil contain VOCs, phenols, heavy metals, and PCBs. The chemicals have also been detected in nearby Lebanon Creek. The contamination threatens Lebanon Creek, the New Lyme Wildlife Area wetlands, and the private drinking water supplies of local residents.

In 1985 the EPA recommended capping, a leachate and groundwater extraction and treatment system, gas venting, fencing, and soil consolidation. All construction activities were completed by 1990. The groundwater system will operate at least until 2005.

Sources: PNPLS-OH; United States Post Office, 4949 Day Road, Jefferson, OH, 44047.

North Sanitary Landfill
Montgomery County
This former gravel pit near Dayton has been operated as a landfill since the 1960s. It has received a variety of municipal and industrial wastes, including waste oils, scrap metal, transformers, slag, rubber production chemicals, paints, and chemical solvents. The landfill has a history of state and local health department violations. Testing has shown that the groundwater and soil contain VOCs, heavy metals, cyanide, and PCBs. The contamination threatens local water supplies that serve about 500,000 people.

Detailed site investigations are under way. The EPA will recommend a final remediation plan after the data have been analyzed.

Source: PNPLS-OH.

Old Mill
Ashtabula County
This former dump site near Rock Creek was used in the 1970s. Various drummed wastes, including solvents, PCBs, and oils, were stored on the property. Leaks and spills contaminated the soil and groundwater with VOCs and heavy metals, including lead. The contamination threatens water wells that supply about 1,500 people within 3 miles of the site.

Response actions included removing drummed waste and contaminated soil (1981–1982) and erecting fencing (1984). In 1985 the EPA recommended removing contaminated soil, installing a groundwater extraction and treatment system, deed restrictions, and connecting affected residents to clean water supplies. These operations were completed by 1989. The groundwater will be treated until at least 2,000.

PRPs voluntarily participated in the removal actions.

Sources: PNPLS-OH; Rock Creek Library, 2988 High Street, Rock Creek, OH, 44084.

Ormet Corporation
Monroe County
Aluminum has been processed on this site near Hannibal since 1958. Various waste products, including potliners, sludges, spent solutions, and wastewaters were stored on-site or discharged into unlined lagoons. Testing has shown that the groundwater and soil contain cyanide, fluorides, PAHs, VOCs, and petrochemicals. Lagoon waters and contaminated groundwater have entered the Ohio River. The contamination also threatens the private and public wells within 3 miles of the site, which serve about 4,500 workers and residents.

Response actions included a site investigation (1987–1994). The EPA recommended a groundwater extraction and treatment system, capping, leachate collection, and the removal and treatment of contaminated soils.

A 1987 AOC with the EPA requires the PRP to undertake the site investigations.

Sources: PNPLS-OH; United States Post Office, Boston Hill Road, Hannibal, OH, 43931.

Powell Road Landfill
Montgomery County
This former gravel pit near Dayton operated as a landfill from 1959 to 1984. It received a variety of municipal and industrial wastes, including construction debris and liquid industrial chemicals. Complaints from local residents prompted inspections by the state, which showed that the groundwater, soil, and air contain VOCs. The contamination threatens public and private wells that serve nearby residents and the Great Miami River.

In 1993 the EPA recommended a final remediation plan of capping, gas collection, excavation and consolidation of contaminated soil, and groundwater and leachate collection and treatment. Construction should be completed by 1997.

PRPs conducted the site investigations.

Sources: PNPLS-OH; Montgomery County Public Library, 215 East Third Street, Dayton, OH, 45402

Pristine, Inc.
Hamilton County
A sulfuric acid plant and a liquid waste recycling firm have operated on this site near Reading. Liquid wastes, including acids, solvents, and degreasers were processed from 1974 to 1981. Liquids and various waste products were stored on-site in drums or discharged on the property. The facility was closed by the State of Ohio in

1981 because of violations. The groundwater and soil contain VOCs, phenols, fluoride, heavy metals, PAHs, and pesticides. The contamination threatens private and public water wells that serve about 13,500 residents.

Response actions included removing liquid wastes, sludges, and contaminated soil (1980–1984). In 1987 the EPA recommended in-situ vitrification, incineration, and soil vapor extraction of contaminated soil; capping; groundwater extraction and air-stripping; razing structures; fencing; deed restrictions; and groundwater monitoring. An interim groundwater treatment system was installed in 1994. Soil vapor extraction will continue until 2003.

A Consent Decree required the PRPs to conduct the removal actions from 1980 to 1984.

Sources: PNPLS-OH; Valley Public Library, 301 West Benson Street, Reading, OH, 45215.

Reilly Tar and Chemical
Tuscarawas County
A coal tar refinery was operated on this site in Dover from 1932 to 1956. Coal tars were left on the ground. The groundwater and soils contain petrochemicals. The contamination threatens the groundwater drinking supplies for about 32,000 local residents.

Response actions included fencing (1988) and removing coal tars (1990). A detailed site investigation is nearing completion.

A UOA (1989) requires the PRPs to conduct the site investigations. The PRPs also conducted the response actions.

Sources: PNPLS-OH; Dover Public Library, 525 North Walnut Street, Dover, OH, 44622.

Republic Steel Corporation Quarry
Lorain County
This former sandstone quarry near Elyria operated as an industrial landfill from 1950 to 1975. It received spent pickle liquors, sulfuric acid derivatives, and wastewaters from a nearby steel mill. Testing has shown that the groundwater, soil, and surface water contain heavy metals, VOCs, oil, grease, pyrene, and phthalates. The contamination threatens the recreational Black River and public and private wells that serve about 2,000 local residents.

In 1988 the EPA recommended excavating contaminated soil and environmental and groundwater monitoring. About 130 cubic yards of contaminated soil were removed. Monitoring is in progress, and the EPA is planning to delete this site from the NPL.

Sources: PNPLS-OH; Elyria Public Library, 320 Washington Avenue, Elyria, OH, 44035.

Rickenbacker Air National Guard
Pickaway and Franklin Counties
This U.S. Air Force base near Lockbourne has been in operation since 1942. Activities included refueling, storage, pesticide research, and maintenance and repair of vehicles and aircraft. Various wastes, including pesticides, metal scrap, oils, sludges, spent batteries, paints, and solvents, were buried or discharged on the 2,100-acre site. Testing has shown that the groundwater and soil contain VOCs, heavy metals, PAHs, and pesticides. The contamination threatens nearby streams and the drinking water source for about 150,000 residents.

At least seven different site investigations are under way. The EPA will recommend a final remediation plan after reviewing the data.

Source: PNPLS-OH.

Sanitary Landfill Co.
Montgomery County
This former landfill near Dayton operated from 1965 to 1980. It received a variety of industrial and municipal wastes, including heavy metal sludges and spent solvents. Testing has shown that the groundwater and soil contain solvents, chromium, copper, cadmium, and lead. The contamination threatens the Great Miami River and the local drinking water aquifer that supplies about 130,000 residents.

In 1993 the EPA recommended capping, deed restrictions, and surface runoff controls. Construction should be under way by 1997.

Sources: PNPLS-OH; Dayton Public Library, 3496 Far Hills Avenue, Kettering, OH, 45429.

Skinner Landfill
Butler County
This landfill near West Chester has been in operation since the 1950s. It has received a variety of municipal and industrial wastes, including demolition debris, heavy metal sludges, solvents, pesticides, and heavy metals. Local wells were found to contain VOCs. Testing showed that the groundwater and soil contain heavy metals, VOCs, pesticides, PCBs, dioxins, and furans. The contamination threatens Mill Creek, Skinner Creek, associated wetlands, and wells that supply about 2,000 people.

A site investigation was completed in 1992. The EPA recommended excavating contaminated soil, capping, soil vapor extraction, and groundwater extraction and treatment. Fencing and provision of clean water hook-ups for affected residents have been completed. The remediation work will begin in 1997.

Sources: PNPLS-OH; Union Township Library, 7900 Cox Road, West Chester, OH, 45069.

South Point Plant
Lawrence County
Ethanol has been manufactured on this site in South Point since 1982. Other chemicals used for explosives and agricultural applications were produced from 1943 to 1981. Various liquid wastes and wastewaters were spilled or discharged on the ground or into lagoons. Surface runoff flows into Solida Creek and the Ohio River. Testing has shown that the groundwater, soil, and surface water contain chloride, nitrate, sulfate, and manganese. The chemicals have penetrated nearby streams. The contamination is a threat to Solida Creek, the Ohio River, and the water wells that supply about 60,000 people

A site investigation is nearing completion and will recommend a final remediation plan.

The PRPs are conducting the site investigation.

Sources: PNPLS-OH; Office of the Mayor, 408 Second Street West, South Point, OH, 45680.

Summit National
Portage County
A former coal mining pit, this site near Deerfield was used as a landfill from 1974 to 1978. It received a variety of municipal and industrial wastes, including heavy metal sludges, solvents, chlorinated solvents, paints, oils, and resins. Liquid wastes overflowed into a stream that flows into Berlin Lake Reservoir. Testing in 1979 revealed that the groundwater and soil contain VOCs, phenols, phthalates, heavy metals, PAHs, and PCBs. The contamination threatens Berlin Lake Reservoir and water wells that serve about 5,000 local residents.

Response actions included removing contaminated soil, tanks, drums, and surface debris (1980). Runoff controls were also built. The EPA recommended a final remediation plan of incinerating contaminated soil, excavating 1,600 buried drums, installing a groundwater pump-and-treat system, capping, deed restrictions, and long-term monitoring. These activities will be completed by 1997.

Legal agreements require the PRPs to conduct the site investigations and remediation work.

Sources: PNPLS-OH; Deerfield Post Office, 1365 State Route 14, Deerfield, OH, 44421.

TRW
Stark County
This former metal casting plant near Minerva used degreasers made of VOCs in the manufacturing of metal products. Waste solutions were discharged in a ditch that flows into South Pond. The groundwater, soil, and sediments contain VOCs and PCBs. The contamination threatens nearby streams, South Pond, and private and municipal water wells that serve about 5,500 people.

Response actions included connecting affected residents to clean water supplies and excavating contaminated soil and sediment (1985). PCB-contaminated soils were stored on-site and capped. In 1986 a groundwater extraction and air-stripping system went into operation and will run at least 30 years.

An AOC (1985) requires the PRPs to clean up the groundwater.

Sources: Minerva Public Library, 677 Linwood Street, Minerva, OH, 44657.

United Scrap Lead
Miami County
A lead battery recycling facility operated on this site in Troy from 1948 to 1980. Crushed battery casings, acid, and contaminated rinsewaters were buried or discharged on the 25-acre property. Testing in the late 1970s showed that the groundwater, surface water, sediments, and soil contain lead and/or arsenic. Contaminants have penetrated the Great Miami River and threaten water wells that serve the local population.

Response actions included excavating contaminated soil and debris (1985) and fencing (1991). In 1988 the EPA recommended soil washing, capping, decontaminating buildings and debris, deed restrictions, and long-term groundwater monitoring. These remedies should be completed by 1997.

PRPs are participating in the site studies and clean-up work, partly through a 1991 AOC.

Sources: PNPLS-OH; Troy-Miami County Library, 419 West Main Street, Troy, OH, 45373.

Van Dale Junkyard
Washington County
This junkyard near Marietta operated as a disposal site in the 1970s. Various wastes, including drums of dyes and organic solvents, were discharged or buried on the property. Testing has shown that the soil and stream and marsh sediments contain VOCs, PAHs, phthalates, and heavy metals. The contamination threatens local groundwater drinking supplies, recreational surface waters, and wetlands.

A site investigation was conducted from 1988 to 1992. In 1994 the EPA recommended consolidating or removing contaminated soils, capping, and long-term environmental monitoring. Soil treatments and capping are in progress.

PRPs are participating in the site clean up.

Sources: PNPLS-OH; Washington County Public Library, 615 Fifth Street, Marietta, OH, 45750.

Wright-Patterson Air Force Base
Greene County
This U.S. Air Force base near Dayton has been in operation since 1941. Activities have included maintenance and repair of aircraft and vehicles and research and development. Solvents, thinners, degreasers, paints, heavy metal sludges, and other related wastes have been discharged or buried on the 8,500-acre site. Testing has shown that the groundwater, surface water, stream sediments, and soil contain VOCs, petrochemicals, alpha and beta radiation, and PAHs. Methane has been detected in the air and soil. The contamination threatens private, base, and municipal wells that serve about 400,000 area residents.

Response actions have included relocating affected residents, excavating drummed waste, installing air-strippers on base wells, and recovering free-product petroleum. A groundwater extraction and air-stripping system was installed in 1991. Capping, gas vents, and leachate control systems will be constructed for landfills on the base. A number of other studies are in progress.

Sources: PNPLS-OH; Greene County Library, One East Main Street, Fairborn, OH, 45324.

Zanesville Well Field
Muskingum County
An area of groundwater that supplies the city of Zanesville was discovered to contain VOCs in 1981. Contaminated wells were shut down. Further investigations identified a local business as the point source. Testing showed that the groundwater and soil contain VOCs and heavy metals. VOCs were also detected in the air. The contamination has affected the drinking water supplies that serve about 40,000 residents within 3 miles of the site.

Buried waste was excavated on the industrial site. In 1991 the EPA recommended a groundwater pump-and-treat system, soil washing, and soil vapor extraction. Construction will be under way by 1997.

EPA-identified PRPs are participating in the remediation work.

Sources: PNPLS-OH; Muskingum County Library, 220 North Fifth Street, Zanesville, OH, 43701.

OKLAHOMA

Compass Industries
Tulsa County
This quarry, located near Tulsa on the Arkansas River, was used as a landfill from 1964 to the 1970s. It received various municipal, commercial, and industrial wastes, including jet fuel, solvents, acids, bleach, benzene, oily sludge, PCBs, and pesticides. The landfill caught fire and burned underground for more than a year in 1982. Seepage and air emissions from the landfill threaten the Arkansas River, nearby homes, and a bald eagle habitat. A number of residences and private wells are within 3 miles of the site.

Response actions included fencing and warning signs (1988). A site investigation completed between 1989 and 1992 recommended a multilayer cap for the landfill and treatment for the contaminated groundwater if necessary. Construction was completed in 1995, and monitoring will continue until 1999.

About 300 PRPs have been identified, including Texaco, Sun Refining, and Standard Royalties Liquidating Trust. These three parties financed the remedial action through a UAO, and the EPA is trying to recover earlier clean-up costs from the PRP group.

Sources: PNPLS-OK; Page Memorial Library, 6 East Broadway, Sand Springs, OK, 74063.

Double Eagle Refinery Company
Oklahoma County
Double Eagle Company's refinery is located in an industrial section of Oklahoma City. Used motor oil was recycled from 1929 to the 1980s by a process of acidulation and filtration. Sampling by the EPA in 1987 indicated that the soil and shallow groundwater are contaminated with lead, xylene, ethylbenzene, and trichloroethane. About 32,000 people reside within 3 miles of the site; Douglas High School is less than 3,000 feet away. The contamination threatens the North Canadian River and area lakes that are sources for drinking water.

A site investigation (1989–1994) recommended stabilization and off-site disposal of the contaminated soil and sludge and groundwater monitoring. The groundwater monitoring system was installed in 1996, and sludge should be removed by 1997.

Out of the 500 companies that used this facility, the EPA has identified 42 PRPs.

Sources: PNPLS-OK; Ralph Ellison Library, 2000 Northeast 23rd Street, Oklahoma City, OK, 73111.

Fourth Street Abandoned Refinery
Oklahoma County
This site is located in an industrial section of northeast Oklahoma City. Waste oil was recycled at the facility from about 1940 to the 1960s. Sampling by the

OKLAHOMA

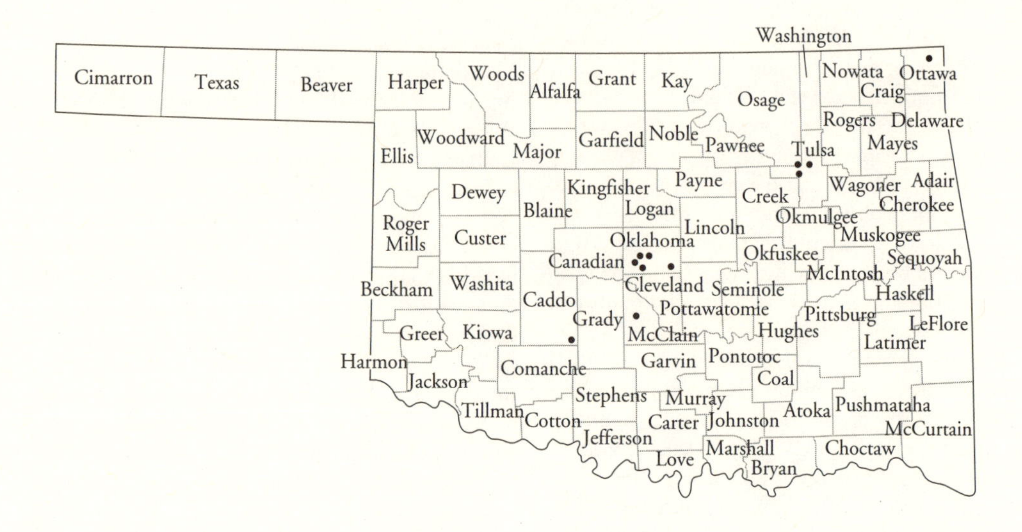

EPA in 1985–1987 revealed that the soil, sludge, and shallow groundwater were contaminated with lead, chrysene, phenanthrene, naphthalene, VOCs, and pesticides. About 1,000 people live within 1 mile of the site; Douglas High School is less than 3,000 feet away. The contamination threatens the North Canadian River and area lakes that are sources for drinking water.

Response actions included an EPA site investigation from 1989 to 1993. The recommended treatment plan includes the stabilization and off-site disposal of contaminated soil and sludge and groundwater monitoring, both to be completed by 1997.

Sources: PNPLS-OK; Ralph Ellison Library, 2000 Northeast 23rd Street, Oklahoma City, OK, 73111.

Hardage/Criner
McClain County
This landfill site is located in a rural agricultural area about 1 mile west of the town of Criner. Various wastes, including drummed liquids, solvents, PCBs, oils, sludges, paint by-products, pesticides, and ink, were deposited in pits or ponds. The soil and groundwater are contaminated with 1,2-dichloroethane, 1,1,2-trichloroethane, tetrachloroethane, trichloroethane, toxaphene, arsenic, and PCBs. The contamination threatens North Criner Creek and the water supplies of about 20 people within a mile of the site.

Response actions included connecting affected residents to clean water supplies (1987) and fencing the site (1988). The remediation plan will consist of restricting access, pumping out and separating liquids for off-site incineration, groundwater extraction and treatment using air-stripping and filtration, and capping.

The EPA filed suit against the operators in 1980. The Hardage Steering Committee, a group of PRPs, undertook site investigations from 1986 to 1988 and proposed a remediation plan different from the EPA's. The EPA won a court case against the PRPs for recovery of past costs, but the court approved the PRP's remediation plan.

Sources: PNPLS-OK; Purcell City Library, 919 North 9th Street, Purcell, OK, 73080.

Mosley Road Sanitary Landfill
Oklahoma County
This landfill is located about 3 miles east of Oklahoma City. During 7 months in 1975 it received 2 million gallons of liquid hazardous waste, including pesticides, solvents, sludges, and emulsions, depositing it in three unlined pits. Soil and groundwater are contaminated with benzene and vinyl chloride. About 60,000 people get their drinking water from private and public wells within 3 miles of the site.

Response actions included the installation of a clay cap (1988). A site investigation from 1989 to 1992 recommended natural attenuation of the contamination plume, groundwater and landfill gas monitoring, upgrading the soil cap, and revegetation. The plan should be initiated by 1997.

About 40 PRPs have been identified, a few of which are financing the site study and reclamation plan through an AOC (1989) and a UAO (1994).

Sources: PNPLS-OK; Ralph Ellison Library, 2000 Northeast 23rd Street, Oklahoma City, OK, 73111.

National Zinc Company
Washington County
This former zinc smelter in Bartlesville has been operating since 1907. Soil and air contamination covering about 8 square miles is from air dispersion of heavy metals (lead and cadmium) and the use of smelter slag for fill on community construction projects. Tests in 1992 showed that about 14 percent of the children in the area have elevated levels of lead in their blood. About 5,000 people reside within 3 miles of the site; nearly 2,000 students and staff are in schools and day care centers in the area.

Response actions included removing contaminated soil from high-access areas in 1992 and 1993. In 1994 the State of Oklahoma recommended the excavation and disposal of contaminated soils from high-risk residential and commercial properties; for lower-risk properties soil excavation, capping, and phosphate treatment were recommended. The operations began in 1995 and are continuing.

Ten PRPs have been identified, three of which are financing the site investigation and clean-up operations through agreements with the State of Oklahoma.

Sources: PNPLS-OK; Bartlesville Public Library, 600 South Johnstone, Bartlesville, OK, 74005.

Oklahoma Refining Company
Caddo County
This former oil refinery in the town of Cyril operated from 1920 to 1984. New owners reopened this facility in 1993. Petroleum refining waste was deposited in a waste pile and about 50 unlined pits and lagoons. Landfarming was also done on-site. EPA tests showed that soil, sludge, and groundwater contain high levels of benzene, toluene, xylene, phenols, arsenic, lead, chromium, acids, and free-product petroleum. About 1,600 people depend on water from public or private wells within 3 miles of the site; the contamination seeps into the recreational Gladys Creek.

Response actions included the removal of drums of hazardous waste, installing netting over ponds to protect wildlife, and fencing (1990–1992). The EPA recommended biological treatment of the wastes and capping. The initiation of these clean-up strategies is expected by 1997. A groundwater pump-and-treat system will be installed after surface remediation is complete.

The PRP, Oklahoma Refinery Company, declared bankruptcy in 1984.

Sources: PNPLS-OK; Cyril City Hall, 202 East Main Street, Cyril, OK, 73029.

Sand Springs Petrochemical Complex
Tulsa County
Located in Sand Springs, this abandoned refinery operated from about 1900 to the 1940s. It has since been used as a recyling center for solvents, transformers, and waste oils. Wastes were deposited in two unlined acid sludge pits. Soils and groundwater are contaminated with trichloroethylene, dichloroethene, and various hydrocarbons. About 15,000 people live in Sand Springs, and there are some private homes and drinking wells within 3,000 feet of the site.

Response actions included the removal of drums and tanks in 1984 by the PRPs under an Administrative Order. The EPA removed 400 drums and fenced the site. A remediation plan consisting of the excavation and incineration of sludges and contaminated soil was conducted from 1993 to 1995. Groundwater monitoring is continuing.

The EPA identified about 1,000 PRPs, the most viable for cost recovery being ARCO. ARCO has conducted the site study and remediation plan through an AO (1984), AOC (1987), and a CD (1989).

Sources: PNPLS-OK; Page Memorial Library, 6 East Broadway, Sand Springs, OK, 74063.

Tar Creek
Ottawa County
This large site includes the towns of Miami, Picher, Cardin, Quapaw, and Commerce in the former Tri-State Mining District, where mining of lead and zinc ore was active from 1900 to 1960. The mines were abandoned in 1970 and allowed to flood. Acidic mine water began discharging into Tar Creek in 1979. Investigations in 1982–1983 showed that the groundwater and mine tailings are contaminated with lead, zinc, and cadmium. Nearly 35 percent of local children were found to have lead in their blood in 1994–1995. A number of high-access areas (playgrounds, schools, and parks) and residential yards have dangerous levels of heavy metals in the soil. The contamination threatens the water supply and health of about 35,000 people in the area.

Response actions included plugging contaminated discharge sites and drilling new water wells (1985). Studies of the mining waste were begun in 1994. The EPA sampled about 2,000 homes to test lead and cadmium levels in the soil. The excavation and removal of the surface contamination was started in 1995 and is ongoing. Groundwater monitoring is also planned.

Six mining companies (including ASARCO, Goldfields, and N.L. Industries) have been identified as viable PRPs. A 1991 CD recovered about $1.3 million from the PRPs for the emergency response actions taken in 1985–1986. The PRPs were issued a Special Notice in 1995 to conduct or finance the removal of the toxic soil, but they refused.

Sources: PNPLS-OK; Miami Public Library, 200 North Main Street, Miami, OK, 74354.

Tenth Street Junkyard
Oklahoma County
This site, located in an industrial area of Oklahoma City, was a landfill, automobile junkyard, and salvage facility from 1950 to 1979. Buried waste includes paint thinners, tires, transformers, and municipal waste. Soils are contaminated with PCBs. About 1,000 people live within 1 mile of the site; some private water wells are also close by.

Drummed waste and old cars were removed in 1985 and the site capped. An EPA site investigation from 1988 to 1993 recommended containment of PCB-laden soils on-site and covering with a permanent cap. The clean-up plan was completed in 1996.

Sources: PNPLS-OK; Ralph Ellison Library, 2000 Northeast 23rd, Oklahoma City, OK, 73111.

Tinker Air Force Base
Oklahoma County
Tinker Air Force Base is located between Oklahoma City and Midwest City (the combined population is about 500,000). Landfills, waste pits, and radioactive disposal sites are located on the 4,277-acre base. Industrial solvents from aircraft maintenance and metal-plating activities were discharged on these sites. The soil and groundwater are contaminated with free-product petroleum, trichloroethylene, and chromium. The underlying aquifer, which supplies water for about 55,000 people, is contaminated.

Response actions included the demolition of buildings, removal of contaminated soil and free-product petroleum, and capping (1985–1988). The U.S. Air Force supplied bottled water to affected residents; Midwest City provided city water hook-ups. Remediation studies from 1982 to 1994 recommended groundwater pump-and-treat systems, soil vapor extraction and thermal destruction for petroleum-rich soils, and pumping toxic liquids from open pits. The clean up is in progress.

Sources: PNPLS-OK; Midwest City Public Library, 8143 West Reno Avenue, Midwest City, OK, 73110.

OREGON

Allied Plating Inc.
Multnomah County
This chrome-plating facility in Portland has been in operation since 1957. Wastewater and sludges containing heavy metals were discharged into a swampy area on the property, which drains into the Columbia Slough. In 1978 heavy metals (copper, nickel, zinc, and chromium) and cyanide were detected in well water that supplies about 20,000 nearby residents. The Columbia Slough flows into the Willamette River.

Response actions included the removal of contaminated soil, surface water, and sediments (1992). The site was deleted from the NPL in 1994. Monitoring is continuing.

Sources: PNPLS-OR; Branford Price Millar Library, Portland State University, 934 Southwest Harrison, Portland, OR, 97207.

East Multnomah County Groundwater Contamination
Multnomah County
This site near Portland is about 3 square miles in size. Tests between 1986 and 1991 showed that the groundwater is contaminated with VOCs, including trichloroethane, an industrial solvent used by many businesses in the area. About 280,000 people use the contaminated groundwater in standby situations only.

Response actions have included installing groundwater extraction and treatment systems and groundwater cutoff trenches (1992–1995). An ongoing site investigation will select a final remediation plan by 1997.

Boeing Company, Portland Plant, Cascade Corporation, and Swift Adhesives have been identified as PRPs by the EPA and are participating in the site studies and clean-up plans.

Sources: PNPLS-OR; Multnomah County Library, 17917 Southeast Stark Avenue, Gresham, OR, 97030.

Fremont National Forest/ White King and Lucky Lass Uranium Mines
Lake County
These two abandoned uranium mines are located in the Fremont National Forest. The open pit mining activities were supervised by the Atomic Energy Commission from 1958 to 1964. The pits are filled with acid water. Piles of ore and waste rock are radioactive. Soil and groundwater are contaminated with radioactive compounds and heavy metals; radon gas and gamma radiation are present at the surface. Runoff drains into Auger Creek, a recreational fishing stream with important wetland components.

Initial response actions consisted of constructing erosion barriers to control runoff. An ongoing site investigation should be completed in 1997, at which time a final remediation plan will be selected.

Sources: PNPLS-OR; Lake County Library, Lake County Courthouse, 513 Center Street, Lakeview, OR, 97630.

Gould, Inc.
Multnomah County
This site in industrial Portland operated from 1949 to 1981 as a secondary lead smelting facility, including recyling lead-acid batteries and zinc alloying. Battery

OREGON

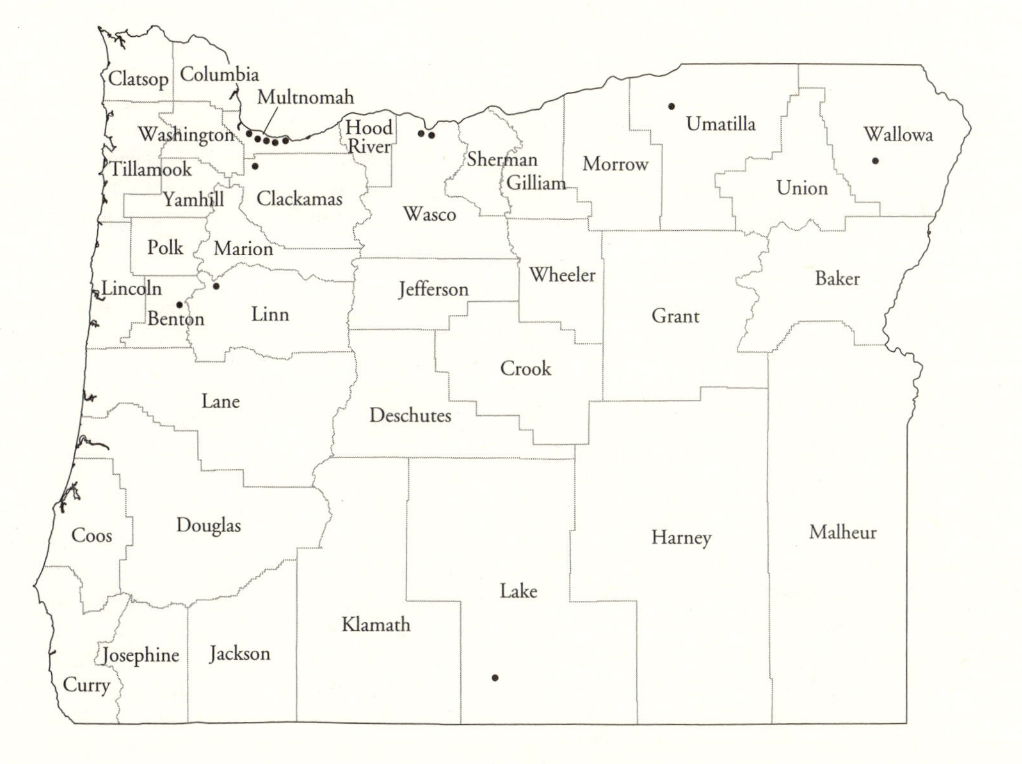

parts were buried on the property, and almost 6 million gallons of acid were dumped into Lake Doane. Groundwater, soil, and lake sediments are contaminated with lead, chromium, arsenic, cadmium, and VOCs. About 10,000 people live within 1 mile of the site, which also flanks the Willamette River.

Initial response actions included excavating battery and smelter waste and removing contaminated soils (1993–1994). An ongoing site investigation should be concluded in 1997 and will recommend a final remediation plan.

A Consent Decree (1989) and Consent Order (1991) between the EPA and N. L. Industries, a PRP, require the company to fund the site investigation and soil clean-up plans. A 1992 UAO with seven other lesser PRPs requires them to contribute toward the clean-up costs.

Sources: PNPLS-OR; Oregon Department of Environmental Quality, Records Management Section, 811 Southwest Sixth Avenue, Ninth Floor, Portland, OR, 97204.

Joseph Forest Products
Wallowa County
Joseph Forest Products was a wood treatment plant that operated 1 mile north of Joseph until 1985. Chromated copper-arsenate solutions were used to preserve lumber. Wastes were stored in a cement pit on the site. Groundwater and soil are contaminated with petroleum, arsenic, chromium, and lead from leaks and spills. The contamination threatens nearby springs, wetlands, the recreational Wallowa River, and the water supplies of 2,000 people in the area.

Response actions included the removal of contaminated soil (1991). Old buildings were razed and additional contaminated soil was excavated in 1993. Groundwater monitoring is continuing.

Sources: PNPLS-OR; Enterprise City Hall, 108 Northeast First Street, Enterprise, OR, 97828.

Martin-Marietta Aluminum Co.
Wasco County
Three different companies operated aluminum processing facilities at this site near The Dalles from at least the 1960s to the present. The soil and groundwater are contaminated with cyanide, fluoride, asbestos, PAHs, and arsenic from operations and landfills (which contained buried waste—asbestos, cathodes, and sludge) on the site. The contamination threatens the water supplies of 14,000 residents in the communities of The Dalles and Chenoweth. The site is also located within the Columbia River flood plain.

Initial response actions included excavating contaminated landfill, covering sludge ponds, and groundwater monitoring (1990). The landfill was capped, a leachate collection installed, and affected residents were connected to The Dalles municipal water system in 1991. The leachate collection system was upgraded in 1992 and 1993. Groundwater monitoring is continuing.

Martin-Marietta, the major PRP, signed a Consent Decree in 1989 that requires it to finance the investigation and remediation plan.

Sources: PNPLS-OR; The Dalles-Wasco County Public Library, 722 Court Street, The Dalles, OR, 97058.

McCormick & Baxter

Multnomah County

This company in north Portland treated wood from 1944 to 1991. Wastes and wood preservatives, including creosote, diesel oil, PCP, and ammonia-based solutions with arsenic, chromium, copper, and zinc, were discharged on the site or into the adjacent Willamette River. Soil and groundwater contain high levels of heavy metals, PAHs, and PCPs. About 12,000 people reside nearby.

Response actions included pumping creosote from extraction wells, removing sludge, and controlling runoff (1995). A detailed site investigation is nearing completion, and a remediation plan is expected by 1997.

The PRP, McCormick and Baxter, filed for bankruptcy in 1988 and cannot fund the clean-up operation.

Sources: PNPLS-OR; St. Johns Library, 7510 North Charleston Street, Portland, OR, 97203.

Northwest Pipe & Casing Co.

Clackamas County

Pipe coating operations were conducted on this 53-acre site from 1956 to 1985. Wastes, including coal tar, tar epoxy, cement mortar, asphalt, primers, and paint by-products were spilled or buried on the property. The groundwater and soil contain high levels of VOCs, PCBs, and PAHs. The contamination flows into the Willamette River, an important fishing resource with habitats for endangered species. About 5,000 people live within 1 mile of the site.

Initial response actions included demolishing the old buildings and fencing the site (1993). A site investigation should be under way by 1997. The EPA is currently searching for PRPs.

Source: PNPLS-OR.

Reynolds Metals

Multnomah County

This aluminum reduction facility on the outskirts of Troutdale was operated from 1941 to 1991. Waste products, including potliners and cryolite, were buried or stored on the property. Elevated levels of PAHs, aluminum, cyanide, fluoride, PCBs, barium, manganese, and copper have been identified in on-site groundwater, soil, and surface water; the contamination has also been identified off-site and threatens the adjacent Columbia and Sandy Rivers.

Response actions included removing contaminated soil and fencing the site (1994). Site investigations to address the other areas of contamination are continuing, and a final remediation plan is expected by 1997.

Reynolds Metals Company has worked cooperatively with the EPA regarding the site investigation and remediation plan.

Sources: PNPLS-OR; Troutdale City Hall, 104 Southeast Kibling Street, Troutdale, OR, 97060.

Teledyne Wah Chang

Linn County

Located in Millersburg, this plant produces rare earth metals and alloys for the nuclear power industry. Production began in 1957, and wastes were buried on-site or poured into unlined sludge lagoons. Groundwater, soil, and sediments in nearby streams contain thorium, uranium, radium, PCBs, and VOCs. Runoff and seepage threaten the recreational Willamette, Truax, and Murder waterways. About 20,000 people live within 3 miles of the site.

Teledyne has intermittently upgraded the facility by installing a wastewater treatment plant and reducing the radiation in waste products (1979). Sludges were excavated, solidified, and removed (1991–1993). Additional EPA-recommended remedies call for groundwater pumping and treatment and excavation of radioactive soil.

Teledyne Wah Chang, through a series of Consent Agreements and Unilateral Orders with the EPA, agreed to finance the site investigation and remediation plan.

Sources: PNPLS-OR; Albany Public Library, 1390 Waverly Drive Southwest, Albany, OR, 97321.

Umatilla Army Depot

Umatilla County

The U.S. Army has operated this site in Hermiston since 1941 as a depot for weapons and chemical warfare agents. Almost 100 million gallons of wastewater from explosives operations were poured into unlined lagoons on the site. The groundwater and soil are contaminated with chemicals from explosives (TNT and cyclonite), lead, chromium, and arsenic. The contamination plume is about 45 acres in size. About 1,000 people live within 3 miles of the site, and local groundwater is used for farming.

A site investigation was completed in 1992 and a final remediation plan recommended in 1994. Activities included removal of contaminated sediments (1994–1996) and groundwater extraction and treatment (1995–1996). Groundwater monitoring is continuing.

Sources: PNPLS-OR; Hermiston Public Library, 235 East Gladys Road, Hermiston, OR, 97838.

Union Pacific Railroad Co.
Wasco County
Located in the city of The Dalles, this site beside the Columbia River operated as a wood treatment plant from 1926 to 1987. Union Pacific Railroad treated railroad ties with copper arsenate, creosote, creosote/fuel, and PCP. Lagoons and chemical spills have polluted the groundwater and soil with creosote, PCP, fuel oil, arsenic, ammonia, VOCs, phenanthrene, and naphthalene. The Columbia River and three aquifers of different depths have been contaminated. The local groundwater is used by about 3,000 residents in the area.

Response actions included containing creosote contamination in the Columbia River (1991). A site investigation is under way and a final remediation plan is expected by 1997.

Union Pacific signed a Consent Order with the State of Oregon in 1989 to finance the site investigation and remediation plan.

Sources: PNPLS-OR; Oregon Department of Environmental Quality, Records Management Section, 811 Southwest Sixth Avenue, 9th Floor, Portland, OR, 97204.

United Chrome Products
Benton County
This former chrome-plating company is located near the town of Corvallis. Electroplating was conducted from 1956 to 1985. Contaminated wastewater was discharged into a dry well on the site and has polluted the soil and groundwater with chromium. The contamination plume extends up to 2 miles away from the site and threatens the Willamette River and the water supply of Corvallis and about 42,000 nearby residents.

Response actions included the removal of contaminated liquid and solid waste (1985). Fencing, provision of clean water hook-ups for affected residents, soil removal, and installation of groundwater pumping and treatment systems were undertaken from 1986 to 1991. Groundwater treatment will continue for at least 2 more years.

Sources: PNPLS-OR; Benton County Public Library, 645 Northwest Monroe Avenue, Corvallis, OR, 97330.

PENNSYLVANIA

A. I. W. Frank/Mustang
Chester County
A variety of industries has operated on this site in Exton since 1962. Activities included manufacturing Styrofoam cups and plates, refrigerators, and freezers and repairing automobiles. Wastes from these operations were discharged or stored on-site. The groundwater and soil contain VOCs. About 77,000 people depend on water drawn from wells within 3 miles of the site. The contamination also threatens recreational Valley Creek.

Response actions have included removing buildings. A detailed site investigation is nearing completion and will recommend a final remediation plan.

Sources: PNPLS-PA; West Whiteland Township Building, 222 North Pottstown Pike, Exton, PA, 19341.

Aladdin Plating
Lackawanna County
Aladdin Plating Company operated on this site near Olyphant from 1947 to 1982. Spent sulfuric acid, chromic acid, cyanide, and nickel, copper, and chromium sludges were discharged into lagoons that regularly overflowed. The EPA discovered that the groundwater and soil contain chromium, lead, and cyanide. Public and private wells within 3 miles of the site serve about 12,000 residents. The contamination also threatens Leggetts Creek, Griffin Creek, and Providence Reservoir.

Response actions included removing drummed waste, demolishing buildings, and installing monitoring wells (1987–1990). Contaminated soil was removed from 1988 to 1992. The groundwater will be monitored for at least 30 years.

Sources: PNPLS-PA; Scott Township Municipal Building, Route 457, Olyphant, PA, 18447.

Ambler Asbestos Piles
Montgomery County
Asbestos-containing waste was dumped on this site located in Ambler from the 1930s to 1974. Waste slurries were also discharged on-site and allowed to dry. In the mid-1970s the EPA determined that the soil, groundwater, surface water, and air contained asbestos. About 10,000 people live within 2 miles of the site. The contamination also threatens Wissahickon Creek.

Response actions included covering the asbestos piles with vegetated soil in 1977. Nearby contaminated playground equipment was removed, the site fenced, and asbestos piles capped in 1984. EPA studies recommended specially engineered covers for the

PENNSYLVANIA

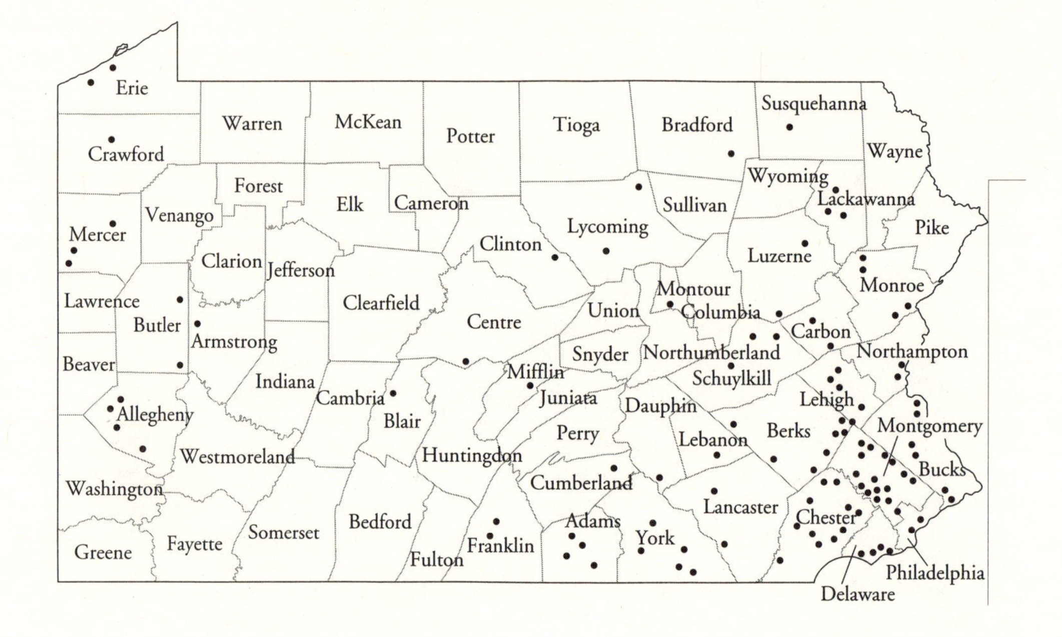

asbestos piles, installation of proper drainage and runoff controls, and better fencing.

The PRPs signed a 1988 Consent Order with the EPA that requires the parties to fund the site investigations and clean-up work. Nicolet Corporation, one of the PRPs, filed for bankruptcy in 1988.

Sources: PNPLS-PA; Wissahickon Valley Library, 209 Race Street, Ambler, PA, 19002.

Amp Inc.
York County
Amp Incorporated near Glen Rock manufactures plastics and polyester using adhesives and lubricants. Testing in 1984 revealed that the groundwater and soil contain VOCs. About 5,000 people within 3 miles of the site use private wells. The contamination also threatens nearby streams and recreational Larkin Pond.

Response actions included the installation of six groundwater extraction wells and two air-stripping towers that keep the contamination plume from migrating off-site. A bedrock flushing infiltration trench, which flushes the contaminants through the bedrock toward the recovery wells, was built in 1991. Additional studies are in progress.

Amp Company has funded the site investigations and remediation work through a 1991 agreement with the State of Pennsylvania and the EPA.

Source: PNPLS-PA.

Austin Avenue Radiation Site
Delaware County
From 1915 to 1925 the radioactive materials from the W. L. Cummings Radium Processing Company were used as fill or mixed with concrete for building. Forty properties in or near Lansdowne are contaminated from radioactive waste and tailings containing radium, thorium, radon, and asbestos.

Response actions included demolishing buildings and conducting emergency soil excavations from private residences. In 1994 the EPA recommended removing all radioactive materials, demolishing or rebuilding all contaminated homes, and removing contaminated soil. This plan will be operational by 1997. A groundwater study is also planned.

Sources: PNPLS-PA; Lansdowne Borough Library, Baltimore Avenue, Lansdowne, PA, 19050.

Avco Lycoming
Lycoming County
This aircraft engine manufacturing plant in Williamsport has been in operation for over 50 years. Varsol, a petroleum solvent, is used extensively on the site. Leaks, spills, and discharges have contaminated the groundwater and soil with VOCs (trichloroethylene). The contamination has penetrated a well field that is 3,000 feet from the site and threatens the drinking water wells for about 3,000 local residents. Pristine trout streams are also nearby.

Response actions included installing air-strippers at the well field. In 1991 the EPA recommended groundwater extraction and treatment through air-stripping.

Avco Lycoming, the PRP, signed a Consent Order (1985) and Unilateral Administrative Order (1992) that require the company to conduct the site investigations and remediation work.

Source: PNPLS-PA.

Bally Groundwater Contamination
Berks County
The State of Pennsylvania discovered organic solvents and VOCs in one of Bally's municipal wells in 1982. Further investigation indicated that the point source was Bally Engineered Structures (BES), a plant that manufactures urethane insulation panels for refrigeration systems. Waste solvents were discharged into unlined lagoons from 1960 to 1965. The groundwater and soil contain VOCs. The contamination threatens the private wells in the area that serve about 3,000 residents.

Response actions included the installation of an air-stripping system for the municipal water supply. In 1989 the EPA recommended a groundwater extraction and air-stripping system for the BES site. Construction should begin by 1997.

BES signed Consent Decrees (1987, 1991) with the EPA that require the company to undertake the site investigations and remediation work.

Sources: PNPLS-PA; Borough Business Office, South Seventh Street, Bally, PA, 19503.

Bell Landfill
Bradford County
This former landfill near Wyalusing operated from at least the 1960s to 1982. It received a variety of municipal and industrial wastes, including ferric hydroxide sludge. Testing in the 1980s by the state revealed that the groundwater and soil contain heavy metals and VOCs. Leachate has penetrated nearby trout streams, including Sugar Run. The contamination also threatens the private water wells of about 1,000 nearby residents.

In 1993 the EPA approved a remediation plan of capping, upgrading the leachate collection system, deed restrictions, removing contaminated soil, installing a gas venting system, and long-term groundwater monitoring. Engineering and design of these remedies is under way.

PRPs identified by the EPA will fund the site investigations and remediation work.

Sources: PNPLS-PA; Terry Township Building, RD 2, P.O. Box 180A, Wyalusing, PA, 18853.

Bendix Flight Systems Division
Susquehanna County
This manufacturer of aircraft instruments near Montrose has been in operation since 1951. Various industrial solvent wastes, including cutting oil and VOCs, were discharged into lagoons on the property. Testing has shown that the soil and groundwater contain VOCs and chloroform. The contamination threatens the private wells of about 1,500 residents and nearby wetlands.

Response actions included draining toxic solutions from the ponds and supplying carbon filters for affected residents. In 1988 the EPA recommended soil vacuum extraction and groundwater extraction and carbon filtration. Construction should begin by 1997.

Consent Orders (1990) require the PRPs to conduct the site investigations and clean-up work.

Sources: PNPLS-PA; Susquehanna Planning Commission, 31 Public Avenue, County Office Building, Montrose, PA, 18801.

Berkley Products Dump
Lancaster County
This former landfill near Denver operated from the 1930s to 1965. From 1965 to 1970 Berkley Products deposited solvents, paint wastes, pigment sludges, and resins in the dump. Testing has shown that the groundwater and soil contain barium, lead, mercury, VOCs, and phthalates. The contamination threatens Cocalico Creek and the local drinking water supply of about 2,000 people.

Buried drums of waste were removed in 1991. A site investigation should be completed by 1997. The EPA will then issue a final remediation plan.

Sources: PNPLS-PA; West Cocalico Township Office, 156B West Main Street, Reinholds, PA, 17569.

Berks Landfill
Berks County
This landfill near Sinking Springs has been in operation since the 1950s. It has received a variety of municipal and industrial wastes, including alkali sludges. In 1985 the EPA discovered contaminants in nearby private wells. The groundwater and soil contain VOCs, lead, and manganese. The contamination threatens nearby streams and wetlands and the drinking water supplies of about 24,000 people within 3 miles of the site.

Response actions included erecting a fence and building clay caps. Leachate and runoff collection systems were built or improved. A site investigation begun in 1991 is nearing completion and will recommend a final remediation plan.

Consent Orders (1986) and Administrative Orders of Consent (1990) with 12 PRPs require the companies to fund the site studies and clean-up work.

Source: PNPLS-PA.

Berks Sand Pit
Berks County
This former sand pit near Mertztown was used as a landfill in the 1960s and 1970s. The site was later developed into a residential area in 1978. In 1983 the EPA discovered contamination in some of the private wells. The soil and groundwater contain VOCs. The contamination threatens Perkiomen Creek and the water supplies of about 1,000 residents.

The EPA excavated a few drums of waste and supplied affected residents with clean water supplies in 1983. A groundwater extraction and air-stripping system was installed in 1993 and will operate until at least 2003.

Sources: PNPLS-PA; Longswamp Township Office, 1010 Main Street, Mertztown, PA, 19539.

Blosenski Landfill
Chester County
This former landfill near Wagontown operated from the 1930s to the 1970s. Various municipal and industrial wastes were received, including solvents, paints, and unidentified liquid wastes. Testing revealed that the groundwater, surface water, and soil contain VOCs, mercury, arsenic, and PAHs. The contamination threatens recreational Indian Spring Run and the private wells for about 600 people in the area.

Response actions included removing drummed waste and a leaking tank truck in 1982. Affected residents were connected to clean water supplies in 1989. Fencing was also erected. About 800 barrels of waste were excavated in 1992. Construction has begun on a groundwater extraction and treatment system that will be operational in 1997. A cap will be built and gas venting may also be installed.

A series of legal agreements between 1971 and 1990 requires the PRPs to conduct the site investigations and remediation work.

Sources: PNPLS-PA; West Caln Township Building, Route 340, Wagontown, PA, 19376.

Boarhead Farms
Bucks County
Boarhead Corporation used this horse-breeding farm near Doylestown in the 1970s to repair equipment and to process and salvage waste materials. Buried wastes include ferrous chloride, ammonia, and sulfuric acid. In 1984 an EPA investigation discovered that the groundwater and soil contain VOCs and heavy metals. The contamination threatens wetlands, the Delaware River, and nearby water wells that serve about 1,000 residents.

Drummed waste was removed by the EPA in 1993. The EPA will complete a detailed site investigation by 1997 and issue a final remediation plan.

Sources: PNPLS-PA; Bucks County Library, 150 South Pine Street, Doylestown, PA, 18901.

Breslube-Penn Incorporated
Allegheny County
Meat-rendering and waste oil recycling operations were conducted on this site near Coraopolis from about 1975–1991. Various wastes were stockpiled on the property or discharged into sludge lagoons. Testing in 1981 revealed that the soils contained high levels of PCBs. Further testing in 1988 revealed that the groundwater and soils contained excessive amounts of VOCs, SVOCs, PCBs, heavy metals, and cyanide. The contamination threatens the drinking water supplies for local residents.

Response actions included excavating and stockpiling contaminated soil (1989–1990). About 5,000 tons of PCB-contaminated material has been removed from the site. Other investigations are under way.

Source: EPA Publication 9320.7-071, Volume 3, No. 1.

Brodhead Creek
Monroe County
This former coal gasification plant in Stroudsburg operated from 1888 to 1944. Waste tar was discharged into two lagoons on the property. In 1981 coal tar was discovered seeping into Brodhead Creek, a recreational trout stream. Groundwater and soil contain coal tar derivatives and PAHs. Brodhead Creek is also polluted. The contamination threatens the drinking water supply for about 500 residents.

Response actions included cutting off the flow of coal tar into Brodhead Creek in 1981. In 1991 contaminated soils were injected with clean water to flush out the coal tar contaminants. Long-term groundwater monitoring is in progress. Other studies will be completed by 1997.

PRPs are cleaning up the site according to a 1987 Consent Order with the EPA and the State of Pennsylvania.

Sources: PNPLS-PA; Stroudsburg Building, Seventh and Sarah Streets, Stroudsburg, PA, 18360.

Brown's Battery Breaking
Berks County
An abandoned battery recycling facility operated on this site in Shoemakersville from 1961 to 1971. Children living in homes built on the site were found to have high levels of lead in their blood. Further tests showed that the groundwater and soil contain lead, nickel, and zinc. The contamination threatens the nearby Schuylkill River and the water supplies of about 1,000 people.

Families were relocated in 1983 during the excavation of contaminated soil and battery casings. More excavations occurred in 1990–1991. Residents and businesses were permanently relocated in 1993. In 1992 the EPA recommended thermal treatment for contaminated soils. Design of this remedy is under way.

Sources: PNPLS-PA; Hamburg Public Library, 35 North Third Street, Hamburg, PA, 19526.

Bruin Lagoon

Butler County

A former petroleum refinery operated on this site near Creek Bruin from the 1940s to the 1970s. Various mineral oil and petroleum wastes were discharged into unlined lagoons on the property. Testing has shown that the groundwater and soil contain sulfuric acid, heavy metals, and hydrogen sulfide. The pollutants have caused fish kills in the nearby Allegheny River. The contamination threatens the drinking water supplies for about 800 local residents.

Response actions included razing structures, removing tanks, and excavating sludges and contaminated soil. Fencing was erected in 1984 and gas vents installed. Free-product petroleum products were removed from the groundwater. Remaining contaminated soil and sludge were stabilized, reburied, and capped. These construction activities were completed in 1992. Long-term monitoring is in progress.

Source: PNPLS-PA.

Butler Mine Tunnel

Luzerne County

This mining tunnel was built in the 1940s to drain acid mine water from underground coal mine workings near Pittston. Hazardous materials were stored in the tunnel. Illegally dumped oil seeped into the mine tunnel and flowed into the Susquehanna River. The groundwater and mine water contain SVOCs and petroleum hydrocarbons. The contamination threatens the water supplies of about 1,500 nearby residents and has negatively impacted the Susquehanna River, an important recreational waterway and drinking water source.

Response actions included installing booms in 1979 and 1985 to collect the oily discharge. A detailed site study (1987–1996) will recommend a final remediation plan.

Seventeen PRPs signed a Consent Order (1987) with the EPA to conduct the site investigations.

Sources: PNPLS-PA; Pittston City Hall, 35 Broad Street, Pittston, PA, 18640.

Butz Landfill

Monroe County

This former landfill site near Tannersville operated from 1963 to 1973. It received a variety of municipal and industrial wastes, including septic sludges and heavy metals. Mercury and VOCs were detected in nearby water wells from 1979 to 1986. Further studies showed that the groundwater and soil at the site contain VOCs and heavy metals (including mercury and chromium). About 7,000 people depend on the local groundwater for drinking purposes. The contamination also threatens nearby recreational waterways.

Response actions included providing affected residents with clean water supplies (1986) and installing air-strippers (1987). In 1992 the EPA recommended the installation of a groundwater pump-and-treat system that is presently being designed.

Sources: PNPLS-PA; Pocono Library, Municipal Building, Route 611, Tannersville, PA, 18372.

C & D Recycling

Luzerne County

This metal reclamation plant near Freeland operated from the 1960s to the 1980s. Sheathed wires and cables were burned on-site to remove the lead and insulation in order to recover the copper wire. Sampling by the state showed that the groundwater and soil contain lead, copper, and other heavy metals. Private wells within 3 miles of the site supply drinking water to about 6,000 residents.

Response actions included removing lead-contaminated soil and debris from the site (1985–1988). A detailed site investigation was completed in 1992. The EPA recommended additional soil and ash removal, which is in progress.

The PRPs have participated in the site studies and remediation work.

Sources: PNPLS-PA; Foster Township Building, 100 Wyoming Street, Freeland, PA, 18224.

Centre County Kepone Contamination

Centre County

A chemical company on this site manufactured the pesticides kepone and mirex from 1958 to 1974. Production wastes were sprayed on fields or stored in unlined lagoons. The chemicals seeped into nearby Spring Creek, and fish contain high levels of kepone. Testing has shown that the groundwater and soil contain VOCs, kepone, mirex, petrochemicals, and PAHs. The contamination also threatens the water supplies of about 2,000 people within 1 mile of the site.

Response actions included removing contaminated material and drums in 1982. An interim groundwater extraction and treatment system was installed. A site investigation was conducted from 1989 to 1995. The EPA recommended continued groundwater treatment, additional soil removal, and groundwater monitoring.

PRPs are required to conduct the field studies through a Consent Order with the EPA.

Sources: PNPLS-PA; Schlow Memorial Library, 100 East Beaver Avenue, State College, PA, 16801.

Commodore Semiconductor Group
Montgomery County
This computer chip manufacturing company in Norristown has been in operation since the 1970s. Waste solvents were leaked into the ground from underground storage tanks. The groundwater and soil contain VOCs. Nearby municipal supply wells were shut down in 1979 because of VOC contamination. About 800,000 people use water drawn from within 3 miles of the site.

Response actions included excavating soil and air-stripping water from contaminated water wells. Affected residents were supplied with carbon filtration systems. In 1992 the EPA recommended continued groundwater treatment and permanent clean water supply hook-ups for residents.

The PRPs signed a Unilateral Order with the EPA in 1993 that requires them to conduct the clean-up work.

Sources: PNPLS-PA; Lower Providence Library, 2765 Egypt Road, Audubon, PA, 19403.

Craig Farm Drum
Armstrong County
Two strip mine pits on this site were used as landfills from 1958 to 1963. A variety of municipal and industrial wastes were buried in drums. The groundwater and soil contain creosotes and VOCs. The contamination threatens the drinking water supply of about 1,800 people as well as Valley Run Creek and the Allegheny River.

Response actions included site investigation studies from 1986 to 1994. The EPA recommended excavation of contaminated soil and the extraction and treatment of groundwater. Engineering plans are in progress.

PRPs have signed a series of legal agreements with the State of Pennsylvania and the EPA that requires the companies to undertake the site investigations and the remediation work.

Source: PNPLS-PA.

Crater Resources
Montgomery County
This former quarry received coking wastes from steel companies near Swedeland from 1918 to 1980. Other wastes included paint cans, ammonia solutions, tars, and various organic compounds. EPA testing in 1992 revealed that the site is contaminated with phenols, PAHs, VOCs, cyanide, and heavy metals. The pollutants threaten nearby water wells that serve about 77,000 people.

A site investigation (1994–1996) will recommend a final remediation plan.

Source: PNPLS-PA.

Crossley Farm
Berks County
This rural farm property was used as a waste disposal site by a local manufacturer from the 1960s to the 1970s. Wastes included drummed trichloroethylene. In 1983 the state discovered that residential wells in the area contained VOCs. Additional testing showed high levels of VOCs in soil and groundwater on Crossley Farm. Wells within 4 miles of the site serve about 5,000 residents.

Response actions included supplying affected residents with clean water supplies or carbon filtration units (1983, 1992). Site contaminants were removed in 1991. A groundwater study will be completed by 1997.

Source: PNPLS-PA.

Croydon TCE Contamination
Bucks County
Groundwater under a 4-square-mile area between Croydon and Bristol is contaminated with trichloroethylene. The source has not been identified. The contamination threatens the private well water for about 200 residents and the Delaware River, a drinking water source for about 18,000 people.

Response actions included connecting affected residents and businesses to the municipal water supply system. In 1990 the EPA recommended the extraction and air-stripping of groundwater. The system became operational in 1995 and will run until 2025.

Sources: PNPLS-PA; Grundy Memorial Library, 680 Radcliffe Street, Bristol, PA, 19007.

Cryochem
Berks County
This metal fabrication plant in Worman has been in operation since 1962. Waste solvents and dyes were discharged into surface drainages that flow into Manatawny Creek. The groundwater and soil contain VOCs. The chemicals have penetrated Ironstone Creek and Manatawny Creek. Residential wells are also contaminated. About 1,200 people depend on the local groundwater for drinking purposes.

Response actions included installing carbon filter units on affected residential water wells (1987). A groundwater extraction and air-stripping system and a soil vapor extraction system will be operational by 1997.

PRPs have been involved in the site studies and remediation work.

Sources: PNPLS-PA; Douglass-Berks Township Building, Douglass Drive, Boyertown, PA, 19512.

Delta Quarries
Blair County
This former landfill operated from the 1960s to 1985. It received a variety of municipal and industrial wastes. The groundwater and soil contain VOCs. The contamination threatens the recreational Little Juniata River and the water wells serving about 1,500 residents.

Response actions included capping the landfill in 1987. The EPA recommended a program of groundwater treatment, deed restrictions, gas venting, and long-term monitoring. The site was fenced in 1992. The groundwater system is being designed.

Consent Orders (1984, 1987) and a Consent Decree (1992) require the PRPs to fund the site investigations and clean-up work.

Source: PNPLS-PA.

Dorney Road Landfill
Lehigh and Berks Counties
This former open-pit iron mine near Allentown was used as a landfill from 1952 to 1978. It received a variety of municipal and industrial wastes, including batteries, sludges, and petroleum products. State inspections in the 1970s showed that the groundwater contains VOCs and heavy metals. The contamination threatens private water wells in the area and recreational wetlands and woodlands.

Response actions included installing erosion controls in 1986. The EPA recommended a clean-up plan of capping, gas collection, groundwater monitoring, deed restrictions, and fencing. Systems should be operational by 1997.

A series of legal agreements with eight PRPs requires the companies to undertake the site studies and remediation work.

Sources: PNPLS-PA; Township Building, 8330 Schantz Road, Breinigsville, PA, 18031.

Douglassville Disposal
Berks County
A waste oil recycling facility operated on this site in Douglassville from 1941 to 1986. Oil wastes and sludges were discharged into unlined lagoons, which overflowed into the Schuylkill River in 1970 and 1972. Examinations have revealed that the groundwater, surface water, and soil contain heavy metals, VOCs, PAHs, and PCBs. Fish contain high levels of PCBs and lead. The contamination threatens the recreational Schuylkill River and the water supplies for about 3,000 residents.

Response actions included removing contaminated soil and drums (1982). In 1990 the EPA recommended removing tanks and contaminated waste, capping lagoon areas, incinerating contaminated soil and filter cakes, and monitoring groundwater. Operations are in progress.

PRPs are participating in the clean-up work according to a 1991 UAO with the EPA.

Sources: PNPLS-PA; Union Township Municipal Building, 177 Center Road, Douglassville, PA, 19518.

Drake Chemical
Clinton County
This former chemical plant in Lock Haven manufactured pesticides, herbicides, and other organic chemicals from the 1960s to 1981. Wastewaters and chemical sludges were stored in unlined lagoons. Testing has shown that the groundwater and soil contain acids, pesticides, and the herbicide fenac. The contamination threatens Bald Eagle Creek, the Susquehanna River, and the drinking water supplies of about 10,000 nearby residents.

Response actions included removing drummed waste, draining tanks, and erecting fencing (1982). A leachate collection system was built in 1986. Buildings were also demolished. Contaminated soil and sludge will be incinerated on-site by 1997. Groundwater extraction and treatment with activated carbon will be operational by 1997–1998.

Source: PNPLS-PA.

Dublin TCE Site
Bucks County
Routine testing in 1986 discovered that groundwater in the Dublin Borough was contaminated with trichloroethylene. The contamination was traced back to an industrial site that housed a variety of businesses, including car restoration and laboratory testing. The contamination threatens the water supply for about 10,000 residents who use private and public wells within 3 miles of the site.

Response actions included supplying clean water supplies to affected residents (1987). A permanent water line was constructed in 1994. A groundwater site investigation will be completed by 1997.

Sources: PNPLS-PA; Dublin Borough Hall, 119 Maple Avenue, Dublin, PA, 18917.

East Mount Zion
York County
The East Mount Zion landfill near York operated from 1955 to 1972. It received a variety of municipal and industrial wastes, including heavy metal sludges. The groundwater and soil contain VOCs and heavy metals. The contamination threatens the private wells of about 200 nearby residents and Rocky Ridge County Park.

In 1990 the EPA recommended capping, erosion and runoff controls, fencing, deed restrictions to limit surface use, and groundwater monitoring. The cap will be completed in 1997.

Sources: PNPLS-PA; Springettsbury Township Building, 1501 Mount Zion Road, York, PA, 17402.

East 10th Street Site

Delaware County

Two companies manufactured rayon and cellophane on this 36-acre parcel in Marcus Hook from about 1910 to 1977. Production wastes, including sludges, were discharged into lagoons. Contents from underground solvent storage tanks were dumped on the ground. The site has since been developed into a business center. Groundwater and soils contain VOCs, asbestos, and heavy metals, including cadmium and mercury. The contamination threatens businesses and residences in the area as well as Marcus Hook Creek.

Response actions included removing debris, transformers, and asbestos from the site and erecting fencing. A detailed site investigation is planned.

A series of legal agreements requires the PRPs to fund the site investigations and clean-up work.

Source: PNPLS-PA.

Eastern Diversified Metals

Schuylkill County

This former wire recycling facility near Hometown operated from 1966 to 1977. Waste insulation was discarded in a huge pile on the property. Wastewater solutions have overflowed from ponds into the Little Schuylkill River, a trout stream. The groundwater and soil contain VOCs, heavy metals, dioxins, PCBs, and phthalates. The contamination threatens the Little Schuylkill River and water wells that serve about 1,400 people.

Response actions included erecting fencing in 1987. Erosion and runoff controls were installed in 1994. A deeper groundwater collection trench was also built. The recycling of the insulation pile should be completed by 1997.

A Consent Agreement (1974), UAO (1987), and UO (1994) require the PRPs to undertake the remediation work.

Sources: PNPLS-PA; Rush Township Municipal Building, Route 54, Hometown, PA, 18252.

Elizabethtown Landfill

Lancaster County

This former sandstone quarry near Elizabethtown operated from 1958 to 1973. It accepted a variety of municipal and industrial wastes, including sludges. In 1985 the EPA discovered leachate seeping from the area. Testing revealed that the groundwater and soil contains VOCs and heavy metals, including manganese and lead. The chemicals have entered Conroy Creek. The contamination threatens the water wells that serve about 13,000 people in the area.

Response actions included capping the site with clay and installing gas vents and a leachate/surface runoff collection system. A detailed site investigation (1990–1997) is nearing completion and will select a remediation plan.

A 1990 Consent Agreement requires the PRPs to undertake the site investigation. The PRPs also conducted the response actions.

Sources: PNPLS-PA; West Donegal Township Building, 7 West Ridge Road, Elizabethtown, PA, 17022.

Fischer and Porter

Bucks County

This company in Warminster manufactures process control equipment. The use of degreasers has contaminated the groundwater with VOCs, which were discovered to have penetrated public water wells in 1979. About 30,000 people use groundwater from within 3 miles of the site for drinking water. The contamination also threatens Pennypack Creek.

Response actions included installing a groundwater extraction and air-stripping system in 1986; the wells are operational again. A permanent water system was installed for affected residents. An EPA site investigation (1992–1997) is nearing completion and will make final recommendations.

The PRPs have participated in site investigations and the response actions.

Sources: PNPLS-PA; Montgomery County Information Center, 120 South York Road, Hatboro, PA, 19040.

Foote Mineral Company

Chester County

This former metal processing facility near Philadelphia manufactured lithium halide and related products from 1943 to 1980s. Various wastewaters, including inorganic fluxes and organic solvents, were discharged into lagoons or burned on the site. Testing indicated that the groundwater and soil contain heavy metals, petroleum hydrocarbons, and VOCs. The chemicals have entered nearby wells. About 42,000 people drink water from well systems within 4 miles of the site.

Response actions included groundwater monitoring and supplying affected residents with clean water supply hook-ups or filters. Underground storage tanks were removed in 1992. Contaminated soil was excavated and biotreated in 1992. A detailed site investigation by the EPA is planned.

An AOC requires the PRPs to conduct the response actions and site investigations.

Source: PNPLS-PA.

Havertown PCP Site
Delaware County
A former wood treatment plant operated on this site near Havertown from 1947 to 1963. Lumber was treated with PCP, and production wastes containing PCP were discharged into a well on the property. The chemicals have entered Naylor's Run and the nearby Delaware River. The groundwater and soil contain PCP, VOCs, and petrochemicals. About 27,000 people live within 1 mile of the property.

Response actions included fencing and runoff controls (1987). An oil and water separator became operational in 1992. Liquid and sludge wastes were drained from storage tanks and incinerated in 1994. Contaminated soil was excavated and covered with a temporary cap in 1995. Groundwater will be collected in a trench and routed to a treatment plant. A deeper groundwater study is in progress.

Source: PNPLS-PA.

Hebelka Auto Salvage Yard
Lehigh County
This former salvage yard near Fogelsville operated from 1958 to 1983. Various scrap automobile parts, tanks, drums, and battery casings were stored on the property. EPA testing in 1985 revealed that the groundwater and soil contain heavy metals, which have entered Iron Run Creek and threaten Lehigh Creek, a trout stream. About 3,300 people use well water drawn from within 3 miles of the site.

In 1989 the EPA recommended the stabilization of lead-contaminated soil and the removal of battery casings (1990–1993). Additional testing indicated that the groundwater did not need to be treated.

A 1992 Consent Decree required the PRPs to remove contaminated soil and battery casings.

Sources: PNPLS-PA; Weisenberg Township Building, Sidestown Road, Fogelsville, PA, 18051.

Heleva Landfill
Lehigh County
Former iron mining pits on this property near Coplay were used as landfills from 1967 to 1981. They received a variety of municipal and industrial wastes, including solvents. Testing showed that the groundwater and soil contain VOCs and heavy metals. The chemicals have entered nearby wells, Coplay Creek, Whitehall Quarry Lake, and the Lehigh River. The contamination also threatens the private water wells of nearby residents.

Response actions included connecting affected residents to clean water supplies (1985–1993). In 1991 the EPA recommended capping, surface water runoff controls, gas venting, and a groundwater extraction and treatment system. The system is presently being designed. Soil investigations are still in progress.

A series of UOs, AOCs, and CAs (1985–1992) requires the PRPs to conduct the site investigations and remediation work.

Sources: PNPLS-PA; North Whitehall Township Building, 600 Levans Road, Coplay, PA, 19803.

Hellertown Manufacturing
Northampton County
A former spark plug manufacturer operated on this site in Hellertown from 1930 to 1982. Various wastes, including cleaners, oils, and heavy metal solutions, were discharged in unlined lagoons on the property. Testing revealed that the groundwater and soil contain VOCs, chromium, and cyanide. The contamination threatens the water supply for about 15,000 residents as well as recreational Saucon Creek.

Response actions included site access restrictions. A site investigation from 1988 to 1991 recommended capping and groundwater extraction and treatment. The groundwater system should be operational in 1997.

Champion Spark Plug Company, the PRP, signed a 1988 Consent Order that requires the company to undertake the site investigations.

Sources: PNPLS-PA; Hellertown Municipal Center, 685 Main Street, Hellertown, PA, 18055.

Henderson Road
Montgomery County
This site near King of Prussia has operated as a waste storage and recycling facility, a vehicle maintenance shop, a parking lot, and a business office center. Part of the property was also used as a landfill. These past activities have contaminated the groundwater, soil, and air with PAHs, chromium, barium, VOCs, and cyanide. The contamination threatens the Upper Merion Reservoir and a public water supply system that serves 800,000 residents.

Response actions included installing carbon adsorption filters on contaminated wells. A groundwater extraction and treatment system and a soil vapor extraction system are operational. A leachate collection system has been constructed. Capping is also planned.

Nine PRPs are participating in the site studies and remediation work according to a 1985 AOC and 1989 CD.

Sources: PNPLS-PA; Upper Merion Library, 175 West Valley Forge Road, King of Prussia, PA, 19406.

Hranica Landfill
Butler County
This former landfill operated from 1966 to 1974. It received a variety of waste, including drums of solvents, paints, and heavy metal sludges. Testing has shown that the groundwater and soil contain heavy metals, VOCs, phenols, and PCBs. The chemicals

have been detected in dairy cows from nearby farms and in nearby springs. The contamination threatens the water supplies of about 4,000 local residents.

Response actions included excavating drummed waste and contaminated soil in 1984. After finishing a site study in 1990, the EPA recommended capping and fencing (1993–1994). Groundwater treatment was not required.

An AOC (1987) and CD (1991) required the PRPs to conduct the site investigations and remediation work.

Source: PNPLS-PA.

Hunterstown Road
Adams County
This former landfill site operated from 1970 to 1980. It received a variety of municipal and industrial wastes, including spent solvents, paint sludges, and asbestos. Testing revealed that the groundwater and soil contain VOCs, heavy metals, and asbestos. About 10,000 residents use private water wells that are within 3 miles of the site. The contamination also threatens nearby streams and wetlands.

Response actions included excavating contaminated soil (1984), fencing (1984–1985), and removing buried drums (1989). Affected residents were connected to the municipal water system. A detailed site investigation recommended installing a groundwater pump-and-treat system, excavating contaminated soil, and capping. Construction should be under way by 1997.

A series of legal agreements with the EPA and State of Pennsylvania requires the PRPs to conduct the site investigations and remediation work.

Source: PNPLS-PA.

Industrial Lane
Northampton County
This former landfill near Easton operated from 1961 to the 1970s. Other industries that have impacted the site are iron mining and iron works operations. The groundwater and soil contain VOCs. The contamination threatens the Delaware and Lehigh Rivers and the private drinking water wells of about 2,000 residents.

Response actions included connecting affected residents to city water supplies (1989) and closing the landfill. A site investigation has recently been completed and will recommend a final remediation plan.

PRPs have participated in the site studies and clean-up work.

Sources: PNPLS-PA; Mary Meuser Memorial Library, 1803 Northampton Street, Easton, PA, 18042.

Jacks Creek
Mifflin County
This scrap metal recycling and refining center in Maitland closed in 1977. Smelting wastes containing heavy metals were piled on the ground. An EPA study in 1984 revealed that the groundwater and soil contain PCBs and heavy metals, including lead. The chemicals have penetrated Jacks Creek. About 1,000 residents use private wells that draw water from within 3 miles of the site.

Response actions included building erosion controls and temporarily capping the smelting waste (1991). The EPA will issue a reclamation plan based on results from a recent site investigation (1990–1995).

PRPs have yet to participate in the remediation work.

Source: PNPLS-PA.

Keystone Landfill
Adams County
This former farm near Hanover operated as a landfill from 1966 to the 1970s. It accepted a variety of industrial and municipal wastes, including heavy metal sludges. EPA tests revealed that the groundwater and soil contain heavy metals, VOCs, and cyanide. About 2,000 people use the local groundwater for drinking purposes.

Response actions included a site investigation that was completed in 1990. The EPA recommended a groundwater pump-and-treat system, capping, gas collection, fencing, and long-term monitoring. Construction should be under way by 1997.

The PRPs signed agreements with the State of Pennsylvania and the EPA (1987, 1991) to conduct the site investigations and remediation work.

Sources: PNPLS-PA; Hanover Public Library, Library Place, Hanover, PA, 17331.

Kimberton Site
Chester County
A manufacturing company that operated on this site near Philadelphia dumped liquid wastes and sludges into unlined lagoons from 1947 to 1959. Testing by the EPA in 1981 revealed that nearby wells, groundwater, and soil contained VOCs, including trichloroethylene. About 500 people live within 1 mile of the property. The contamination also threatens nearby waterways, including recreational French Creek.

Response actions included removing drummed waste and contaminated soil (1984). Affected residents were connected to city water supplies in 1992. A groundwater extraction and air-stripping system became operational in 1993 and will operate for 30 years.

PRPs have conducted all the response actions.

Sources: PNPLS-PA; East Pikeland Township Building, Rappsdam Road, Phoenixville, PA, 19460.

Lackawanna Refuse
Lackawanna County
This former coal strip-mining property near Old Forge operated as a landfill from 1973 to 1978. The pits received a variety of industrial and municipal wastes, including paints and sludges. Testing in 1980

by the EPA revealed that the groundwater and soil contain nitrate, heavy metals, VOCs, dieldrin, and PCBs. Leachate has entered nearby waterways, including the recreational St. John's Creek and Lackawanna River. Wildlife and aquatic life in the area are contaminated with these chemicals.

Response actions included fencing (1983), removing drummed waste and contaminated soil, and installing a leachate collection system and synthetic cap (1989). Subsequent sampling has shown that the contamination has been reduced to acceptable standards, and the site will be deleted from the NPL.

PRPs were sued by the EPA and contributed funds toward the site studies and clean-up work.

Sources: PNPLS-PA; Old Forge Borough Council, 312 South Main Street, Old Forge, PA, 18518.

Letterkenny Army Depot, PDO Area
Franklin County
This U.S. Army site near Chambersburg has been in operation since 1942. Activities include the maintenance and repair of vehicles and missiles. Various wastes, including chlorinated organic solvents, were discharged into landfills on the property. Testing in the 1980s revealed that the groundwater and soil contain VOCs and heavy metals. Nearby Rocky Spring Lake has been polluted with chloroform and trichloroethylene. The contamination threatens the drinking water supplies of about 17,000 people who live within 5 miles of the base.

Response actions included removing contaminated soil (1990). A site investigation concluded in 1994 that a groundwater extraction and treatment system is required to clean the groundwater. Construction is expected to be under way by 1997.

Sources: PNPLS-PA; Letterkenney Public Affairs Office, Room SDSLE-CY, Chambersburg, PA, 17201.

Letterkenny Army Depot, Southeast Area
Franklin County
This U.S. Army site near Chambersburg has been in operation since 1947. Activities include the maintenance and repair of vehicles and missiles. Various wastes, including chlorinated organic solvents, were discharged into landfills on the property. Testing in 1983 revealed that the soil and groundwater contain VOCs and heavy metals. The contamination threatens the local drinking water supplies for 10,000 people within 4 miles of the site.

Response actions included providing clean water supplies to affected residents (1987–1989). Low-temperature thermal treatment of contaminated soil was completed in 1995. A groundwater study will be completed by 1997.

Sources: PNPLS-PA; Letterkenny Public Affairs Office, Room SDSLE-CY, Chambersburg, PA, 17201.

Lindane Dump
Allegheny County
This former landfill in Harrison Township received a variety of municipal and industrial wastes from 1900 to 1950, including the pesticide lindane. The site was later developed into a county park. Testing in 1984 showed that the groundwater and soil contain various pesticides. About 13,000 people live near the site. The contamination threatens the Allegheny River, a drinking water source for nearby communities.

Response actions included installing a leachate collection system. In 1992 the EPA recommended a final remediation plan consisting of capping and improving the leachate collection system. These improvements should be completed by 1997.

A 1993 Consent Decree requires the PRPs to fund the remediation work, and they have participated in all phases of the site clean up.

Source: PNPLS-PA.

Lord-Shope Landfill
Erie County
This former landfill near Girard operated from 1959 to 1979. It received a variety of municipal and industrial wastes, including organic solvents, oils, acids, and sludges. Testing by the EPA showed that the soil and groundwater contain VOCs and heavy metals (barium and arsenic). The contamination threatens the local drinking water supplies for about 5,000 residents as well as Elk Creek, a recreational stream that is also used for agricultural purposes.

Response actions included removing drummed waste, pumping out leachate, and capping the site (1983). In 1990 the EPA recommended in-situ soil vapor extraction and a groundwater pump-and-treat system using metal pretreatment and air-stripping technologies. Construction is under way.

Consent Orders (1982, 1987) and a Consent Decree (1991) require the PRPs to conduct the site investigations and remediation work.

Sources: PNPLS-PA; Wilcox Library, 8 Main Street, Girard, PA, 16417.

Malvern TCE Contamination
Chester County
Groundwater contamination in the town of Malvern was detected in the 1980s. Further investigations showed that the point source for contamination by VOCs is the Chemclene Corporation, which processes industrial cleaning solvents, hydraulic fluid, and hydrogen peroxide. Soils also contain PCBs. The contamination threatens the drinking water supplies of about 14,000 residents.

Response actions included removing buried drums and contaminated soils. Affected residents were supplied with water filters. An EPA site investigation will be completed by 1997.

Chemclene, the PRP, signed a 1988 Consent Order with the EPA to fund the site investigations.

Source: PNPLS-PA.

McAdoo Associates

Schuylkill County

This former coal mining site near McAdoo was used as a waste storage facility from 1975 to 1979. Various liquid wastes were held in underground tanks on the property. Testing has shown that the soil and groundwater contain heavy metals and VOCs. The contamination threatens the water supplies for about 5,000 residents within 1 mile of the site.

Response actions included removing tanks (1980), excavating contaminated soil (1982), and additional tank and debris removal (1988–1989). The site was capped in 1992 after all contaminated soils were removed. A groundwater treatment plant was constructed in 1995.

PRPs identified by the EPA signed a 1988 Consent Decree that requires the companies to fund the remediation work.

Sources: PNPLS-PA; Hazleton Public Library, McAdoo Branch, 515 Kelayres Road, McAdoo, PA, 18237.

Metal Banks

Philadelphia County

This former metal reclamation plant near Philadelphia operated from 1968 to 1972. Copper was stripped from transformers, and PCB-contaminated waste oil was discharged on the property. The oil migrated into the Delaware River in 1972, 1977, and 1978. Testing has shown that the groundwater and soil on the site contain PCBs. The contamination has impacted the Delaware River, a drinking water source for nearly two million residents.

Response actions included separating free-product oil from the groundwater (1981–1989). A detailed site investigation should be completed by 1997.

In 1980 the EPA sued Metal Banks of America, the PRP. Twenty other PRPs were later identified by the EPA. A 1991 Consent Order requires the PRPs to conduct the site investigations.

Sources: PNPLS-PA; Philadelphia Library, 2228 Cottman Avenue, Philadelphia, PA, 19149.

Metropolitan Mirror and Glass

Schuylkill County

This company manufactured mirrors on this site in the town of Frackville from 1959 to 1982. Various wastes, including silver solutions, thinners, and other solvents, were discharged into on-site lagoons. Some of these chemicals were detected in Frackville's municipal water supply in 1986. Testing has shown that the soils and groundwater contain heavy metals (including mercury, lead, and aluminum) and VOCs. Water wells within 4 miles of the site serve about 1,000 nearby residents.

Response actions include a site investigation (1994–1997) that will recommend a final remediation plan.

The EPA has identified a number of PRPs that are expected to participate in the site studies and remediation work.

Source: PNPLS-PA.

Middletown Air Field

Dauphin County

This U.S. Air Force base near Harrisburg operated until 1966. Maintenance activities on aircraft contaminated the soils and groundwater with VOCs and heavy metals (lead). Nearby water wells were closed in 1983 because of excessive levels of VOCs. Water wells within 3 miles of the site serve about 20,000 residents. The contamination also threatens Swatara Creek and the Susquehanna River.

Response actions included removing sludges and hazardous liquids (1984), constructing a groundwater extraction and air-stripping facility (1990), land use restrictions, and long-term groundwater monitoring (1994–1997). A site investigation regarding contaminated soils will be completed by 1997.

Sources: PNPLS-PA; Middletown Public Library, 20 North Catherine Street, Middletown, PA, 17057.

Mill Creek Dump

Erie County

This former wetland near Erie was used as a landfill for over 40 years. Various municipal and industrial wastes were buried in the marsh, including solvents and bulk liquids. Testing has shown that the groundwater and soils contain VOCs, PAHs, PCBs, and heavy metals. The contamination threatens the water supplies of nearby residents, Marshall's Run, and recreational wetlands.

Response actions included fencing (1983), removing drummed waste (1983–1986), and constructing groundwater collection trenches (1990). A groundwater pump-and-treat system became operational in 1992. Soils were capped in 1995–1996.

A 1992 UAO requires 19 PRPs to undertake the remediation work.

Sources: PNPLS-PA; Millcreek Township Building, 3608 West 26th Street, Erie, PA, 16506.

Modern Sanitation Landfill

York County

This farm near York operated as a landfill from the 1940s to 1979. It received a variety of municipal and industrial wastes, including sludges and solvents. Testing by the State of Pennsylvania and the EPA showed that the soil and groundwater contain VOCs. The contamination threatens over 300 wells within 2 miles of the site and agricultural and recreational water sources, including Kreutz Creek, a trout stream.

Response actions included constructing a leachate control system and a groundwater pump-and-treat system. In 1991 the EPA recommended continuing groundwater treatment, capping, fencing, and long-term groundwater monitoring. Construction will be under way by 1997.

A Consent Decree with Modern Sanitation, the PRP, requires the company to reimburse all past clean-up costs and fund the remediation plan.

Sources: PNPLS-PA; Windsor Township Building, 400 Bahms Mill Road, Red Lion, PA, 17356.

Moyers Landfill
Montgomery County
This former landfill near Eagleville operated from 1940 to 1981. It received a variety of municipal and industrial wastes, including PCBs, paints, radioactive materials, and solvents. Leachate has penetrated Skippack Creek. The groundwater and soils contain heavy metals, VOCs, and PCBs. Trout in surrounding streams contain high levels of PCBs. The contamination threatens the local water supplies of about 800 people.

Response actions included a site study by the EPA (1985) that recommended surface runoff controls, capping, gas venting, a leachate collection system, and long-term groundwater monitoring. Construction activities (1992–1997) are nearing completion.

Sources: PNPLS-PA; Lower Providence Township Building, 100 Parklane Drive, Eagleville, PA, 19403.

MW Manufacturing
Montour County
MW Manufacturing operated a scrap wire recycling facility on this site near Danville in the 1960s and 1970s. Insulation was stripped from the wires and treated with chlorinated solvents, which were discharged on the site. The current owner, Warehouse 81, also strips and stores copper wire in piles on the site. The groundwater and soils contain VOCs and heavy metals. The contamination threatens the groundwater drinking source for about 1,200 local residents as well as nearby waterways, including Mauses Creek, a trout stream.

Carbon waste piles and PCB waste were removed in 1990–1992. In 1992 the EPA recommended a groundwater extraction and treatment system, which became operational in 1995. Affected residents were also connected to the city water supply system. Contaminated soils will be incinerated on-site.

A 1993 UO requires EPA-identified PRPs to undertake the groundwater treatment plan, but they have been reluctant to comply with orders from the EPA or the State of Pennsylvania.

Source: PNPLS-PA.

Naval Air Development Center
Bucks County
This U.S. Navy research and development center near Warminster has been in operation since 1944. Research programs include the development of antisubmarine warfare systems. Various wastes, including paints, solvents, oils, and electroplating sludges were buried or discharged on-site. The groundwater and soils contain VOCs and heavy metals. The chemicals have been detected in residential and commercial water sources. The contamination threatens the Delaware River and the water supplies of about 100,000 people.

Response actions included connecting affected residents to clean water supplies (1993), constructing an interim groundwater pump-and-treat system (1993–1994), and ongoing site investigations. A decision regarding contaminated soils is expected by 1997.

Sources: PNPLS-PA; Warminster Free Library, 1076 Emma Lane, Warminster, PA, 18974.

North Penn Area 1
Montgomery County
This area of groundwater contamination near Philadelphia was discovered in 1979. Investigations revealed that VOCs from various nearby dry cleaning and textile businesses had contaminated wells, groundwater, and soils on the point source properties. Approximately 9,000 people live within 1 mile of the site. The contamination also threatens nearby recreational Skippack Creek.

Response actions included closing contaminated wells and a site investigation. In 1994 the EPA recommended soil excavation and groundwater pump-and-treat remedies that should be completed by 1997.

The EPA has identified nine PRPs that are expected to participate in the remediation work.

Sources: PNPLS-PA; Borough of Souderton Building, 331 West Summit, Souderton, PA, 18964.

North Penn Area 2
Montgomery County
Groundwater in this part of Hatfield was discovered to be polluted with VOCs in 1983. Further investigations identified the former Ametek Company, which used degreasers and solvents in the production of springs and reels, as the point source for the contamination. Waste solutions containing trichloroethylene were held in unlined lagoons. Groundwater and soils contain VOCs, which threaten the drinking water source for about 70,000 residents within 3 miles of the site.

Response actions included removing contaminated soil (1987) and sampling area wells (1989–1990). Detailed site investigations (1988–1996) will recommend a final remediation plan.

Ametek Company, a PRP, has conducted the initial response actions.

Sources: PNPLS-PA; EPA Region 3, 9th Floor, 841 Chestnut Street, Philadelphia, PA, 19107.

North Penn Area 5

Montgomery County

Routine testing in 1983 discovered that the groundwater under this 35-acre site near Philadelphia is contaminated with VOCs (trichloroethylene and trichloroethane). The contamination is thought to be derived from local industry, including American Electronics Laboratories. Wells within 3 miles of the site supply drinking water for about 100,000 residents.

Response actions included removing contaminated soil and building a groundwater pump-and-treat system using air-stripping (1981). Site investigations that will recommend a final remediation plan in 1997–1998 are in progress.

American Electronics Laboratories, an EPA-identified PRP, signed a 1981 Consent Order to fund the groundwater treatment plan.

Source: PNPLS-PA.

North Penn Area 6

Montgomery County

The groundwater under this 200-acre site in Lansdale is contaminated with VOCs. The contamination is thought to be related to metal processing and paint manufacturing industries in the area. At least 25 properties have been identified as being potential point sources. Many operations used solvents and degreasers that were stored in leaking underground tanks. Wells within 3 miles of the site supply drinking water for about 100,000 residents.

Response actions included connecting affected residents to clean water supplies (1989–1993). Site investigations that will recommend a final remediation plan are in progress.

The EPA-identified PRPs have participated in the response actions.

Sources: PNPLS-PA; Lansdale Public Library, Susquehanna Avenue and Vine Street, Lansdale, PA, 19446.

North Penn Area 7

Montgomery County

Groundwater under this 650-acre site northwest of Philadelphia is contaminated with VOCs thought to be derived from local metal fabrication businesses. Solvents and degreasers containing trichloroethylene were stored in on-site tanks that leaked. About 90,000 people use the local groundwater within 3 miles of the site for drinking purposes. The contamination also threatens Wissahickon Creek.

Response actions included removing contaminated soil and leaking tanks and installing a groundwater extraction and air-stripping system (1982–1987). An EPA site investigation that will recommend a final remediation plan in 1997–1998 is in progress.

PRPs have participated in the response actions.

Sources: PNPLS-PA; Upper Gwynedd Township Building, Parkside Place, North Wales, PA, 19454.

North Penn Area 12

Montgomery County

A 20-acre area of VOC-contaminated groundwater near Worcester was discovered in 1979 through routine sampling. Further investigations showed that the VOCs were generated by a business that manufactured electric motors and used VOCs as degreasers. Waste oils and solvents were stored in underground tanks. Local private and public wells within 3 miles of the site are used by about 16,000 residents. Other wells are used for agricultural and livestock watering purposes.

Response actions included connecting residents to city water supplies or providing carbon filters. Ongoing site investigations should be finished by 1997.

Two PRPs operating under a 1989 Consent Order with the EPA declared bankruptcy. The EPA is completing the site investigation with federal funds.

Sources: PNPLS-PA; Worcester Township Hall, 1721 Valley Forge Road, Worcester, PA, 19490.

Novak Sanitary Landfill

Lehigh County

This former landfill near Allentown operated from the 1950s to 1984. It received a variety of municipal and industrial wastes, including demolition debris, solvents, organic wastes, and heavy metal sludges. Testing has shown that the groundwater and soils contain VOCs and heavy metals, including barium. The contamination threatens drinking water wells that serve about 18,000 residents, as well as recreational Jordan Creek and its wetlands.

Response actions included connecting affected residents with clean water supplies (1985). An EPA site investigation (1988–1993) recommended capping the landfill, installing leachate and gas collection systems, and restoring wetlands. The design of these remedies is in progress.

PRPs signed a 1988 ACO that requires the companies to undertake the site investigations.

Sources: PNPLS-PA; Parkland Community Library, 4422 Walbert Avenue, Allentown, PA, 18104.

Occidental Chemical Corporation

Montgomery County

Various businesses have operated on this site near Pottstown since the 1930s, including manufacturing of aircraft engines, tires, and polyvinyl chloride (PVC) resins. Wastes, including oils, metal filings, rubber, and sludges, were buried in a landfill on the property until 1985. The groundwater and soils contain VOCs. The contamination threatens the recreational Schuylkill River and the drinking water supplies for about 30,000 residents.

Response actions included a site investigation (1989–1993). The EPA recommended groundwater extraction and treatment through air-stripping and carbon adsorption, removing PVC sludges and contaminated soils, groundwater monitoring, and erosion controls. Construction began in 1994.

A Consent Order (1989) and Unilateral Order (1994) require the PRPs to conduct the site investigation and remediation work.

Sources: PNPLS-PA; Pottstown Public Library, 500 High Street, Pottstown, PA, 19464.

Ohio River Park

Allegheny County

This former landfill near Coraopolis operated from the 1930s to the 1950s. It received a variety of municipal and industrial wastes, including sludges and pesticides. Testing has shown that the groundwater and soil contain VOCs, beryllium, and manganese. The contamination threatens the drinking water of 1,300 residents on Neville Island.

Response actions included fencing, shoreline stabilization, site studies, and the removal of selected waste. A recently completed field study (1991–1994) will recommend a final remediation plan.

Sources: PNPLS-PA; Coraopolis Memorial Library, State and School Streets, Coraopolis, PA, 15108.

Old York Landfill

York County

This former landfill in York operated as a landfill from 1961 to 1975. It received a variety of municipal and industrial wastes, including VOCs and heavy metals. Testing in 1981 by the EPA found that the groundwater and soils contained VOCs, iron, magnesium, and beryllium. Nearby wells were also polluted. About 2,000 people depend on water drawn from wells within 3 miles of the site.

Response actions included a detailed site investigation. In 1991 the EPA recommended a soil cover, removal of contaminated soils, and the installation of a groundwater extraction and treatment system.

A 1987 Consent Order and 1992 Unilateral Administrative Order (1992) require six PRPs to fund the remediation work.

Sources: PNPLS-PA; Springfield Village Library, 35-C North Main Street, Jacobus, PA, 17407.

Osborne Landfill

Mercer County

This former strip mine near Grove City was used as a landfill from the 1950s to 1978. It received a variety of municipal and industrial wastes, including paints, slags, sludge, and petrochemicals. Testing has shown that the groundwater and soils contain heavy metals, VOCs, PCP, and PCBs. The contamination threatens nearby Swamp Run, Wolf Creek, associated wetlands, and the water supplies of the 8,000 residents of Grove City.

Response actions included fencing the site and removing drummed waste and contaminated soil (1983). The EPA conducted a site investigation from 1988 to 1990 and recommended a slurry wall, clay cap, and revegetation. A groundwater extraction and treatment system will be operational by 1997. Studies regarding wetland mitigation are in progress.

Cooper Industries, a PRP, voluntarily cleaned the site in 1983 and funded the site investigation. Subsequent Consent Orders and Administrative Orders with Cooper and other PRPs require the companies to conduct the remediation work.

Source: PNPLS-PA.

Palmerton Zinc Pile

Carbon County

A zinc smelter operated on this site near Palmerton from 1900 to 1980. The New Jersey Zinc Company created a 32-million-ton waste pile that is 200 feet high and over 10,000 feet in length. Smelter fumes have polluted the valley and defoliated Blue Mountain. Acid and heavy metals have migrated into Aquashicola Creek. Testing has shown that the air is contaminated with lead, cadmium, and zinc. The soils and groundwater also contain these metals, which have been detected in neighborhood yards and gardens. Children in Palmerton have elevated levels of lead and cadmium in their bloodstreams. Farm animals and wildlife also suffer from heavy metal accumulations in their tissues. The contamination threatens the health and drinking water supplies of the 8,000 residents in the area.

Response actions included neutralizing the acid mine drainage with lime and constructing better erosion controls (1983). Revegetation was undertaken from 1987 to 1991. Other studies regarding the waste pile and contaminated soils and groundwater are in progress.

A Consent Order (1985) and Consent Decree (1989) require the PRPs to undertake site investigations and remediation work. The PRPs have refused to comply; the EPA has taken over the work and will sue the PRPs to recover clean-up costs.

Sources: PNPLS-PA; Palmerton Library, 402 Delaware Avenue, Palmerton, PA, 18071.

Paoli Rail Yard

Chester County

The Paoli Rail Yard in Paoli has repaired and serviced commuter trains since the 1960s. Liquid wastes included oils containing PCBs. Sampling by the EPA in 1984–1985 revealed that the soils and groundwater contain high levels of PCBs. Workers on-site have elevated amounts of PCBs in their bloodstreams. The contamination has also penetrated recreational Valley Creek. About 1,500 people live within 1 mile of the site.

Response actions included constructing erosion controls, excavating contaminated soils, and cleaning up neighborhood yards (1986–1987). In 1992 the EPA recommended removal and treatment of PCB-contaminated soil (1995–1996).

PRPs have participated in some of the reclamation work.

Sources: PNPLS-PA; Paoli Library, 18 Darby Road, Paoli, PA, 19301.

Publicker Industries

Philadelphia County

Liquor and industrial alcohol was processed on this site in Philadelphia from 1912 to 1985. Fuel oil was also stored on the premises. The last owner declared bankruptcy in 1986, leaving behind tanks, drums, chemicals, and millions of gallons of hazardous waste. Testing revealed that the groundwater and soils contain VOCs, PCBs, and heavy metals. The contamination threatens the Delaware River, peregrine falcon nesting sites, and the drinking water supplies for about 190,000 residents.

Response actions included removing hazardous materials and solid waste (1987–1988). Additional surface debris was removed in 1989–1990. Asbestos material was removed from 1992 to 1994. Groundwater studies are in progress.

Consent Orders (1987, 1988) require the PRPs to undertake some of the remediation work. The EPA sued other PRPs in 1990 for the recovery of past costs.

Source: PNPLS-PA.

Raymark

Montgomery County

This manufacturing site in Hatboro has produced rivets, fasteners, and other machined items since 1947. Various wastes, including degreasers and electroplating sludges, were discharged into on-site lagoons. VOCs were also stored in aboveground tanks. Nearby municipal wells are contaminated with VOCs. Investigations revealed that soils and groundwater at the Raymark property contain trichloroethylene (TCE). Almost one million residents depend on water drawn from within 3 miles of the site. The contamination also threatens recreational Pennypack Creek.

Response actions included excavating sludge from the lagoons (1972–1973). In 1991 the EPA recommended capping, soil vapor extraction, and groundwater extraction and air-stripping. All systems are operational.

Sources: PNPLS-PA; Union Library, 243 South York Road, Hatboro, PA, 19040.

Recticon/Allied Steel

Chester County

Various manufacturing businesses, including metal fabrication and silicon wafer production, have operated on this site in Parker Ford since the 1970s. The widespread use of TCE and other degreasers has contaminated the soils and groundwater with VOCs, which have penetrated nearby wells. About 17,000 people depend on water drawn from wells within 3 miles of the site. The contamination has also migrated into the Schuylkill River, another drinking water source.

Response actions included removing contaminated soil (1981) and interim groundwater pumping (1981). Carbon filtration units were supplied to affected residents. A site investigation was conducted from 1991 to 1993. In 1993 the EPA recommended groundwater treatment, removal of contaminated soil, and provision of permanent clean water supplies to affected residents. Construction should be under way by 1997.

PRPs have participated in the site investigations and response actions through Consent Orders (1981, 1990) and a UAO (1994) with the EPA.

Sources: PNPLS-PA; East Coventry Township Building, 855 Ellis Woods Road, Pottstown, PA, 19464.

Resin Disposal

Allegheny County

This former landfill in Jefferson Borough operated from 1949 to 1964. It received a variety of municipal and industrial wastes, including organic solvents, resins, old filters, sludges, and resin oils. Testing has shown that the groundwater and soil contain VOCs and heavy metals, including lead and arsenic. The contamination threatens the Monongahela River and the local water supplies for about 25,000 people.

A leachate collection system was installed in 1973. In 1991 the EPA recommended a final remediation plan of capping and separating free-product oil and petroleum. A groundwater study is still in progress.

PRPs have undertaken site studies and remediation work according to a Consent Order (1987) and a Consent Decree (1992) with the EPA.

Source: PNPLS-PA.

Revere Chemical Company

Bucks County

This former chemical company near Doylestown processed acids, metals, and plating wastes in the 1960s. Various liquid wastes containing chromic acid, copper sulfate, sulfuric acid, ammonia, and other heavy metals were discharged in unlined lagoons on the property. Testing has shown that the groundwater and soils contain VOCs, PAHs, phthalates, and heavy metals. The contamination threatens Rapp Creek, the Delaware River, and the drinking water supplies for over 2,000 nearby residents.

Response actions included removing drummed waste (1970–1971, 1991) and heavy metal sludges (1984). In 1993 the EPA recommended soil vapor extraction. Groundwater studies are in progress.

PRPs have conducted site studies and removal actions according to a 1988 Consent Order and 1991 Administrative Order.

Sources: PNPLS-PA; Bucks County Library, 150 South Pine Street, Doylestown, PA, 18901.

River Road Landfill

Mercer County

This former landfill near Sharon operated from 1962 to 1986. It received a variety of municipal and industrial wastes, including metals, PCBs, asbestos, and waste rinsewaters. Groundwater and soils contain VOCs, PCBs, phenols, and heavy metals. The chemicals have penetrated the Shenango River. The contamination also threatens the drinking water supplies for about 9,000 nearby residents.

Response actions included closing the landfill (1987). Ongoing site investigations should be completed by 1997 and will recommend a final remediation plan.

A 1990 Consent Order requires the PRPs to undertake the site studies.

Sources: PNPLS-PA; Buhl-Henderson Library, 11 North Sharpsville Avenue, Sharon, PA, 16146.

Rodale Manufacturing Company

Lehigh County

Wiring devices and other electrical parts were manufactured on this site in Emmaus from the 1950s to 1986. Electroplating wastes and contaminated rinsewaters containing TCE, oils, cyanide, and phosphates were discharged into wells on the site. Testing has shown that the groundwater and soil contain heavy metals, VOCs, oils, and cyanide. Municipal wells have also been contaminated. About 21,000 people depend on water drawn from wells within 3 miles of the site.

Response actions included the installation of air-strippers on contaminated wells (1990). Ongoing site investigations should be completed by 1997 and will recommend a final remediation plan.

Source: PNPLS-PA.

Route 940 Dump

Monroe County

This site near Pocono Summit was used as a drummed waste storage and disposal property in the 1970s. Testing revealed that the soils and groundwater contain VOCs and heavy metals. The contamination threatens recreational Indian River Creek and the water supplies for about 4,200 people.

Response actions included removing drums and stockpiling contaminated soil (1983). More debris was removed in 1984. The stockpiled soil was treated with a soil shredder in 1988. Long-term monitoring is continuing.

A 1987 Consent Order requires the PRPs to undertake the site investigation.

Source: PNPLS-PA.

Saegertown Industrial Area

Crawford County

VOCs were detected in the town of Saegertown's water supply in 1980. Further investigations showed that the source is a 100-acre area of businesses involved in repairing and cleaning railroad cars, manufacturing ceramic capacitors and rubber products, and fabricating steel. Various waste solutions and rinsewaters, including VOCs, PAHs, and heavy metals, were discharged on-site. The local groundwater within 3 miles of the site serves about 1,200 residents.

Response actions included fencing. A site investigation was completed from 1990 to 1992. In 1993 the EPA recommended a final treatment plan of on-site incineration of contaminated soil and groundwater extraction and treatment. Construction will be under way by 1997.

Consent Orders (1990, 1993, 1994) require the PRPs to undertake the site investigations and remediation work.

Sources: PNPLS-PA; Saegertown Area Library, 320 Broad Street, Saegertown, PA, 16433.

Shriver's Corner

Adams County

This former landfill near Shriver's Corner received drummed wastes, including VOCs, sludges, solvents, and paints, in the 1970s. The groundwater and soils contain VOCs (toluene and xylene). About 5,000 residents depend on water drawn from wells within 3 miles of the site.

Response actions including supplying affected residents with carbon filtration units. Drummed waste and contaminated soil was removed in 1984. Recommendations from a groundwater and soil study that was completed in 1994–1995 are pending.

Consent Orders (1984, 1987) require Westinghouse Corporation, the PRP, to conduct the site investigations.

Source: PNPLS-PA.

Stanley Kessler

Montgomery County

Stanley Kessler has degreased and repacked welding wire at this site near King of Prussia since 1963. Liquid wastes, including spent acids and bases and VOC degreasers, were stored or discharged on the property. Testing in 1979 revealed that the soils and groundwater contain VOCs, which have also penetrated Upper Merion Reservoir. The contamination threatens the drinking water supplies of nearly 800,000 residents.

Response actions included removing liquid wastes and contaminated soil (1981) and the installation of an interim groundwater recovery and treatment system (1984). A site investigation completed between 1991 and 1994 recommended treating the groundwater with activated carbon and long-term monitoring. Construction will be under way by 1997.

A 1991 Consent Decree requires the PRPs to conduct the site investigations and pay for past cleanup costs.

Sources: PNPLS-PA; Upper Merion Township Library, 175 West Valley Forge Road, King of Prussia, PA, 19406.

Strasburg Landfill

Chester County

This former landfill site near Strasburg operated in the 1970s and 1983. It received a variety of municipal and industrial wastes, including VOCs and heavy metal sludges. Testing in 1983 revealed that the soil and groundwater are contaminated with VOCs and heavy metals. The chemicals have penetrated nearby wells, Briar Run Creek, and Brandywine Creek. Private and public wells within 3 miles of the site serve about 1,000 residents.

Response actions included capping and installing a leachate collection system (1983). Affected residents were connected to clean water systems (1989). In 1992 the EPA recommended a new cap, a gas venting system, and a subsurface leachate collection system. A leachate air-stripper was built in 1991. Recommendations from a groundwater study are expected in 1998.

A 1989 UAO requires the PRPs to fund the site studies and remediation work.

Sources: PNPLS-PA; Taylor Memorial Library, 216 East State Street, Kennett Square, PA, 19348.

Taylor Dump

Lackawanna County

This former coal strip mine near Scranton was used as a landfill from 1964 to 1968. It received a variety of municipal and industrial wastes, including organic chemicals and heavy metals. Testing has shown that the groundwater and soils contain VOCs, chlordane, phthalates, PCBs, PAHs, and heavy metals. Air samples from the site contain anomalous levels of VOCs and chlordane. The contamination threatens the health and drinking water supplies of about 10,000 nearby residents. Nearby ponds have also been contaminated.

Response actions included removing solid, liquid, and drummed waste from the site (1983). Another removal action was conducted in 1987. The EPA recommended removing contaminated soils and building a soil cover, which was completed in 1988. These remedies improved the quality of the groundwater and treatment was not required. Monitoring will continue through 1999. The EPA will soon recommend this site for deletion from the NPL.

Sources: PNPLS-PA; Taylor Borough Municipal Building, 122 Union Street, Taylor, PA, 18517.

Tobyhanna Army Depot

Monroe County

This U.S. Army facility in Tobyhanna has been in operation since the 1950s. Used as a production warehouse and staging facility, various wastes were burned, buried, or discharged on the 1,300-acre site. Testing has shown that the groundwater and soils contain VOCs, heavy metals, and PCBs. The contamination threatens the drinking water source for about 4,000 nearby residents as well as wetlands and streams.

Response actions included supplying affected residents with clean water supplies (1987–1991). PCB-contaminated soils were also removed. Ongoing site investigations will address groundwater remedies in 1997.

Sources: PNPLS-PA; Coolbaugh Township Municipal Building, 5500 Memorial Boulevard, Tobyhanna, PA, 18466.

Tonolli Corporation

Carbon County

A lead smelter and lead battery recycling plant operated on this site near Nesquehoning from 1974 to

1985. Batteries were crushed to recover lead and plastic. Tonolli Corporation went bankrupt in 1985, the same year testing revealed that the soil and groundwater contain arsenic, cadmium, lead, and chromium. The chemicals have also penetrated Nesquehoning Creek. The contamination threatens the water supplies for about 17,000 residents within 3 miles of the site.

Response actions included pumping out the lagoons and stabilizing the site surface (1989). In 1992 the EPA recommended removing battery wastes, excavating contaminated soils and creek sediments, and installing a groundwater pump-and-treat system. Construction will be under way by 1997.

The EPA has identified 46 PRPs. A Consent Order (1989), Unilateral Order (1991), and Administrative Order (1993) require the companies to conduct the site investigations and remediation work.

Sources: PNPLS-PA; Nesquehoning Borough Office, 123 Catawissa Street, Nesquehoning, PA, 18240.

Tyson's Dump

Montgomery County

This former sandstone quarry near King of Prussia was used as a landfill from 1962 to 1973. It received a variety of municipal and industrial wastes, including organic solvents and septic sludges. In 1983 the EPA discovered that the groundwater and soils contain VOCs. The contamination threatens the recreational Schuylkill River, its wetlands, and the water source for about 26,000 residents.

Response actions included constructing a leachate control system, fencing, and installing a soil cap (1983). A soil vacuum extraction system became operational in 1988, and other soil treatments are being considered. A groundwater extraction and air-stripping system was constructed in 1989.

A 1988 Consent Decree requires the PRPs to undertake the remediation work.

Sources: PNPLS-PA; Upper Merion Library, 175 West Valley Forge Road, King of Prussia, PA, 19406.

UGI Columbia Gas Plant

Lancaster County

This former gas manufacturing plant in Columbia operated from 1853 to 1948. Operations on the site produced tar and purifier wastes, which occasionally overflowed into the Susquehanna River. In 1985 it was discovered that the groundwater, soil, and sediments contain VOCs, heavy metals, PAHs, and cyanide. Sediments in a downstream stretch of the Susquehanna River contain PAHs and cyanide. The contamination threatens the Susquehanna River and private drinking wells in the area that serve about 1,000 people.

Response actions included removing tar sludges in 1987. A detailed site investigation is planned.

Sources: PNPLS-PA; Columbia Public Library, Columbia, PA, 17512.

USN Ships Parts Center

Cumberland County

Beginning in the 1940s activities on this U.S. Navy base near Mechanicsburg included ship repair, engineering, and storage of ammunition and metal ores. Various wastes, including construction rubble, medical supplies, gas mask canisters, paints, varnishes, oils, asbestos, transmission fluids, antifreeze, and PCB solvents, were buried on the 824-acre site. The groundwater and soils contain heavy metals, pesticides, VOCs, PCBs, and PAHs. Wells within 4 miles of the base supply drinking water to about 9,000 residents and workers.

Response actions included the bioremediation of VOC-contaminated soils. Site investigations are under way.

Sources: PNPLS-PA; Mechanicsburg Library, 51 West Simpson Street, Mechanicsburg, PA, 17055.

Wade ABM

Delaware County

This former rubber recycling facility near Chester operated from 1950 to the 1970s; the site later became an illegal industrial waste disposal site. Toxic chemicals, including PCBs, acids, and cyanide, were dumped on the ground or stored on-site in drums. Inspections revealed that the groundwater and soils contain heavy metals, PCBs, resins, VOCs, and PCBs. The contamination threatens the local drinking water supply and nearby wildlife and wetland habitats.

Response actions included removing wastes and contaminated soils in 1981–1982. Additional removals occurred between 1987 and 1989, when the site was deleted from the NPL. The site is still being monitored.

PRPs have contributed toward the site investigations and remediation work.

Source: PNPLS-PA.

Welsh Landfill

Chester County

This former landfill in Chester County near Honey Brook operated from 1970 to 1977. It received a variety of municipal and industrial wastes, including organic solvents, drummed waste, oils, and abandoned cars. Testing in the 1980s showed that nearby wells were contaminated with VOCs and heavy metals. Groundwater and soils on the site contain VOCs, lead, zinc, mercury, and cadmium. The contamination threatens Honey Brook and the private drinking wells that serve about 1,000 people.

Response actions included removing drummed waste and contaminated soil (1985) and providing safe drinking water to affected residents (1989). In 1990 the EPA recommended capping the landfill and restricting land use. Conclusions from recently completed groundwater survey are pending.

UAOs (1991, 1993) require the PRPs to conduct the removal actions.

Sources: PNPLS-PA; Honey Brook Library, Pequea Avenue, Honey Brook, PA, 19344.

Westinghouse Electric, Sharon Plant
Mercer County
This facility in Sharon produced transformers from 1922 to 1984. PCBs used in the transformer fluids were spilled or leaked into the ground. Wastewater entering the Shenango River carried anomalous concentrations of PCBs. Testing revealed that the groundwater and soils contain PCBs, solvents, and dioxins. The Shenango River is also polluted. The contamination threatens the drinking water supplies for about 75,000 residents.

Response actions included removing contaminated solutions (1976) and skimming free-product oil from the groundwater (1994–1995). Detailed site investigations are in progress.

A 1985 Administrative Order and 1994 UAO require Westinghouse to fund the site studies and remediation work.

Sources: PNPLS-PA; Henderson Community Library, 11 North Sharpsville, Sharon, PA, 16146.

Westinghouse Elevator Company Plant
Adams County
The Westinghouse Corporation built elevators on this site near Gettysburg in the 1970s and 1980s. Chlorinated solvents were used in the painting and degreasing processes. Waste solvents and sludges were stored on-site and then removed. In 1983 the state discovered that nearby wells contain VOCs. The groundwater and soils are contaminated with high levels of VOCs. The contamination threatens recreational Rock Creek and the drinking water for about 14,000 residents.

Response actions included removing contaminated soil (1983), connecting affected residents to clean water supplies (1984), and building a groundwater extraction and air-stripping system (1984). In 1992 the EPA recommended groundwater air-stripping and removal of free-product petroleum.

A Consent Order (1988) and UO (1992) require the PRPs to conduct the site investigations and cleanup work.

Sources: PNPLS-PA; Adams County Public Library, 59 East High Street, Gettysburg, PA, 17325.

Whitmoyer Laboratories
Lebanon County
Veterinary pharmaceuticals, aniline, and soluble arsenic compounds were manufactured or stored on this site near Myerstown from 1934 to 1984. Liquid wastes including arsenic were discharged into unlined lagoons. Arsenic pollution in the area was first detected in 1964. Testing showed that the groundwater and soils contain VOCs and arsenic. The contamination threatens the private water sources for about 5,000 nearby residents, agricultural water sources, and Tulpehocken Creek.

Response actions included excavating arsenic sludges (1964), providing clean water to affected owners (1987), and removing abandoned chemical wastes (1988–1989). Buildings were demolished and removed in 1995. Contaminated soil will be treated through fixation and capping. A groundwater extraction and treatment system became operational in 1994.

PRPs have contributed toward the site investigation and remediation work.

Sources: PNPLS-PA; Whitmoyer Library, 199 North College Street, Myerstown, PA, 17067.

William Dick Lagoons
Chester County
This disposal site near Wagontown consisted of three unlined lagoons that received wastes, including petroleum, latexes, resins, and various rinsewaters, from the 1950s to 1970. Testing in 1987 revealed that nearby wells contain trichloroethylene. The groundwater and soils on the site contain PAHs, pesticides, and VOCs. The contamination threatens nearby Birch Run and private water wells that serve about 1,500 residents.

Response actions included supplying affected residents with clean water supplies (1987) and fencing the site. The construction of a groundwater extraction and treatment system will be under way by 1997. Contaminated soils will be thermally treated, backfilled, and capped on-site.

Consent Agreements (1987, 1988) require the PRPs to fund the response actions and site investigations.

Sources: PNPLS-PA; West Caln Township Building, Route 340, Wagontown, PA, 19376.

Willow Grove Air Station
Montgomery County
This U.S. Navy base in Willow Grove began operations in the 1920s. Wastes from aircraft support and

maintenance activities were discharged or buried on the 1,000-acre property. Sampling has shown that the groundwater and soil contain PCBs, VOCs, and jet fuel. The contamination threatens local drinking water supplies and nearby wetlands.

Response actions included removing storage tanks and contaminated soil (1991). Detailed site investigations are in progress.

Source: PNPLS-PA.

York County Landfill
York County
This landfill near York, operating since 1974, has received a variety of municipal and industrial wastes, including VOCs, solvents, and sludges. Testing in 1983 revealed that the groundwater and soils contain VOCs. The contamination threatens the local drinking water supplies for about 2,500 residents.

Response actions included groundwater monitoring and supplying clean water supplies to affected residents (1984). A groundwater extraction and air-stripping system was also installed. The EPA is considering other groundwater treatment plans.

Consent Agreements (1984, 1987) require the PRPs to participate in the site studies and response actions.

Sources: PNPLS-PA; Mason Dixon Public Library, Main Street, Stewartsville, PA, 17363.

PUERTO RICO

Barceloneta Landfill
Florida County
This active landfill near Florida Afuera consists of a limestone sinkhole depression that is being filled with municipal and industrial refuse, including sludges. Testing has shown that the groundwater and soils contain VOCs and heavy metals. The contamination threatens the drinking source for about 12,000 people, agricultural wells, and an on-site recreational swimming and fishing stream.

A site investigation begun in 1988 is nearing completion and will recommend a final remediation plan.

An AOC (1990) requires eight PRPs to conduct the site investigation.

Sources: PNPLS-PR; Barceloneta City Hall, Barceloneta, PR, 00617.

Fibers Public Supply Wells
Guayama County
This municipal well field was discovered to be contaminated with halogenated solvents in 1982. The point source is thought to be a former synthetic fiber plant that discharged solvent-contaminated rinsewaters into unlined lagoons on the site. The groundwater and soil contain VOCs and halogenated solvents. The contamination threatens the local drinking water supplies for about 40,000 people.

Response actions consisted of closing the contaminated wells and removing contaminated soil (1994). A groundwater pump-and-treat system should be operational by 1997.

Administrative Orders (1985, 1986, 1987) and Consent Decrees (1992) require four PRPs to conduct the site studies and remediation work and to reimburse the EPA for past costs.

Sources: PNPLS-PR; Guayama Public Library, Guayama, PR, 00655.

Frontera Creek
Humacao County
Various industries near Junquito discharged liquid wastes into Frontera Creek from 1971 to 1981. Chemicals detected in the stream include pesticides and mercury. Livestock deaths led to an EPA investigation that revealed that local residents had high levels of mercury in their bloodstreams. Five hundred residents were permanently evacuated. The contamination threatens local drinking water supplies and the recreational and commercial fishing activities in Frontera Creek and coastal lagoons.

A site investigation was conducted from 1986 to 1991. The EPA recommended excavating contaminated soils and sediments. The clean up was completed in 1994–1995. Long-term monitoring is under way.

An AOC (1986) and Consent Decree (1992) required the PRPs to conduct the site investigation and remediation work.

Sources: PNPLS-PR; Office of the Mayor, Humacao City Hall, Humacao, PR, 00661.

GE Wiring Devices
Juana Diaz County
Mercury light switches were produced on this site near Juana Diaz from 1957 to 1969. Mercury and mercury-contaminated solid waste were discarded on the premises. An investigation showed that the groundwater and soil contain mercury; mercury vapors have also been detected at the surface. The contamination threatens the San Jacaquas River and the health and drinking water supplies of about 10,000 nearby residents.

PUERTO RICO

Quebradillas
Barceloneta
Cataño
Aguadilla
Isabela
Camuy
Hatillo
Arecibo
Manatí
Vega Baja
Vega Alta
Dorado
Toa Baja
San Juan
Carolina
Loíza
Aguada
Rincón
Moca
S. Sebastián
Florida
Toa Alta
Bayamón
Truj. Alto
Rio Grande
Luquillo
Morovis
Corozal
Guaynabo
Canóvanas
Fajardo
Culebra
Añasco
Lares
Utuado
Ciales
Naranjito
Aguas Buenas
Gurabo
Las Marías
Mayagüez
Jayuya
Orocovis
Comerio
Caguas
Juncos
Las Piedras
Nuguabo
Ceiba
Maricao
Adjuntas
Barranquitas
Cidra
San Lorenzo
Humacao
Hormigueros
San Germán
Sabana Grande
Villalba
Aibonito
Cayey
Vieques
Cabo Rojo
Yauco
Peñuelas
Ponce
Coamo
Yabucoa
Patillas
Juana Díaz
Lajas
Guayanilla
Santa Isabel
Salinas
Guayama
Arroyo
Maunabo
Guánica

Response actions consisted of installing runoff controls in 1982. In 1988 the EPA recommended soil treatment and disposal and long-term groundwater monitoring. A hydrometallurgical technique was selected for the removal of mercury from the soils and should be operational by 1997. Other studies are in progress.

An AOC requires General Electric, the PRP, to conduct the site investigations and remediation work.

Sources: PNPLS-PR; Mayor's Office, Calle Degetan #35, Juana Diaz, PR, 00665.

Juncos Landfill

Juncos County

This former landfill near Juncos received a variety of municipal and industrial wastes, including mercury-bearing thermometers. Testing revealed that the groundwater and soils contain VOCs and heavy metals, including chloroform and mercury. Leachate is escaping from the site. The contamination threatens the water supplies of about 10,000 local residents and nearby streams.

Response actions included fencing the landfill and installing a temporary soil cap (1984). A site investigation was conducted from 1984 to 1991. The EPA recommended a permanent cap that should be completed by 1997. Groundwater monitoring will determine if natural attenuation of the contamination plume is occurring.

A Consent Order (1984), Administrative Order (1984), and UAO (1992) require seven PRPs to finance the site investigations and remediation work.

Sources: PNPLS-PR; Juncos Public Library, Apartado 2306, Calle Alagarin Final, Juncos, PR, 00666.

Naval Security Group Activity

Toa Baja County

This U.S. Navy communications station is located near the town of Sabana Seca. Solid and liquid wastes from base activities, including paints, pesticides, oils, solvents, and acids, were discharged or buried on the site. Testing has shown that the groundwater and soils contain heavy metals (arsenic and lead), pesticides (chlordane), and PCBs. The contamination threatens recreational and commercial fishing grounds, the San Pedro Marsh coastal wetland, and the local water supplies for about 50,000 residents and base personnel.

Response actions included erecting fencing (1988). A series of site investigations undertaken in 1993 are nearing completion and will recommend a final remediation plan.

Sources: PNPLS-PR; Jamie Fondella Garriga Public Library, Toa Baja, PR, 00659.

RCA Del Caribe

Barceloneta County

Aperture masks for color televisions have been manufactured on this site in Barceloneta since 1971. Waste ferric chloride solutions were held in lined lagoons. Sinkholes developed that ruptured the lagoons and drained the wastewater into the underlying rocks. The wastewater is thought to be present in the groundwater and soils. The suspected contamination threatens local drinking water supplies for about 13,000 residents and agricultural water sources.

A site investigation completed in 1994 determined the site was not a threat to human health. Monitoring is continuing, and the site will be recommended for deletion from the NPL.

General Electric Company, the PRP, performed the site investigations.

Sources: PNPLS-PR; Office of the Mayor, City Hall, Barceloneta, PR, 00617.

Upjohn Facility

Barceloneta County

Upjohn Corporation manufactures pharmaceutical drugs and chemicals on this site in Barceloneta. A leaking underground storage tank contaminated the groundwater and soil with carbon tetrachloride in 1982 and polluted local wells. Containment actions were conducted from 1982 to 1986. The contamination is a threat to nearby wetlands, Río Grande de Arecibo, Río de Manati, and the drinking water supplies for about 3,000 residents.

Response actions included drilling new wells, connecting affected residents to clean water supplies, and installing extraction wells to contain the contamination plume. In 1988 the EPA recommended groundwater extraction and air-stripping and long-term monitoring. This system will be operational by 1997.

A Consent Order (1987) and UAO (1988) requires Upjohn to perform the site investigations and clean-up work.

Source: PNPLS-PR; Office of the Mayor, City Hall, Barceloneta, PR, 00617.

V & M/Albaladejo

Vega Baja

A metal recycling operation in the Almirante Norte Ward of Vega Baja burned cables, batteries, and other electrical equipment on this site to recover metal. Sampling in 1989 indicated that the soils contain high levels of antimony, cadmium, copper, arsenic, lead, and silver. Sulfuric acid is also thought to be present. The contamination threatens nearby streams and drinking water wells that supply about 53,000 residents.

This site was proposed to the NPL in 1996; a final decision is expected in 1997–1998.The EPA has recommended excavating the contaminated soil. Additional studies are planned.

Source: EPA Publication 9320.7-071.

Vega Alta Wells
Vega Alta County
An area of groundwater in the Vega Alta Well Field was discovered in 1983 to be contaminated with VOCs. The polluted wells were shut down. Further testing has shown that the groundwater, soils, and stream sediments contain VOCs. The contamination is a threat to nearby streams and the drinking water supplies for about 28,000 people.

An air-stripper was installed in 1984. In 1987 the EPA recommended connecting affected residents to clean water supplies and expanding the groundwater extraction and air-stripping system. The system became operational in 1993 but was ineffective because the flow of the plume had changed. New studies are in progress.

UAOs (1989, 1990) require five PRPs to conduct the site investigations and remediation work.

Sources: PNPLS-PR; City Hall, Apartado 292, Vega Alta, PR, 00762.

RHODE ISLAND

Central Landfill
Providence County
This landfill in Johnston has been in operation since the 1970s and receives about 85 percent of the state's municipal, industrial, and hazardous wastes, including liquid latex waste, acids, oils, and spent VOC solvents. Testing in the 1980s showed that the groundwater and soils contain VOCs, arsenic, beryllium, cadmium, lead, manganese, and vanadium. The pollutants have penetrated neighboring wells, waterways, and wetlands. About 4,000 people who live within 3 miles of the site depend on private groundwater wells.

Response actions included connecting affected residents to clean water supplies and installing a landfill gas collection system. Contaminated residential properties have been bought by the landfill owner. In 1994 the EPA recommended capping and a groundwater extraction and treatment system. Remedies for off-site contamination are still being studied.

The PRP signed a Consent Order with the EPA (1987) to conduct the site investigations and response actions.

Sources: PNPLS-RI; Mohr Memorial Library, 1 Memorial Drive, Johnston, RI, 02919.

Davis Landfill
Providence County
This former landfill site near Gloucester and Smithfield operated from 1974 to 1982. It received a variety of municipal and industrial wastes, including heavy metals and pesticides. Testing in the 1980s revealed that the groundwater, surface water, and soils contain excessive amounts of VOCs, PAHs, pesticides, and heavy metals. The contamination threatens the private water wells of about 5,000 people within 3 miles of the site.

An EPA site investigation (1990–1997) is in progress and will recommend a final remediation plan in 1997–1998.

Sources: PNPLS-RI; Smithfield Public Library, 50 Esmond Street, Esmond, RI, 02917.

Davis Liquid Waste
Providence County
This liquid waste disposal facility near Smithfield operated in the 1970s. It received paint and metal sludges, oils, acids, pesticides, solvents, phenols, halogens, fly ash, and laboratory pharmaceuticals. In 1978 testing showed that the groundwater and soils contain VOCs and heavy metals, including arsenic and lead. The contamination threatens the private drinking water supplies of about 5,000 residents and nearby wetlands.

Response actions included removing drums and supplying affected residents with temporary clean water supplies (1985–1986). A permanent water supply system was later constructed. EPA studies recommended on-site incineration of contaminated soils and groundwater extraction and treatment through air-stripping and carbon filtration (1997–2007).

After initial resistance to EPA requests, the PRP has agreed to conduct certain removal actions on the site.

Sources: PNPLS-RI; Smithfield Town Hall, 64 Farnum Pike, Smithfield, RI, 02917.

Davisville Naval Construction Battalion Center
Washington County
This U.S. military installation in North Kingstown was established in 1951 as a base for U.S. Navy construction forces. Wastes from various activities, including solvents, paint thinners, degreasers, PCBs,

RHODE ISLAND

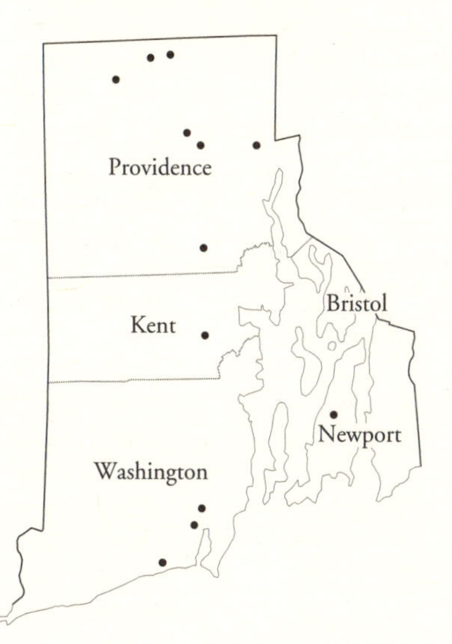

sewage sludge, fuel oil, contaminated rinsewaters, and calcium hypochlorite, were buried or discharged in pits or lagoons on the base. The groundwater and soils contain heavy metals, PAHs, PNAs, solvents, PCBs, and fuel oil. Wells located within 3 miles of the site serve about 27,000 residents. The contamination also threatens Narragansett Bay and Allen Harbor.

Response actions included removing contaminated soil and structures (1991–1992). A number of site investigations to be completed by 1997 will address different areas of contamination on the base.

Sources: PNPLS-RI; North Kingstown Free Library, 100 Boone Street, North Kingstown, RI, 02852.

Landfill and Resource Recovery
Providence County
This former landfill in North Smithfield operated from 1927 to 1985. It received a variety of wastes, including drummed organic chemicals, sludges, and demolition materials. About 3,000 residents within 3 miles of the site depend on private drinking water wells. The groundwater and soil contain arsenic, lead, and VOCs. VOC emissions have also been detected in the air. The contamination threatens recreational Trout Brook and Slatersville Reservoir.

Response actions included capping the landfill and installing a gas venting system. Erosion controls were also built. Long-term groundwater and air monitoring is under way.

A CO (1985) and AO (1990) require the PRPs to fund the remediation work.

Sources: PNPLS-RI; Municipal Annex Building, 85 Smithfield Road, North Smithfield, RI, 02895.

Newport Naval Education and Training Center
Newport County
This 1,000-acre site near Newport has been used by the U.S. Navy as a refueling center since 1900. Part of the base was also used as a landfill and received acids, paints, solvents, and oils from the 1950s to the 1970s. Sludges were dumped on the ground or burned on-site. Testing has shown that the groundwater and soils contain petroleum products, heavy metals, VOCs, and PCBs. The contamination threatens Narragansett Bay and private drinking water and agricultural wells in the area. About 10,000 people live within 3 miles of the site.

Response actions included pumping out underground storage tanks. EPA investigations have recommended capping the landfill and installing a groundwater extraction and treatment system. A number of other studies are still in progress.

Sources: PNPLS-RI; Newport Public Library, Aquidneck Park, Newport, RI, 02840; Middletown Free Library, Middletown, RI, 02842.

Peterson/Puritan
Providence County
This former landfill and industrial site in the towns of Cumberland and Lincoln began operations in 1959. The Peterson/Puritan plant manufactured aerosol consumer products. The municipal water supply systems for Cumberland and Lincoln were

discovered to contain VOCs and were shut down. Further testing found that the groundwater and soil at the Peterson/Puritan site contain chlorinated solvents, VOCs, phthalates, PCBs, and heavy metals. About 12,000 people live within 4 miles of the contamination plume, which also threatens the Blackstone River.

Response actions included erecting fencing and removing contaminated soil and drums (1992). In 1993 the EPA recommended soil vapor extraction and groundwater extraction and treatment. Construction was begun in 1996. Additional site studies are in progress.

Peterson/Puritan was sued by the town of Lincoln in 1984 and helped fund the city's new water supply system. A 1987 AO requires the company to undertake the site investigations on the property.

Sources: PNPLS-RI; Cumberland Public Library, 1464 Diamond Hill Road, Cumberland, RI, 02864.

Picillo Farm

Kent County

This former pig farm operated as an illegal waste disposal site in the 1970s. It was discovered in 1977 when a serious explosion occurred. The State of Rhode Island and the EPA cleaned the site intermittently between 1977 and 1988. Testing revealed that the groundwater and soil contain VOCs, phenols, and PCBs. About 2,000 people who live within 3 miles of the site depend on the local groundwater for drinking purposes. The contamination also threatens Roaring Brook, the Moosup River, Great Cedar Swamp, Whitford Pond, and nearby cranberry bogs.

Response actions included removing contaminated soil and various waste from 1977 to 1988. In 1993 the EPA recommended soil vapor extraction and the extraction and air-stripping of groundwater. Construction of these remedies should be under way by 1997.

The EPA has identified 12 PRPs. Legal agreements require these companies to undertake the site investigations and remediation work.

Sources: PNPLS-RI; Coventry Public Library, 1672 Flat River Road, Coventry, RI, 02816.

Rose Hill Landfill

Washington County

This former landfill near South Kingstown operated from 1967 to 1983. It received a variety of municipal and industrial wastes, including sewage sludge. The groundwater and soils contain VOCs and heavy metals. The contamination threatens Saugatucket River, Saugatucket Pond, Mitchell Brook, and the private water wells of about 17,000 people.

Response actions included connecting affected residents to city water supplies. Ongoing EPA site investigations to be completed by 1997 will recommend a final remediation plan.

Sources: PNPLS-RI; South Kingstown Public Library, 1057 Kingstown Road, Peace Dale, RI, 02883.

Stamina Mills

Providence County

This textile mill near North Smithfield operated from the 1900s to 1977. A trichloroethylene spill occurred in 1969 on-site and was never cleaned up. Nearby wells were discovered to be contaminated with VOCs in 1981. Additional investigations revealed that the groundwater and soils contain VOCs, dieldrin, heavy metals, and PAHs. The contamination threatens the local drinking water supply for about 3,000 people, wetlands, and the Branch River.

Affected residents were connected to city water supplies between 1981 and 1987. Fencing was erected in 1986. Hazardous solutions were pumped from storage tanks on the property in 1990. The EPA recommended soil vacuum extraction, the excavation of contaminated soil, capping, and the extraction and treatment of groundwater with ultraviolet light and hydrogen peroxide. Construction is expected to begin in 1997.

The PRP is required to remediate the site through a 1991 AO with the EPA.

Sources: PNPLS-RI; North Smithfield Public Library, 20 Main Street, Slatersville, RI, 02876.

West Kingston Town Dump

Washington County

This former landfill site near South Kingstown operated from the 1930s to 1987. It received a variety of waste, including paints, pesticides, and sludges. Testing has shown that the groundwater and soils contain heavy metals and VOCs, which have penetrated nearby wells. About 30,000 people use the local groundwater for drinking purposes. The contamination also threatens the Chipuxet River and Hundred Acre Pond.

Waste removals were conducted in 1987 and 1992–1993. A detailed site investigation has been initiated.

Source: PNPLS-RI.

Western Sand & Gravel

Providence County

This former gravel pit near Burrillville operated as a landfill from 1975 to 1979. It received a variety of wastes, including liquid chemical waste and sewage sludges. Testing revealed that the groundwater and soils contain VOCs and heavy metals, and nearby private wells have been contaminated. About 600 residents use the local groundwater for drinking purposes. The contamination also threatens Tarkiln Brook and Slaterville Reservoir.

Response actions included pumping out liquid chemical and septic waste in 1980. Affected residents were connected to clean water supplies. The area was capped and fenced in 1988. A groundwater study in 1991 recommended natural attenuation of the contamination plume and groundwater monitoring.

About 45 PRPs signed a Consent Decree with the EPA that requires them to fund the site investigations and remediation work.

Sources: PNPLS-RI; Burrillville Town Hall, 105 Harrisville Main Street, Harrisville, RI, 02830.

SOUTH CAROLINA

Aqua-Tech

Spartanburg County

This former landfill near Greer operated from 1940 to 1991. It received a variety of municipal, hazardous, and industrial wastes, including drummed chemicals, gas cylinders, and laboratory materials. Hazardous materials were also improperly stored on the site. Testing in the 1980s showed that the groundwater, surface water, and soils contain VOCs, heavy metals, phosgene, and radioactive compounds. The contamination threatens recreational Maple Creek, the South Tyger River, and private wells that supply nearby residents.

Response actions included removing about 7,000 drums of waste, storage tanks, gas cylinders, and other debris (1991–1994). A detailed site investigation is planned.

UAOs required about 100 PRPs to conduct the emergency removal actions at the site.

Sources: PNPLS-SC; Jean M. Smith Library, 505 Pennsylvania Avenue, Greer, SC, 29651.

Beaunit Corporation

Greenville County

This former textile plant near Fountain Inn operated from 1952 to 1977. Various liquid wastes, including spent dyes, were discharged into an unlined lagoon on the site. The lagoon overflowed into Howard Branch in the 1970s. An examination by the state revealed that the soil and creek sediments contained PCBs, VOCs, and heavy metals. The contamination threatens the local groundwater used for drinking purposes by about 1,500 nearby residents.

A site investigation was completed from 1992 to 1994. Recommendations for clean up are pending.

An AOC (1992) requires five PRPs to undertake the site investigations.

Sources: PNPLS-SC; Greenville County Library, 400 North Main Street, Fountain Inn, SC, 29644.

Carolawn

Chester County

This former waste storage and recycling facility near Fort Lawn operated in the 1970s. The owner declared bankruptcy in 1980. Inspections revealed that liquid chemical wastes were stored in leaking drums and tanks. Waste sludges and other liquid wastes were discharged into unlined lagoons. Testing revealed that the groundwater, soils, and nearby stream sediments contain VOCs, heavy metals, and chloroform. Neighboring private wells are also contaminated. Runoff from the site flows into the Catawba River. The contamination threatens nearby streams and wetlands and the local drinking water supplies for about 3,000 people.

Response actions included removing contaminated sludges and soils (1981–1982). In 1985 affected residents were connected to clean water supplies. In 1989 the EPA recommended installing a groundwater extraction and treatment system. Soil treatment studies are nearing completion. Construction of the groundwater treatment system should be completed by 1997.

An AOC and a 1991 Consent Decree requires the PRPs to fund the site investigations and clean-up work.

Sources: PNPLS-SC; Lancaster County Public Library, 313 South White Street, Lancaster, SC, 29720.

Elmore Waste Disposal

Spartanburg County

This former waste disposal area near Greer operated from 1975 to 1977. Various liquid wastes were buried or stored in drums and tanks on the site. Testing by the state revealed that the soils, stream sediments, and groundwater contain VOCs and heavy metals. The contamination threatens Wards Creek, the South Tyger River, and drinking water supplies of nearby residents.

Response actions included the partial clean up of the site in 1977. Liquids, drums, and contaminated soil were removed by the state in 1986. A site investigation was conducted between 1991 and 1992. More contaminated soil was removed in 1994, and a groundwater pump-and-treat system is being constructed.

SOUTH CAROLINA

A 1977 Consent Order required the PRP to clean up the site.

Sources: PNPLS-SC; Greer Branch Library, 113 School Street, Greer, SC, 29651.

Geiger
Charleston County
Waste oil was processed and incinerated on this site near Rantowles from 1969 to 1974. Oils were held in unlined lagoons that occasionally overflowed. Testing has shown that the groundwater and soils contain lead, chromium, PCBs, and petrochemicals. The contamination threatens the local groundwater drinking supplies for the towns of Rantowles and Hollywood, agricultural wells, and streams.

Response actions included a site investigation. The EPA recommended a groundwater pump-and-treat system, which is expected to be operational by 1997. Contaminated soils will be solidified and reburied on the site.

Sources: PNPLS-SC; Hollywood Town Hall, 6316 Highway 162, Hollywood, SC, 29449.

Golden Strip Septic Tank Service
Greenville County
Septic wastes, electroplating sludges, and other industrial liquid wastes were discharged into unlined lagoons from 1960 to 1975 on this site near Greenville. Testing showed that the groundwater and soils contain heavy metals, including lead, zinc, arsenic, cadmium, and chromium. The contamination threatens water wells that serve about 1,000 local residents, as well as wetlands and Gilder Creek.

In 1991 the EPA recommended the excavation, solidification, and reburial of contaminated soil and sludge; deed restrictions; and long-term groundwater monitoring. The soil treatment should be completed by 1997.

A Consent Decree requires the PRPs to conduct the clean-up work.

Sources: PNPLS-SC; Greenville Public Library, 300 College Street, Greenville, SC, 29601.

Helena Chemical Landfill
Allendale County
Pesticides, containers, and various production wastes from the Helena Chemical Company were buried on this site near Fairfax from the 1960s to 1978. Testing has shown that the groundwater, soils, and nearby stream sediments contain high levels of pesticides. The contamination threatens private and public wells within 3 miles of the site, which supply about 2,600 people.

Response actions included removing some of the buried waste in 1984, followed by capping. More contaminated soils were removed in 1992. In 1993 the EPA recommended the excavation and treatment of contaminated soils, installation of a groundwater pump-and-treat system, and the reconstruction of affected wetlands. Construction will be under way by 1997.

Consent Orders (1981, 1984) and a UAO (1994) require the Helena Chemical Company to conduct the site investigations and clean-up work.

Sources: PNPLS-SC; City Hall, Town of Fairfax, Highway 278 and Laurens Avenue, Fairfax, SC, 29827.

Kalama Specialty Chemicals
Beaufort County
Various chemicals including herbicides were produced from 1973 to 1979 on this site near Beaufort. Liquid chemicals and wastewaters were stored in unlined lagoons that occasionally overflowed. An explosion resulted in a large spill of VOCs. Groundwater, surface water, and soils contain lead and VOCs. The contamination threatens coastal wetland habitats and the water supplies of about 16,000 residents.

A detailed site investigation was conducted from 1988 to 1994. The EPA recommended groundwater extraction and treatment, the solidification and reburial of contaminated soils, and long-term groundwater monitoring. Construction should be under way by 1997.

A Consent Order (1988) and Consent Decree (1994) require the PRPs to conduct the site investigations and remediation work.

Sources: PNPLS-SC; Beaufort County Library, 710 Craven Street, Beaufort, SC, 29902.

Koppers Company
Florence County
This active wood treatment plant near Florence has been in operation since at least the 1960s. Wood timbers are treated with creosote, PCP, and chromated copper arsenate. Various liquid wastes have been sprayed on fields to evaporate or stored in unlined lagoons. Complaints from nearby residents revealed that private wells were contaminated with creosote. Testing has shown that the groundwater, surface water, and soils contain PAHs, PCP, arsenic, mercury, oil, and grease. The contamination threatens agricultural wells, nearby streams, and the local drinking water supplies for about 1,500 residents.

Response actions included connecting affected residents to clean water supplies. An interim groundwater pump-and-treat system was also installed. Recommendations from a site study (1988–1995) are pending.

The PRPs have conducted the response actions.

Sources: PNPLS-SC; Florence County Library, 319 South Irby Street, Florence, SC, 29501-4795.

Koppers Company, Charleston Plant
Charleston County
This former wood treatment facility near Charleston operated from 1925 to 1978. Timbers were soaked in creosote, drip-dried, and stored on the property. Contaminated surface runoff flowed into nearby streams and marshes. Testing has shown that the soil, groundwater, and drainage ditches contain PAHs and other creosote by-products. The contamination threatens water supplies for about 95,000 people, the Ashley River, and Charleston Harbor, a recreational and commercial fishing area.

A site investigation was conducted from 1992 to 1996. Final recommendations from the EPA for a remediation plan are pending.

Sources: PNPLS-SC; Charleston County Public Library, 404 King Street, Charleston, SC, 29402.

Leonard Chemical Company
York County
Hazardous wastes were treated and stored on this site near Catawba from the 1960s to 1982. Materials included solvents, VOCs, inks, sludges, and filters. Leaks and spills have contaminated the groundwater and soil with heavy metals and VOCs. The chemicals threaten agricultural wells, streams, and the local drinking water for about 6,000 residents.

Response actions included removing drummed wastes and contaminated soil in 1983. An ongoing site investigation is nearing completion and will make recommendations for a final remediation plan.

The PRPs have participated in the site investigation and the response actions.

Sources: PNPLS-SC; York County Library, 138 East Black Street, P.O. Box 10032, Rock Hill, SC, 29731.

Lexington County Landfill
Lexington County
This former gravel pit near Cayce began operations as a landfill in the 1960s. It received a variety of munic-

ipal and industrial wastes. In 1987 the EPA detected heavy metals, VOCs, and pesticides in the groundwater and soil. The contamination threatens public and private water wells within 3 miles of the site. The groundwater is also used for agricultural purposes.

A detailed site study was completed in 1994. The EPA recommended upgrading the gas venting system, extracting and treating groundwater and leachate, and long-term environmental monitoring. The remedies are presently being designed.

Sources: PNPLS-SC; Cayce–West Columbia Library, 1500 Agusta Road, West Columbia, SC, 29073.

Medley Farm Drum Dump

Cherokee County

Various chemicals, sludges, and other wastes were stored illegally in drums and lagoons on this farm near Gaffney from 1973 to 1978. Contaminated runoff has entered Thickety and Jones Creeks. Testing by the EPA indicated that the groundwater and soils contain VOCs, PCBs, and pesticides. The contamination threatens the local drinking water supplies for about 200 people and nearby streams and wetlands.

Response actions included removing drummed waste and contaminated soil in 1983. Lagoon liquids were treated on-site and discharged. A site investigation was completed from 1988 to 1991. The EPA recommended a final remediation plan of extracting and air-stripping groundwater, soil vapor extraction, and groundwater monitoring. The extraction systems should be operational by 1997.

An AOC (1988) and Consent Decree (1991) require the PRPs to conduct the site investigations and remediation work.

Sources: PNPLS-SC; Cherokee County Library, 300 East Rutledge Street, Gaffney, SC, 29340.

Palmetto Recycling Inc.

Richland County

Lead was recovered from old batteries on this site near Columbia from 1979 to 1982. Various wastes on the property included acids and battery casings. Testing showed that the soils contain lead, chromium, barium, and cadmium. The contamination threatens the local groundwater used for domestic and agricultural purposes, as well as recreational Cane Creek and Broad River.

Response actions included removing contaminated soil and surface water in 1984–1985. A detailed site investigation has been initiated.

The PRP conducted the initial removal actions in 1984–1985.

Sources: PNPLS-SC; Northeast Regional Library, 7490 Parklane Road, Columbia, SC, 29223.

Palmetto Wood Preserving

Lexington County

This former wood treatment facility near Columbia operated from 1963 to 1985. Wood was treated with fluoride-chromate-arsenate-phenol and acid-copper-chromate solutions. Lumber was drip-dried and stored on the premises. Testing has shown that the groundwater and soils contain heavy metals and PCP. The compounds have also been detected in nearby wells. The contamination threatens wells that serve about 2,500 local residents.

Response actions included connecting affected residents to clean water supplies (1985–1990). Contaminated soil was excavated, treated, solidified, and reburied from 1988 to 1989. A groundwater extraction and treatment system will be operational by 1997.

Sources: PNPLS-SC; Lexington County Administration Building, 212 South Lake Drive, Lexington, SC, 29072.

Para-Chem

Greenville County

Chemicals and adhesives have been manufactured on this site near Simpsonville since 1965. Various organic and inorganic wastes were buried on the 100-acre site from 1975 to 1979. In 1984 the state determined that the groundwater and soil contained VOCs and heavy metals. The contamination has penetrated Big Durban Creek and threatens wells that supply drinking water to about 1,500 residents.

Response actions included removing buried waste and contaminated soil (1987). An interim groundwater extraction and treatment system was also installed. In 1993 the EPA recommended excavating and biologically treating sludge and continuing groundwater treatment and monitoring.

A Consent Order (1985) and Consent Decree require the PRP to undertake the removal actions and remediation work.

Sources: PNPLS-SC; Fountain Inn Library, 400 North Main Street, Fountain Inn, SC, 29681.

Parris Island Marine Corp Depot

Beaufort County

This U.S. Army training facility is centered on Parris Island, about 4 miles south of Beaufort. Activities on the base have generated a variety of wastes that have been buried on the site, including pesticides, PCB-contaminated oils, mercury, other heavy metals, sludges, and spent solvents. Liquid wastes and surface runoff have entered Ribbon Creek and nearby tidal marshes. The groundwater, soils, and stream sediments contain mercury, lead, and a variety of inorganic chemicals. Aquatic life, including oysters, contain high levels of PCBs.

A detailed site investigation is planned.
Source: PNPLS-SC.

Rochester Property
Greenville County
A packaging company illegally buried production wastes on this property near Travelers Rest in 1971–1972. Wastes included drummed wood glue and print binder residues. The site came to the attention of the State of South Carolina in 1982 when an oozing substance was discovered during construction on a neighboring property. Subsequent testing showed that the groundwater and soils contain VOCs, phthalates, and heavy metals. The contamination threatens the drinking water supplies for about 2,500 local residents.

Response actions included removing drummed waste, debris, and contaminated soil in 1990. A site investigation was conducted in 1992. The EPA recommended an air-stripping groundwater treatment system that is presently being built.

The PRPs have conducted the removal actions, site studies, and remediation work.
Sources: PNPLS-SC; Greenville County Library, 310 South Main Street, Travelers Rest, SC, 29690.

Rock Hill Chemical Company
York County
Various solvents from paint and textile industries were distilled on this property near Rock Hill in the 1960s. Sludges from storage tanks were placed on the ground and covered with construction debris. An examination by the state in 1985 discovered rusting tanks and the sludge pile. Testing revealed that the groundwater and soils contain VOCs, heavy metals, and PCBs. The contamination threatens the Catawba River and water sources that serve about 6,000 residents.

Response actions included removing sludges and contaminated soil between 1986 and 1989. In 1994 the EPA recommended a groundwater extraction and treatment system.
Sources: PNPLS-SC; York County Library, 138 East Black Street, Rock Hill, SC, 29731.

Sangamo Weston
Pickens County
Sangamo Weston manufactured electric capacitors on this site in Pickens from 1955 to 1976. PCBs were used as the nonconducting fluid. Various wastes, sludges, and spent chemicals were stored or discharged on the 224-acre site. Testing has shown that the groundwater and soils contain VOCs and PCBs, which have contaminated Twelve Mile Creek and recreational Lake Hartwell. The contamination threatens nearby waterways and the drinking water source for about 15,000 people.

Response actions included removing contaminated soil (1975) and installing fencing and a temporary cap (1986). In 1994 the EPA recommended environmental monitoring of Twelve Mile Creek, low-temperature thermal treatment of contaminated soil, and groundwater extraction and treatment. Soil remediation is in progress.

Consent Orders (1986, 1987) have required the PRP to undertake the site investigations.
Sources: PNPLS-SC; Pickins Public Library, Easley Branch, 110 West First Avenue, Easley, SC, 29640.

Savannah River
Aiken County
Operated by the U.S. Department of Energy, this 300-square-mile property near the South Carolina/Georgia border has produced military nuclear materials and fuels. Five nuclear reactors are on-site. Various solid and liquid chemical and radioactive wastes were stored, spilled, discharged, or buried at nearly 500 locations on the sprawling base. Sampling has shown that the groundwater and soils contain VOCs, heavy metals, and radionuclides (including plutonium). Radionuclides have been detected in Savannah River Swamp, a sensitive wetland. The contamination threatens Upper Three Runs Creek, the recreational Savannah River, and wells that supply water to employees and nearby residents.

Response actions included cleaning up old dump sites and removing solid and liquid wastes and contaminated soil. In 1990 sludges were solidified and capped. Other remedies have included the solidification and reburial of contaminated soil, capping, deed restrictions, and groundwater extraction and air-stripping. Numerous site studies are still in progress.
Sources: PNPLS-SC; U.S. Department of Energy Reading Room, University of South Carolina, 171 University Parkway, Aiken, SC, 29801; Gordon Library, Savannah State College, Tompkins Road, Savannah, GA, 31404.

SCRDI Bluff Road
Richland County
A waste storage and recycling company operated on this site near Columbia in the 1970s and 1980s. Various toxic and flammable wastes were stored in tanks, drums, or lagoons on the property. Surface runoff enters Myers Creek and Congaree Swamp. Testing has shown that the soils and groundwater contain VOCs, pesticides, PCBs, and heavy metals. The contamination threatens the drinking water supplies of nearby residents.

Response actions included removing nearly 8,000 drums of liquid chemical waste in 1982. A site investigation was conducted between 1984 and 1990. The EPA recommended soil vapor extraction and the extraction and air-stripping of groundwater. The soil vapor system became operational in 1994–1995.

An AOC (1988) and Consent Decree (1991) require PRPs to conduct the site investigations and remediation work. The PRPs also voluntarily conducted the removal actions in 1982.

Sources: PNPLS-SC; Richland County Library, Landmark Square Shopping Center, 6864 Garners Ferry Road, Columbia, SC, 29209.

SCRDI Dixiana

Lexington County

This former recycling facility near Cayce operated in the 1970s. Various liquid wastes such as paints, acids, oils, phenols, dyes, and solvents were stored on the site in drums or tanks. Mishandling of the wastes and leaking drums contaminated the groundwater and soils with VOCs. The contamination threatens water wells that supply about 1,300 residents.

Response actions included installation of a groundwater extraction and treatment system using air-stripping, which became operational in 1992. Monitoring is continuing.

Sources: PNPLS-SC; Cayce–West Columbia Library, 1500 Augusta Road, West Columbia, SC, 29037.

Shuron

Barnwell County

Ophthalmic vision lenses were produced on this site near Barnwell from 1958 to 1992. Production wastes, including grinders, glass, polishers, oils, sludges, and asbestos, were discharged into unlined lagoons or buried in pits on the property. Some wastes were dumped into wetlands. Sampling by the state in 1991 revealed that the soil and groundwater contain VOCs, phthalates, and heavy metals. The contamination threatens Turkey Creek, recreational Salkehatchie River, associated wetlands, and wells that supply drinking water to about 6,000 residents.

This site was proposed to the NPL in 1996; a final decision is expected in 1997–1998. Site investigations are planned.

The PRP filed for bankruptcy in 1992.

Source: EPA Publication 9320.7-071.

Townsend Saw Chain Company

Richland County

Recording equipment and chain saws have been manufactured on this site near Pontiac since 1964. Various liquid wastes, including paints and metal-plating sludges, were discharged onto the ground. Testing in the 1980s showed that the groundwater and soils contain excessive levels of heavy metals and VOCs. The contamination threatens Woodcreek Lake, several streams and wetlands, and private and public wells within 3 miles of the site, which serve thousands of residents.

Response actions included connecting affected residents to clean water supplies and constructing a groundwater extraction and treatment system (1982). A subsequent site investigation recommended expanding the groundwater treatment system (1995).

Various legal agreements have required the PRPs to participate in the site investigations and remediation work.

Sources: PNPLS-SC; Richland County Public Library, 7490 Park Lane Road, Columbia, SC, 29223.

Wamchem

Beaufort County

This chemical and dye manufacturing company operated on a small island in a salt marsh near Burton. Liquid wastes were discharged into the marsh and McCalleys Creek from 1959 to 1971; after 1971 wastes were held in lagoons or sprayed on fields. Testing has shown that the groundwater and soils contain VOCs. The chemicals have entered nearby streams and wetlands that are home to various endangered species. The contamination threatens the local drinking water supplies for about 2,000 nearby residents.

Response actions included a detailed site investigation. The EPA recommended groundwater extraction and air-stripping, low-temperature thermal treatment of contaminated soil, and groundwater monitoring. Contaminated soil was removed in 1994. The groundwater system should be operational by 1997.

A Consent Decree requires the PRPs to undertake the site investigations and remediation work.

Sources: PNPLS-SC; Beaufort County Library, 710 Craven Street, Beaufort, SC, 29902.

Annie Creek Mine
Lawrence County
Gold was mined and milled at this site on Annie Creek, located in Black Hills National Forest near the town of Lead, from 1907 to 1916. Mine tailings were treated with cyanide and piled at the headwaters of the creek. Testing in 1989 showed that the groundwater, soils, surface water, and sediments in Annie and Spearfish Creeks contain elevated amounts of arsenic. The contamination has damaged the waters, wetlands, and habitats of these two streams.

Response actions included capping the exposed tailings and building runoff controls (1994). Groundwater and environmental monitoring is in progress. In 1994 the EPA determined that the response actions were sufficient to mitigate the contamination.

The site investigation was performed by the PRP.

Sources: PNPLS-SD; Lawrence County Registrar of Deeds, 90 Sherman Street, Deadwood, SD, 57732.

Ellsworth Air Force Base
Meade and Pennington Counties
This U.S. Air Force base near Rapid City was established in 1942. The support and maintenance of aircraft commands have generated a variety of solid and liquid wastes, including solvents, degreasers, oils, fuels, grease, heavy metals, and pesticides, which were buried or discharged on the 4,800-acre property. EPA tests in the 1980s showed that the groundwater and soils contain VOCs, arsenic, and chromium. The contamination threatens nearby streams and water wells that serve about 2,000 local residents.

Response actions included providing affected residents with clean water supplies (1991). Interim cleanup actions include soil vapor extraction and construction of two groundwater extraction and treatment systems (1995). Other studies are in progress.

Sources: PNPLS-SD; Rapid City Public Library, 610 Quincy Street, Rapid City, SD, 57701.

Whitewood Creek
Lawrence, Meade, and Butte Counties
This site along Whitewood Creek consists of an 18-mile stretch of floodplain that has been contaminated by gold mining operations. Tailing were dumped in the creek from the 1870s to 1977. Testing by the EPA has shown that the groundwater, surface water, soils, and creek sediments contain arsenic, cadmium, lead, cyanide, selenium, sulfates, lead, and zinc. In 1975 about 50 cattle died from arsenic poisoning after eating contaminated corn. The contamination has damaged the habitats and ecosystems of Whitewood Creek and surrounding farmland. The local drinking water supplies for the town of Whitewood are also at risk.

Response actions included capping, removing contaminated soil from residential properties, deed restrictions, and environmental monitoring (1990–1994). Although groundwater monitoring will continue, the EPA is in the process of deleting the site from the NPL.

Homestake Mining Company, a PRP, agreed to conduct the site studies and remediation work and to reimburse past costs according to agreements with the EPA (1982, 1991).

Sources: PNPLS-SD; Lawrence County Registrar of Deeds, 90 Sherman Street, Deadwood, SD, 57732.

Williams Pipe Line Company
Minnehaha County
This company buried various solid and liquid wastes, including leaded stillbottoms and sludge, oils, pesticides, and solvents, on this site in Sioux Falls in the 1970s. In 1987 the EPA determined that the groundwater and soils contain petroleum by-products, VOCs, PAHs, pesticides, arsenic, and other heavy metals. The contamination threatens the Big Sioux River, the Skunk River, and water wells that service about 100,000 residents. Response actions included a site investigation (1992–1994) that recommended groundwater monitoring to ensure that natural attenuation is occurring.

AOs (1991, 1995) require the PRP to conduct the site study and groundwater monitoring.

Sources: PNPLS-SD; Sioux Falls Public Library, 5300 West 12th Street, Sioux Falls, SD, 57107.

SOUTH DAKOTA

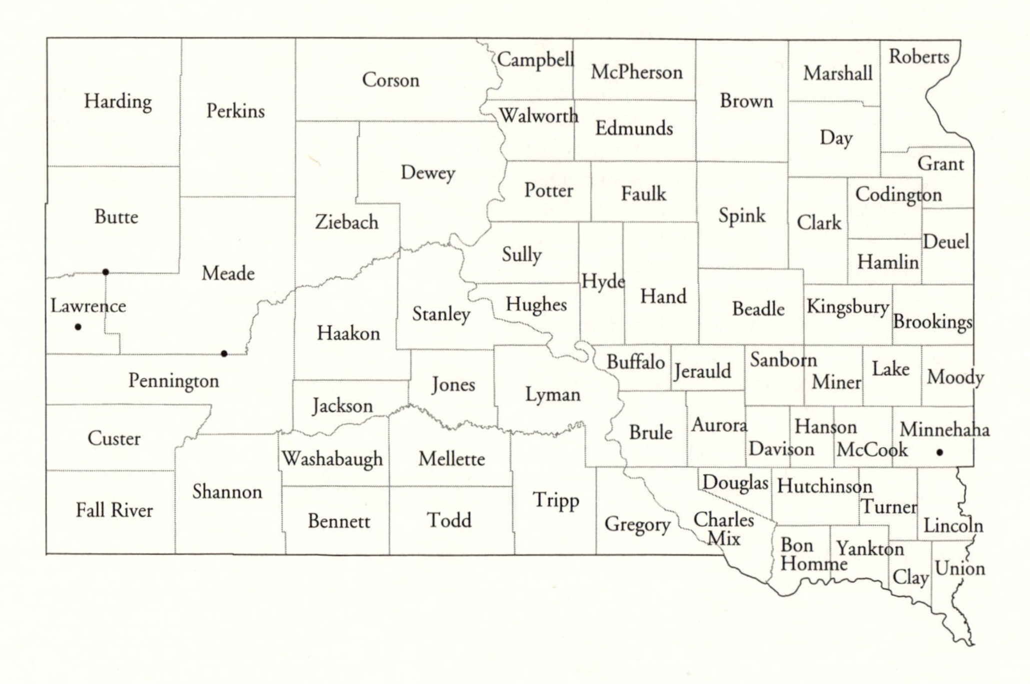

TENNESSEE

American Creosote Company
Madison County
Wooden timbers were treated with preservatives at this site near Jackson from the 1930s to 1981. The process consisted of soaking lumber in creosote and PCP. Wastewaters were discharged into the Forked Deer River or on-site lagoons. The lagoons occasionally overflowed onto neighboring properties. Groundwater and soils contain VOCs and PAHs, as well as heavy metals from previous operations. The contamination threatens water well systems that serve about 60,000 residents, wetlands, and the Forked Deer River.

Response actions consisted of removing sludge and contaminated soil, capping, fencing, and treating contaminated surface water (1983–1988). A groundwater study completed in 1993 recommended no further action.

Sources: PNPLS-TN; Jackson-Madison County Library, 433 Lafayette Street, Jackson, TN, 38301.

Amincola Dump
Hamilton County
This former dump near Chattanooga operated from 1970 to 1973. It received mostly construction and demolition debris and creosote-soaked railroad ties. Testing has shown that the groundwater and soils contain PAHs and heavy metals, including chromium. The contamination threatens the Tennessee River and water well systems that supply about 150,000 people.

Response actions included removing contaminated soil, backfilling, and groundwater monitoring (1993).

A Consent Decree with the EPA (1991) requires the PRPs to undertake site studies and remediation work.

Sources: PNPLS-TN; Chattanooga Hamilton County Library, 1001 Broad Street, Chattanooga, TN, 37402.

Arlington Packaging
Shelby County
Pesticides were manufactured and processed on this site near Arlington from 1971 to 1978. Various chemicals and toxins were stored, leaked, or spilled on the premises. Testing by the EPA in 1983 revealed that the soils and groundwater contain pesticides and pesticide by-products, which have penetrated local wells. The contamination threatens the drinking water supplies for the towns of Arlington and Gallaway, wetlands, and the recreational Loosahatchie River Canal.

Drummed wastes and chemicals, debris, and contaminated soil were removed in 1983. Neighboring

TENNESSEE

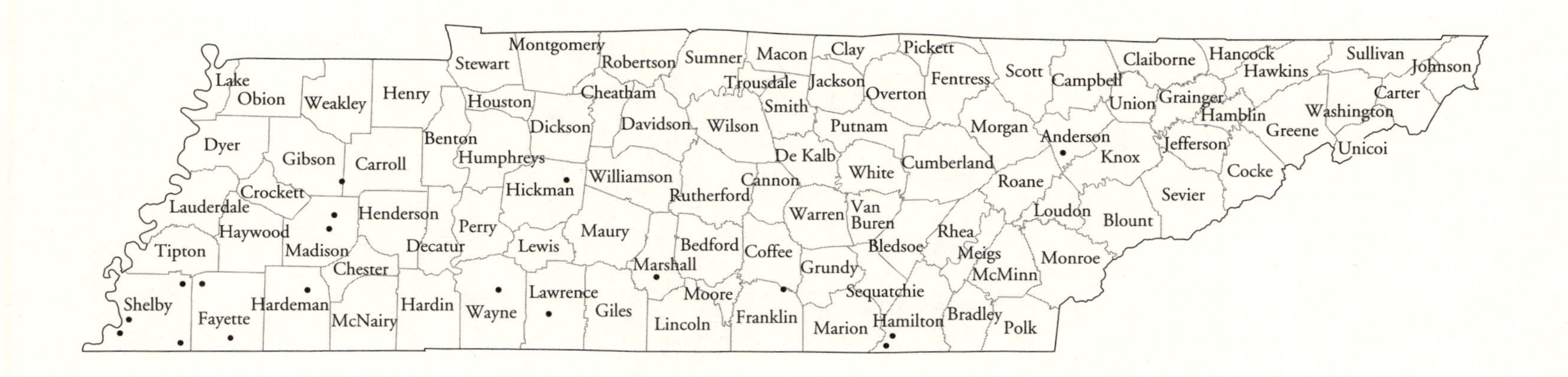
Lake
Obion
Weakley
Henry
Stewart
Montgomery
Robertson
Sumner
Macon
Clay
Pickett
Scott
Campbell
Claiborne
Hancock
Hawkins
Sullivan
Johnson
Carter
Washington
Unicoi
Greene
Hamblin
Grainger
Union
Fentress
Overton
Jackson
Trousdale
Smith
Houston
Cheatham
Dyer
Gibson
Carroll
Benton
Humphreys
Dickson
Davidson
Wilson
Putnam
Morgan
Anderson
Knox
Jefferson
Cocke
Sevier
Crockett
Lauderdale
Haywood
Tipton
Madison
Henderson
Hickman
Williamson
Rutherford
Cannon
De Kalb
White
Cumberland
Roane
Loudon
Blount
Warren
Van Buren
Bledsoe
Rhea
Meigs
McMinn
Monroe
Perry
Decatur
Chester
Lewis
Maury
Marshall
Bedford
Coffee
Grundy
Sequatchie
Shelby
Fayette
Hardeman
McNairy
Hardin
Wayne
Lawrence
Giles
Lincoln
Moore
Franklin
Marion
Hamilton
Bradley
Polk

residential yards were decontaminated in 1990. In 1991 the EPA recommended on-site thermal desorption of contaminated soil, excavation and carbon adsorption of contaminated groundwater, solidification of arsenic-contaminated soil, and groundwater monitoring. Construction is in progress.

A 1992 UAO requires the PRPs to perform the remediation work.

Sources: PNPLS-TN; Arlington Public Library, 11968 Walker Street, Arlington, TN, 38002.

Arnold Engineering Development Center
Coffee and Franklin Counties
This U.S. Air Force site near Manchester and Tullahoma conducts testing and training in the fields of aerodynamics, propulsion, and space simulation. Wastes generated from various operations, including acids, electroplating solutions, PCBs, fuels, and VOC degreasers, were buried in landfills on the 32,000-acre base. The groundwater, surface water, and soils contain PCBs and heavy metals. PCBs occur in sediments in Rollins/Rowland Creek, Bradley Creek, and Brumalow Creek and in aquatic life. The contamination threatens the drinking water supplies for about 4,000 workers and residents.

Detailed site investigations are planned.

Source: PNPLS-TN.

Carrier Air Conditioning Company
Shelby County
Carrier Air Conditioning Company manufactures air conditioners on this site in Collierville. Wastewaters contaminated with trichloroethylene and paint sludges were leaked or discharged into lagoons on the property. The groundwater and soils contain VOCs (trichloroethylene). TCE has also been detected in municipal wells. The contamination threatens the drinking water supply for about 13,000 residents.

Response actions included removing contaminated soil (1980). An air-stripping unit was installed in 1990 to clean the water from contaminated city wells. In 1992 the EPA recommended soil vapor extraction and continued groundwater air-stripping.

Carrier Company, the PRP, signed an AO and a UAO (1992) with the EPA that require the company to conduct the site investigations and clean-up work.

Sources: PNPLS-TN; Memphis Shelby County Library, 91 Walnut Street, Collierville, TN, 38017.

Chemet Company
Fayette County
Antimony oxide was produced on this site located near Moscow from 1978 to 1987. Raw materials and production chemicals were stored on-site in drums and bags. Sampling in the 1980s revealed that the soils, surface water, and possibly groundwater contain excessive levels of arsenic, antimony, and lead. The contamination threatens the health of nearby residents and elementary school students.

Response actions included erecting fencing (1990) and removing contaminated soil (1990–1991). Detailed site investigations are planned.

A 1990 court order required the PRP to erect fencing and remove hazardous waste from the site.

Sources: PNPLS-TN; Lagrange-Moscow Elementary School, 15655 Highway 57 East, Moscow, TN, 38057.

Gallaway Pits
Fayette County
This former gravel pit near Gallaway was used as a landfill during the 1970s and 1980s. It received a variety of municipal and industrial wastes, including pesticides and other liquids. Testing has shown that the soil, surface water, and groundwater contain pesticides (including chlordane and toxaphene). The contamination threatens local water supply wells that serve about 1,000 residents.

Response actions included excavating and stabilizing contaminated soil, capping, and long-term groundwater monitoring. Construction activities have been completed, and monitoring has indicated that clean-up goals are being achieved. The site will be nominated for deletion from the NPL.

Sources: PNPLS-TN; Gallaway City Hall, 607 Watson Drive, Gallaway, TN, 38036.

ICG Iselin Railroad
Madison County
This railroad yard in Jackson has had a succession of owners since 1906. Activities on the site have included rail car and engine maintenance, steel fabrication, fuel storage, and coal-fired power generation. Waste products included lead batteries, railroad tracks, fuels, acids, and degreasers. Testing has shown that the soil and groundwater contain heavy metals and VOCs. The chemicals have penetrated Jones Creek and threaten the water wells that supply about 32,000 residents.

EPA investigations (1994–1997) are nearing completion and will recommend a final remediation plan.

Sources: PNPLS-TN; Jackson-Madison County Library, 433 East Lafayette, Jackson, TN, 38301; Washington-Douglas Headstart, 409 Iselin Street, Jackson, TN, 38301.

Lewisburg Dump
Marshall County
This former quarry near Lewisburg operated as a landfill from the 1950s to 1979. It received a variety of municipal and industrial wastes, including inor-

ganic chemicals and solvents. Testing has shown that the soil and groundwater contain plastic phthalates, copper, lead, and VOCs. The contamination threatens Big Rock Creek and water wells that supply about 200 nearby residents.

Response actions included site investigations (1987–1992). The EPA recommended removing debris, capping, and long-term groundwater monitoring. The site will be deleted from the NPL if clean-up goals continue to be met.

PRPs have participated in the site studies and clean-up work.

Sources: PNPLS-TN; Marshall County Memorial Library, 310 Farmington Pike, Lewisburg, TN, 37091.

Mallory Capacitor Company
Wayne County
This site in Waynesboro was the former location of an electrical capacitor company active between 1969 and 1984. Various production wastes, including PCB oils and degreasers, were stored in tanks on the site. Leaks, spills, and discharges on the surface contaminated the soils and groundwater with VOCs and PCBs. The chemicals threaten wetlands, springs, the recreational Green River, and local water wells that serve about 1,000 people.

Response actions included removing contaminated soil (1988–1989). A site investigation was conducted from 1989 to 1991. The EPA recommended a groundwater extraction and treatment system that should be operational by 1997.

The PRPs have participated in the site studies and remediation plan according to an AO and UAO with the EPA.

Sources: PNPLS-TN; Wayne County Public Library, U.S. Highway 64, East Waynesboro, TN, 38485.

Memphis Defense Depot
Shelby County
This 640-acre U.S. Army depot in Memphis has been in operation since 1942. Its main purpose is to supply army units with clothing, food, medical supplies, electronics, industrial chemicals, and fuels. Various wastes such as oil, grease, thinners, paints, pesticides, degreasers, and methyl bromide have been buried on the site. Testing has shown that the groundwater and soils contain arsenic, lead, chromium, nickel, VOCs, PAHs, cadmium, zinc, and pesticides. The contaminated has entered nearby wells and streams and threatens water supplies for about 150,000 people. Fish in Lake Danielson contain toxic amounts of pesticides and PCBs.

Detailed site investigations to be completed by 1998 will recommend a final remediation plan.

Sources: PNPLS-TN; Memphis/Shelby County Public Library, Main Branch, Government and Law Section, 1850 Peabody Avenue, Memphis, TN, 38104; Memphis/Shelby County Public Health Department, Pollution Control Division, 814 Jefferson Avenue, Memphis, TN, 38106; Cherokee Public Library, 3300 Sharp Avenue, Memphis, TN, 38111.

Milan Ammunition Plant
Carroll and Gibson Counties
The U.S. Army has operated this ammunition manufacturing and storage base near Milan since 1942. Munitions have also been deactivated by removing TNT using hot water and steam. The wastewater was discharged into unlined ponds on the 22,500-acre base. The groundwater, surface water, soils, and nearby streams contain explosives derivatives, cadmium, mercury, lead, VOCs, and nitrates. The contamination threatens agricultural wells, private and public wells within 5 miles of the base that serve about 14,000 residents, and recreational waterways.

Response actions included dredging the settling ponds (1971), capping (1984), and removing lead-based paint from structures. The EPA began an intensive investigation in 1990. Final recommendations will include additional capping and installation of a groundwater extraction and treatment system.

Sources: PNPLS-TN; Fields Library, 1075 A East Van Hook Street, Milan, TN, 38358.

Murray-Ohio Dump
Lawrence County
This former industrial landfill near Lawrenceburg accepted a variety of wastes, including paint and electroplating sludges, from 1963 to 1982. Seepage has been observed entering nearby drainages and streams. The groundwater and soils are thought to contain chromium, nickel, zinc, other heavy metals, and VOCs. Chromium has been detected in Shoal Creek. The contamination threatens Shoal Creek and the private water wells that serve about 3,000 people within 3 miles of the site.

The site was capped in 1981. An investigation (1990–1994) recommended upgrading the cap, deed restrictions, and groundwater monitoring.

The PRP has conducted the site investigations and on-site capping.

Sources: PNPLS-TN; Lawrenceburg Public Library, 519 East Garnes Road, Lawrenceburg, TN, 38464.

North Hollywood Dump
Shelby County
This former municipal dump near Memphis operated from the 1930s to 1980. It received a variety of materials, including sodium hydrochloride production wastes, heavy metal sludges, and pesticide-bear-

ing sludges. Testing by the EPA in 1980 revealed that the groundwater, surface water, and soils contain chlordane, endrin, other pesticides, lead, copper, and arsenic. The contamination threatens the drinking water supplies for about 10,000 people living within 3 miles of the site and an elementary school.

Response actions included constructing leachate control systems and fencing (1980). Surface chemical wastes were removed in 1981. An investigation from 1982 to 1990 recommended a clay cap, drainage of contaminated ponds, deed restrictions, and groundwater monitoring. Construction is under way.

Various agreements (1984, 1988, and 1991) require the PRPs to fund the site investigations and remediation work.

Sources: PNPLS-TN; Memphis-Shelby County Library, 1850 Peabody Avenue, Memphis, TN, 38104.

Oak Ridge Reservation
Anderson County
This U.S. Department of Energy facility near Oak Ridge consists of 37,000 acres that contain nearly 300 contaminated sites. Operations included research and development on nuclear reactors, chemical and biological programs, uranium-235 enrichment processes, and nuclear weapons components. Resulting wastes have been buried on-site in drums or stored in tanks and buildings on the property. Leakage from buried or stored waste has contaminated the soils and groundwater with heavy metals (mercury and lead), organic compounds, VOCs, nitrates, and radionuclides (including cesium-137). The contamination threatens the water supplies of about 44,000 people along the Tennessee River as well as Poplar Creek, Clinch River, Bear Creek, and Lower Watts Bar Reservoir on the Tennessee River.

Response actions have included dredging sediments from Watts Bar Dam Reservoir, capping landfill areas (1993), removing mercury-contaminated tanks and sediments (1991–1993), removing drummed waste and sludges (1993–1995), and excavating and removing cesium- and strontium-contaminated soil. A number of ongoing site investigations are nearing completion and will recommend final remediation plans.

Sources: PNPLS-TN; Oak Ridge Public Library, 1401 Oak Ridge Turnpike, Oak Ridge, TN, 37830.

Tennessee Products
Hamilton County
This former coke plant in Chattanooga operated until the 1950s. Various wastes, including coal tars, were dumped into Chattanooga Creek or piled on its banks. The soils, sediments, surface water, wetlands, and fisheries in Chattanooga Creek are contaminated with PAHs. Fish and wildlife may also contain elevated levels of PAHs and other tar derivatives.

Response actions included fencing parts of Chattanooga Creek (1993). A detailed site investigation is planned.

Sources: PNPLS-TN; Alton Park Community Health Center, 100 East 37th Street, Chattanooga, TN, 37410.

Velsicol Chemical Corporation
Hardeman County
Velsicol Chemical Corporation disposed of large amounts of pesticides and VOCs in the Hardeman County landfill near Toone between 1964 and 1973. The groundwater and soils contain VOCs that have penetrated nearby water wells. The contamination threatens water wells that supply thousands of local residents.

Response actions included providing affected residents with clean water supplies, capping, regrading, and site revegetation (1979–1980). A groundwater study (1989–1991) recommended groundwater extraction, settling, air-stripping, and carbon adsorption treatment. This system should be operational by 1997.

The PRP, Velsicol Chemical Company, has conducted the site investigations and remediation work according to several agreements with the EPA.

Sources: PNPLS-TN; Velsicol Hardeman Library, 213 North Washington Street, Bolivar, TN, 38008.

Wrigley Charcoal Plant
Hickman County
This former charcoal plant in Wrigley operated from the late 1800s to the 1960s and manufactured charcoal briquettes, iron products, and wood alcohol. A 1985 inspection revealed tar pits, old drums, and leachate seeping into Mill Creek. Testing showed that the groundwater and soils contain VOCs, phenols, and PAHs. Mill Creek is also contaminated. The contamination threatens the public and private water wells that serve about 6,000 local residents.

Response actions included stabilizing the tar pits (1988), rerouting a stream, and excavating tar waste (1989). In 1991 the EPA recommended removing waste debris and tanks, tar-contaminated soils, and wood-tar sludges. Other studies, including groundwater remedies, are in progress.

Sources: PNPLS-TN; Hickman County Public Library, 120 West Swann Street, Centerville, TN, 37033.

Air Force Plant No. 4

Tarrant County

This site is owned by the U.S. Air Force and operated by Lockheed Corporation. Located near White Settlement, a suburb of Fort Worth, the plant has tested military aircraft since 1941. Solid and liquid wastes, including solvents, spent chemicals, and plating solutions, were dumped in landfills and pits. Tests have shown that the soil and shallow groundwater are contaminated with trichloroethylene. The contamination threatens Lake Fort Worth and the drinking water supply of the 13,000 residents of White Settlement.

Response actions included the removal of contaminated soil. A site investigation study should be completed by 1997, with recommendations for a clean-up plan.

The U.S. Air Force will mitigate the site.

Sources: PNPLS-TX; Fort Worth Public Library, Central Branch, 300 Taylor Street, Fort Worth, TX, 76102.

ALCOA/Lavaca Bay

Calhoun County

The ALCOA plant near Point Comfort and Port Lavaca has processed bauxite, cryolite, and chlorine-alkali chemicals since 1948. A smelter circuit was removed in 1980. Chrome plating was also conducted. Bauxite, aluminum fluoride, and carbon paste are presently being processed on-site. Past activities discharged mercury-rich wastewater into Lavaca Bay and contaminated about 50,000 acres of marine estuary environment and commercial fishing grounds. Bauxite residue, sludge, spent solvents, cyanide, asbestos insulation, and PCBs were disposed of on-site. Soils and groundwater are contaminated with mercury and other chemicals. About 13,000 people live in the area.

The Texas Department of Health Response closed Lavaca Bay to fishing in 1988.

Three PRPs have been identified. ALCOA signed an AO with the EPA in 1994 to fund the site investigation and feasibility study.

Sources: PNPLS-TX; Calhoun County Library, 200 West Mahan, Port Lavaca, TX, 77979.

Baily Waste Disposal

Orange County

Located near the town of Bridge City, this landfill operated from 1950 to the 1970s. Various municipal and industrial wastes were accepted. Soils and groundwater are contaminated with organic chemicals, chloroform, benzene, phenols, pyridenes, chlorinated hydrocarbons, and heavy metals, including arsenic. About 8,000 people within 3 miles of the site depend on local groundwater for drinking water. The contamination has also penetrated a salt marsh.

Response actions included fencing the site (1984). A site investigation recommended the removal and solidification of soils and contaminated marsh sediments in 1988. The remediation process began in 1993 and is continuing.

Twenty-two PRPs have been identified. Consent Decrees (1990, 1996) require the companies to fund 97 percent of the site investigation and clean-up cost.

Sources: PNPLS-TX; Orange Public Library, 220 North Fifth Street, West Orange, TX, 77630.

Bio-Ecology Systems, Inc.

Dallas County

Located in Grand Prairie, this industrial solid waste processing facility operated from 1972 to 1978. Bio-Ecology Systems was licensed to decontaminate and bury a variety of waste. Soils and shallow groundwater are contaminated with VOCs and heavy metals. Private drinking water wells are within 1 mile of the site; Grand Prairie's municipal wells are within 3 miles.

Response actions included site investigations from 1982 to 1985. Contaminated soil and storage tanks were removed in 1985 and the site fenced. A remediation plan consisted of constructing a new on-site landfill. Contaminated soils have been excavated, contained, and covered with grass (1986–1993).

A principal PRP, Bio-Ecology Systems, declared bankruptcy in 1978. Minimum settlements are being negotiated with about 80 remaining PRPs.

Sources: PNPLS-TX; EPA Region 6 Library, 1445 Ross Avenue, Suite 1200, Dallas, TX, 75202.

Brio Refining, Inc.

Harris County

Brio Refining operated a petrochemical site near the city of Friendswood from 1957 to 1982. Styrene and vinyl chloride tar were stored in unlined lagoons. Waste solvents, vinyl chloride, chlorinated solvents, fuel oil, metallic catalysts, sludges, and tars were deposited in pits on the property. Soils and shallow groundwater are contaminated with 1,2 dichloroethane, vinyl chloride, fluorene, pyrene, copper, and anthracene. These chemicals have also been identified in air releases. The groundwater and surface runoff have contaminated Clear Creek, a fishing stream. About 5,000 people live within 1 mile of the site.

Response actions included fencing the site, removing contaminated sludges and solids, and covering the pits with plastic (1985). A site investigation recommended a clean-up plan of barrier wells, groundwater pump-and-treatment systems, free-product recovery, and incineration of contaminated soil. The incineration remedy has yet to be initiated.

TEXAS

Dallam
Sherman
Hansford
Ochiltree
Lipscomb
Hartley
Moore
Hutchinson
Roberts
Hemphill
Oldham
Potter
Carson
Gray
Wheeler
Deaf Smith
Randall
Armstrong
Donley
Collingsworth
Parmer
Castro
Swisher
Briscoe
Hall
Childress
Hardeman
Bailey
Lamb
Hale
Floyd
Motley
Cottle
Foard
Wilbarger
Cochran
Hockley
Lubbock
Crosby
Dickens
King
Knox
Baylor
Yoakum
Terry
Lynn
Garza
Kent
Stonewall
Haskell
Throckmorton
Gaines
Dawson
Borden
Scurry
Fisher
Jones
Shackelford
Stephens
Andrews
Martin
Howard
Mitchell
Nolan
Taylor
Callahan
Eastland
El Paso
Hudspeth
Culberson
Loving
Winkler
Ector
Midland
Glasscock
Sterling
Coke
Runnels
Coleman
Ward
Crane
Upton
Reagan
Irion
Tom Green
Concho
McCulloch
Reeves
Jeff Davis
Pecos
Crockett
Schleicher
Menard
Sutton
Presidio
Brewster
Terrell
Val Verde
Edwards
Kinney
Maverick

Wichita
Archer
Clay
Montague
Cooke
Grayson
Fannin
Lamar
Red River
Bowie
Young
Jack
Wise
Denton
Collin
Hunt
Delta
Titus
Franklin
Hopkins
Cass
Morris
Camp
Palo Pinto
Parker
Tarrant
Rockwall
Dallas
Kaufman
Rains
Wood
Upshur
Marion
Van Zandt
Harrison
Gregg
Erath
Hood
Johnson
Ellis
Somervell
Henderson
Smith
Rusk
Panola
Comanche
Bosque
Hill
Navarro
Cherokee
Shelby
Brown
Hamilton
Freestone
Anderson
Nacogdoches
Mills
Limestone
McLennan
San Augustine
Coryell
Houston
Sabine
San Saba
Lampasas
Falls
Leon
Angelina
Bell
Robertson
Trinity
Newton
Burnet
Madison
Mason
Llano
Williamson
Milam
Walker
Polk
Tyler
Jasper
Kimble
Brazos
San Jacinto
Grimes
Burleson
Gillespie
Blaco
Travis
Lee
Montgomery
Hardin
Washington
Kerr
Hays
Bastrop
Liberty
Orange
Real
Kendall
Waller
Austin
Harris
Comal
Caldwell
Fayette
Jefferson
Bandera
Chambers
Colorado
Guadalupe
Bexar
Ft. Bend
Gonzales
Lavaca
Uvalde
Medina
Galveston
Wharton
Wilson
Brazoria
De Witt
Jackson
Zavala
Frio
Atascosa
Karnes
Victoria
Matagorda
Goliad
Calhoun
Dimmit
La Salle
Live Oak
Bee
Refugio
McMullen
Aransas
San Patricio
Jim Wells
Webb
Nueces
Duval
Kleberg
Zapata
Jim Hogg
Brooks
Kenedy
Starr
Willacy
Hidalgo
Cameron

Over 30 PRPs have been identified. A PRP task force, through a CO (1985), CD (1989), and AO (1991), will finance some of the site investigation and remediation plan. A class action lawsuit by affected citizens forced the PRPs to buy properties in an adjacent subdivision.

Sources: PNPLS-TX; San Jacinto College, South Campus, 13735 Beamer Road, Houston, TX, 77089.

Crystal Chemical Company

Harris County

This site is located in a residential and light industrial section of Alief, a suburb of Houston. Herbicides were manufactured from 1968 to 1981. Ingredients and finished products were stored or spread on the ground. The soil and groundwater are contaminated with arsenic, and the contamination has moved beyond the site boundary. Twenty thousand people and over 20 water wells are located within 1 mile of the site.

Response actions included draining the pits, filling in contaminated ponds, building a temporary cap, and fencing (1981–1983). Site investigations were completed from 1984 to 1992. The final remediation plan consists of groundwater excavation and treatment through ion exchange, excavation and consolidation of contaminated soil (including off-site areas), and multilayer capping. These remedies should be under way by 1997.

Thirteen PRPs have been identified. A Consent Order (1987), Administrative Order (1992), and Unilateral Administrative Order (1992) require the PRPs to contribute to the costs of the site investigation and clean up.

Sources: PNPLS-TX; Jungman Public Library, 5830 Westheimer Road, Houston, TX, 77057.

Crystal City Airport

Zavala County

The municipal airport at Crystal City was used as a crop-dusting base until 1982. The soils are contaminated with pesticides (DDT and toxaphene), herbicides, fungicides, and arsenic. The site is located within a few miles of private homes, water wells, and a high school.

Response actions included consolidating sludge (1983–1984) and repairing fencing (1988). A site investigation recommended on-site consolidation of contaminated material and capping (1987). The remedies were completed in 1990. A 5-year monitoring period has just been completed, and the site will likely be deleted from the NPL.

Sources: PNPLS-TX; Crystal City Public Library, 101 East Dimmit Road, Crystal City, TX, 78839.

Dixie Oil Processors, Inc.

Harris County

This site, about 2 miles north of the city of Friendswood, has been operated by Intercoastal Chemical Company since 1969. Activities have included recycling and processing copper, hydrocarbons, and oil; oil washing; and blending chemical solvents. Soils and groundwater are contaminated with ethylbenzene, hexachlorobenzene, and copper. Housing developments and municipal water wells are located within a few miles of the site. The contaminated groundwater flows into the Mud Gulley waterway.

The most contaminated soils were excavated in 1984. Site investigations were completed in 1988 and recommended a program of additional soil removal and monitoring. Excavation was completed in 1993. Monitoring is continuing.

A CO (1986), AOC (1989), and UAO (1991) with a group of 12 PRPs require them to contribute to the cost of the site investigation and clean-up work.

Sources: PNPLS-TX; San Jacinto College, South Campus, 13735 Beamer Road, Houston, TX, 77089.

French Ltd.

Harris County

Originally a former sand mine, this site near Crosby was licensed as a petrochemical waste disposal site from 1966 to 1973. A variety of wastes, including sludges, were dumped in the sand pit. EPA testing from 1980 to 1983 indicated that the soils and groundwater are contaminated with VOCs, phenols, heavy metals, and PCBs. Air emissions have also been detected. The local groundwater is used for residential and farming purposes. The site is also located in the floodplain of the San Jacinto River.

Response actions included fencing and pumping out contaminated sludge (1980–1983). The site was flooded in 1989, and the EPA supplied bottled water to affected residents. Injection wells were installed to drive back the contamination plume. A site investigation concluded in 1988 that the groundwater should be pumped out and treated with biological and carbon adsorption technologies (1992–1995). Contaminated soil and sludge will be treated using in-situ biodegradation. Groundwater monitoring will continue for 30 years.

About 76 viable PRPs have been identified by the EPA. Legal actions, including a Consent Decree in 1990, require the PRPs to finance the site investigations and remediation plan.

Sources: PNPLS-TX; Crosby Public Library, 135 Hare Road, Crosby, TX, 77532.

Geneva Industries/Fuhrmann Energy
Harris County
A petrochemical production company and a salvage company used the site, located in a light industrial section of Houston, from 1967 to 1985. Biphenyls were manufactured from 1967 to 1984; equipment was stripped and recycled from 1984 to 1985. The soils and groundwater are contaminated with PCBs, PNAs, and trichloroethylene. About 35,000 people and numerous wells are located within 1 mile of the site.

Site investigations recommended groundwater extraction and carbon adsorption treatment and off-site disposal of contaminated soil and drummed waste (1986). The groundwater system is currently operational.

The EPA is trying to recover costs from 5 PRPs.

Sources: PNPLS-TX; M.D. Anderson Library, University of Houston, 4800 Calhoun Boulevard, Houston, TX, 77204.

Highlands Acid Pit
Harris County
Located near the town of Highlands, this disposal site was operational in the 1950s. Sludges and a variety of industrial waste acids were deposited in unlined pits on the property. Soils and groundwater are contaminated with toluene, benzene, phenols, xylene, sulfate, arsenic, cadmium, lead, and manganese. The pits are on a peninsula in the San Jacinto River and are periodically flooded. The peninsula is gradually sinking into the river, and portions of the site are underwater. About 5,000 people live within a few miles of the site; a number of private wells are located there as well.

Response actions included fencing the site (1984, 1985). Site investigations from 1984 to 1987 recommended excavation and off-site disposal of waste and toxic soil and a 30-year groundwater monitoring program.

Sources: PNPLS-TX; Houston Central Library, Government Documents, 500 McKinney Street, Houston, TX, 77002.

Koppers Company
Bowie County
Located near downtown Texarkana, this wood treatment facility was operated by a variety of owners from 1903 to 1962. EPA tests in 1980–1981 showed that the soil and shallow groundwater are contaminated with PAHs. Buildings were razed in 1962 and a residential area built on part of the site. About 25,000 people live within 4 miles of the area. Creosote seeps have been found in adjacent Wagner Creek.

The worst soil contamination was excavated and stored on-site in drums (1985). The area was fenced, and Koppers replaced the contaminated yards of 24 homes (1985–1986). A site investigation was completed in 1988, and the remediation plan was revised from 1990 to 1992. The residential area was bought out in 1992 and the affected families relocated; abandoned housing was demolished in 1994. The recommended clean-up plan consisting of off-site soil disposal and groundwater treatment with carbon adsorption or fluidized carbon has yet to be initiated.

A series of legal actions, including a UAO in 1993, requires the PRP to undertake the site investigation and remediation plan.

Sources: PNPLS-TX; Texarkana City Hall, 320 Texas Boulevard, Texarkana, TX, 75501.

Lone Star Army Ammunition Plant
Bowie County
This site is owned by the U.S. Army and was used to detonate explosives. Munitions were also loaded and assembled on-site. EPA tests in 1993–1994 indicated that the soils and groundwater are contaminated with explosives such as tetryl and with chromium, lead, and mercury. Although the site is located in a sparsely populated area, domestic water wells are located nearby. The contamination has migrated into Elliot Creek, about 1,000 feet away.

Ongoing site investigations need to be completed before a final clean-up plan can be recommended.

The U.S. Army is the responsible party and has agreed to finance the site investigation and remediation work.

Sources: PNPLS-TX; Texarkana Public Library, Texarkana, TX, 75501.

Motco Inc.
Galveston County
This waste recycling facility operated from 1958 to 1968 and was reopened for a few years in the early 1970s. Recycling wastes and sludges were discharged into seven unlined pits and storage tanks on the property. Tests showed that the soil and groundwater are contaminated with styrene tar, VOCs, and heavy metals. About 3,000 people live within 1 mile of the site. The contamination also threatens a coastal marsh system.

Emergency response actions between 1980 and 1985 included pumping out contaminated water, repairing dike containment systems, removing soil, capping, and pulling out storage tanks. A site investigation (1985–1993) recommended a groundwater pump-and-treat system, recovering and incinerating free-product petrochemicals, and removal or incineration of sludge, tar, and soil (1993–1996).

Twenty PRPs have been identified. Various legal actions against them have recovered a share of the emergency action costs. UAOs in 1991 and 1992 against seven PRPs required them to finance the clean-up plan.

Sources: PNPLS-TX; College of the Mainland Library, 1200 Amburn Road, Texas City, TX, 77591.

North Cavalcade Street

Harris County

Located in a commercial area of Houston, this former wood treatment plant operated from 1946 to 1961. The site is now occupied by two warehouses built in the 1980s. The soil and shallow groundwater are contaminated with PAHs and benzene. About 50,000 people live in the area. The shallow groundwater contamination does not immediately threaten the community water supply, which is drawn from a deeper aquifer.

A site investigation recommended the biological degradation of contaminated soil and groundwater extraction and treatment using carbon adsorption (1988). Contaminated soil has been stockpiled for biological treatment, which should be under way by 1997. The groundwater treatment system has not been as effective as anticipated and is being redesigned.

Sources: PNPLS-TX; Houston Central Library, Government Documents, 500 McKinney Street, Houston, TX, 77002.

Odessa Chromium No. 1

Ector County

This chrome-plating operation located in an industrial section of Odessa was active from the 1960s to the 1970s. Tests have shown that the groundwater is contaminated with hexavalent chromium. The contamination plume is approximately 20 acres in size and has entered about 20 nearby wells. Over 200 municipal and private water wells are located within a mile of the site; 3,500 residents live close by.

Response actions included developing alternate water supplies for the affected residents. Site investigations completed between 1986 and 1988 recommended the extraction and electrochemical treatment of contaminated groundwater. The system is installed and in operation.

Four PRPs were identified, three of which have sufficient funds to contribute to the remediation of the site. A settlement with Bell Petroleum/Regal International in 1990 recovered past and future costs. Negotiations continue with Sequa Company, who refused to obey a 1989 Unilateral Administrative Order.

Sources: PNPLS-TX; Ector County Library, 321 West Fifth Street, Odessa, TX, 79761.

Odessa Chromium No. 2

Ector County

This chrome-plating facility located in an industrial section of Odessa operated from the 1960s to the 1970s. Tests have shown that the groundwater is contaminated with hexavalent chromium. The contamination plume is over 40 acres in size and has penetrated at least 14 nearby wells. About 3,500 people and over 400 private and municipal water wells are located within 1 mile of the site.

Response actions included connecting affected residents to clean water supplies. Site investigations completed between 1986 and 1988 recommended groundwater extraction and treatment through electrochemical technology or ion exchange. The system is installed and operational.

Four PRPs have been identified by the EPA. They have agreed to contribute to the cost of the site studies and remediation plan.

Sources: PNPLS-TX; Ector County Library, 321 West Fifth Street, Odessa, TX, 79761.

Pantex Plant (USDOE)

Carson County

The U.S. Army and Department of Energy have operated this facility near Amarillo since 1942. Nuclear operations, including assembling and disassembling nuclear weapons, were started in 1950. Waste materials were burned or buried in unlined pits on the property; liquid wastes were poured into unlined ditches or ponds. Soil and groundwater are contaminated with acetone, furan, toluene, trichloroethylene, butanone, bromoform, arsenic, chromium, barium, lead, mercury, and silver. The site is located in an agricultural area, and the contamination threatens private wells and Amarillo's municipal wells located within 4 miles of the base. The groundwater is also used for irrigating crops and watering livestock.

Final remediation actions will be recommended when the ongoing site investigations are completed.

Sources: PNPLS-TX; EPA Region 6 Library, 1445 Ross Avenue, Dallas, TX, 75202.

Petro Chemical Systems, Inc.

Liberty County

This petrochemical waste disposal site near Liberty operated from about 1970 to 1979. Soil and groundwater are contaminated with naphthalene, chrysene, fluorene, benzene, styrene, and lead. A local road, which was covered with waste oil, is also contaminated. In 1974 the disposal site was subdivided into a residential neighborhood. A number of shallow wells in the area are used for drinking water.

Response actions included fencing (1986) and digging up and resurfacing the road (1988). Site studies from 1988 to 1991 recommended soil vapor extraction, catalytic oxidation of organic contamination, runoff controls, and landfill caps. Pilot studies should be completed by 1997. Nearby and on-site residents will be temporarily relocated during the work.

Viable PRPs include ARCO, DuPont, Exxon, Lubrizol, and Tenneco Polymers. An out-of-court settlement by ARCO with the EPA and Department of Justice required ARCO to pay about $17 million toward the estimated $27 million total clean-up cost. Settlements are also being negotiated with other PRPs.

Sources: PNPLS-TX; Liberty Municipal Library, 1710 Sam Houston Avenue, Liberty, TX, 77575.

RSR Corporation
Dallas County
Located in West Dallas, this site is an abandoned smelter. Waste, including slag and battery parts, was disposed of on-site. EPA tests confirmed that smokestack emissions have led to widespread soil contamination by lead, arsenic, and cadmium. About 10 percent of the children living within 3,000 feet of the smokestack showed signs of lead poisoning. About 17,000 people live in the area.

Emergency response actions included removing contaminated soil from the yards of over 400 homes (1993–1994), demolishing 167 apartment complexes, excavating lead-contaminated soil (1994–1995), and removing 500 drums of battery acid, slag, and laboratory chemicals (1995). Site investigations are continuing on the smelter and landfill locations, and recommendations for clean up are expected by 1997–1998.

PRPs include RSR Corporation, Quemetoc Metals, Murmur Corporation, and the city of Dallas.

Sources: PNPLS-TX; Dallas Public Library, West Branch, 2332 Singleton Boulevard, Dallas, TX, 75212.

Sheridan Disposal Services
Waller County
This site near Hempstead processed waste oil and solvents from various businesses from 1963 to 1973. The liquid wastes were stored in ponds and lagoons, landfarmed, and incinerated. EPA tests in 1985–1986 showed that the soil and groundwater are contaminated with VOCs, benzene, toluene, PCBs, and heavy metals. The contamination threatens the wells of a number of nearby residences. Contaminated runoff and oil have been observed in the Brazos River.

Response actions included fencing the site (1986), periodic pumping of the lagoons (1987), and conducting a biotreatment pilot study (1991). EPA site investigations completed in 1988–1989 recommended biodegradation to detoxify the contaminated material. An erosion control system was installed to protect the Brazos River. Sludges will be treated, stabilized, and capped on-site. Groundwater will be monitored to confirm that natural attenuation is breaking down the contamination plume.

About 150 PRPs have been identified. Seventy have negotiated minimum settlements with the EPA; 38 others have formed the Sheridan Site Trust to help finance the clean-up operations. A legal action by nonsettling PRPs has suspended the clean-up work.

Sources: PNPLS-TX; Waller County Library, 2331 Eleventh Street, Hempstead, TX, 77445.

Sikes Disposal Pits
Harris County
Located near Crosby, this 185-acre landfill operated during the 1960s. A variety of waste was dumped in the unlined sand pits, mostly petrochemical waste, sealed drums, organic sludge, and phenolic tars. The soil and groundwater are contaminated with a variety of chemicals, including heavy metals (arsenic, mercury, cadmium, chromium, and zinc). About 10,000 people and a number of water wells are located within 2 miles of the site. The site is also close to the San Jacinto River and Jackson Bayou.

Response actions included excavating tar (1983) and fencing (1989). Site investigations led to a recommendation in 1986 to incinerate all contaminated material, including groundwater. Incineration procedures were conducted from 1990 to 1994. Remediation is complete, and the site has been seeded.

The EPA will pursue cost recovery from 11 PRPs.

Sources: PNPLS-TX; Crosby Public Library, 135 Hare Road, Crosby, TX, 77532.

Sol Lynn Industrial Transformers
Harris County
This site located in an industrial part of Houston processed used wire and transformers from 1965 to 1975. Solvents and other chemical solutions containing PCBs were dumped on-site. Soils and groundwater are contaminated with trichloroethylene and PCBs. About 2,000 people live within 1 mile of the site; four city water wells and four private wells are within 3 miles of the site.

Response actions included fencing (1989). Drummed waste was also removed. Site investigations and feasibility studies (1988–1992) recommended air stripping of groundwater and off-site disposal of contaminated soil. The groundwater treatment system was installed in 1993. Soil and groundwater remediation are in progress.

Three PRPs have been identified. Consent Decrees (1990, 1992) require the viable PRP to finance the clean-up work.

Sources: PNPLS-TX; Houston Central Library, Government Documents, 500 McKinney Street, Houston, TX, 77002.

South Cavalcade Street
Harris County
Located in the city of Houston, this site was a former wood treatment plant from 1910 to 1962. The buildings were dismantled in 1962, and the property has since been commercially developed. The soil and groundwater are contaminated with creosote, creosote derivatives, PAHs, and metal salts. About 4,500 people reside within 1 mile of the site.

Response actions included a site investigation study (1988) that called for a installing groundwater treatment system using filtration and activated carbon adsorption, soil washing, and in-situ soil flushing to remove creosote derivatives. Construction of these systems began in 1995.

Four PRPs have been identified. Beazer East Inc. is financing the current clean-up work through an AOC and a CD (1991). In 1991 Beazer East paid the EPA $500,000 for past costs. Other PRPs have negotiated minimum settlements.

Sources: PNPLS-TX; Houston Central Library, Government Documents, 500 McKinney Street, Houston, TX, 77002.

Stewco Inc.
Harrison County
Stewco operated a truck washing facility at this site near Waskom from 1976 to 1983. The site is located in a sparsely populated area where the nearest homes and water wells are about half a mile away. Wastes on-site include phthalate sludge, DDT sludge, and PCE. Sludges were held in unlined ponds. Monitoring wells indicate that the groundwater is slightly contaminated with VOCs.

Response actions included fencing, pumping out the evaporation ponds, excavating sludges, backfilling, and capping (1984–1985). A site investigation concluded in 1988 that no further action was needed. Monitoring is continuing, and the site will likely be deleted from the NPL.

Stewco declared bankruptcy in 1983. Five other PRPs have been identified and may be targeted for cost recovery.

Sources: PNPLS-TX; Waskom City Hall, 304 Texas Avenue, Waskom, TX, 75692.

Texarkana Wood Preserving Company
Bowie County
This former wood treatment plant in Texarkana operated from 1909 to 1984. Creosote sludges were held in unlined ponds on the property. Tests have shown that the soil and groundwater are contaminated with creosote derivatives, PNAs, PCP, and dioxins. The contamination has migrated into nearby Day's Creek. About 200 people live within 3 miles of the site.

Response actions included controlling surface runoff and fencing the site (1985–1986). The worst surface contamination was removed by the EPA in five operations between 1986 and 1990. In 1992 a site investigation recommended groundwater extraction and treatment and extraction, incineration, and reburial of contaminated soil and sludge on-site. Local residents objected to the incineration method and compelled Congress to authorize additional studies, which should be completed by 1997.

Fourteen PRPs have been identified for cost recovery.

Sources: PNPLS-TX; Texarkana City Hall, 320 Texas Boulevard, Texarkana, TX, 75501; Texarkana Public Library, Texarkana, TX, 75501.

Tex-Tin Corporation
Galveston County
A former tin-copper smelter, this site is located in a mixed industrial/residential area of Texas City. Wastes were disposed of in ponds, acid ponds, slag heaps, and a landfill on the 170-acre property from 1941 to 1991. Tests have shown that soil and groundwater are contaminated with VOCs, arsenic, chromium, copper, cadmium, lead, nickel, radionuclides, antimony, barium, beryllium, mercury, selenium, and radium. Three different levels of groundwater have been contaminated. Gamma radiation has been measured at the surface. About 22,000 people live within 4 miles of the site.

A site investigation from 1990 to 1993 recommended an NPL listing. This was challenged by Tex-Tin Corporation in Federal Appeals Court, which ruled in favor of Tex-Tin's removal from the NPL. Additional field testing has been done, and the Texas Natural Resource Conservation Commission and the EPA proposed the site for the NPL again in 1996. A final ruling is expected in 1997–1998.

A group of 130 PRPs conducted the original site investigation through an AOC (1990).

Sources: PNPLS-TX; Moore Memorial Library, 1701 Ninth Avenue North, Texas City, TX, 77590; EPA Publication 9320.7-071.

Triangle Chemical Company
Orange County
Located outside Bridge City, this site manufactured and processed chemicals from 1970 to 1981. Liquids were stored in aboveground storage tanks. The soil and groundwater are contaminated with VOCs, eth-

ylbenzene, benzene, and chlorobenzene. About 100 homes are within 3,000 feet of the site.

Response actions included fencing and the consolidation and off-site disposal of contaminated soil (1982). A site investigation completed in 1985 recommended deep-well injection and mechanical aeration of contaminated soil. This work was finished in 1990. Groundwater monitoring is continuing.

The owner abandoned the site and declared bankruptcy in 1981. Four other PRPs are not considered viable for cost recovery.

Sources: PNPLS-TX; Orange Public Library, 220 North Fifth Street, West Orange, TX, 77630.

United Creosoting Company
Montgomery County
United Creosoting Company operated this wood treatment facility in Conroe from 1946 to 1972. The property has since been redeveloped with businesses and about 100 homes. Soils and shallow groundwater are contaminated with PCP and creosote derivatives. About 13,000 people live within 2 miles of the site.

Response actions included a temporary cap. An investigation and feasibility study concluded in 1989 that the site should be remediated with the extraction of free-product creosote derivatives from groundwater, off-site incineration of the separated liquids, and reburial of treated soils on site. Residential yards have been treated and restored. Other actions are in progress.

Ten PRPs will be evaluated for liability and cost recovery.

Sources: PNPLS-TX; Montgomery County Library, 400 North San Jacinto, Conroe, TX, 77301.

UTAH

Hill Air Force Base
Davis and Weber Counties
Aircraft are repaired and maintained on this U.S. Air Force base near Ogden. Various wastes, including electroplating solutions, sludges, oils, and fuel, were buried or discharged on the 6,665-acre property. The groundwater and soils contain VOCs, free-product petrochemicals, and heavy metals. The contamination threatens local agricultural and drinking water supplies.

Response actions included capping (1984, 1986), construction of slurry walls (1985), and installation of interim groundwater extraction and treatment systems. A number of other studies that are nearing completion will recommend final remediation plans.

Sources: PNPLS-UT; Davis County Library, 155 North Wasatch Drive, Layton, UT, 84041.

Kennecott Tailings, South Zone
Salt Lake County
This area of open-pit copper mining near Bingham has been in operation since the 1860s. Surface runoff from piles of waste rock and leaching systems has contaminated Bingham Creek with heavy metals, including lead and arsenic. The groundwater, soils, and stream sediments contain heavy metals, acids, and sulfates. The contamination threatens Bingham Creek, the Salt Lake Valley, and the drinking water supplies of about 75,000 people.

Response actions included removing piles of waste rock and tailings, excavating contaminated soil from neighborhood yards, dredging Bingham Creek and Large Bingham Reservoir, and capping (1991–1994). Groundwater studies in progress will be completed by 1998.

An AOC (1991) and UO (1993) require the PRPs to conduct the removal and capping remedies.

Sources: PNPLS-UT; Magna Public Library, 8339 West 3500 South, Magna, UT, 84044.

Kennecott Tailings, North Zone
Salt Lake County
Located on the shores of Great Salt Lake, this copper milling and smelter operation has been active since 1906. Unregulated smelting activities toxified the surrounding area with metal-laden emissions and dust. Waste solutions and sludges were held in unlined ponds. Groundwater and soils contain heavy metals, including arsenic and selenium. The contamination threatens wetlands, Great Salt Lake, and the water supplies for about 16,000 residents.

Response actions have included cleaning up surface debris and contamination, including about 1 million cubic yards of flue dust. Other studies that will address water treatment are in progress.

Sources: PNPLS-UT; Magna Public Library, 8339 West 3500 South, Magna, UT, 84044.

Midvale Slag
Salt Lake County
This former copper-lead smelter near Midvale operated from 1902 to 1992. Large piles of slag and various smelter wastes remain on the property beside the Jordan River. They contain high levels of heavy metals, including arsenic, cadmium, and lead, which have contaminated the groundwater and soils. The

UTAH

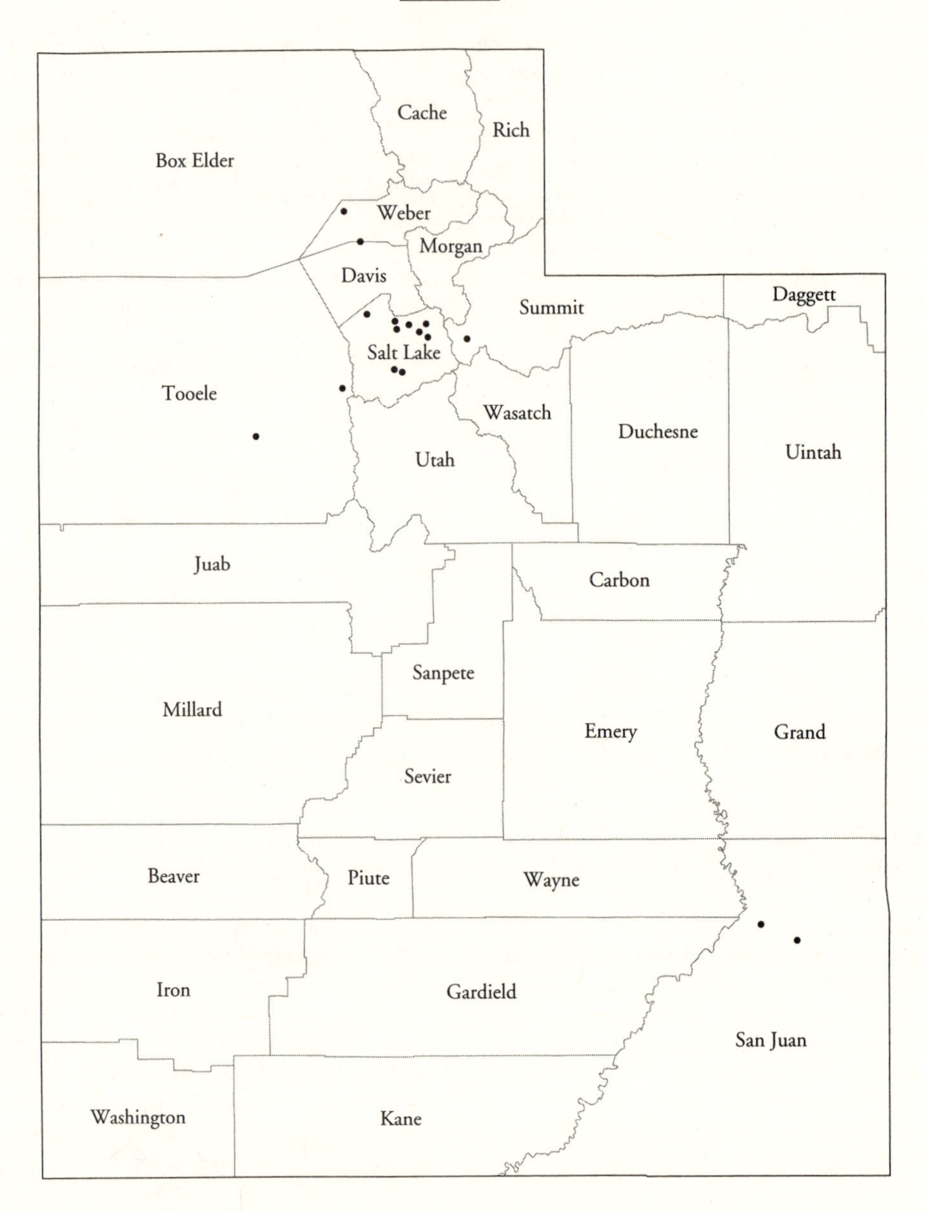

contamination threatens local drinking water supplies and the Jordan River.

Response actions included removing abandoned chemicals and fencing the site (1990). A site investigation was completed in 1994. The EPA will recommend the solidification and reburial of smelter waste, removal of contaminated soil, and capping. Other studies are in progress.

Sources: PNPLS-UT; Ruth Vine Tyler Library, 315 Wood Street, Midvale, UT, 84047.

Monticello Mill
San Juan County
The U.S. Department of Energy supervised this former uranium-vanadium ore milling facility, located near Monticello, which operated from 1942 to 1961. Airborne particles from the vast tailings piles have contaminated about 400 acres. The groundwater and soils contain uranium, thorium, radium, radon, and other radioactive decay products. The contamination threatens Montezuma Creek and the

health and drinking water supplies of about 3,000 residents.

Response actions included a site investigation. In 1990 the DOE recommended removing and stockpiling about 3 million cubic yards of tailings and radioactive waste. Contaminated soil and tailings will also be removed from surrounding properties. All removal actions should be completed by 1998. Groundwater studies are still in progress.

Sources: PNPLS-UT; Abajo Building, 21st and Main Streets, Monticello, UT, 84535.

Monticello Radioactive-Contaminated Properties
San Juan County
A number of properties in Monticello totaling about 4 square miles are contaminated with radioactive waste—a result of tailings from a vanadium-uranium ore mill that were used in the 1960s as backfill or to make concrete. The soils and some concrete structures contain uranium and emit radon gas. The contamination threatens the health of about 2,000 residents in the city of Monticello.

Response actions included a site investigation. In 1989 the EPA recommended removing contaminated materials from the residential properties. To date about 300 of the 410 properties have been cleaned up. This work will continue until about 1997–1998.

Sources: PNPLS-UT; Abajo Building, 21st and Main Streets, Monticello, UT, 84535.

Murray Smelter
Salt Lake County
This former ASARCO lead smelter near Murray operated from 1902 to 1949. Smokestack emissions transported heavy metals over a large area, contaminating soil and residences. Heavy metal–bearing slag piles were left on the property and later used as road base, gravel, and fill. The groundwater and soils contain barium, antimony, cadmium, mercury, lead, selenium, zinc, and arsenic. The contamination threatens Little Cottonwood Creek and public and private wells that supply about 150,000 people.

A site investigation is planned.

Sources: PNPLS-UT; EPA Region 8 Superfund Records Room, 999 18th Street, Suite 500, Denver, CO, 80202; Murray Public Library, 166 East 5300 South, Murray, UT, 84107.

Ogden Defense Depot
Weber County
This U.S. military facility near Ogden, an important supply distribution center for the U.S. Army, has been in operation since the 1940s. Various chemical wastes, including methyl bromide and mustard gas, were stored or buried on the 1,000-acre site. Testing has shown that the groundwater and soils contain VOCs, heavy metals, dioxins, furans, chlordane, and PCBs. The contamination threatens local water supplies, streams, and Pineview Reservoir.

Response actions included removing contaminated soil and defusing buried explosives (1988). The EPA recommended a groundwater extraction and air-stripping system that became operational in 1992. Contaminated soils were removed in 1994. Other studies are in progress.

Sources: PNPLS-UT; Weber County Library, 2464 Jefferson Avenue, Ogden, UT, 84401.

Petrochem
Salt Lake County
This former oil refinery and petroleum recycling center near Salt Lake City operated from 1953 to 1988. Hazardous wastes were also stored on the site. Leaking drums, tanks, and furnaces have contaminated the groundwater and soils with PAHs, phthalates, pesticides, PCBs, furans, dioxins, and heavy metals, including lead and mercury. VOCs have also been detected in air samples. The contamination threatens agricultural wells, wetlands that shelter the threatened bald eagle and endangered peregrine falcon, and private and public wells within 4 miles of the site, which serve about 30,000 residents.

Response actions included removing or stabilizing hazardous wastes on the site (1988). Fencing was also erected. A site investigation (1992–1995) will recommend a final remediation plan.

The EPA identified nearly 500 PRPs. The PRPs formed a steering committee that helped negotiate settlements with the EPA. A 1992 AOC requires the PRPs to conduct the site investigations.

Sources: PNPLS-UT; Marriott Library, University of Utah, Salt Lake City, UT, 84112.

Portland Cement
Salt Lake County
Cement kiln dust and old chromate bricks were discarded in large piles on this site near Salt Lake City until 1983. Heavy metals from these materials, including lead, chromium, molybdenum, cadmium, and arsenic, were spread by wind and surface runoff. Testing has shown that the groundwater and soils contain heavy metals and that the groundwater is acidic. The contamination threatens the health of about 12,000 people within 1 mile of the site, the Jordan River Canal, City Drain Pond, and associated wildlife.

Response actions included fencing and the periodic application of a dust suppressant. In 1990 the EPA recommended removing the kiln dust, bricks, and contaminated soils. The removal actions are in progress. A groundwater decision is pending.

A cash settlement was negotiated with the PRP, Lone Star Industries.

Sources: PNPLS-UT; Salt Lake City Library, 577 South 900 West, Salt Lake City, UT, 84101.

Richardson Flat Tailings

Summit County

Spent tailings piles from the Keetley Ontario Mine and United Park City Mines were generated in the mid-1900s on this site near Park City. The most recent mining activities occurred from 1975 to 1981. Tailings materials are slumping into Silver Creek, a cold water fishery. Testing has revealed that groundwater, soils, tailings, and surface water contain a variety of heavy metals, including mercury, silver, zinc, calcium, lead, copper, arsenic, and cadmium. Contaminated dust from the tailings piles is also being carried by wind. The contamination is a threat to Silver Creek, its wetlands and wildlife habitats, and the water supplies of about 5,000 nearby residents.

Detailed site investigations are in progress.

Source: PNPLS-UT.

Rose Park Sludge Pit

Salt Lake County

This former sludge disposal pit near Salt Lake City was filled with petroleum wastes from the 1920s to 1957. Wastes included acidic refinery sludges that contain PAHs and sulfur dioxide. The site was later developed into a recreational park with ball fields, tennis courts, and a golf course. The contamination is a threat to the health of local residents and to groundwater supplies.

Response actions included constructing a clay cap and slurry wall (1983) and revegetation (1984). Groundwater monitoring began in the late 1980s and will continue until 2015. The response actions have remediated the groundwater, and the EPA is considering deleting this site from the NPL.

Sources: PNPLS-UT; Utah State Department of Environmental Quality, Division of Environmental Response and Remediation, 168 North 1950 West, First Floor, Salt Lake City, UT, 84116.

Sharon Steel (Midvale Tailings)

Salt Lake County

This former copper, lead, and zinc mill and smelter in Midvale operated from 1905 to 1971. Ore processing resulted in the generation of about 10 million tons of waste rock tailings. Lead-contaminated sand from the tailings was used in local gardens and playground sandboxes. Testing has shown that the groundwater and soils contain lead, iron, arsenic, zinc, magnesium, and cadmium. The heavy metals have also entered the recreational Jordan River and its wetlands. The contamination threatens the health and water supplies for the greater Salt Lake City area.

Response actions included fencing (1989) and razing structures on the site (1992). In 1993 the EPA recommended groundwater monitoring and the consolidation and capping of contaminated soils and tailings. Construction is under way. Since 1993 contaminated soil from about 400 residences has been excavated and the yards replaced with clean fill and landscaped. The treatment of additional yards should be completed by 1997.

Sources: PNPLS-UT; Ruth Vine Eyler Library, 315 Wood Street, Midvale, UT, 84047.

Tooele Army Depot (North Area)

Tooele County

This U.S. Army property near Tooele is one of the largest ammunition storage and equipment and vehicle maintenance bases in the United States. Various wastes, including sludges, heavy metals, degreasers, solvents, chemical agents, fuels, and munitions, have been buried or discharged in unlined lagoons since 1942. Testing has shown that the groundwater and soils contain VOCs, chromium, PCBs, explosives derivatives, nitrates, and lead. The contamination is a threat to water wells that supply about 16,000 residents.

Response actions included constructing an interim groundwater and air-stripping system (1994), removing drummed waste (1994), and excavating areas of contaminated soil (1995–1997). Other studies are in progress.

Sources: PNPLS-UT; Tooele City Public Library, 47 East Vine Street, Tooele, UT, 84074.

Utah Power & Light

Salt Lake County

A coal gasification plant operated on this site near Salt Lake City from 1873 to 1908. The property was later used by Utah Power and Light Company to treat utility poles with creosote and to store barrels that once contained VOCs, degreasers, and solvents. Spills, leaks, and discharges from these activities contaminated the soil and groundwater with PAHs, coal tars, VOCs, cyanide, lead, herbicides, and pesticides. The contamination threatens local drinking water supplies and streams.

Response actions included removing about 50,000 empty barrels (1988). A site investigation was conducted from 1990 to 1993. The EPA recommended soil vapor extraction and the excavation of coal tars and creosote-contaminated soils for recycling into road asphalt. Groundwater monitoring will continue to assure that the contamination plume is naturally attenuating. Clean-up work should be finished by 1997.

AOCs (1990, 1993) and a Consent Decree require the PRPs to conduct the site investigations and remediation work.

Sources: PNPLS-UT; Utah Department of Environmental Quality, 1950 West North Temple, Salt Lake City, UT, 84114.

Wasatch Chemical Company
Salt Lake County
Pesticides, herbicides, and other industrial chemicals were manufactured on this site near Salt Lake City from the 1950s to the 1990s. Various wastes and sludges were stored in steel drums and tanks or discharged into a concrete-lined holding pond and possibly into a ditch system that connects to Great Salt Lake. Groundwater and soils contain VOCs, herbicides, pesticides, and dioxins. The contamination threatens Great Salt Lake and the drinking water supplies for about 90,000 nearby residents.

Response actions included removing tanks, drums, and cylinders and installing a temporary cap (1986). A site investigation (1988–1993) recommended the consolidation and in-situ vitrification of sludge and contaminated soil and the extraction and air-stripping of groundwater. Treatment systems are either operating or being constructed.

A Consent Order (1986) and Consent Decrees (1988, 1991) require the PRPs to conduct the site investigations and clean-up work.

Sources: PNPLS-UT; Salt Lake City Library, 577 South 900 West, Salt Lake City, UT, 84101.

VERMONT

Bennington Municipal Sanitary Landfill
Bennington County
A former gravel pit, this landfill operation was active from 1969 to 1985. A variety of municipal and industrial waste was accepted, including organic solvents, which were discharged into an unlined lagoon on the property. Soil and groundwater are contaminated with lead, PCBs, arsenic, and VOCs (benzene and xylene). Surface runoff flows into a nearby wetland and trout fishing stream. About 2,000 people who use private wells live within 3 miles of the site.

Response actions included building runoff and leachate control systems, consolidating contaminated soil in the landfill, soil capping, and fencing (1994). A site investigation (1991–1996) will recommend a final remediation plan shortly.

The PRP is financing the site investigation, and will participate in the clean-up actions.

Sources: PNPLS-VT; Bennington Public Library, 101 Silver Street, Bennington, VT, 05201.

BFI Landfill
Windham County
A former gravel pit, this site near Bellow Falls operated as an unlined landfill from 1968 to 1991. Various municipal and industrial wastes were accepted, including heavy metals, pesticides, and VOCs. Contamination was noticed in nearby residential wells as early as 1979. Soils and groundwater contain VOCS, chromium, copper, and lead. Surface runoff and the contaminated groundwater have penetrated area wells and threaten the water source of about 3,000 people within 1 mile of the site; the Connecticut River is also threatened.

Response actions included connecting affected residents to city water, installing a gas collection system (1989), and constructing runoff and leachate control systems (1993). A site investigation (1992–1994) concluded that the contamination plume will naturally attenuate over time, using the controls already installed. Long-term groundwater monitoring will continue. In 1994 controls were expanded and the landfill capped.

The two PRPs, BFI, Inc., and Disposal Specialists, Inc., have agreed to finance the site investigations and clean-up work through an Administrative Order on Consent (1992).

Sources: PNPLS-VT; Rockingham Free Public Library, 65 Westminster Street, Bellow Falls, VT, 05101.

Burgess Brothers Landfill
Bennington County
This gravel pit near Bennington was operated as a dump from the 1940s to the 1970s; it is still active as a sand pit and salvage yard. Union Carbide Corporation buried battery manufacturing waste, including lead sludge, in unlined pits, Tests by the EPA indicate that the soil and groundwater contain high levels of heavy metals (mercury and lead) and VOCs (vinyl chloride and trichloroethylene). About 14,000 people live within 3 miles of the site, and many use local private and public wells. The contamination also threatens Barney Brook, the Waloomsic River, and adjacent wetlands in the Green Mountain National Forest.

VERMONT

A site investigation begun in 1991 should be completed in 1997. The report will include recommendations for a final remediation plan.

The PRPs are financing the site investigation and feasibility study.

Sources: PNPLS-VT; Bennington Public Library, 101 Silver Street, Bennington, VT, 05201.

Darling Hill Dump

Caledonia County

This unlined landfill located near Lyndonville operated from 1952 to 1989. It received a variety of municipal and industrial wastes, including metal, demolition materials, liquid industrial wastes, plating solutions, alkali degreasers, and organic solvents. Soil and groundwater are contaminated with low levels of VOCs. Private and city wells within 3 miles of the site serve about 4,000 residents. The contamination also threatens the recreational Passumpsic River.

Response actions included the installation of a carbon filtration system on the city water system in Lyndonville. A 1992 report by the EPA indicated no further actions were necessary. Groundwater monitoring is continuing.

Consent Orders signed in 1989 required the PRPs to perform the site investigation and install the carbon filtration system.

Sources: PNPLS-VT; Town Hall, Town of Lyndon, 24 Main Street, Lyndonville, VT, 05851.

Old Springfield Landfill

Windsor County

This landfill in the town of Springfield operated from 1947 to 1968. It accepted a variety of municipal and industrial wastes, including liquid waste and sludges. The site was later developed into a mobile home park. EPA tests in 1976 showed that the soil and groundwater are contaminated with VOCs (benzene and vinyl chloride), PCBs, and PAHs. About 500 people, many with private wells, live within 1 mile of the site. Nearby springs and wells have also been contaminated. Leachate seepage and surface runoff migrate into the recreational Black River and Seavers Brook.

Response actions included connecting affected residents to city water supplies (1984) and purchasing homes on the site (1990). In 1992 the EPA recommended construction of groundwater, runoff, and leachate collection systems; installation of a groundwater extraction and treatment system and gas venting system; and capping. All were finished in 1993. Groundwater monitoring is continuing.

A 1984 AO required the PRPs to furnish alternative water supplies. The PRPs financed the various site studies and clean-up work through two CDs (1990, 1991) with the EPA.

Sources: PNPLS-VT; Springfield Public Library, Main Street, Springfield, VT, 05156.

Parker Sanitary Landfill
Caledonia County
This solid waste landfill near Lyndon operated from 1972 to 1992. A variety of municipal and industrial waste was received, including metal-plating solutions, oil, electroplating sludges, paint waste, chlorinated solvent sludges, and metallic salts. The liquid waste was poured on the ground or into unlined trenches. Testing in 1984 revealed that the soil and groundwater are contaminated with VOCs, including trichloroethylene. Nearby wells and streams were also polluted. Over 125 wells, including a municipal well field that serves about 3,200 people, are located within 3 miles of the site.

Response actions included connecting affected residents to city water. A site investigation completed in 1995 recommended landfill capping and groundwater extraction and treatment.

The EPA is negotiating with PRPs to fund the clean-up work.

Sources: PNPLS-VT; Cobleigh Public Library, 70 Depot Street, Lyndonville, VT, 05851; Lyndonville State College Library, Lyndonville, VT, 05851.

Pine Street Canal
Chittenden County
This site is located in Burlington. From 1908 to 1966 a coal gasification plant operated near the Pine Street Canal and a former wetland known as Maltex Pond. Waste solutions and oil from the plant were poured into the Maltex Pond area. Petrochemical seepage was later detected in Pine Street Canal. Tests show that the soil and groundwater are contaminated with PAHs, cyanide, VOCs (benzene, toluene, and xylene), and heavy metals (lead). A number of businesses and private homes are within 2 miles of the site. The site is highly accessible and prone to flooding.

Response actions included fencing, removing coal tar, and capping the filled-in Maltex Pond wetland with clay (1985). Detailed EPA studies (1985–1992) recommended a clean-up plan that has since been withdrawn in favor of additional studies.

PRPs have been identified for possible cost recovery or participation in the remediation work.

Sources: PNPLS-VT; Fletcher Free Public Library, 235 College Street, Burlington, VT, 05401; University of Vermont Library, Burlington, VT, 05401.

Tansitor Electronics, Inc.
Bennington County
Located in a rural part of Bennington, Tansitor Inc. builds electronic capacitors. Liquid wastes, oils, and acid sludges were either discharged on the ground or into a nearby waterway. Tests by the state showed that soils and groundwater are contaminated with boron, VOCs, and heavy metals. Drinking water wells within 3 miles of the site serve about 1,500 residents in Vermont and New York. The runoff and seepage have contaminated recreational Brown's Creek.

Site investigations are in progress and will determine a recommended remediation plan.

Sources: PNPLS-VT; Bennington Public Library, 101 Silver Street, Bennington, VT, 05201.

VIRGIN ISLANDS (U.S.)

Island Chemical Corporation
St. Croix
Pharmaceutical chemicals, perfumes, flavorings, polishes, inks, and resins are manufactured on this site in St. Croix. Various wastes were held in a processing pit on the property, or discharged into the River Gut. Testing in 1985–1986 revealed that the groundwater and soils contain phthalates, VOCs, chloroform, PAHs, heavy metals, and pesticides. The contamination threatens local drinking water supplies and the River Gut.

Response actions included removing drummed chemicals (1989–1991). A site investigation is planned.

Source: PNPLS-VI.

Tutu Well Field
Tutu
Contaminated groundwater was discovered in part of the Tutu Well Field in the 1980s. Well sampling revealed a 108-acre area of groundwater contamination with high levels of VOCs. Point sources identified to

VIRGIN ISLANDS (U.S.)

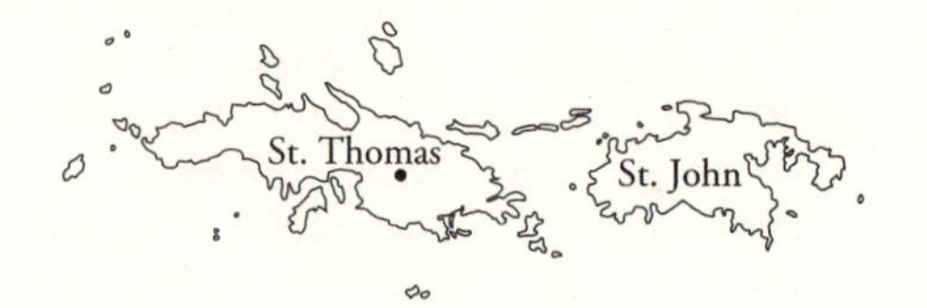

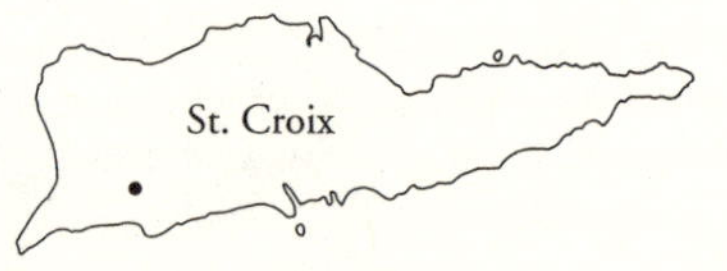

date include gasoline stations, auto repair shops, and dry cleaners that used VOC-containing solutions, degreasers, strippers, and cleaning fluids. The contamination threatens Turpentine Run, Mangrove Lagoon, the Caribbean Sea, and the local drinking water supplies for about 2,000 residents.

Response actions included closing the polluted wells, connecting affected residents to clean water supplies, and long-term monitoring. A site investigation (1992–1996) will recommend a final remediation plan.

Administrative Orders (1987), UAOs (1990), and AOs (1992) require three PRPs to conduct the site investigations and clean-up work.

Source: PNPLS-VI.

VIRGINIA

Abex Corporation
Portsmouth
Abex Corporation was a brass and bronze foundry that operated in Portsmouth from 1928 to 1978. Various solid and liquid foundry wastes, including lead-bearing furnace sands, were buried or dumped on the property. In 1984 testing by the EPA showed that the soil, groundwater, and air are contaminated with lead, copper, and tin. Heavy metals were also detected in nearby homes. The contamination threatens the health of about 10,000 people who live or work within 1 mile of the site.

Response actions included paving the site and erecting fencing (1988). Contaminated soil was removed in 1992. Site investigations from 1989 to 1992 recommended demolition of the foundry structures and the removal of the deeper contaminated soil. This work should be under way by 1997.

Legal agreements between the EPA and Abex require the company to participate in the site studies and remediation work.

Sources: PNPLS-VA; Portsmouth Public Library, 601 Court Street, Portsmouth, VA, 23704.

VIRGINIA

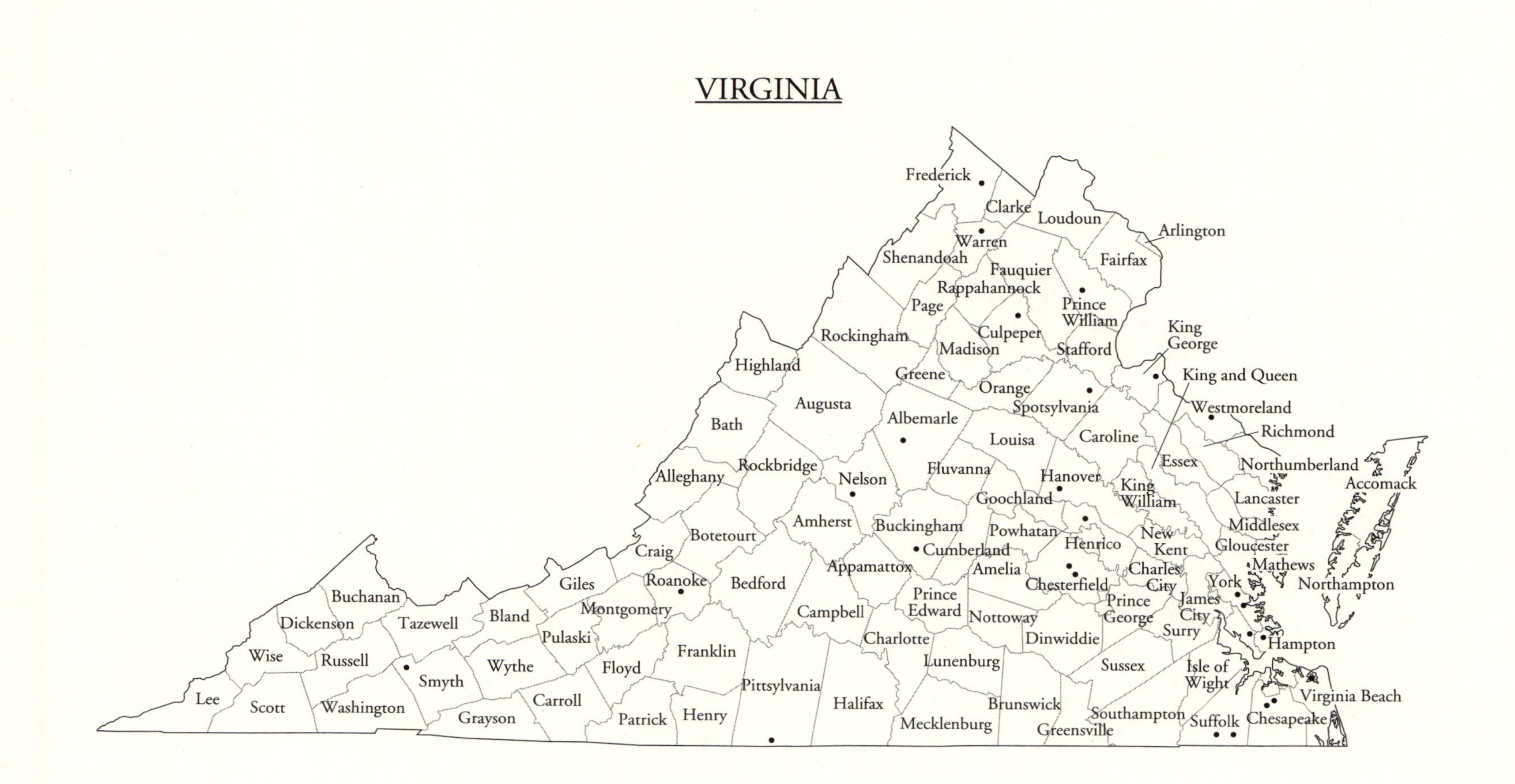
Frederick
Clarke
Loudoun
Arlington
Warren
Shenandoah
Fairfax
Fauquier
Rappahannock
Page
Prince William
Culpeper
Rockingham
Madison
Stafford
King George
Highland
Greene
King and Queen
Orange
Augusta
Spotsylvania
Westmoreland
Albemarle
Bath
Louisa
Caroline
Richmond
Essex
Northumberland
Rockbridge
Alleghany
Fluvanna
Nelson
Hanover
King William
Accomack
Goochland
Lancaster
Amherst
Buckingham
Powhatan
Middlesex
Botetourt
Henrico
New Kent
Gloucester
Craig
Cumberland
Appamattox
Amelia
Charles City
Mathews
Giles
Roanoke
Bedford
Chesterfield
York
Northampton
Buchanan
Prince Edward
Prince George
James City
Montgomery
Campbell
Dickenson
Tazewell
Bland
Nottoway
Surry
Pulaski
Charlotte
Dinwiddie
Hampton
Franklin
Wise
Russell
Lunenburg
Wythe
Floyd
Sussex
Isle of Wight
Smyth
Pittsylvania
Lee
Carroll
Halifax
Brunswick
Virginia Beach
Scott
Washington
Henry
Southampton
Grayson
Patrick
Mecklenburg
Suffolk
Chesapeake
Greensville

Arrowhead Associates
Westmoreland County
An electroplating company operated on this site near Montross from 1966 to 1979. Other companies have assembled and packed cosmetic cases since 1981. Various wastes from the electroplating operations were discharged or buried on the site. Testing has shown that the soils and groundwater contain high levels of VOCs, heavy metals, and cyanide. Private water wells within 3 miles of the site serve about 1,200 people. The contamination also threatens a coastal wetland and nearby recreational waterways.

Response actions included removing drummed wastes, tanks, and contaminated soils, sludges, and wastewater (1986–1990). In 1991 an EPA investigation recommended a remediation plan of groundwater extraction and treatment through air-stripping and carbon adsorption along with soil vapor extraction. Construction is expected to be under way by 1997.

Scovill Corporation, the PRP, signed Consent Orders (1986, 1989) and a Consent Decree (1994) that require the company to fund the site investigations and remediation work.

Sources: PNPLS-VA; Assistant County Administrator, Westmoreland County, Social Services Building, Peachgrove Lane, Montross, VA, 22520.

Atlantic Wood Industries
Portsmouth
This former wood treatment plant near Portsmouth operated from 1926 to 1992. Timber was treated with creosote and PCP. Liquid wastes were either discharged on the property or held in leaking tanks. Testing has shown that the groundwater and soils contain creosote, PCP, and heavy metals. About 15,000 people live or work within 1 mile of the site. The contamination has penetrated the Elizabeth River. Oysters in the river contain high levels of creosotes.

Response actions included removing contaminated water from leaking storage tanks (1982). Contaminated soils were removed in 1995. Ongoing site investigations will address remedies for groundwater and soil contamination.

A 1987 Consent Order requires the PRPs to fund the site investigations and remediation work.

Sources: PNPLS-VA; Portsmouth Public Library, 601 Court Street, Portsmouth, VA, 23704.

Avtex Fibers
Warren County
Companies on this site near Front Royal manufactured rayon materials from 1940 to 1989. Various solid and liquid wastes were buried or discharged in pits on the property. In 1982 the state detected carbon disulfide in a nearby well. The present owner, Avtex Fibers, was shut down by the state in 1989 for illegally discharging wastes into the Shenandoah River. The groundwater and soils at the plant contain high levels of carbon disulfide, phenol, heavy metals, sodium, and PCBs. The contamination threatens the Shenandoah River and the private water wells of about 1,500 people in the area.

Response actions included supplying clean water to affected residents (1984). In 1989 toxic chemicals and solutions were removed from the plant, and the site was stabilized. Additional removal actions, including excavating sludges and PCB-contaminated soil, were conducted from 1994 to 1996. A groundwater extraction and treatment system will also be installed. Ongoing studies will address the clean up of the Shenandoah River.

A series of legal agreements (1986–1993) between the EPA and PRPs requires the companies to fund the site studies and clean-up work.

Sources: PNPLS-VA; Samuels Public Library, 538 Villa Avenue, Front Royal, VA, 22630.

Buckingham County Landfill
Buckingham County
This landfill site near the town of Buckingham operated from 1962 to 1983. It received a variety of municipal and industrial wastes, including chemical solutions from the furniture manufacturing industry. Studies in the 1980s showed that the groundwater and soil contain VOCs. The contamination threatens local water wells that supply about 1,200 residents within 3 miles of the site.

In 1994 the EPA recommended soil vapor extraction, capping, groundwater monitoring, and possibly groundwater extraction and treatment should the contamination plume expand.

Sources: PNPLS-VA; Buckingham County Library, Route 2, Box 41B, Dillwyn, VA, 23936.

C & R Battery Company
Chesterfield County
This company recycled lead from old batteries from the 1970s to 1985. Testing in 1982 showed high levels of lead in the groundwater and soil. Lead was also detected in drainage ditches leading to the James River. The contamination threatens nearby wetlands, the James River, and private water wells within 3 miles of the site, which serve about 1,200 people. High levels of lead were also found in some employees who worked in the plant.

Response actions included treating acidic soil with lime, excavating and storing soil, capping, and fencing (1986). In 1990 the EPA recommended a final remediation plan of stabilizing contaminated soil, closing the acid pond, and revegetation. Long-term groundwater monitoring has shown that the contamination plume is naturally attenuating.

Enforcement actions by the State of Virginia between 1979 and 1984 led to site reclamation and the construction of a treatment plant. C & R Battery declared bankruptcy in 1985.

Sources: PNPLS-VA; Chesterfield Public Library, 9501 Lori Road, Chesterfield, VA, 23832.

Chisman Creek
York County
This former gravel pit was filled with fly ash from the Yorktown Power Generating Station from 1957 to 1980. Testing in 1980 showed that the groundwater and soil contain nickel, vanadium, selenium, and sulfate. Nearby residential wells were contaminated with vanadium and shut down. The contamination threatens the local drinking water supplies of about 1,000 people as well as nearby wetlands.

Response actions included connecting affected residents to clean water supplies, covering pits, and installing groundwater barriers. Erosion controls were completed in 1989. A groundwater extraction and treatment system was also installed, which has significantly lowered metal values in the groundwater.

A Consent Decree was signed with Virginia Power Company to fund the site investigations and clean-up work.

Sources: PNPLS-VA; York County Public Library, 8500 George Washington Parkway, Yorktown, VA, 23692.

Culpeper Wood Preservers
Culpeper County
This wood treatment plant near Culpeper has operated since at least the 1970s. Lumber is soaked in a chromated arsenate solution and left to dry. Drippings have soaked into the ground. Wastewater was held in an unlined lagoon that occasionally overflowed into neighboring surface waters. The soil and groundwater contain copper, arsenic, and chromium. The contamination threatens nearby recreational waterways and water wells that serve about 2,000 local residents.

Response actions included removing contaminated soil and improving surface controls (1981). Ongoing site investigations by the EPA are nearing completion and will recommend a final remediation plan in 1996–1997.

Two Consent Agreements and a Consent Order require the PRPs to participate in the site studies and reclamation work.

Sources: PNPLS-VA; Culpeper Town and County Library, 605 South Main Street, Culpeper, VA, 22701.

Defense General Supply Center
Chesterfield County
This U.S. military property near Richmond has been in use since 1942. Activities have included hazardous waste disposal, fire-fighting training, acid neutralization, and managing, storing, and processing military supplies. Testing has shown that the groundwater and soils contain VOCs, PAHs, heavy metals, and pesticides. The contamination threatens recreational Kingsland Creek and the drinking water supplies for nearby residents.

Response actions included soil vacuum extraction (1993–1995). A groundwater extraction and air-stripping/carbon adsorption treatment system should be constructed by 1997. Other studies are in progress.

Sources: PNPLS-VA; Chesterfield Public Library, 9501 Lori Road, Chesterfield, VA, 23832.

Dixie Caverns County Landfill
Roanoke County
This former landfill near the town of Salem operated from 1965 to 1976. It received a variety of municipal and industrial wastes, including metal, sludge, and fly ash. Runoff from the site flows into the Roanoke River. Testing has shown that the soils contain PAHs and other organic chemicals. The contamination threatens the Roanoke River and the private drinking water supplies of about 2,500 nearby residents.

Drummed waste and contaminated soil were removed by the EPA in 1983. Fly ash was removed and treated off-site in 1995–1996.

Sources: PNPLS-VA; Roanoke County Public Library, Glenvar Branch Library, 8917 Daugherty Road, Salem, VA, 24153.

First Piedmont Rock Quarry
Pittsylvania County
This former rock quarry was used as a landfill from 1970 to 1972. It received a variety of waste, including toxic liquids from the Goodyear Tire & Rubber Company. The groundwater and soil contain high amounts of heavy metals (arsenic, chromium, iron, and manganese). The metals have also been found in nearby wells. The contamination threatens recreational Lawless and Fall Creeks and drinking water supplies of about 2,000 nearby residents.

Response actions included detailed site investigations from 1987 to 1991. The final remediation plan consists of capping, leachate collection, and removing contaminated soils. Construction is expected to be completed by 1997.

First Piedmont, Corning Glass Works, and Goodyear Tire and Rubber, the PRPs, signed a CO

(1987) and AO (1992) with the EPA that require the companies to fund the clean-up work.

Sources: PNPLS-VA; Pittsylvania County Public Library, 24 Military Drive, Chatham, VA, 24531.

Fort Eustis

York County

This U.S. Army base in the city of Newport News began operations in 1918. Activities have included training troops and supporting rail, marine, and amphibious transportation systems. Equipment and vehicles were also repaired and maintained. Various solid and liquid wastes were disposed of in pits and landfills on the 8,300-acre property. Testing between 1987 and 1990 revealed that the soil and groundwater contain pesticides (chlordane and DDT), PCBs, heavy metals, and PAHs. The contaminants have penetrated nearby waterways, including Bailey's Creek, Brown's Lake, and Milstead Island Creek. The contamination threatens the health of about 20,000 people who live or work in the area and the wetlands and tidal marshes of the James River and Chesapeake Bay.

Response actions included capping landfill areas (1994). Detailed site investigations that will recommend a final remediation plan are in progress.

Source: PNPLS-VA.

Greenwood Chemical Company

Albemarle County

The Greenwood Chemical Company manufactured industrial chemicals, pesticides, and pharmaceuticals from 1945 to 1985. Wastes were buried on site in drums or discharged into unlined ponds. Tests by the EPA showed that the soil and groundwater contain VOCs, heavy metals, and cyanide. The contamination threatens Stockton Creek and nearby water wells that serve about 1,700 residents.

Response actions included removing buried drums (1987), excavating sludge, and razing buildings (1992–1993). Site investigations by the EPA recommended stabilization or off-site incineration of contaminated soil and extraction and UV/oxidation treatment of groundwater. Construction should be completed by 1997. Additional soil studies are in progress.

Sources: PNPLS-VA; Jefferson-Madison Library, 201 East Market Street, Charlottesville, VA, 22553.

H & H Burn Pit

Hanover County

This landfill site near Farrington was used by Haskell Chemical Company from 1960 to 1976. Haskell burned liquid manufacturing wastes in a shallow pit. In 1984 the EPA detected PCBs on neighboring properties. Testing showed that the groundwater and soils contain excessive levels of VOCs (vinyl chloride), PCBs, barium, chromium, and beryllium. The contamination threatens nearby recreational waterways and wetlands and water wells within 3 miles of the site, which serve 2,500 people.

Response actions included removing contaminated soil and drums in 1982. A recently completed site investigation by the EPA will issue a final remediation plan by 1997–1998.

The PRPs, Haskell Chemical Company and H & H Incorporated, conducted the removal actions.

Sources: PNPLS-VA; Pamunkey Regional Library, 102 South Railroad Avenue, Ashland, VA, 23005.

L. A. Clarke & Son

Spotsylvania County

This wood treatment facility near Fredericksburg operated from 1937 to 1988. Timbers were treated with creosote. Creosote and creosote-tainted wastewater leaked, spilled, or was discharged onto the ground or into unlined drainage ditches. Excess water was sprayed on dirt roads to control dust. Testing has shown that the soil, groundwater, and surface water contain creosote derivatives, benzene, and PNAs. The contamination threatens the drinking water for about 1,600 residents within 1 mile of the site as well as recreational Massaponax Creek, West Vaco Pond, and Ruffins Pond.

An EPA study completed in 1988 recommended in-situ soil flushing and soil biodegradation (landfarming). Buildings were demolished in 1993. Removal of sludge from the lagoon should be completed by 1997. A groundwater remedy is still being studied.

The EPA signed a CD (1989) with PRP RF & P Railroad to conduct the first phase of the clean-up work.

Sources: PNPLS-VA; County Administrator's Office, 9104 Courthouse Road, Spotsylvania, VA, 22553.

Langley Air Force Base

Hampton County

This U.S. Air Force base and the adjacent Langley Research Center cover about 4,000 acres. Airfield and aeronautical research has been conducted at the site since 1917. Various wastes, including oils, solvents, pesticides, paint wastes, PCB- and PCT-containing fluids, batteries, photographic chemicals, compressors, and heavy metals, were buried or discharged in various locations on the property. Soils and groundwater contain heavy metals, phenols, PCBs, PCTs, oils, mercury, and pesticides. The contamination threatens the health of about 15,000 workers and the recreational Black River and Chesapeake Bay. Important fisheries and crab beds have already been damaged.

Detailed site investigations are in progress and will recommend a final remediation plan.

Sources: PNPLS-VA; Hampton Central Library, 4207 Victoria Boulevard, Hampton, VA, 23669; Poquoson Public Library, 774 Poquoson Avenue, Poquoson, VA, 23662.

Marine Corps Combat Development Command
Prince William, Stafford, and Fauquier Counties
This U.S. military facility near Quantico began operations in 1917. It has been used primarily as a training facility and for research and development on military weapons and defense systems. Various wastes, including oils from transformers, paints, acids, solvents, fuel, and electroplating sludges, were buried on the property. Testing has shown that the soils and groundwater contain VOCs, PCBs, pesticides, and heavy metals, especially arsenic. The contamination threatens the health of base personnel and has penetrated the Potomac River.

Site investigations and groundwater monitoring are in progress and will recommend a final remediation plan.

Source: PNPLS-VA.

Naval Surface Warfare Center
King George County
This U.S. Navy facility near Dahlgren has been in operation since 1918. Its primary purpose is to research, develop, and test weaponry. Various wastes, including explosives derivatives, fuels, mercury, and transformer oils, were buried or discharged into landfills on the 2,700-acre base. A testing range is also on the property. A sampling program revealed that the soil and groundwater contain mercury, PCBs, pesticide derivatives, and PAHs. The contamination has penetrated recreational Hideaway Pond and the Potomac River. The contamination also threatens Gambo Creek, the Potomac River, and the drinking water supplies for about 7,000 residents.

Response actions included removing contaminated soil in 1994. Detail site investigations are in progress and will recommend a final remediation plan.

Sources: PNPLS-VA; Smoot Memorial Library, Route 3, King George, VA, 22485.

Naval Weapons Station, Yorktown
York County
This 10,500-acre U.S. military facility began operations in 1918 to maintain, produce, and store weapons, explosives, and other military equipment. Various wastes, including carbon batteries, inert explosives, electric shop hardware, fly ash, mine casings, transformers, burn pad residues, and heavy metals, were buried on the property. Testing in the 1980s showed that the groundwater and soils contain VOCs, heavy metals (cadmium and mercury), TNT, and other explosives compounds. The contamination threatens the recreational York River, Chesapeake Bay, Whiteman Swamp, and the health of about 3,300 military personnel.

Response actions included removing various hazardous debris from the on-site landfills (1992–1995). Contaminated soil from the transformer area was excavated in 1994. Other site investigations are in progress.

Sources: PNPLS-VA; York County Library, 8500 George Washington Highway, Yorktown, VA, 23692.

Rentokil Company
Henrico County
This former wood treatment plant near Richmond operated from 1956 to 1990. Lumber was treated with creosote, PCP, xylene, ammonium phosphates, sulfates, and chromated copper-arsenate solutions. Chemical wastes were held in an unlined lagoon from 1956 to 1974. Testing by the EPA in the 1980s revealed that the soil and groundwater contain excessive levels of PCP, creosote derivatives, heavy metals, and dioxin. The contamination threatens nearby wetlands, recreational North Run, and the private water supplies of about 400 local residents.

Response actions included connecting affected residents to clean water supplies (1987), removing sludge (1987) and structures (1991), and building runoff controls (1992). In 1993 the EPA recommended excavation of contaminated sediments, off-site incineration, capping, excavation and low-temperature thermal desorption of soil, construction of slurry walls, and groundwater monitoring. Construction should be under way by 1997.

PRPs Virginia Properties and Rentokil signed Consent Orders (1987, 1992, 1994) to conduct the remediation work.

Sources: PNPLS-VA; Henrico County Public Library, 1001 North Laburnum Avenue, Richmond, VA, 23223.

Rhinehart Tire Fire Dump
Frederick County
This tire dump near Winchester caught fire in 1983. Over five million tires created a smoke plume that polluted the air over four states. Hot oil from the burning rubber and firefighting chemicals flowed into Massey Run. Testing has shown that the groundwater and soils contain heavy metals, VOCs, oils, and PAHs. The chemicals have penetrated Massey Run,

Hogue Creek, and the Potomac River. The contamination threatens the drinking water supplies for thousands of residents in the region.

Response actions included removing oily waste from the burned tires in 1983. Runoff controls were also constructed. The EPA recommended a final remediation plan of runoff controls and collection of free-product petroleum. Other studies in progress will address groundwater clean up.

An AO (1984) and ACO (1989) with the PRPs require the parties to participate in the site clean up.

Sources: PNPLS-VA; Handley Library, 100 West Piccadilly Street, Winchester, VA, 22601.

Saltville Waste Disposal Ponds
Smyth and Washington Counties
This site consists of two former wastewater ponds near Saltville that held mercury-laden wastewaters from a nearby chemical company. The soils and groundwater are contaminated with mercury, which has damaged the North Fork of the Holston River. The contamination threatens the private water wells of nearby residents and the Holston River, which hosts two endangered aquatic species.

Response actions included dredging the river to remove mercury-contaminated sediments (1982). Contaminated soil was removed in 1991. Runoff controls and diversion ditches were built in 1991. A water treatment plant will be operational by 1997. Other studies are in progress.

Legal agreements with the PRPs required the companies to dredge the Holston River and monitor various criteria from 1982 to 1988. A 1988 Consent Decree requires them to fund the clean-up work.

Sources: PNPLS-VA; Saltville Town Hall, Saltville, VA, 24370.

Saunders Supply Company
Suffolk County
This former wood treatment plant near Chuckatuck operated from at least the 1960s to 1991. Lumber was treated with PCP, fuel oil, and chromated copper arsenates. Wastes were leaked, spilled, or burned on-site. In 1984 the EPA detected high levels of arsenic, chromium, PCP, copper, and dioxins in the groundwater and soil. The contamination has penetrated Godwin's Mill Pond Reservoir, a drinking water source for about 30,000 Suffolk residents.

Response actions included removing contaminated soil and silt and constructing a groundwater pump-and-treat system. In 1991 the EPA recommended low-temperature thermal desorption of soil, surface clean up, and long-term groundwater monitoring. Construction should be under way by 1997.

The PRPs conducted the response actions.

Sources: PNPLS-VA; Suffolk Public Library, 443 West Washington Street, Suffolk, VA, 23434.

Sewells Point Naval Complex
Norfolk
This U.S. Navy base near Norfolk supports various naval operations, including ship maintenance and repair. Halogenated solvents, acids, paint wastes, electroplating solutions, petroleum, oils, lubricants, asbestos, degreasers, solvents, and lubricants are some of the common waste products that have been buried or discharged on the 4,631-acre base. Recent testing has shown that the groundwater and soils contain VOCs, heavy metals, PCBs, pesticides, and SVOCs. The contamination threatens local drinking water supplies, streams, wetlands, the Elizabeth River, Willoughby Bay, Chesapeake Bay, and important coastal habitats and fishing grounds.

This site was proposed to the NPL in 1996; a final decision is expected in 1997–1998. Additional studies are planned.

Source: EPA Publication 9320.7-071.

Suffolk City Landfill
Suffolk County
This former landfill in Suffolk operated from 1967 to 1984. It received a variety of municipal waste, including highly toxic pesticides. The EPA detected pesticides in the groundwater and soils. The contamination threatens Great Dismal Swamp and the drinking water of about 3,000 local residents.

Response actions included the installation of a leachate collection system (1991). A detailed site investigation revealed that the contamination has degraded naturally and is no longer a threat. The EPA has recommended the deletion of this site from the NPL.

Sources: PNPLS-VA; Suffolk Public Library, 443 West Washington Street, Suffolk, VA, 23434.

U.S. Titanium
Nelson County
This former titanium mining and processing site near Piney River operated from 1931 to 1971. Waste products, including ferrous sulfate and heavy metals, were discharged into settling ponds or stored in waste piles on-site. Runoff from the ferrous sulfate piles killed fish in the Piney and Tye Rivers in the late 1970s. Testing has revealed that the groundwater and soils contain aluminum, iron, nickel, zinc, cadmium, and titanium. The contamination threatens the Piney and Tye Rivers and the water supplies of about 200 local residents.

Site investigations (1986–1989) recommended extracting and treating the iron-rich acidic groundwater, treating sulfate-contaminated soil with lime, and erecting erosion controls. Remediation should be completed by 1997.

American Cyanamid Company, a PRP, signed a Consent Agreement (1986) that requires it to conduct the site investigations and clean-up work.

Sources: PNPLS-VA; Nelson County Memorial Library, Route 29, South Lovingston, VA, 22949.

WASHINGTON

Alcoa Smelter
Clark County
This site began operations as an aluminum smelter in 1940. Located along the Columbia River, the Aluminum Company of America (ALCOA) deposited production waste containing fluoride, heavy metals, and cyanide on-site. Tests in the early 1980s revealed cyanide and fluoride in the groundwater. Soils contain high levels of aluminum. About 50,000 people live within 3 miles of the site, and nearby farms are irrigated with the same groundwater.

Response actions included the excavation and removal of contaminated soil, capping, erection of erosion controls, and fencing (1992). Groundwater monitoring is continuing.

ALCOA, the PRP, financed the remediation plan.

Sources: PNPLS-WA; Washington Department of Ecology, 300 Desmond Drive, Lacey, WA, 98503.

American Crossarm & Conduit Company
Lewis County
This site, located in the town of Chehalis, was operated as a wood treatment facility from 1948 to 1983. Utility pole crossarms were pressure-treated with creosote and later with PCP. Solid waste was also stored on the property. In 1987–1988 the EPA found that the groundwater and soil were contaminated with diesel fuel, PCP, creosote, and dioxins. Runoff from the site drains into Dillenbaugh Creek and the Chehalis River. About 200 homes are located near the site.

Response actions included the incineration of waste material (1988–1989), fencing (1992), and removal of contaminated drums, tanks, and concrete (1992). Soil excavation, removal of waste oil, and the installation of groundwater monitoring wells were completed from 1992 to 1996.

In 1986 the EPA required American Crossarm & Conduit Company to finance site investigations and part of the clean-up costs.

Sources: PNPLS-WA; Timberland Regional Library, 76 Northeast Park Street, Chehalis, WA, 98532.

American Lake Gardens
Pierce County
This site is part of residential Tacoma. Because of a homeowner's complaint, the EPA sampled water supplies in the area in 1983 and discovered toxic levels of VOCs (trichloroethylene and dichloroethylene) and metals. The contamination is derived from abandoned disposal areas on the adjacent McChord Air Force Base. Three thousand people live in the community, some of whom use private wells.

Response actions included drilling a network of groundwater monitoring wells in 1985. The U.S. Air Force provided bottled water to affected residents until they could be connected to a municipal water source. A USAF-sponsored site investigation and remediation plan consisted of installing a groundwater extraction and treatment system and restrictions on groundwater use (1993). Groundwater monitoring is continuing.

Sources: PNPLS-WA; McChord Air Force Base Library, Building 765, 62 CSG/SS1, Tacoma, WA, 98438.

Bangor Naval Submarine Base
Kitsap County
This site covers nearly 7,000 acres near the town of Silverdale. An area of wastewater disposal on the property has contaminated groundwater with VOCs, plastics, acids, pesticides, and heavy metals. Burning and detonation exercises also left the explosives TNT and cyclonite (RDX) in the soils and groundwater. About 4,000 people live within 3 miles of the facility.

Response actions began in 1990 when the U.S. Navy installed a groundwater extraction and treatment system to contain the contamination plume. Other actions included excavation and bioremediation of contaminated soils (1993–1995). Long-term groundwater monitoring will continue.

Sources: PNPLS-WA; Kitsap Regional Library, 1301 Sylvan Way, Bremerton, WA, 98310; Bangor Naval Submarine Base Library, Building 2500, Silverdale, WA, 98315.

Bangor Ordnance Disposal
Kitsap County
This hazardous waste site on the Bangor Naval Submarine Base was used as an explosives test range. Over 2 million pounds of explosives were detonated between 1965 and 1973. The groundwater, soil, and surface water are contaminated with lead, TNT, and cyclonite. Farmland and wetlands adjoin the site.

WASHINGTON

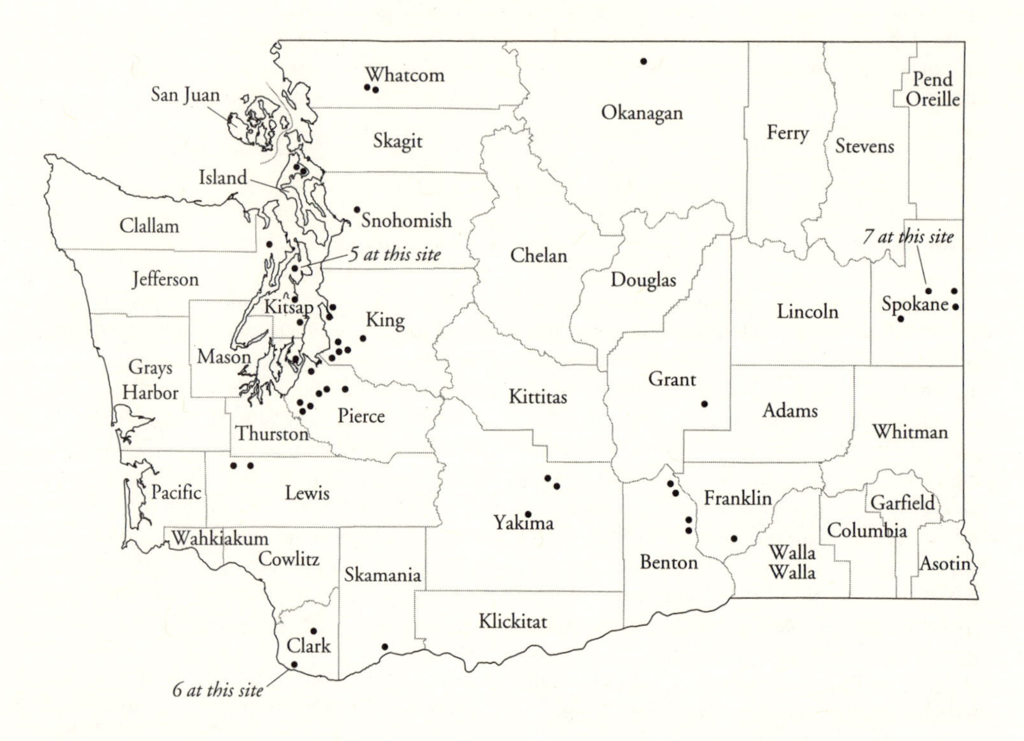

Local wells are used by nearly 4,000 people within 3 miles of the site.

The U.S. Navy began controlling contaminated runoff in 1983 and initiated a remediation plan in 1991 that consisted of constructing a leachate treatment and water distribution system (1991–1993) and soil washing (1994–1996).

Sources: PNPLS-WA; Kitsap Regional Library, 1301 Sylvan Way, Bremerton, WA, 98310.

Bonneville Power Administration, Ross Complex

Clark County

This 200-acre site near Vancouver is the control center for the generation and transmission of electricity in the northwestern United States. Laboratory wastes, oil, PCBs, heavy metals, and PCB-bearing electrical parts were buried in various disposal sites on the property. Tests in 1987–1988 showed that groundwater and soils were contaminated with VOCs, PCBs, PCP, PAHs, lead, dichloroethylene, and chloroform. Vancouver's 105,000 residents depend on drinking wells that are within 3 miles of the site, and recreational waterways are within 500 feet.

Response actions included excavation and removal of contaminated soil, bioremediation of contaminated soil, and capping of dumps (1991–1996). Groundwater monitoring is continuing.

Sources: PNPLS-WA; Vancouver Community Library, 1007 East Mill Plain Boulevard, Vancouver, WA, 98663.

Boomsnub/Airco Gases

Clark County

This site in the city of Vancouver houses an active metal-plating plant (Boomsnub Company) and a gas manufacturing company (Airco Gases). EPA tests in 1991 showed that Boomsnub has polluted the grounds with chromium through improper handling and spills. Airco's operations contributed high levels of VOCs. The contamination threatens the water source for more than 55,000 people and is migrating away from the site.

Emergency response actions included installing a groundwater extraction and treatment system to contain the contamination plume (1991–1994). Excavation and treatment of contaminated soil has also been initiated. A final remediation plan has yet to be selected.

Source: PNPLS-WA.

Centralia Municipal Landfill

Lewis County

This unlined landfill trenching operation in the town of Centralia operated from 1958 to 1994. The site received a variety of city and industrial wastes, including PCB-contaminated soil, paint wastes, pesticides, and sulfur wastes. An inadequate leachate collection system occasionally overflowed onto adjacent property and into Salzer Creek. The groundwater has high levels of manganese, sodium, and iron. The landfill threatens the aquifer that supplies the drinking water to about 12,000 people living within 3 miles of the site as well as the nearby spawning grounds of trout and salmon.

Response actions included fencing (1991), a temporary cap and gas control system (1991), a permanent cap (1994), and a permanent gas collection and treatment system (1994). Groundwater monitoring is continuing.

The PRPs identified by the EPA signed Consent Decrees in 1991 and 1993 that require them to finance the remediation plan.

Sources: PNPLS-WA; Washington Department of Ecology, Records Center, 510 Desmond Drive SE, Olympia, WA, 98503.

Colbert Landfill
Spokane County
Located 2 miles north of Colbert, this landfill accepted commercial and municipal wastes from 1968 to 1986. Included were organic solvents from electronics manufacturing companies and solvent wastes from the U.S. Air Force. Testing of nearby private wells in 1980 showed that the water contained solvents. The groundwater and soil are also contaminated with VOCs, including methylene chloride and trichloroethane. About 1,500 people live within 3 miles of the site; the Little Spokane River is less than 3,000 feet from the site.

Response actions included supplying bottled water and city water hook-ups to affected residents (1984–1985). Groundwater monitoring wells were drilled in 1987. A pilot groundwater extraction and treatment well was tested in 1991, and the final system was installed in 1995. A landfill cap will be constructed in 1997.

Sources: PNPLS-WA; North Spokane Branch Library, 44 East Hawthorne Road, Spokane, WA, 99218.

Commencement Bay, Near Shore/Tide Flats
Pierce County
This large site is an active industrial center in Tacoma that is approximately 12 square miles in size. There are about 500 sources of contamination, including paper mills, chemical factories, smelters, coal gasification companies, a metal recycling facility, and oil refineries. Groundwater, surface water, soils, and sediments contain high levels of VOCs, PNAs, PCBs, mercury, and heavy metals (copper, lead, and arsenic). About 162,000 people live in Tacoma, and the pollutants threaten the important fishing grounds offshore.

Response actions include the dredging, treatment, and disposal of marine sediments (1989 to the present). ASARCO Mining Company closed its copper, lead, and arsenic smelting facilities in 1987 and has been working with the public and the EPA on a final remediation plan that includes soil removal, capping, and runoff control systems. ASARCO has removed contaminated soil from over 200 residential properties since 1989 in an ongoing program. Tar pits will be excavated and reclaimed, and groundwater monitoring wells have been installed (1987–1994).

The EPA is working with a number of PRPs that will help finance the extensive remediation plan.

Sources: PNPLS-WA; Tacoma Public Library, 1102 Tacoma Avenue South, Tacoma, WA, 98402.

Commencement Bay, South Tacoma Channel
Pierce County
Located in Tacoma, this site covers about 3 square miles and includes an industrial sector, the Tacoma Municipal Landfill, and the city's drinking water wells. The landfill received a variety of industrial and municipal wastes, including hazardous substances. EPA studies found that one of the city's wells was contaminated by waste oil and solvents from operations of the Burlington Northern Railroad and the Time Oil Company. Groundwater and soil are contaminated by VOCs (methylene chloride, toluene, benzene, and xylene), heavy metals, PAHs, and PCBs. At least 24,000 people live within 1 mile of the site, and the contaminated groundwater flows toward Leach Creek and its surrounding wetlands.

Response actions have included pumping and air-stripping groundwater (1983), installing a carbon adsorption groundwater treatment system (1988), removing underground storage tanks and contaminated soils (1990–1992), constructing a vapor extraction system for contaminated soil (1992), installing a gas extraction and burning system (1986–1992), and capping landfills (1992–1995). Groundwater monitoring will continue for at least 10 years.

The EPA has signed several AOs and CDs with PRPs, most notably Burlington Northern Railroad, the city of Tacoma, and Time Oil Company, that require them to finance the site investigations and remediation plan.

Sources: PNPLS-WA; Tacoma Public Library, 1102 Tacoma Avenue South, Tacoma, WA, 98402.

Fairchild Air Force Base
Spokane County
This site about 12 miles west of Spokane has operated as a U.S. Air Force base since 1942. Four major waste

disposal areas, including industrial waste lagoons, have received drums of carbon tetrachloride, paint wastes, and plating sludges. Testing in 1986–1987 showed that the groundwater and soils are contaminated with VOCs (including trichloroethylene), petroleum, SVOCs, inorganic compounds, and heavy metals (cadmium, zinc, and lead). About 20,000 people within 3 miles of the air base use private wells, and important wetlands are located nearby.

Response actions included providing affected residents bottled water and hook-ups to safe water supplies (1989–1990), excavating contaminated soils (1992), installing groundwater extraction and treatment systems (1993–1994), landfill capping (1994–1995), and constructing a passive fuel recovery system that recovers free-product petroleum from the groundwater (1995). Additional groundwater extraction wells may be installed.

Sources: PNPLS-WA; Spokane Public Library, 906 West Main Street, Spokane, WA, 99201.

FMC Corporation
Yakima County
Located in the town of Yakima, this site was a pesticide manufacturing facility from 1951 to 1986. FMC Corporation buried pesticides, including DDT, diazinon, and dieldrin, in a pit on the site. Tests in 1987 discovered that the groundwater and soils were contaminated with these pesticides. Area groundwater is used by nearly 10,000 people living within a mile of the site and by farmers for watering crops and livestock.

Response actions included removing contaminated soil (1989–1992) and incinerating it on-site (1992–1994).

The EPA signed documents with FMC Corporation that required it to finance the site investigation and remediation plan.

Sources: PNPLS-WA; Yakima Regional Library, 102 North 3rd Street, Yakima, WA, 98901.

Fort Lewis Landfill No. 5
Pierce County
This U.S. Army facility near Tacoma has repaired and maintained aircraft, vehicles, and weapons on this site since 1917. Wastes dumped in on-site landfills included caustic paint-stripping solutions, battery acid, and municipal wastes. Groundwater testing in 1987 showed high amounts of heavy metals and organic compounds. About 50,000 people live on the army base, and municipal water wells are about 2 miles away.

Response actions included landfill caps (1990). An army study concluded that the site did not pose a threat to human health or the environment (1992), and the site was deleted from the NPL listing in 1995. Groundwater monitoring is continuing.

Sources: PNPLS-WA; Pierce County Library, 6300 Wildaire Road Southwest, Tacoma, WA, 98499.

Fort Lewis Logistics Center
Pierce County
This U.S. Army facility near Tacoma has repaired and maintained aircraft, vehicles, and weapons on this site since 1917. The Logistics Center is a 650-acre industrial complex with a contamination zone that is 2 miles long and nearly half a mile wide. Contaminants include VOCs and PAHs. The zone is migrating towards a residential area and the town of Tillicum. Both near-surface and deeper aquifers and Lynn Lake contain VOCs (trichloroethylene). About 50,000 people live on the army base.

Response actions included installing a groundwater extraction well field (1991–1993) and an air-stripping treatment plant (1994–1995). An air-sparging system to treat groundwater and a vapor extraction system to treat soils became operational in 1996. Other contaminated soils are being washed or treated by low-temperature desorption (1995–1997).

Sources: PNPLS-WA; Pierce County Library, 6300 Wildaire Road Southwest, Tacoma, WA, 98449.

Frontier Hard Chrome, Inc.
Clark County
This former chrome-plating facility operated from 1955 to 1983 in the city of Vancouver. Wastewater was initially discharged into the sewer system and later into disposal sites on the property, including a dry well. Tests showed that the groundwater and soils are contaminated with VOCs and heavy metals, including chromium, iron, and nickel. The contamination threatens the Columbia River, nearby businesses and homes, and the water supply for 10,000 city residents.

A site investigation has recommended soil removal and treatment, capping, and installation of a groundwater extraction and treatment system. The EPA is continuing additional groundwater monitoring. The clean up has not begun because of restricted funds.

Sources: PNPLS-WA; Washington Department of Ecology, Woodland Square, 4415 Woodview Drive Southeast, Lacey, WA, 98504.

General Electric Company
Spokane County
This former service shop in Spokane repaired and cleaned electrical transformers from 1961 to 1980. Waste oils were stored on-site. An inspection by the Washington Department of Ecology discovered high amounts of PCBs in the soil on the property. About 200,000 people live within 3 miles of the site, which is also less than 1,500 feet from the Spokane River.

Response actions included razing buildings on the site and installing groundwater monitoring wells (1990). Treatment using in-situ vitrification of contaminated soil (a technology designed by General Electric) was approved after successful field trials and will be installed by 1997.

General Electric Company, through a Consent Decree with the State of Washington, will finance the remediation plan.

Sources: PNPLS-WA; Washington Department of Ecology, North 4601 Monroe Street, Suite 100, Spokane, WA, 99205.

Greenacres Landfill
Spokane County
Located near Liberty Lake, this municipal dump was operated from 1951 to 1972. A variety of municipal, industrial, and agricultural wastes were buried there. State examiners found a nearby well to be contaminated with chlorinated organic solvents in 1978. Other chemicals in the groundwater include VOCs, acid, and heavy metals. The site is a threat to aquifers that supply water to about 350,000 people.

A site investigation called for groundwater monitoring through 1996 and possibly the construction of a landfill cap in 1998. The groundwater data will be evaluated to see if other additional treatments are necessary.

Spokane County and other PRPs signed a Consent Decree with the State of Washington to finance the site investigation study.

Sources: PNPLS-WA; Washington Department of Ecology, North 4601 Monroe Street, Suite 100, Spokane, WA, 99205.

Hamilton Island Landfill
Skamania County
This unlined landfill is owned by the U.S. Army Corps of Engineers. From 1977 to 1982 the pit received wastes from construction related to the Bonneville Dam and from the demolition of a town. Paints and degreasers were also buried there. Groundwater and soils are contaminated with heavy metals (arsenic and cadmium) and VOCs. About 1,000 people depend on wells within 4 miles of the site, which is located within a National Scenic Area and beside the Columbia River.

Response actions included fencing the site (1987). The U.S. Army conducted an investigation (1993–1995) that concluded that no clean-up action was needed. The site was removed from the NPL list in 1995.

Sources: PNPLS-WA; North Bonneville City Hall, Community Library, North Bonneville, WA, 98639.

Hanford 100-Area
Benton County
The Hanford site covers about 11 square miles and is part of the Hanford Nuclear Reservation, managed by the U.S. Department of Energy. Built in the 1940s, it made plutonium for nuclear weapons; there are nine nuclear reactors near the Columbia River. The soils and groundwater are contaminated with radioactive wastes, including strontium, cobalt, chromium, and tritium. Although the groundwater is not used for drinking water, it does discharge into the Columbia River, an important fishing resource with a number of municipal water supply intakes along its course.

Response actions included the disposal of millions of cubic yards of radioactive soil by the DOE in the 1980s. Radioactive liquid waste sites are being treated through pump-and-treat methods and removal of contaminated soil. Remediation plans for buried solid radioactive waste are still being studied.

An Interagency Agreement and Consent Order were signed by the EPA, the State of Washington, and the DOE in 1989 to develop and execute a remediation plan for the site.

Sources: PNPLS-WA; U.S. Department of Energy Reading Room, Washington State University, 100 Sproat Road, Room 130 West, Richland, WA, 99352.

Hanford 200-Area
Benton County
The Hanford 200-Area covers 215 square miles of the Hanford Nuclear Reservation. Owned by the U.S. Department of Energy, this part of the reservation was used to process nuclear materials, including plutonium. Over 1 billion cubic yards of radioactive waste were buried in more than 200 disposal sites in the area. Initial sampling in 1987 revealed that the groundwater and soils are polluted with strontium, iodine, plutonium, tritium, uranium, cyanide, and carbon tetrachloride. The groundwater flows into the Columbia River, an important fishing resource and water source for the cities of Richland, Pasco, and Kennewick, with 100,000 residents combined.

Response actions included the removal of carbon tetrachloride from soils using a large-scale vapor extraction system (1992-present) and from groundwater using a pump-and-treat system (1996-present). Plans for recovering and treating liquid radioactive waste and groundwater are still being studied.

In 1989 the EPA, State of Washington, and DOE agreed to conduct site investigations and develop a remediation plan.

Sources: PNPLS-WA; U.S. Department of Energy Reading Room, Washington State University, 100 Sproat Road, Room 130 West, Richland, WA, 99352.

Hanford 300-Area
Benton County
One of four areas on the U.S. Department of Energy's Hanford Nuclear Reservation, the 300-Area was used for the manufacturing of fuel for nuclear reactors.

There are over 25 radioactive waste dump sites on the property. The soils and groundwater are contaminated with uranium, nitrate, trichloroethane, PCBs, copper, strontium, and uranium. The contamination plume is about 2 square miles in area and flows into the nearby Columbia River, a major fishing resource and water source for the cities of Richland, Pasco, and Kennewick, with 100,000 residents combined.

Response actions included the exhuming of drums of hexane and kerosene (1991–1992). As part of a pilot study, contaminated soils were removed in 1991. Ongoing groundwater and soil studies will lead to the selection of final treatment plans by 1997–1998.

An Interagency Agreement and Consent Order between the EPA, the State of Washington, and the DOE called for the joint study of the site and development of the remediation plan.

Sources: PNPLS-WA; U.S. Department of Energy Reading Room, Washington State University, 100 Sproat Road, Room 130 West, Richland, WA, 99352.

Hanford 1100-Area

Benton County

One of four NPL sites on the Department of Energy's Hanford Nuclear Reservation, the 1100-Area was used mostly for maintenance operations. A pit in the area received up to 15,000 gallons of battery acid, and petroleum products and antifreeze may have leaked into the groundwater. Soil and groundwater were found to be contaminated with VOCs, nitrates, sodium, sulfate, heavy metals, and PCBs. The contaminated groundwater flows into the Columbia River, a major fishery resource and water supply source for the cities of Richland, Pasco, and Kennewick, with 100,000 residents combined. The Yakima River also borders the site.

Response actions included a site investigation that was completed in 1993. Excavation and disposal of contaminated soil and landfill capping should be completed by 1997. Groundwater monitoring will continue.

The EPA, DOE, and State of Washington entered into an agreement regarding how to fund the site investigation and clean-up plan.

Sources: PNPLS-WA; U.S. Department of Energy Public Reading Room, Washington State University, 100 Sproat Road, Room 130 West, Richland, WA, 99352.

Harbor Island

King County

The island is in Elliot Bay at the mouth of the Duwamish River, about 1 mile from Seattle. Previous industries such as smelting and shipbuilding have contaminated the groundwater and soils with benzene, ethylbenzene, xylene, mercury, cadmium, lead, zinc, PAHs, PCBs, and petroleum. About 10,000 people live within 1 mile of the island, which is located in an important wildlife habitat.

Response actions included removing drums of metal-plating solutions (1991) and contaminated soil (1991). Future remediation plans include detoxifying marine sediments, excavating and treating contaminated soil, capping, extracting petroleum from groundwater, and monitoring the groundwater.

The EPA is negotiating with PRPs to finance the clean-up plan.

Sources: PNPLS-WA; EPA Region 10 Record Center, 7th Floor, 1200 Sixth Avenue, Seattle, WA, 98101.

Hidden Valley Landfill

Pierce County

Originally a gravel pit, this landfill near Puyallup began operating in 1967. A variety of municipal and industrial wastes were accepted, including bulk liquids and heavy metal sludges. The older portion of the landfill was unlined. Tests in the mid-1980s indicated that the groundwater is contaminated with heavy metals, VOCs, and nitrates. Within a few miles of the site are housing developments, a gun club, municipal water wells, and wetlands.

The only response action has been a preliminary evaluation of the site by the EPA, which concluded that public health and the environment were not at immediate risk. A final remediation plan has yet to be selected.

Land Recovery, Inc., the PRP, signed a Consent Decree with the State of Washington to finance the initial study. Another CD will be issued to require the company to oversee the construction of the remediation plan.

Sources: PNPLS-WA; Washington Department of Ecology, Records Center, 7272 Cleanwater Lane, Olympia, WA, 98504.

Jackson Park Housing Complex

Kitsap County

Located 2 miles northwest of Bremerton, this 300-acre tract includes housing for U.S. Navy personnel and other military facilities. From 1904 to 1959 the property was an ammunition depot. Chemical derivatives from explosives contaminated the soil and groundwater with ammonium picrate, dyes, arsenic, cadmium, chromium, copper, lead, nickel, zinc, trinitrotoluene, dinitrotoluene, and benzene compounds. These chemicals have traveled into nearby Ostrich Bay, an important wildlife habitat and commercial fishing area.

Response actions included the removal of contaminated soil from the site. A final remediation plan has yet to be selected.

Sources: PNPLS-WA; Central Library, Bremerton, WA, 98337.

Kaiser Aluminum Mead Works

Spokane County

Located near the town of Mead, this 240-acre property is an aluminum reduction facility. Waste materials and wastewater were disposed of on-site. Tests on nearby private wells in 1978 revealed concentrations of cyanide. The groundwater and soil are contaminated with fluoride and cyanide. About 6,000 people are served by the local water system. The contamination plume may also affect the Little Spokane River.

Response actions included dispensing bottled water, drilling new wells, or providing city water hook-ups to affected residents. A site investigation report finished in 1993 is still being reviewed by the State of Washington. A final remediation plan has yet to be selected.

Kaiser Aluminum Mead Works, the PRP, agreed to finance the study and remediation plan.

Sources: PNPLS-WA; Washington Department of Ecology, North 4601 Monroe Street, Suite 100, Spokane, WA, 99205.

Lakewood Site

Pierce County

This 1-square-mile commercial area is located in the town of Lakewood. Two city wells were found to be contaminated with chlorinated organic compounds in 1981. The groundwater and soil contain trichloroethylene and tetrachloroethylene, common components in degreasers and solvents. The source of the contamination plume is a commercial dry cleaner.

Response actions included excavating contaminated sludge (1984–1985) and excavating and treating contaminated soil through vapor extraction (1992). A groundwater extraction and treatment system was also installed. Groundwater monitoring is ongoing.

Because two PRPs could not afford to fund the clean up, the State of Washington placed a lien on their properties in 1991.

Sources: PNPLS-WA; Pierce County Library, 6300 Wildaire Road Southwest, Tacoma, WA, 98499.

McChord Air Force Base/Wash Rack

Pierce County

This U.S. Air Force base covering about 4,600 acres just south of Tacoma. Operations at a former aircraft washing facility, which used chemical solvents to remove dirt, oil, and grease, contaminated the soil and groundwater. A plume of floating fuel about 300,000 square feet in area lies above the groundwater, which is polluted with benzene. About 16,000 residents live within 3 miles of the site; 300 wells are within 5 miles of the site.

Response actions included studies from 1992 to 1994 that called for natural attenuation of the plume. Data from groundwater monitoring indicated that attenuation is occurring.

Sources: PNPLS-WA; McChord Air Force Base Library, Building 765, 62 CSG/SSL, Tacoma, WA, 98438.

Mica Landfill

Spokane County

Located near the town of Mica, this landfill was used by Spokane County Utilities Co. to dispose of various wastes, including baghouse dust and asbestos, from 1971 to 1991. The groundwater is contaminated with VOCs, heavy metals, and phenols. Methane gas is also present. Within 3 miles of the site are 4,000 residents, two municipal wells, over 120 private wells, and numerous waterways.

Response actions included construction of a permanent cap and gas and leachate collection systems (1995). Groundwater monitoring is continuing to determine if additional actions are needed.

Sources: PNPLS-WA; Spokane Public Library, 906 West Main Avenue, Spokane, WA 99201; Spokane County Utilities Division, West 1026 Broadway Avenue, Spokane, WA, 99260.

Midway Landfill

King County

This gravel quarry was operated as an unlined landfill from 1966 to 1983 by the city of Seattle. Buried wastes include coolant, oily sludges, preservative distillates, dyes, and paint waste. Groundwater and soils are contaminated with heavy metals, PCBs, and VOCs (benzene and vinyl chloride). Combustible gas has been detected in buildings 3,000 feet away. About 8,000 people live near the site, and the Green River is 1 mile away.

Response actions included installing gas extraction wells (1985–1986), fencing, capping the landfill (1990), and installing runoff and drainage control systems (1991). Groundwater and gas monitoring is continuing.

The city of Seattle signed a Consent Decree with the State of Washington in 1990 to finance the remediation plan.

Sources: PNPLS-WA; King County Library, 212 Second Avenue North, Kent, WA, 98032.

Moses Lake Well Field

Grant County

Routine sampling in 1988 showed that two of Moses Lake's municipal drinking water wells were contaminated with trichloroethylene. A number of private wells are also in the area. Likely sources of the contamination are Larson Air Force Base, a waste treatment plant, and a burn pit at a fire-fighting training school.

Response actions included blending contaminated water with clean water to meet federal standards for TCE levels. The EPA and the U.S. Army Corps of Engineers are presently conducting detailed site investigations, including the construction of groundwater monitoring wells. A remediation plan will be selected after the data have been evaluated.

The EPA has identified a number of PRPs to share in the cost of the clean-up plan.

Sources: PNPLS;WA; Moses Lake Community Library, 418 East Fifth Avenue, Moses Lake, WA, 98837.

Naval Air Station, Whidbey Island, Ault Field
Island County
The naval station covers nearly 7,000 acres on Whidbey Island. Aircraft and vehicle maintenance, engine testing, painting, and pest control activities generated wastes including solvents, heavy metals, paints, and PCP. The wastes were dumped in four landfills that are on top of shallow and sea-level aquifers that provide water for 21,000 residents within 3 miles of the site. Groundwater and soils are contaminated with VOCs, PCBs, heavy metals, pesticides, PAHs, and dioxin. Nearby lakes and wetlands are used for recreation and irrigation.

Response actions included the installation of pump-and-treat groundwater systems, provision of city water hook-ups for affected residents, excavation and disposal of contaminated soil and sediments (1992–1995), and a landfill cap (1996–1997).

Sources: PNPLS-WA; Oak Harbor Public Library, 3075 300th Avenue West, Oak Harbor, WA, 98277.

Naval Air Station, Whidbey Island, Sea Plane Base
Island County
This naval station covers nearly 7,000 acres on Whidbey Island. Wastes from general activities, including solvents, zinc chromate, thinners, acid, lead-based sealants, and paint wastes were dumped in five disposal sites. Soils and groundwater were contaminated with lead, arsenic, pesticides, and PAHs. The contaminants are a threat to recreational lakes on-site and to coastal wetland habitats.

Response actions included investigations by the U.S. Navy (1993) and the excavation and removal of contaminated soil (1994). The site was deleted from the NPL in 1995. Monitoring is continuing.

Sources: PNPLS-WA; Oak Harbor Public Library, 3075 300th Avenue West, Oak Harbor, WA, 98277.

Naval Undersea Warfare Engineering Station
Kitsap County
This site outside Keyport was originally used by the U.S. Navy as a torpedo testing range. Various naval activities have contaminated the site with VOCs, petroleum, and heavy metals from disposals and spills. An unlined landfill in a salt marsh is in contact with groundwater. The polluted groundwater discharges into creeks, bays, and estuaries that are used for recreation and commercial fishing. High levels of VOCs and metals have been found in shellfish. About 250 homes within 3 miles of the site are serviced by private or municipal water wells.

Response actions included removal of chromium-tainted soils (1992). A remediation plan for the landfill sites will be chosen when ongoing investigations have been completed.

Sources: PNPLS-WA; Kitsap Regional Library, 1301 Sylvan Way, Bremerton, WA, 98310; Poulsbo Library, 700 Northeast Lincoln Street, Poulsbo, WA, 98370.

North Market Street
Spokane County
This 50-acre site about 1 mile north of Spokane was formerly a petroleum refinery and is currently a petroleum storage facility. Waste oil was deposited in unlined lagoons on the property. Groundwater and soil are contaminated with VOCs and petroleum products. An aquifer provides drinking water for about 200,000 people within 3 miles of the site.

The first phase of the site investigation and feasibility study was completed in 1994. A final remediation plan has yet to be selected.

Sources: PNPLS-WA; Hillyard Branch Library, North 4001 Cook Street, Spokane, WA, 99207.

Northside Landfill
Spokane County
Located in northwestern Spokane, this 345-acre site was operated as a landfill from 1931 to 1991. (A new waste disposal cell has now been installed and the site is operating as a sanitary landfill.) Prior to 1992, it received a variety of municipal, commercial, and industrial wastes, including sludges. The groundwater and soils are contaminated by organic solvents, tetrachloroethylene, and trichloroethylene. The contamination plume has penetrated the underlying aquifer, the drinking water source for the city of Spokane.

Response actions included providing city water hook-ups for affected residents (1983), closing landfill units, capping, groundwater treatment and monitoring, installing a gas collection system, fencing, and landscaping (1990–1993). Groundwater monitoring is continuing.

Sources: PNPLS-WA; Spokane Engineering Services Division, West 808 Spokane Falls Boulevard, Room 318, Spokane, WA, 99201.

Northwest Transformer
Whatcom County
This facility was used as a salvage yard from 1958 to 1985; equipment was dismantled and reclaimed, waste fuel was burned, and transformer oils were

drained into seepage pits. In 1985 county health officials found PCBs in soil and private wells near the site. The surrounding area is both residential and agricultural. Farms use the groundwater for irrigation, and about 200 people live within 1 mile of the site.

Response actions included removing contaminated soil, debris, and liquids and fencing (1985). A landfill cap was constructed, and groundwater monitoring is ongoing.

About 100 PRPs signed an AO with the EPA in 1990 to finance the site investigation and remediation plan.

Sources: PNPLS-WA; Everson Community Library, 104 Kirsh Drive, Everson, WA, 98247.

Northwest Transformer, South Harkness Street

Whatcom County

This transformer facility operated from 1958 to 1987 in the town of Everson. A state inspector found high levels of PCBs, arsenic, cadmium, and lead in the soils in 1985. About 2,200 people live within 3 miles of the site; over 10,000 people use groundwater from within 3 miles of the site.

An investigation was completed in 1994. Buildings were razed and the contaminated soil removed (1994–1995). Groundwater monitoring is continuing.

PRPs financed the remediation plan.

Sources: PNPLS-WA; Everson Community Library, 104 Kirsh Drive, Everson, WA, 98247.

Old Inland Pit

Spokane County

This former gravel pit in Spokane operated from about 1977 to 1983. It received wastes, including baghouse dust, heavy metals, and VOCs, from the Inland Asphalt Company and the Spokane Steel Foundry. Soils and groundwater are contaminated with heavy metals and organic solvents. The threatened aquifer supplies drinking water to about 30,000 residents in the area.

An initial evaluation by the State of Washington indicated that the site posed no immediate risks; additional studies will lead to the selection of a final remediation plan.

Sources: PNPLS-WA; Washington Department of Ecology, North 4601 Monroe Street, Suite 100, Spokane, WA, 99205.

Old Navy Dump/Manchester Annex

Kitsap County

The Manchester naval facility and laboratory is on Puget Sound, about 1 mile north of Manchester. The site was used by the U.S. Navy for building and repairing submarine equipment. Waste products were buried in landfills on the property. Soils and runoff water are contaminated with lead, mercury, cadmium, PCBs, petroleum, and asbestos. These contaminants threaten the groundwater, local wells, and the sensitive wetland habitats in Clam Bay.

A site investigation study is under way, and the U.S. Army Corps of Engineers will undertake the final clean-up effort.

Sources: PNPLS-WA; Manchester Town Library, Manchester, WA, 98353.

Pacific Car & Foundry Company

King County

From 1907 to 1988 this site manufactured vehicles, railroad cars, and anodes in an industrial section of the town of Renton. Waste products, including sand, wood, metal, paints, oils, and solvents, were dumped in a marsh and buried. Tests in 1986 revealed that the soils and groundwater are contaminated with heavy metals, petroleum products, PCBs, and PAHs. The contamination threatens nearby streams and wetlands and the drinking water of about 37,000 people within 3 miles of the site.

Initial response actions included removing contaminated soil (1987). The remediation plan consists of excavation and treatment of soils through bioremediation and groundwater monitoring (1992–1997).

Pacific Car & Foundry Company, the designated PRP, signed Consent Decrees in 1989 and 1991 with the State of Washington to finance the site investigation and clean-up plan.

Sources: PNPLS-WA; Renton Public Library, 100 Mill Avenue South, Renton, WA, 98055.

Pacific Sound Resources

King County

Located in West Seattle, this site operated as a wood treatment plant from 1909 to 1994. Leakages and spills have allowed creosote, PCP, chemonite, heavy metals (arsenic, chromium, copper, and zinc), PAHs, and inorganic wood preservatives to contaminate the soil and groundwater. Surface runoff flows into Elliot Bay, an important commercial fishing resource and fish migration route. About 175,000 people live within 4 miles of the site.

Response actions included the immediate removal of contaminated soil (1990) and fencing. Arsenic-laden soils were removed in 1995. The PRP, Pacific Sound Resources, is conducting a site investigation (1994–1997) that will result in the selection of a final remediation plan.

Legal actions by the EPA against Pacific Sound Resources require it to finance the site investigation and clean up.

Sources: PNPLS-WA; Seattle Public Library, 2306 42nd Avenue Southwest, Seattle, WA, 98116.

Pasco Sanitary Landfill
Franklin County
This active landfill is located about 1 mile northeast of Pasco. An open dump was operated from 1956 to 1971. Another part of the site accepted hazardous waste from 1972 to 1981 and took in about 50,000 drums of hazardous substances, including paints, resins, herbicides, sludges, and pesticide containers. The groundwater is contaminated with VOCs (trichloroethylene, toluene, and xylene). The site threatens the water supply for about 11,000 people living within 3 miles of the site.

The initial response action is a site investigation that was started in 1992. Groundwater monitoring wells have been installed. A final remediation plan has yet to be determined.

An AO between the State of Washington and 29 PRPs requires them to finance the site investigation study.

Sources: PNPLS-WA; Washington Department of Ecology, North 4601 Monroe Street, Suite 100, Spokane, WA, 99205.

Pesticide Lab
Yakima County
The site is an agricultural research laboratory that is managed by the U.S. Department of Agriculture in the town of Yakima. Wastes from pesticide mixing and manufacturing were discharged on-site into a septic tank system. Testing in 1990 indicated that the groundwater contained low levels of pesticide compounds. The local groundwater is used by about 60,000 people in the area.

In 1992 the septic tank system and surrounding soils were removed. The site was deleted from the NPL in 1993. Monitoring is continuing.

Source: PNPLS-WA.

Port Hadlock Detachment
Jefferson County
Owned by the U.S. Navy, this facility is on Indian Island near Port Townsend. Operated mostly as a munitions storage area, related wastes were deposited in landfills and pits on the property. Soil and groundwater on parts of the island are contaminated with lead, cadmium, pesticides, and PCBs. Marine sediments and clams are also contaminated. The contamination threatens offshore waters, special wildlife habitats, and a variety of marine species.

Response actions included the removal of contaminated soil and sediments in 1994. Landfills were capped and erosion control systems installed (1995). Groundwater monitoring is continuing, and additional remedies may be required.

Sources: PNPLS-WA; Port Hadlock Library, Ness Corner Road and Cedar Avenue, Port Hadlock, WA, 98339.

Puget Sound Naval Shipyard Complex
Kitsap County
This active U.S. navy facility in Bremerton, on Puget Sound, has been in operation since 1891. Its main activity has been the construction and mooring of naval ships and storing of supplies and equipment. Hazardous waste from these activities was disposed of on-site in dumps or holding areas. The groundwater and soils are contaminated with petroleum, heavy metals, PCBs, VOCs, and SVOCs. The contamination flows into Sinclair Inlet, an important recreational and fishing waterway.

Initial actions included the removal of lead-contaminated soil (1994). A final remediation plan will be selected after reviewing the results of a detailed site investigation (1992–1996).

Sources: PNPLS-WA; Kitsap Regional Library, 1301 Sylvan Way, Bremerton, WA, 98310.

Queen City Farms
King County
This site about 2 miles from Maple Valley manufactured compost and used ponds for the disposal of wastes from 1955 to 1964. Sampling by the EPA in 1980 found the ponds to be contaminated with VOCs and heavy metals (arsenic). Soils contain PCBs and heavy metals. About 8,000 people and over 100 private and public water wells are within 3 miles of the site.

Initial response actions included the removal of liquid and solid wastes (1985–1986). Construction of an upgradient water diversion system, capping, and grading were also completed. More soil and debris were removed in 1988 and 1990 and groundwater monitoring wells installed. Constructing a vertical barrier wall and enlarging the cap should be under way by 1997.

The EPA signed agreements with Queen City Farms and Boeing Co. in 1985, 1991, and 1994 that require the PRPs to finance the site investigation and remediation plan.

Sources: PNPLS-WA; Maple Valley Public Library, 23730 Maple Valley Road, Maple Valley, WA, 98038.

Seattle Municipal Landfill
King County
This landfill facility in Kent was used by the city of Seattle from 1968 to 1986 and accepted municipal and industrial wastes, including sand-blasting grit and sludges. Testing in 1984 showed that the groundwater contained heavy metals and VOCs and that landfill gas contained VOCs (toluene, xylene, and trichloroethylene). Well systems within 3 miles of the site serve over 18,000 nearby residents.

Response actions from the late 1980s to 1991 included the installation of leachate collection and gas control systems and fencing. A permanent cap and improvements to the leachate collection system were completed in 1995. Groundwater monitoring is continuing.

The city of Seattle, through a 1987 Consent Agreement, financed the site investigation and remediation plan.

Sources: PNPLS-WA; Kent Library, 232 South Fourth Street, Kent, WA, 98032.

Silver Mountain Mine
Okanogan County
This abandoned gold and silver mine operated from 1928 to the 1960s. In the 1980s the mine tailings were treated with cyanide to remove the remaining gold. The project was abandoned in 1983, and the cyanide-contaminated tailings and holding basin were not reclaimed. Groundwater and soils are contaminated with cyanide and arsenic. Less than five people live within 3 miles of the mine, although local wells are used for farming.

Response actions included the removing the water in the basin, covering the tailings, and fencing the site (1985). The mine tailings were later capped (1992). Revegetation was completed in 1993.

Sources: PNPLS-WA; Okanogan County Health District, 237 North 4th Street, Okanogan, WA, 98840.

Spokane Junkyard/Associated Properties
Spokane County
Located in a light commercial area of the city of Spokane, this site took in a variety of materials between the 1940s and 1983, ranging from cars to appliances and transformers. A recycling circuit was also operated on the properties. EPA tests in 1987–1989 discovered high levels of PCBs and heavy metals (mercury, cadmium, and lead) in the soil. Water supply wells for about 165,000 people are within 4 miles of the site.

Response actions included the removal of 140 drums of hazardous waste and 140 cubic yards of asbestos (1987). A site evaluation was completed in 1995, and a final remediation plan will be selected by 1997. About 200,000 square feet of contaminated soil remains on the site.

Sources: PNPLS-WA; Hillyard Public Library, North 4005 Cook Street, Spokane, WA, 99207.

Toftdahl Drum Site
Clark County
Located in Brush Prairie, this facility was used for cleaning industrial waste drums. The unusable drums were crushed and buried on site. The EPA removed six of these drums in 1983. Testing showed that the groundwater, soil, and surface water had high levels of heavy metals and PCBs. About 6,000 people live within 3 miles of the site, and runoff drains into Morgan Creek.

Response actions included surveys that located more drum burial sites (1984). These drums and the contaminated soil were later removed. Groundwater monitoring is scheduled through 1998. The site was removed from the NPL in 1988.

Sources: PNPLS-WA; EPA Hazardous Waste Records Center, 1200 Sixth Avenue, Seattle, WA, 98101.

Tulalip Landfill
Snohomish County
This landfill on the Tulalip Indian Reservation is bordered by wetlands. The Seattle Disposal Company dumped about 4 million cubic yards of industrial, hospital, and city wastes in a landfill site on the reservation from 1964 to 1979. In 1987 the EPA ordered the reservation to collect and remove contaminated leachate, which it refused to do. A subsequent inspection in 1990 showed that waste materials, including demolition debris, were still being dumped on the property. The groundwater, surface water, soils, and wetland waters are contaminated with lead, copper, cadmium, chromium, PCBs, and VOCs (toluene and xylene). A number of wells within 4 miles of the site supply drinking water to about 8,000 people, and the runoff drains into the flanking wetlands and Puget Sound, a sensitive estuary habitat.

A detailed site investigation that will determine the final remediation plan is under way.

The EPA has identified over 6,000 potentially responsible parties. To date, the EPA has made 95 settlement offers, to which 65 parties have agreed.

Sources: PNPLS-WA; Marysville Library, 6120 Grove Street, Marysville, WA, 98270.

Vancouver Water Station No. 1
Clark County
This well field in the city of Vancouver pumps nearly 20 million gallons of water per day to residents in Vancouver and Clark County. Routine testing in 1988 revealed low levels of VOCs (tetrachloroethylene) in some of the wells.

Site inspections by the EPA in 1990–1991 failed to identify a source of the contamination. Vancouver built five air-stripping towers in 1993 to clean the groundwater. A detailed site investigation that will determine a final remediation plan is continuing.

Source: PNPLS-WA.

Vancouver Water Station No. 4
Clark County
This well field in the city of Vancouver near the Columbia River supplies drinking water to about 110,000 residents. Routine testing in 1988 revealed low levels of VOCs (trichloroethylene and perchloroethylene) in the water.

The city of Vancouver built two air-stripping towers to clean the groundwater (1992). A detailed site investigation that will determine a final remediation plan is in progress.

Source: PNPLS-WA.

Western Processing Company

King County

The Western Processing Company processed and recycled a variety of products, including animal by-products, yeast, and industrial wastes (oil, electroplating solvents, battery acid, and pesticides). Toxic materials were improperly stored in drums, bulk tanks, uncovered piles, and lagoons on-site. The company was closed by court order in 1983 after about 25 years of operation. Groundwater and soils are contaminated with phenols, heavy metals, VOCs, and PCBs. About 10,000 people live within 3 miles of the site.

Response actions included removing a variety of toxic materials and debris (1983), building a runoff collection and treatment system (1984), excavating contaminated soil (1985), and constructing a groundwater extraction and treatment system (1988). Contaminated sediments in Mill Creek were removed in 1994. Groundwater monitoring is continuing.

Consent Decrees signed in 1984 and 1986 require the PRPs to finance the site investigation and remediation plan and to maintain monitoring for 30 years.

Sources: PNPLS-WA; EPA Region 10 Hazardous Waste Records Center, 1200 Sixth Avenue, Seattle, WA, 98101.

Wyckoff Co./Eagle Harbor

Kitsap County

This site housed a former wood treatment facility that operated from 1905 to 1988 and a shipyard in Eagle Harbor on Bainbridge Island. Lumber was pressure-treated with creosote and PCP. For most of the period sludge waste was buried on site, and wastewater was discharged into a seepage basin. The groundwater and soils are contaminated with VOCs, PAHs, PCPs, creosote, heavy metals, dioxins, petroleum by-products, and furans. The contaminated groundwater has polluted recreational Eagle Harbor and its fish and shellfish. About 2,000 people live within a mile of the site, and the town of Eagle Harbor depends on a local aquifer for drinking water.

Initial response actions included removing toxic materials from the site and installing groundwater treatment systems (1988–1990). An area of mercury-contaminated soils was removed and capped in 1992. Ongoing efforts include capping contaminated bottom sediments in Eagle Harbor, expanding the groundwater extraction and treatment systems, and possibly constructing a barrier wall around the site (1994-present).

Legal agreements between the EPA and the Wyckoff Company (1984, 1988, 1991) require the PRP to finance the site studies and remediation plan.

Sources: PNPLS-WA; Bainbridge Branch Library, 1270 Madison North, Bainbridge Island, WA, 98110.

Yakima Plating Company

Yakima County

This facility in Yakima has electroplated car bumpers since 1962. Wastewater was legally discharged into a drain system on the property. EPA tests in 1986 showed low levels of copper, lead, and zinc in the soil. The local groundwater is the water source for about 100,000 people within 3 miles of the site.

Response actions included the removal of drums of plating wastes, contaminated soils, and groundwater monitoring (1991). The site has been deleted from the NPL. Monitoring is continuing.

Sources: PNPLS-WA; Yakima Regional Library, 102 North Third Street, Yakima, WA, 98901.

WEST VIRGINIA

Allegany Ballistics Laboratory

Mineral County

Established in 1942 near Cresaptown, Maryland, this U.S. Navy facility researches, develops, and tests rocket propellants and motors for rockets and weapons. Various waste materials, including heavy metals, VOCs, acids, sludges, paints, photographic solutions, and thinners have been buried in landfills on the base. Local drinking water wells are threatened by the contamination, which has already penetrated the Potomac River.

Response actions included shutting down contaminated wells in 1981. Site investigations are under way.

Source: PNPLS-WV.

Fike Chemical

Kanawha and Putnam Counties

This chemical plant near Nitro produced specialty chemicals until 1988. Hazardous production wastes were discharged into unlined lagoons, storage tanks, or drums. Leakage and spills have contaminated the groundwater and soils with VOCs, cyanide, and dioxin, which can also be carried in the air. The chemicals threaten the drinking water source for about 10,000 residents and the recreational Kanawha River.

WEST VIRGINIA

An emergency clean up removed drummed waste and tanks in 1988. More clean ups occurred in 1993, including the removal of cyanide. Buried tanks were cleaned and removed in 1994–1995. Groundwater site investigations are in progress.

Sources: PNPLS-WVA; Nitro Public Library, 1700 Park Avenue, Nitro, WV, 25143.

Follansbee Site
Brooke County
This coal tar processing plant near Follansbee has been in operation since 1914. Various production wastes, including sludges, VOCs, and heavy metals, were stored or discharged on-site into lagoons and tanks. Testing has shown that the groundwater and soils contain high levels of PAHs, VOCs, and heavy metals. The contamination threatens the drinking water supplies of about 10,000 residents, numerous springs and seeps, and the Ohio River.

Response actions included installing a groundwater trench and pumps (1983). A recent investigation (1990–1996) will recommend a final remediation plan.

A Consent Decree (1984) and Administrative Order on Consent (1990) require Koppers Industries and Wheeling-Pittsburgh Steel to undertake site investigations and remediation work.

Sources: PNPLS-WV; Follansbee City Building, Main Street, Follansbee, WV, 26037.

Ordnance Works Disposal Areas
Monongalia County
This 670-acre disposal area along the Monongahela River has received wastes from a number of industries in the Morgantown area since 1941, including by-products from coke, ammonia, and methanol production that were buried or held in lagoons. Testing has shown that the groundwater and soils contain heavy metals, PCBs, and PAHs. The contamin-

ation threatens local drinking water sources and the Monongahela River.

Response actions included removing drummed waste in 1984. In 1989 the EPA recommended capping, soil bioremediation or solidification, and air monitoring. Investigations into the contamination levels in the industrial complex are in progress.

PRPs have participated in the site studies and clean-up work according to a 1990 Consent Order.

Sources: PNPLS-WV; Morgantown Public Library, 373 Spruce Street, Morgantown, WV, 26505.

Sharon Steel Corporation (Fairmont Coke Works)
Marion County
Coke was produced from coal on this site near Fairmont from about 1920 to 1979. Other by-products included tars, phenols, ammonium sulfate, coke oven gas, and VOCs. Wastes such as coal tars and a coal dust known as "breeze" were left on the 107-acre property. Sludges and other wastewaters were held in unlined ponds. Contaminated surface runoff flows into nearby streams that drain into the Monongahela River. Testing has shown that the groundwater and soil contain cyanide, heavy metals, PAHs, and other coal tar derivatives. The contamination threatens local drinking water supplies, streams, wetlands, and associated ecosystems.

The PRP, Sharon Steel Corporation, stopped operations in 1979 as a result of EPA lawsuits. Sharon Steel has yet to remediate the site. In 1993 the EPA conducted a removal action to stabilize the property. It was proposed as an NPL site in 1996; a final decision is expected in 1997–1998. Other studies and remediation work are planned.

Source: EPA Publication 9320.7-071.

West Virginia Ordnance
Mason County
This former U.S. Army ordnance center near Point Pleasant manufactured explosives from 1942 to 1945. The area is now a recreational wildlife management area. Past operations have contaminated the groundwater and soils with TNT, trinitrobenzene, asbestos, arsenic, lead, and beryllium. The contamination threatens the 11,000 people who use the property recreationally every year, nearby private water wells, and the various wildlife and wetland habitats.

Response actions included burning TNT residues, building a soil cover, removing asbestos, and flashing other explosives (1989). The cap was also upgraded. Groundwater pump-and-treat systems have been installed or are being considered in six areas on the 8,300-acre property.

Sources: PNPLS-WV; Mason County Public Library, Sixth and Viand Streets, Point Pleasant, WV, 25550.

WISCONSIN

Algoma Municipal Landfill
Kewaunee County
The city of Algoma used this 13-acre site as a landfill facility from 1969 to 1983. Illegal substances such as paint wastes, lacquers, thinners, and asbestos were also buried there. In 1984 anomalous levels of heavy metals and VOCs were detected in the groundwater. Aquifers that supply drinking water to nearly 5,000 residents underlie the unlined pit, and about 200 people with private wells live within 1 mile of the site.

Initial response actions (1991–1993) included gas and groundwater monitoring and the construction of a clay cap. Interim response actions consist of ongoing groundwater and gas monitoring.

The city of Algoma and several companies signed an Administrative Order on Consent (1988) and a Consent Decree (1991) with the EPA to finance the site investigation and remediation plan.

Sources: PNPLS-WI; Algoma Public Library, 406 Fremont Street, Algoma, WI, 54201; Algoma City Hall, 416 Fremont Street, Algoma, WI, 54201.

Better Brite Plating Company Chrome and Zinc Shops
Brown County
Located near the town of De Pere, the Better Brite Plating Company performed chrome and zinc metal plating from 1963 to 1989. Over 20,000 tons of metal-laden solutions are believed to have leaked from plating tanks buried on the 2-acre property. The worst violation was a 1979 spill of 2,200 gallons of chromium-rich plating solution and rinsewater. In 1979 state examiners discovered extensive chromium contamination in the soil around the shops and in the yards of nearby homes. Follow-up investigations confirmed yellow, chromium-rich water running from the property into the city sewer. The pollutants (zinc, chromium, other heavy metals, cyanide, and VOCs) threaten the drinking water supply for 46,000 people in the towns of De Pere, Allouez, and Ashwaubenon. Chemicals in the soil and surface water have entered the underlying groundwater.

Initial response actions in 1979 consisted of installing groundwater monitoring wells and a surface

WISCONSIN

water collection system to control runoff. Polluted soil, chrome and zinc acids, toxic liquids, and cyanide sludge were removed in 1986. The buildings were razed and the area capped with low-permeability clay. In 1991 a sump-pump water treatment system was installed to extract contaminated groundwater. Interim actions consist of monitoring the water treatment system and controlling surface runoff.

For 10 years Better Brite ignored orders by the Wisconsin Department of Natural Resources (DNR) to clean up the site and finally abandoned it after declaring bankruptcy.

Sources: PNPLS-WI; Brown County Public Library, De Pere Branch, 380 Main Avenue, De Pere, WI, 54115.

City Disposal Corporation Landfill
Dane County
From 1966 to 1977, this unlined 24-acre landfill received a variety of hazardous waste products, including liquid industrial wastes (sealed and unsealed), organic chemicals, solvents, oils, and other petrochemicals. EPA-sponsored groundwater testing and soil sampling revealed VOC contamination. The site, operated by Waste Management of Wisconsin, was closed in 1977. Over 5,000 people live within 3 miles of the landfill, 160 of whom are within a mile of the site and use private water wells. Runoff from the site drains into Badfish Creek and also threatens Grass Lake, a special habitat for sandhill cranes.

After the initial investigation was completed in 1992, EPA authorities recommended building a new cap and installing a groundwater pump-and-treat system that is expected to be operational by 1997.

An Administrative Order on Consent (1987) and Unilateral Administrative Order (1993) between Waste Management of Wisconsin, the PRP, and the EPA requires the company to investigate and remediate the site.

Sources: PNPLS-WI; Dunn Town Hall, 4156 County Trunk Highway B, McFarland, WI, 53558.

Delavan Municipal Well No. 4
Walworth County
This site is defined as the aquifer that was used by Delavan Well No. 4 until it was shut down in 1982 because it contained high levels of VOCs, including trichloroethylene. Area soils are also contaminated. The city of Delavan still occasionally uses the water from Well No. 4, blending it with clean water to dilute the contaminants to safe levels. Over 3,000 people, a number of businesses, and a school have used the contaminated water.

State and federal officials supervised the installation and monitoring of groundwater extraction and soil venting systems in 1990. A final remediation plan is expected by 1997.

Sources: PNPLS-WI; Aram Public Library, 404 East Walworth Avenue, Delavan, WI, 53115.

Eau Claire Municipal Well Field
Eau Claire County
VOC contamination in groundwater from a well field that covers over 500 acres was discovered by the Wisconsin Department of Natural Resources (DNR) and the EPA in 1981. The field supplies drinking water to about 58,000 residents in Eau Claire County. The contamination plume, nearly 3 miles in length, originates from the facilities of National Presto Industries. Chemicals include trichloroethane, dichloroethene, and tetrachloroethene.

Immediate response actions by the EPA in 1984 were the installation of an air-stripping unit to remove VOCs from pumped groundwater, which was then sent to a municipal water treatment plant. In 1992 additional extraction wells were installed, and homes with polluted wells were connected to the municipal water supply.

National Presto Industries will reimburse the EPA for about 95 percent of its costs according to a 1993 settlement with the EPA and U.S. Department of Justice.

Sources: PNPLS-WI; L. E. Phillips Memorial Library, 400 Eau Claire Street, Eau Claire, WI,54701; Eau Claire City Hall, 203 S. Farwell Street, Eau Claire, WI, 54701.

Fadrowski Drum Disposal
Milwaukee County
The Fadrowski Drum Disposal site was originally a landfill that was operated by Edward Fadrowski from 1970 to 1981. The site covers 20 acres on the eastern edge of the town of Franklin. From 1983 to 1993 the EPA exhumed hundreds of drums of hazardous waste, including VOCs (especially toluene), lead, chromium, arsenic, lubricant sludges, and DDT. Contaminants in the groundwater include mercury, benzene, chromium, barium, and cyanide. Soil samples show high levels of PAHs, inorganic compounds, and VOCs (especially toluene). Runoff flows into a pond and wetlands that are used for swimming and recreation. About 18,000 residents drink water from aquifers within 3 miles of the site.

EPA studies from 1987 to 1991 recommended a remediation plan that consists of removing drums, capping the waste disposal area, installing erosion controls and storm sewers, pumping out and backfilling the pond, and fencing the site (1993–1995). A final clean-up plan has yet to be determined.

Sources: PNPLS-WI; Franklin Public Library, 9229 West Loomis Road, Franklin, WI, 53132.

Hagen Farm
Dane County
The Hagen Farm site is about 1 mile east of Stoughton. Originally a gravel pit, it became a disposal site for the city of Stoughton, Uniroyal Plastics Corporation, and other industries from about 1963 to 1966. It was later filled in and used as a pasture for sheep. An examination by the DNR and Uniroyal between 1980 and 1982 found VOCs (tetrahydrofuran, benzene, ethylbenzene, toluene, and xylene) in nearby residential wells, groundwater, and soils. Stoughton's city wells and eight private wells are located within 2 miles of the landfill.

Response actions between 1992 and 1994 included fencing, capping the main disposal area, installing an in-situ vapor extraction system, and using a biological treatment process known as fixed film.

The State of Wisconsin sued Uniroyal and Waste Management of Wisconsin, the present owner, to undertake the site investigation and remediation plan.

Sources: PNPLS-WI; Stoughton Public Library, 304 South Fourth Street, Stoughton, WI, 53589; Dunkirk Town Hall, County Trunk Highway North, Stoughton, WI, 53589.

Hechimovich Sanitary Landfill
Dodge County
This former commercial landfill, located about 2 miles south of Mayville, operated from 1959 to 1986. Paint sludges, cutting oils, and spent organic solvents were dumped in unlined pits. The DNR shut down the op-

eration in 1980. EPA tests found VOCs in two nearby water wells in 1984. The landfill threatens nearby recreational wetlands and the drinking water of about 8,000 people in the towns of Mayville and Horicon.

Initial response actions included the installation of a landfill cap and gas collection system. The extent of the groundwater contamination is being studied, and a final remediation plan will be selected by 1997.

Sources: PNPLS-WI; Mayville Public Library, 111 North Main Street, Mayville, WI, 53050.

Hunts Disposal Landfill

Racine County

This former sand and gravel pit, located on the outskirts of Caledonia, operated as a landfill from 1959 to 1974. A variety of industrial and city wastes were buried there, including four 10,000-gallon tanks containing arsenic acid sludge. An inspection by the DNR in 1975 revealed that the site was poorly covered and that contaminants were seeping into the groundwater. The soil and groundwater contain high levels of VOCs, arsenic, and lead. The county purchased the 84 acres with plans to convert it into a regional park. The 1,500 residents within 1 mile of the site use private wells. Groundwater flows away from the site toward the Root River, a recreational waterway.

Initial actions (1990–1995) consisted of repair of erosion damage, sealing of leachate seeps, and revegetation. Long-term plans include installing groundwater and gas collection and treatment systems and constructing a slurry wall to control runoff.

Waste Management, Inc., and other PRPs signed a Consent Decree with the EPA in 1992 to finance the site investigation and remediation plan.

Sources: PNPLS-WI; Caledonia Town Hall, 6922 Nicholson Road, Caledonia, WI, 53108.

Janesville Ash Beds

Rock County

This 5-acre parcel is located on the north side of Janesville and has been owned and operated by the city since the 1950s. Five separate unlined ash beds received millions of gallons of industrial liquids and sludges every year between 1974 and 1983. The beds were removed and the facility closed in 1985. VOCs (benzene, acetone, tetrachloroethene, and trichloroethene) have been detected in the groundwater, and methane gas is being released at the surface. VOCs have also been identified in the nearby Rock River. Residential neighborhoods are within a few hundred feet of the site.

The remediation plan for the ash beds consisted of restricted land use, restricted use of groundwater between the site and the Rock River, groundwater pumping, and installation of an air-stripping groundwater treatment system (1992–1995).

The various PRPs undertook the site investigation study. A Consent Decree (1992) between the parties, the EPA, and the State of Wisconsin calls for the parties to finance the remediation plan.

Sources: PNPLS-WI; Janesville Public Library, 316 South Main Street, Janesville, WI, 53545.

Kohler Company Landfill

Sheboygan County

The Kohler Company Landfill site is about 40 acres in size. Still operational, it began receiving foundry and manufacturing wastes from the Kohler Company in the 1950s. The wastes included enamel powder, brass lint, plating sludges, solvents, paint by-products, hydraulic oils, and sludges from wastewater settling lagoons. VOCs, phenols, PAHs, PCBs, and heavy metals have contaminated the groundwater and soil and have penetrated the Sheboygan River, which is only a few hundred feet away. The river sediments are thought to be toxic to human and aquatic life. Approximately 1,600 people live within a 3-mile radius.

Response actions (1985–1995) included site studies by Kohler Company and the EPA, closing and capping the landfill, restricting access, and building a drainage system to capture runoff and contaminated leachate. A final remediation plan has yet to be determined.

The State of Wisconsin, the EPA, and Kohler Company signed an AOC in 1985 that calls for the company to finance the site investigations.

Sources: PNPLS-WI; Kohler Public Library, 230 School Street, Kohler, WI, 53044.

Lauer I Sanitary Landfill

Waukesha County

This site, which was closed in 1973, began receiving industrial and city wastes in the 1950s. Pollutants in the soil and groundwater include benzene, toluene, cyanide, and zinc, which may affect nearby crops and farm livestock. Contaminated leachate from the capped landfill leaked into the Menominee River in 1974. Nearly 24,000 people live within 3 miles of the site.

A long-term remediation plan is being developed that will include fixing the landfill cap and installing gas and leachate collection systems.

An agreement between Waste Management of Wisconsin and the State of Wisconsin requires the company to finance the site investigation and remediation plan.

Sources: PNPLS-WI; Maude Shunk Library, W156 N8486 Pilgrim Road, Menomonee Falls, WI, 53051.

Lemberger Landfill

Manitowoc County

Originally a gravel pit and open dump from 1940 to 1970, the site was used by Lemberger Landfill Incorporated from 1970 to about 1983. A variety of city and industrial wastes, including thousands of tons of fly ash, were buried at the site. About 3,000 people live within 3 miles of the landfill.

Nearby residents complained that leachate was seeping in their yards. Testing in 1985 showed that the groundwater, surface water, and soils are polluted with VOCs (vinyl chloride and methylene chloride), phenols, heavy metals (cadmium and zinc), SVOCs, pesticides, PCBs, and inorganic compounds. Nearby farmland, crops, livestock, and the recreational Branch River also carry these contaminants.

Initial response actions (1984–1995) included drilling new water wells for affected residents. The remediation plan will consist of restricting access to the site, restricting groundwater use, closing and capping the landfill, building a slurry wall to control runoff, and installing a groundwater extraction and treatment system.

The EPA, the Wisconsin DNR, and the PRPs signed a Consent Decree (1992) to finance the remediation plan.

Sources: PNPLS-WI; Manitowoc Public Library, 808 Hamilton Street, Manitowoc, WI, 54220; Whitelaw Village Hall, 232 E. Menasha Avenue, Whitelaw, WI, 54247.

Lemberger Transport and Recycling
Manitowoc County

About 1 million gallons of paint sludge and tar were buried in this unlined landfill between 1970 and 1976. Other wastes include aluminum dust and PCBs. Neighboring farmers have plowed up drummed waste and bulk waste. Wisconsin DNR studies in 1985 indicated that the groundwater and soil are contaminated with VOCs, phenols, heavy metals (lead, chromium, and aluminum), pesticides, and PCBs. The site is less than a mile from the recreational Branch River, and the nearly 3,000 residents within 3 miles of the site derive their water from private wells.

Nearly 1,500 drums of waste were removed from the site, which was then covered by a composite clay cap. The final remediation plan calls for restricted groundwater use, groundwater monitoring, and the installation of a groundwater extraction and treatment system (1991–1997).

Certain parties have signed an AOC (1993) with the EPA to finance the site investigation and remediation plan.

Sources: PNPLS-WI; Manitowoc Public Library, 808 Hamilton Street, Manitowoc, WI, 54220; Whitelaw Village Hall, 232 E. Menasha Avenue, Whitelaw, WI, 54247.

Madison Metropolitan Sewerage District Lagoons
Dane County

The site is a 135-acre lagoon, or sludge pond, located near wetlands near Nine Springs Creek. The city of Madison has deposited sludge from a nearby sewer treatment plant on the site since 1942. The sludge contains high levels of PCBs. Breaks in the dikes led to the release of thousands of gallons of liquid sludge, which killed fish in Nine Springs Creek and the Yahara River. A mobile home park is within a quarter-mile of the lagoons, and about 94,000 people within 3 miles of the site use well water.

The site is currently being studied, and a long-term remediation plan will be selected by 1997.

Sources: PNPLS-WI; Madison Public Library, 201 West Mifflin Street, Madison, WI, 59703; Madison Metropolitan Sewage District, 1610 Moorland Road, Madison, WI, 53713.

Master Disposal Service Landfill
Waukesha County

This landfill, near the town of Brookfield, operated from 1977 to 1982. It is located on a plateau in the middle of wetlands near the Fox River; some of the wetlands were also filled with solvents, adhesives, oils, paints, and other industrial and foundry wastes. Sampling by the Wisconsin DNR has shown that the groundwater is contaminated with VOCs (benzene, toluene, and xylene), chlorinated solvents, and heavy metals (barium and manganese). City wells within 3 miles of the landfill supply drinking water to about 10,000 residents.

The remediation plan included capping the site and installing a groundwater pump-and-treat system and a gas venting system (1990–1996). The site will continue to be monitored.

Master Disposal Service signed an agreement with the State of Wisconsin to finance the site investigation and remediation plan.

Sources: PNPLS-WI; Brookfield Public Library, 1900 Calhoun Road, Brookfield, WI, 53005.

Mid-State Disposal Landfill
Marathon County

Located about 4 miles northeast of the town of Stratford, this landfill was operated from 1970 to 1979. A variety of commercial, industrial, and municipal wastes were buried there, including asbestos, solvents, paper mill and paint sludges, pesticides, and metals. Handling violations led to the release of 200,000 gallons of toxic leachate into Rock Creek. Dangerous levels of heavy metals (copper, iron, and arsenic), VOCs (benzene, methylene chloride, and vinyl chloride), phthalates, and dieldrin have been detected in the groundwater, surface water, and soils. Surface runoff drains into Rock Creek and the Big Eau Pleine River. Ten homes are located within a 1-mile radius of the landfill.

The State of Wisconsin sued Mid-State in 1979, which forced the closure of the site. Clean-up actions from 1988 to 1994 consisted of restricting groundwater access, fencing the site, gas collection and monitoring, installing on-site leachate and runoff control systems, capping sludge, constructing new soil covers, and long-term groundwater monitoring.

A citizens' group sued Mid-State Disposal, Weyerhauser, and the Wisconsin DNR in 1980 for improper disposal of hazardous waste.

Sources: PNPLS-WI; Marathon County Library, 300 Larch Street, Stratford, WI, 54484.

Moss-American

Milwaukee County

Operations at this northwestern Milwaukee site began in 1921 when the Moss Tie Company began treating railroad ties with creosote. A later owner, Kerr-McGee, operated the site until it was closed in 1976. Contaminants in the groundwater and soil include PAHs, BTX compounds, and free-product creosote. Runoff from the site has contaminated the Little Menomonee River. People wading in the river in 1971 suffered chemical burns, which led to investigations and lawsuits against Kerr-McGee. About 10,000 people within 1 mile of the site use city wells for drinking water.

Initial response actions included dredging the river to remove contaminated sediment and stripping away oil-saturated soil at the site (1973–1978). A final remediation plan consisting of rerouting the Little Menomonee River, installing a groundwater pump-and-treat system, and soil washing and bioremediation using bacteria will be operational by 1997.

Sources: PNPLS-WI; Mill Road Library, 6431 North 76th Street, Milwaukee, WI, 53223.

Muskego Sanitary Landfill

Waukesha County

Located near the town of Muskego, operations at this site consisted of a gravel pit, an animal rendering plant, and sewage wastewater lagoons. In addition to animal carcasses and blood, waste oils and paint were also buried in the dump. The facility was closed in 1981. The following year the Wisconsin DNR determined that the groundwater was contaminated with VOCs and heavy metals (lead and chromium). Nearby wetlands and the Fox River are threatened by the contaminated groundwater and surface runoff. A number of private wells in the area were also contaminated.

Response actions (1985–1994) included connecting affected homes to the city of Muskego's water supply and removing contaminated soil and drums of liquid waste from the site. After extensive study, a remediation plan that included building a landfill cap and installing leachate and gas collection systems was selected.

An AO (1987) and a UAO (1992) between the EPA, the State of Wisconsin, and Wisconsin Waste Management and 40 other PRPs calls for the parties to finance the site investigation and the remediation plan.

Sources: PNPLS-WI; Muskego Public Library, South 8200 Racine Avenue, Muskego, WI, 53150.

N. W. Mauthe Company, Inc.

Outagamie County

The 2-acre property, owned by N. W. Mauthe Company and operated by Wisconsin Chromium Corporation, was the site of a chromium electroplating facility from 1946 to 1976. The soil and groundwater contain anomalous levels of chromium and VOCs. Contaminated groundwater leaked into residential basements, storm sewers, and the Fox River in 1982. About 11,000 people live within 3 miles of the site.

Initial actions (1982) consisted of installing a runoff and shallow groundwater collection system, excavating and shipping contaminated soil, fencing, and paving with asphalt to minimize the contamination of rainwater. The final remediation plan (1992–1994) included extracting and treating the contaminated groundwater.

Sources: PNPLS-WI; Appleton Public Library, 225 North Oneida Street, Appleton, WI, 54911.

National Presto Industries, Inc.

Eau Claire and Chippewa Counties

The U.S. government maintained weapons and manufacturing facilities on this 325-acre site near the city of Eau Claire in the 1940s. National Presto Industries bought the property in 1947 and made consumer goods and parts for weapons and aircraft until 1980. During that period wastewater was discharged into seepage pits, a gravel pit, and constructed lagoons. Groundwater is contaminated with VOCs and heavy metals, and the contamination plume extends for 3 miles to the Eau Claire Well Field, another Superfund site. VOCs are present in Lake Hallie, about a mile north. The water supply for the City of Eau Claire (with a population of 54,000) and the private wells near the site are threatened by the contamination.

Initial response actions (1989–1993) consisted of supplying bottled water to those with threatened water supplies, fencing the site, and pumping wastewater from the lagoons. A permanent alternative drinking water source was provided in 1992. Groundwater extraction wells to contain the plume began operating in 1994. Other techniques that may be used are soil vapor extraction, waste consolidation, and capping.

A CO (1986), a UO (1989), and an AOC (1993) between the State of Wisconsin and National Presto Industries calls for the PRP to finance the site investigation and remediation plan.

Sources: PNPLS-WI; Chippewa Falls Public Library, 105 West Central Road, Chippewa Falls, WI, 54729; Hallie Town Hall, Route 9, Hagen Road, Chippewa Falls, WI, 54729.

Northern Engraving Company

Monroe County

This site is owned by a company that manufactures metal accessories for automobiles. From about 1960 to 1980, wastewater, sludge, and metal finishing by-products were deposited in dumps, seepage pits, and sludge lagoons. Groundwater and soil are contaminated by VOCs (trichloroethylene) and heavy metals (copper, chromium, iron, zinc, and nickel), and fluoride.

Initial response actions (1988) included removing or stabilizing sludge and soil, covering the lagoon, and installing groundwater monitoring wells. The surface clean up has reduced the contamination in the groundwater, and monitoring will continue.

The PRPs signed a Consent Decree with the EPA and the State of Wisconsin that requires the companies to fund the site investigation and remediation plan.

Sources: PNPLS-WI; Sparta Public Library, West Main and Court Streets, Sparta, WI, 54656.

Oconomowoc Electroplating Company, Inc.

Dodge County

The Oconomowoc Electroplating Company, which operated from 1957 to 1992, is located beside Davy Creek near the town of Ashippun. Untreated wastewater laden with heavy metals was discharged into the wetlands beside the creek. Sludges were later placed in unlined lagoons. Plating wastes and chemical spills have corroded the concrete floor in the plant and contaminated the ground. Pollutants in the groundwater, surface water, and soils include VOCs, heavy metals (arsenic, cadmium, copper, lead, and chromium), and cyanide. These threaten water supplies and may be present in toxic levels in wetland wildlife. About 1,500 people live within 3 miles of the plant.

Response actions (1987–1991) included fencing, removing drums of waste, and razing the plant. Future remediations will include a installing groundwater pumping and treatment system, removing the lagoons, and excavating the contaminated soil from the site and wetlands (1993–1997).

The Oconomowoc Electroplating Company declared bankruptcy in 1991.

Sources: PNPLS-WI; F & M Bank, North 533, Highway 67, Ashippun, WI, 53003.

Old Janesville Landfill

Rock County

This abandoned gravel pit occupies 18 acres on the north side of Janesville and is owned by the city. It was used as an unlined municipal sanitary landfill from 1963 to 1978, receiving industrial drummed wastes, solvents, oils, paints, paint thinners, and sludge-ash mixtures. The site was closed in 1978 and covered with clay. Methane gas has been detected in the air. VOCs and heavy metals such as arsenic, barium, lead, iron, manganese, and cadmium have contaminated the soil and groundwater and have leaked into the Rock River. Residential areas are within a few hundred feet of the landfill site.

EPA response actions (1991–1995) have consisted of restricting land use; installing a landfill gas and flaring system, a new landfill cap, groundwater extraction wells, and an air-stripper groundwater treatment system; and continued monitoring.

The PRPs have agreed to remediate the site.

Sources: PNPLS-WI; Janesville Public Library, 316 South Main Street, Janesville, WI, 53545.

Omega Hills North Landfill

Washington County

From 1977 to 1982 this landfill site near metropolitan Milwaukee accepted about 30,000 tons of hazardous waste, including heavy metals, solvents, liquid wastes, and asbestos. The site closed in 1989. A built-in leachate collection system malfunctioned, and about 200 million tons of liquid waste accumulated under the landfill. The shallow groundwater is contaminated with VOCs (benzene, toluene, vinyl chloride, and trichloroethylene), zinc, nickel, arsenic, cadmium, phthalates, cyanide, gases, petrochemicals, and pesticides. About 42,000 people live within 3 miles of the site and use private and municipal wells in the area.

Response actions have included fencing the site; building leachate, runoff, and gas collection systems; and installing a pretreatment plant for collected leachate. A final remediation plan has yet to be selected.

Source: PNPLS-WI.

Onalaska Municipal Landfill

La Crosse County

Originally a sand and gravel pit in the 1960s, this site was operated as a landfill for the town of Onalaska from 1969 to 1980. Wastes dumped in the pit include about 2,500 drums of liquid waste, a 500-gallon tank truck containing paint residues, and industrial chemicals such as naphtha and toluene. The groundwater and soil are contaminated with VOCs (trichloroethylene), toluene, naphtha, free-product petroleum and hydrocarbons, and barium. The groundwater flows away from the site towards the recreational Black River, Lake Onalaska, and the Upper Mississippi Valley Wildlife Refuge. Although it is located in a rural agricultural area, it is close to residential homes and a recreational club. About 300 people live within 1 mile of the landfill.

Response actions included capping the site and replacing a contaminated private well (1982). An insitu bioremediation and groundwater pumping and treatment system was constructed in 1994. A gas venting system will also be installed.

Sources: PNPLS-WI; La Crosse County Public Library, Onalaska Branch, 741 Oak Avenue South, Onalaska, WI, 54650; Holmen Library, 103 State Street, P.O. Box 539, Holmen, WI, 54636.

Penta Wood Products

Burnett County

This former wood treatment facility near the town of Siren operated from 1953–1992. Lumber was treated with PCP, fuel oil, ammonia-copper-zinc-arsenate, and ammonia-copper-arsenate. Testing in 1988 showed that various spills, leaks, and drips have contaminated the soil and groundwater with PCP, arsenic, zinc, and copper. The contamination threatens the private and municipal water supplies of about 3,000 residents within 4 miles of the site.

Response actions included removing contaminated soil and waste solutions from the lagoons (1992). Drummed wastes were also removed. Additional studies are under way.

The PRP, Penta Wood Products, declared bankruptcy in 1992.

Source: EPA Publication 9320.7-071, Volume 3, No. 1.

Refuse Hideaway Landfill

Dane County

The Refuse Hideaway Landfill, located in a rural area about 2 miles west of Middleton, accepted a variety of industrial, commercial, and city wastes from 1974 to 1988. These included barrels of glue and paint, paint stripper sludge, and VOCs (vinyl chloride, methylene chloride, and acetone). A year after closing the landfill, the owner declared bankruptcy. A Wisconsin DNR investigation revealed that the landfill was poorly capped and that nearby wells were contaminated by a toxic plume nearly a mile in length. About 15,000 people reside within 4 miles of the site.

Response actions included installing treatment systems on contaminated wells and constructing methane gas and leachate collection systems (1988–1991). A final clean-up plan has yet to be selected.

Source: PNPLS-WI.

Ripon City Landfill

Fond du Lac County

This site is in a rural, agricultural area about a mile northwest of Ripon. The privately owned property was leased by Ripon as a landfill site from 1967 to 1983. The groundwater is contaminated with VOCs, including benzene, xylene, and vinyl chloride. The cities of Ripon and Green Lake (with a combined population of 13,000) pump drinking water from wells within 4 miles of the site, and about 2,000 people in the area use private wells.

Site investigations began in 1984, and a final remediation plan has yet to be selected.

Source: PNPLS-WI.

Sauk County Landfill

Sauk County

This site was operated as a landfill by Sauk County from 1973 to 1983. Among the municipal and industrial wastes buried there were lead- and cadmium-bearing dusts from a local foundry. EPA inspections in 1985 noted the presence of methane gas and found that VOCs and heavy metals had contaminated the groundwater. About 900 people use private wells within 3 miles of the site.

Response actions included building a clay cap over the landfill (1983), installing a gas collection system, and monitoring groundwater (1994–1995). A groundwater pump and treatment system may be installed later.

Sauk County, in a 1991 agreement with the State of Wisconsin, agreed to finance the site investigation.

Sources: PNPLS-WI; Reedsburg Public Library, 345 Vine Street, Reedsburg, WI, 53959; Baraboo Public Library, 230 Fourth Avenue, Baraboo, WI, 53913.

Schmalz Dump

Calumet County

Located 500 feet from Lake Winnebago, this site was an unauthorized dumping ground for various businesses in the 1960s and 1970s. Refuse included car parts, pulp chips and ash from paper mills and utility companies, and PCB-bearing demolition debris from construction projects. Adjacent wetlands were also filled in. Groundwater and soils are contaminated with heavy metals (lead and chromium) and PCBs. About 60 homes and businesses are within 1,000 feet of the dump, and Lake Winnebago is a major water source for the city of Appleton, whose population is about 60,000.

Response actions included security fencing (1985), removing 3,500 cubic yards of contaminated soil (1988), installing a seeded soil cap (1994), and groundwater monitoring.

Agreements between the EPA and two of the eight PRPs required them to pay for the costs of the soil cap and an additional two million dollars.

Sources: PNPLS-WI; University of Wisconsin Center, Fox Valley Library, 1478 Midway Road, Menasha, WI, 54952.

Scrap Processing Company

Taylor County

This site has operated as a scrap processing facility and salvage yard. Between 1955 and 1981 lead and acid batteries were crushed on-site to recover lead, which produced about 400,000 gallons of acid that ran into an unlined pond. Surface water containing acid and heavy metals has migrated into the recreational Black River, a popular fishing stream. Soils contain anomalous levels of heavy metals (lead, barium, copper, and zinc) and PCBs. A number of homes, including those in a mobile home park, are located beside the site and use private wells. To date these wells have not shown any contamination of the groundwater.

Initial response actions included draining the holding pond (1984) and removing the contaminated soil from the property (1986). An ongoing EPA investigation will determine if additional clean up is necessary.

A 1983 EA by the State of Wisconsin against Scrap Processing, the PRP, forced the company to finance the initial clean-up work.

Sources: PNPLS-WI; Medford Public Library, 104 E. Perkins Street, Medford, WI, 54451.

Sheboygan Harbor and River

Sheboygan County

This site downstream from Tecumseh Products Company, a die casting plant in Sheboygan Falls, consists of the lower Sheboygan River and Sheboygan Harbor on Lake Michigan. Affected communities are Sheboygan Falls, Kohler, and Sheboygan. High PCB levels have been detected in fish, wildlife, surface water, and river sediments since 1977. Heavy metals (copper, lead, zinc, arsenic, and chromium) are also present. Anomalous levels of PCBs have been detected in lake sediments less than a mile from the drinking water intakes in Lake Michigan for the Sheboygan area.

Initial response actions included dredging contaminated sediments from the river and testing different biotreatment techniques (1989–1991). Detailed investigations are ongoing and will result in the selection of a final treatment plan.

Consent Orders (1986, 1990) between Tecumseh Products Company and the EPA require the company to finance the investigations and the removal of the contaminated river sediments.

Sources: PNPLS-WI; Mead Public Library, 710 Plaza Eight, Sheboygan, WI, 53801; Sheboygan City Hall, 828 Center Avenue, 2nd Floor, Sheboygan, WI, 53801.

Spickler Landfill

Marathon County

The Spickler site operated as a landfill from 1970 to 1976. Various municipal and industrial wastes, including asbestos dust and mercury sludge, were buried in the pit. Leachate has seeped from the site onto adjacent private property. The landfill is perched on top of an aquifer that provides well water to 2,000 homes within 3 miles of the site. Site investigations in 1984 have shown that the groundwater in the aquifer is contaminated with VOCs (toluene) and heavy metals (barium).

Response actions have included the construction of clay caps and leachate and gas collection systems (1992–1995). Ongoing groundwater monitoring will determine if a groundwater treatment system is needed.

Weyerhauser Industries, BASF Inc., and Weinbrennar Company, through a CO, AOC (1992), and UAO (1994) with the State of Wisconsin and the EPA, are required to finance the site investigation and remediation plans.

Sources: PNPLS-WI; Spencer Village Hall, 117 East Clark Street, Spencer, WI, 54479.

Stoughton City Landfill

Dane County

Operated as a landfill by the city of Stoughton from 1952 to 1982, this site accepted a variety of industrial and municipal wastes, including plastic and rubber by-products, solvents, liquid chemicals, and vinyl plastics. Stoughton planned to build a park on the closed site until the Wisconsin DNR discovered high levels of VOCs and arsenic in the groundwater in 1983. Soils contain PAHs, phthalates, zinc, cadmium, and lead. The dump is bordered by wetlands and the Yahara River, and about 10,000 people live within 3 miles of the site.

The remediation plan calls for a multilayer cap, fencing, restricted groundwater use, and groundwater pumping and treatment. Construction is expected to be under way by 1997.

A 1988 Consent Order between the EPA and Uniroyal, Inc., and the city of Stoughton called for the parties to fund the investigation. The EPA is now funding the clean up because Uniroyal has declared bankruptcy and Stoughton does not have the necessary funds.

Sources: PNPLS-WI; Stoughton Public Library, 304 South Fourth Street, Stoughton, WI, 53589; Stoughton City Hall, 381 East Main Street, Stoughton, WI, 53589.

Tomah Armory

Monroe County

The city of Tomah operated this site as an unlined dump from the early 1950s to 1955. Among the municipal and industrial wastes deposited were solvents and heavy metals from a polyethylene plant in Tomah. The groundwater, soil, and surface water are contaminated with VOCs and heavy metals (lead and chromium). About 9,500 people live within 3 miles of the dump, and the recreational Lemonweir River is about 500 feet away.

Response actions have been initiated with a site investigation (1993–1996). A final remediation plan has yet to be selected.

Sources: PNPLS-WI; Tomah Public Library, 716 Superior Avenue, Tomah, WI, 54660.

Tomah Fairgrounds

Monroe County

This site was operated by the city of Tomah as an unlined dump from 1953 to 1959. Among the industrial and municipal wastes received were chemicals and heavy metals from a polyethylene plant in Tomah. The dump was later covered, seeded, and developed as the Tomah fairgrounds. The groundwater, soil, and surface water contain high levels of VOCs, chromium, and lead. About 4,100 people live within 1 mile of the site. Recreational Lake Tomah is about 400 feet away downslope.

A site investigation was begun in 1993. The results will be used to determine a final remediation plan.

Sources: PNPLS-WI; Tomah Public Library, 716 Superior Avenue, Tomah, WI, 54660.

Tomah Municipal Sanitary Landfill

Monroe County

The city of Tomah operated this site as a licensed, unlined landfill from 1959 to 1979. Industrial waste included over 1,500 barrels containing heavy metals (including barium), solvents, ethyl acetate, and trichloroethylene. An EPA investigation in 1984 found that the groundwater and sediments in nearby Deer Creek are contaminated by VOCs, cadmium, chromium, and lead. Tomah's 7,300 residents get their water from city wells within 3 miles of the site. Some of the private wells beside the landfill have been contaminated.

Response actions have included drilling new wells for affected homes and connecting other residences to municipal water (1993). An ongoing site investigation was started in 1994.

Sources: PNPLS-WI; Tomah Public Library, 716 Superior Avenue, Tomah, WI, 54660.

Waste Management of Wisconsin Brookfield Sanitary Landfill

Waukesha County

This abandoned gravel pit was operated by Waste Management of Wisconsin as a sanitary landfill from 1969 to 1981. Testing by the EPA in 1985 showed anomalous levels of VOCs (vinyl chloride) and cyanide in the groundwater. About 11,000 people obtain water from wells within 3 miles of the site, which is also less than a mile from recreational wetlands and streams.

A detailed site investigation is presently under way.

Source: PNPLS-WI.

Waste Research & Reclamation Company

Eau Claire County

Contamination at this site was caused by inadequate waste handling procedures at the Waste Research and Reclamation Company, which has been in operation since 1981 and recycles hazardous waste, blends fuel, and transports hazardous waste for disposal (including flammables and fluorinated and chlorinated solvents). About 150 people live within a mile of the site, which is also close to Lowes Creek and the Chippewa River. The groundwater and soil is contaminated with VOCs.

A detailed site investigation has recommended that the site be removed from the Superfund program and be handled under the Resource Conservation and Recovery Act (RCRA).

Sources: PNPLS-WI; Phillips Memorial Library, 400 Eau Claire Street, Eau Claire, WI, 54701.

Wausau Groundwater Contamination

Marathon County

Tests on Wausau's groundwater wells near the Wisconsin River in 1983 found that three were contaminated with VOCs. Most of the city's 33,000 residents live within 3 miles of the contaminated wells.

Response actions included the installation of carbon filters and air strippers on affected wells (1984), groundwater pumping and air-stripping (1990), and soil vapor extraction (1992). Vapor phase carbon units may also be installed to treat the gases produced by the soil vapor extraction.

A Consent Decree (1990) between the EPA and the PRPs requires the parties to finance part of the remediation plan.

Sources: PNPLS-WI; Marathon County Public Library, 400 First Street, Wausau, WI, 54401.

Wheeler Pit

Rock County

The abandoned Wheeler pit, about 2 miles east of Janesville, was used as a landfill from 1956 to 1974. General Motors Corporation disposed of paint, wastewater sludges, and coal ashes from its assembly plant in Janesville. Tests by GMC and the Wisconsin DNR in 1981 showed that the groundwater was contaminated with VOCs (including trichloroethylene), zinc, arsenic, barium, iron, manganese, and chromium. Heavy metals (including cadmium and lead) and SVOCs were found in soils. Janesville's water supply wells, the Rock River, and about 51,000 people are located within 3 miles of the site.

Response actions included fencing, installing a multilayer cap, restricting groundwater and land use, and long-term groundwater monitoring (1991–1993). Groundwater monitoring is continuing.

Sources: PNPLS-WI; Janesville Public Library, 316 South Main Street, Janesville, WI, 53545.

WYOMING

Baxter/Union Pacific
Albany County
Railroad ties were treated with creosote oil, asphalt oil, and PCP on this property near Laramie from 1886 to 1983. Contaminated wastewaters were discharged into an unlined lagoon on the property until the mid-1950s. In 1981 the EPA discovered that the groundwater and soils contain PAHs and PCBs. The contamination is flowing away from the Laramie River, the main source of water for the 24,000 residents of Laramie. Contact with on-site contaminants is still a threat.

Response actions included the following interim measures: moving the channel of the Laramie River further away from the site, building a slurry wall, and extracting and treating contaminated groundwater (1986–1990). In 1991 the EPA recommended building a hydraulic slurry wall, recovering free-product creosote oil, and capping. These activities will continue until about 1998.

The PRPs have participated in the site studies and remediation work.

Source: PNPLS-WY.

F. E. Warren Air Force Base
Laramie County
This U.S. Air Force base near Cheyenne was established in 1947 and became a Strategic Air Command center in 1958. Various wastes from base activities, including spent solvents, degreasers, acids, fuels, oils, and petrochemicals, were stored, leaked, spilled, discharged, or buried on the 5,900-acre property. The groundwater and soils contain VOCs, fuel, and petrochemical by-products. The contamination threatens Crow and Diamond Creeks and the private water wells of about 2,500 residents.

Response actions included removing spent acids and contaminated soil and sludge (1986, 1989). A number of site investigations are under way.

Sources: PNPLS-WY; Laramie Public Library, 2800 Central Avenue, Cheyenne, WY, 82001.

Mystery Bridge
Natrona County
An oil refinery and natural gas processing station operated in this area near Evansville from the 1950s to the 1980s. Liquid wastes were stored in an unlined pond from 1965 to 1984. Toluene and other VOCs were stored on site in drums and tanks. The area has since been developed into a residential neighborhood. Tests in 1986 revealed that the groundwater and soils contain VOCs and hydrocarbons. The contamination is a threat to local groundwater supplies.

Response actions included supplying affected residents with clean water supplies (1987–1988), removing drummed waste and sludges (1988), and building interim groundwater pump-and-treat and soil vapor extraction systems (1988). In 1990 the EPA recommended constructing three groundwater extraction and air-stripping systems for three contamination plumes; so far two are in operation.

Consent Orders (1988) require three PRPs to conduct the site investigations and remediation work.

Source: PNPLS-WY.

WYOMING

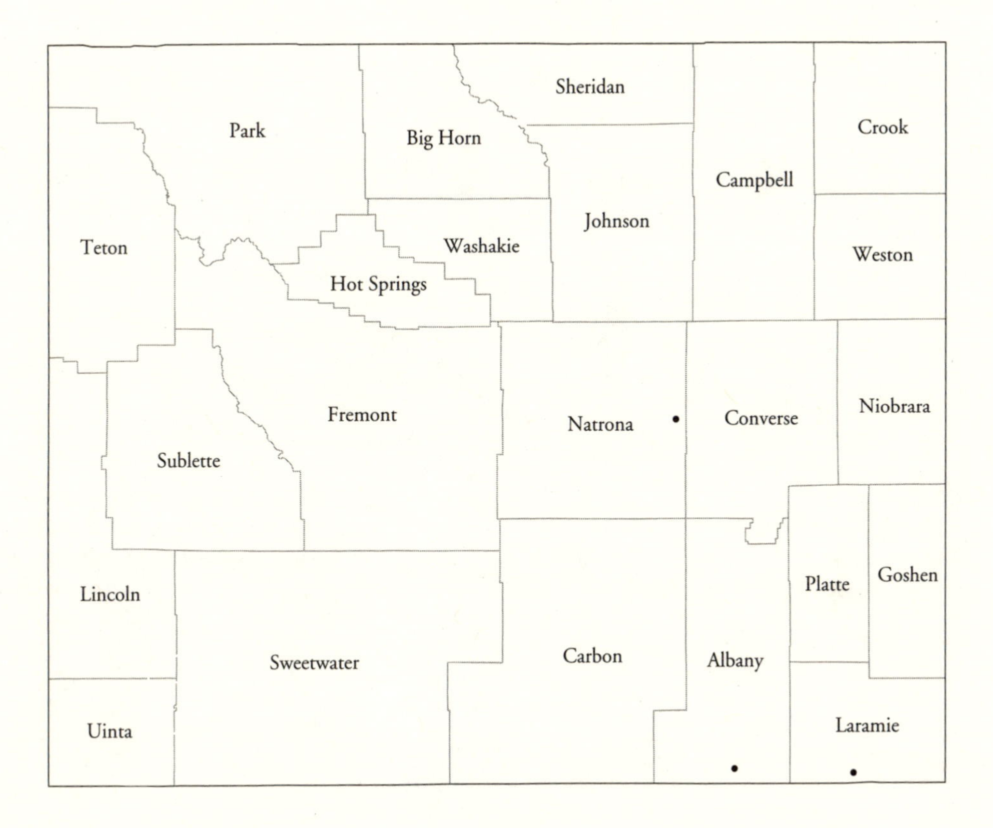
Park
Big Horn
Sheridan
Crook
Campbell
Johnson
Teton
Washakie
Weston
Hot Springs
Fremont
Natrona
Converse
Niobrara
Sublette
Lincoln
Platte
Goshen
Sweetwater
Carbon
Albany
Uinta
Laramie

APPENDIX A

The Distribution of Superfund Sites by State and Type (including 13 proposed NPL sites)

State	*Landfill*	*Industrial*	*Federal*	*Mining*	*Total*
Alabama	0	10	3	0	13
Alaska	0	3	5	0	8
Arizona	2	5	3	0	10
Arkansas	4	8	0	0	12
California	10	58	19	8	95
Colorado	2	6	3	7	18
Connecticut	6	9	1	0	16
Delaware	7	11	1	0	19
Florida	16	37	6	0	59
Georgia	6	7	1	0	14
Guam	1	0	1	0	2
Hawaii	0	1	3	0	4
Idaho	0	7	1	3	11
Illinois	12	26	3	0	41
Indiana	17	16	0	0	33
Iowa	8	9	1	0	18
Kansas	3	7	2	1	13
Kentucky	9	6	1	0	16
Louisiana	3	15	1	0	19
Maine	4	5	3	0	12
Maryland	4	6	4	0	14
Massachusetts	5	19	7	0	31
Michigan	25	46	2	1	74
Minnesota	11	26	3	0	40
Mississippi	0	5	0	0	5
Missouri	3	13	3	2	21
Montana	0	5	0	4	9
Nebraska	0	8	2	0	10
Nevada	0	0	0	1	1
New Hampshire	7	10	1	0	18
New Jersey	40	65	4	0	109
New Mexico	1	7	0	4	12
New York	25	51	5	0	81
North Carolina	0	21	2	0	23
North Dakota	1	1	0	0	2
Ohio	14	20	5	0	39
Oklahoma	4	5	1	1	11

(continues)

The Distribution of Superfund Sites by State and Type *(continued)*

State	*Landfill*	*Industrial*	*Federal*	*Mining*	*Total*
Oregon	0	11	1	1	13
Pennsylvania	42	53	7	2	104
Puerto Rico	3	6	1	0	10
Rhode Island	8	2	2	0	12
South Carolina	4	20	2	0	26
South Dakota	0	1	1	2	4
Tennessee	6	8	4	0	18
Texas	5	22	2	0	29
Utah	0	6	3	7	16
Vermont	6	2	0	0	8
Virgin Islands (U.S.)	0	2	0	0	2
Virginia	7	11	7	1	26
Washington	13	27	17	1	58
West Virginia	1	3	2	0	6
Wisconsin	25	17	0	0	42
Wyoming	0	2	1	0	3
TOTAL	370	747	147	46	1,310

APPENDIX B: COMMON CONTAMINANTS FROM WASTES IN SUPERFUND SITES

VOCs (Volatile Organic Compounds)
acetone
acetylene
benzene
bromoform
BTX compounds (benzene-toluene-xylene)
butanone
carbon disulfide
carbon tetrachloride
chlorinated aromatic compounds
chloroform
dibromo-3-chloropropane
dichloroethene
dichlorobenzene
dichloroethane
dichloroethylene
dimethyl ether
dinitrotoluene
ethyl acetate
ethylbenzene
ethylene
ethylene dibromide (fumigant)
freon
halocarbons (for example, trichloroethane)
heptane
hexane
hexanone
isopropyl alcohol
methane
methanol
methyl ethyl ketone
methylene chloride
methylisobutylketone (MIBK)
monochlorobenzene
naphtha
perchloroethylene (PCE)
phosgene gas (chemical warfare agent)
pyridines
styrene
tetrachloroethene
tetrachloroethylene
tetrahydrofurans
toluene
trichloroethane
trichloroethylene (TCE)
vinyl chloride
xylene

SVOCs (Semi-Volatile Organic Compounds)
acenaphthylene
aniline
anthracene
arochlor (a form of PCB)
benz(a)anthracene
benzidine
benzo(a)pyrene
benzonitrile
carcinogenic hydrocarbon (CPAH)
chlorinated dibenzo-dioxin
chlorinated dibenzo-furan
chrysene
creosote
crystal violet (purple dye)
cyclonite (RDX) (explosive)
dibenzofuran
dichlorobenzene
dinitrotoluene (DNT) (explosive)
ethylene glycol (antifreeze)
fluorene
fluoroanthene
furans
hexachlorobutadiene
hexachlorophene
isophorone
kerosene
MOCA (industrial hygiene analyte)
naphthalene
nitroguanidine
pentachlorophenol (PCP)
phenanthrene
phenol
phthalates (common plastics by-product)
polybrominated biphenyl (PBB)
polychlorinated biphenyl (PCB)

polychlorinated terphenyl (PCT)
polycyclic aromatic hydrocarbon (PAH)
polymethyl methacrylate
polynuclear aromatic hydrocarbon (PNA)
polyvinyl chloride
pyrene
tetrachlorodiobenzo(p)dioxin (TCDD)
tetryl (explosive)
thiophene
toluene diamine
toluene disocyanate
tributyltin (TBT)
trinitrobenzene
trinitrotoluene (TNT) (explosive)
trischloropropyl phosphate (flame retardant)

Herbicides
2,4-D
2,4,5-T
2-4-5-TP (silvex)
atrazine
fenac
silvex (2-4-5-TP)

Pesticides
aldrin
atrazine
benzene hexachloride (lindane)
chlordane
DDD
DDE
DDT
diazinon
dicamba
dieldrin
dinoseb
endrin
fenac
heptachlor
hexachlorobenzene
kepone
lewisite (chemical warfare agent with DDT)
lindane (benzenehexachloride)
methyl parathion
mirex
pentachlorophenol (PCP)
thiocarbamates
toxaphene

Inorganic Compounds
aluminum
americium (radioactive)
ammonium picrate
antimony (heavy metal)
arsenic (heavy metal)
asbestos
barium (heavy metal)
beryllium (heavy metal)
bismuth (heavy metal)
boron
cadmium (heavy metal)
carbon-14 (radioactive)
chemonite (copper-arsenic-zinc salt)
chromic acid
chromium (heavy metal)
cobalt (heavy metal)
copper (heavy metal)
cupric ammonium carbonate
cyanide
ferric ferrocyanide
iodine
lead (heavy metal)
manganese (heavy metal)
mercury (heavy metal)
molybdenum (heavy metal)
nickel (heavy metal)
nitrate
plutonium (radioactive)
radionuclides (radioactive)
radium (radioactive)
radium-226
radon gas
radon-222 (radioactive)
selenium (heavy metal)
silver (heavy metal)
sodium chloride
sodium hydroxide
sodium hypochlorite
strontium (heavy metal)
sulfate
sulfur monochloride
sulfuric acid
technetium-99 (radioactive)
thiocyanate
thionyl chloride (corrosive)
thorium (radioactive)
tributyltin (TBT) (paint additive)
tritium (radioactive)
uranium (radioactive)
vanadium pentoxide
zinc (heavy metal)

APPENDIX C: GLOSSARY OF COMMON TERMS

Administrative Order (AO): See Unilateral Administrative Order (UAO).

Administrative Order on Consent (AOC): A voluntary legal agreement between the EPA and potentially responsible parties (PRPs) that outlines the requirements for site studies or clean-up work and does not require the approval of a judge. Also called a Consent Order.

Air-Sparging: The process by which air is forced into groundwater, enhancing bioremediation and increasing volatilization of contaminants.

Air-Stripping: The process by which an airstream is forced through VOC-contaminated groundwater. The VOCs are evaporated into the air; the air is then treated to remove the VOCs.

Aquifer: A rock or sediment that is permeable, contains water, and can transmit economical quantities of water to wells.

Baghouse Dust: The accumulation of tiny particulate matter from an airstream that passes through a fiberglass baghouse filter.

Bioremediation (Biodegradation): The use of microorganisms (bacteria) to digest contaminants, breaking them down into nonhazardous components.

Bioventing: The introduction of oxygen into contaminated soil by injection or extraction to stimulate the biodegradation of the soil.

Capping: A process that covers an area of contamination with clay, soil, or a geosynthetic membrane that prevents rainwater from entering the zone of contamination.

Carbon Adsorption: A remediation techique that forces contaminated groundwater through tanks that contain activated carbon. The contaminants adhere to the carbon.

Catalytic Oxidation: A common remediation technique for fuel contamination in which organic compounds and VOCs are destroyed by passing the contaminated air stream through a catalyst at lower temperatures (about 1500°F).

Chromated Copper Arsenate: An insecticide made from copper, chromium, and arsenic that is used extensively as a wood preservative.

Clarifier: A vessel that allows contaminants such as heavy metal compounds to settle from the groundwater, making it clearer.

Consent Decree (CD): A legal agreement between the EPA and the potentially responsible party (PRP) that generally details the clean-up actions as outlined in the Record of Decision. A consent decree is issued by a judge and is subject to a public comment period.

Consent Order (CO): A voluntary legal agreement between the EPA and potentially responsible parties (PRPs) that outlines the requirements for site studies or clean-up work and does not require the approval of a judge. Also called an Administrative Order on Consent.

Contamination Plume: An area of groundwater contamination with definable limits that can usually be shown to originate from one or more point sources.

Creosote: Wood preservation chemicals, including PAHs and PNAs, derived from the distillation of tar.

Dechlorination: The removal of chlorine from contaminated soil or groundwater. This usually involves the breakdown of chlorinated VOCs and the release of chlorine to form derivative compounds that have no chlorine. It is similar to the dehalogenation process for these compounds, which uses alkaline polyethylene glycolate.

Degreasers: Cleaning agents used in manufacturing, especially for cleaning metal parts. VOC degreasers include TCE.

Dioxins: Compounds that generally result from insufficient combustion of chlorinated compounds.

Enforcement Action (EA): Any action from a state or federal agency that requires a PRP to provide information or comply with state or federal regulations.

Electrochemical Treatment: A remediation technique that applies an electrical current to contaminated groundwater, which drives certain contaminants, especially metals, to anodes or cathodes.

Fixed Film: A remediation technique used in structures contaminated by radioactivity to capture radioactive emissions.

Flocculation: The process by which solid particles in water are increased in size by chemical or biological actions so that they can be more easily separated.

Free-Product: Liquid-phase organic material that does not dissolve in water, such as petroleum products. Free-product generally forms a layer on top of the water table.

Free-Product Recovery or Extraction: The pumping or passive separation and collection of undissolved liquid-phase organics from the groundwater (usually petroleum or creosote).

Fugitive Dust: Dust that is not captured by control or filtration devices.

Furan: A flammable liquid derived from wood oils or produced synthetically that is used in the manufacture of organic compounds.

Gas Collection, Venting, or Flaring: The release of gas, generally methane, from landfills by allowing the pressurized gas to vent through drill holes or by applying a vacuum. The gas is commonly burned or flared.

Gasification: The process of converting soft coal into gas, which results in large amounts of coal tar waste.

Greensand Treatment: The removal of iron and other heavy metal particles from groundwater by passing the groundwater through a glauconitic sand filter.

Groundwater: The water contained in the interconnected pores in sediment or in rock fractures below the water table.

Groundwater Treatment Wall: A trench that is filled with iron fillings or other reactive compounds that will react with the contaminants in the groundwater. As the groundwater passes through the wall, the contaminants are removed through a chemical reaction in the wall.

Halogenated Compounds: VOCs or SVOCs that typically contain chlorine or other elements from the halogen group. Frequently used as degreasers and in the dry cleaning industry, they are a special concern because they are so reactive.

Heavy Metals: Metallic elements, including iron, copper, lead, zinc, chromium, magnesium, barium, cesium, selenium, nickel, cobalt, mercury, cadmium, and strontium.

Injection Wells: Wells that pump treated water into the water table.

In-Situ: A technique that does not require the removal or extraction of the contaminated soil, sediment, or groundwater; the material is treated in place.

Interagency Agreement: An agreement between different federal agencies, such as the EPA and the U.S. Army Corps of Engineers, that usually provides funding for a joint project.

Ion Exchange: The exchange of one compound in contaminated groundwater for another that is less hazardous.

Landfarming: The application of contaminated waste, usually soils, to the land surface and periodically tilling to help aerate and biodegrade the waste.

Leachate: The water that percolates downward through landfills and leaches soluble compounds from the waste. Leachate accumulates at the bottom of landfills and contains a wide range of contaminants.

Leaking Underground Storage Tank (LUST) Program: A federal program funded by gasoline taxes that oversees the clean up of contamination from underground storage tanks that have leaked petrochemicals into the surrounding soil and groundwater.

Low-Temperature Thermal Treatment, Aeration, or Desorption: A treatment that heats contaminated soil to temperatures between 200° and 600°F, volatilizing the organic contaminants.

Natural Attenuation: The reduction of contaminant levels in groundwater through the natural processes of dilution, volatilization, biodegradation, and adsorption.

Polychlorinated Biphenyl (PCB): A class of compounds derived from oils, lubricating oils, transformers, and machine shop operations.

Polycyclic Aromatic Hydrocarbons (PAHs): A class of organic compounds derived from heavy fuels, coals, and decaying organic matter, often from incomplete combustion.

Pentachlorophenol (PCP): A synthetic petrochemical that is actually a pesticide but is commonly used in wood treatment.

Phenols: Very poisonous organic compounds derived from petroleum refining and from the manufacture of plastics, resins, textiles, and dyes.

Phosphate Treatment: A remediation technique for soils contaminated with lead, zinc, and other heavy metals.

Phthalates: A group of compounds derived from plastics and plasticizers.

PNPLS: Acronym listed in the source note for each entry that refers to annual EPA reports for each state titled "Superfund: Progress at National Priority List Sites."

Polynuclear Aromatic Hydrocarbons (PNAs): A class of compounds that are highly reactive and commonly found in creosotes.

Potentially Responsible Party (PRP): A business operation or individual determined by the EPA to be responsible for contamination at a Superfund site. A "viable" PRP is one that has sufficient funds to finance a significant part of the clean up.

Pump-and-Treat: A general term for the common groundwater remediation strategy of drilling extraction wells, pumping out the contaminated groundwater, and treating the groundwater using a number of different techniques.

Record of Decision (ROD): A public document that explains the remediation techniques that have been selected to clean up a site.

Sand Filtration: The removal of solids from contaminated water by passing the water through a sand filter.

Semi-volatile Organic Compounds (SVOCs): A class of compounds similar to VOCs but that vaporize less readily.

Slurry Wall: An underground wall or trench filled with bentonite clay that is designed to change the direction of flow of the contaminated groundwater.

Soil Flushing: The injection of water or water containing a cleanser into the contaminated soil. The contaminants are removed into the water, which is then extracted and treated.

Soil Vapor Extraction: The excavation of contaminated soil, to which a vacuum is applied through a network of pipes that draws off contaminating vapors. Alternatively, the area is capped, a vacuum is applied through a series of wells, and the vapors are removed from between soil particles.

Soil Venting: See Bioventing.

Soil Washing: Similar to soil flushing, except the contaminated soil is dug out and treated or washed to remove the contaminants.

Solidification/Stabilization: The binding of contaminants and soil through chemical reactions that render the contaminants inert. The solidified material is then reburied.

Sump-Pump Treatment: The use of a central pit in landfills to collect and remove leachate.

Thermal Destruction (Incineration): The use of very high temperatures (about 2,000°F) to volatilize and combust organic contaminants in hazardous waste.

Ultraviolet Oxidation: Destruction of organic contaminants in groundwater by passing the groundwater through a tank that subjects the water to ultraviolet radiation, ozone, or hydrogen peroxide.

Unilateral Administrative Order (UAO): An order from the EPA to potentially responsible parties (PRPs) that requires them to undertake clean-up actions at a site. A UAO does not require the approval of a judge.

Volatile Organic Compounds (VOCs): A class of compounds that readily evaporate and are commonly used for solvents, degreasers, paints, thinners, and fuels.

Vitrification: The process by which contaminated soils are subjected to high temperatures that melt them into a glassy substance that is inert and resists leaching.

Water Table: The upper surface of the zone of saturation in the top of the groundwater, or subsurface.

APPENDIX D: ADDRESSES FOR EPA REGIONAL OFFICES

Region 1: Connecticut, Maine, Massachusetts, New Hampshire, Rhode Island, and Vermont
Waste Management Division, HAA-CAN-1
John F. Kennedy Federal Building
Boston, MA 02203-2211
(617) 565-3420

Region 2: New Jersey, New York, Puerto Rico, and Virgin Islands (U.S.)
Emergency and Remedial Response Division
290 Broadway, 19th Floor
New York, NY 10007-1866
(212) 637-3000

Region 3: Delaware, District of Columbia, Maryland, Pennsylvania, Virginia, and West Virginia
Site Assessment Section, 3HW33
841 Chestnut Building
Philadelphia, PA 19107
(215) 566-5000

Region 4: Alabama, Florida, Georgia, Kentucky, Mississippi, North Carolina, South Carolina, and Tennessee
Waste Management Division
100 Alabama Street S.W.
Atlanta, GA 30303
(404) 562-8357

Region 5: Illinois, Indiana, Michigan, Minnesota, Ohio, and Wisconsin
Waste Management Division
Metcalfe Federal Building
77 West Jackson Boulevard, 19th Floor
Chicago, IL 60604-3507
(312) 353-2000

Region 6: Arkansas, Louisiana, New Mexico, Oklahoma, and Texas
Hazardous Waste Management Division, 6H-M
1445 Ross Avenue, 12th Floor
Dallas, TX 76202-2733
(214) 665-6444

Region 7: Iowa, Kansas, Missouri, and Nebraska
Waste Management Area
726 Minnesota Avenue
Kansas City, KS 66101
(913) 551-7000

Region 8: Colorado, Montana, North Dakota, South Dakota, Utah, and Wyoming
Hazardous Waste Management Division, 8HWM-SR
999 18th Street, Suite 500
Denver, CO 80202-2466
(303) 312-6312

Region 9: American Samoa, Arizona, California, Guam, Hawaii, Nevada, and Northern Marianas
Waste Management Division, H-1
75 Hawthorne Street
San Francisco, CA 94105
(415) 744-1730

Region 10: Alaska, Idaho, Oregon, and Washington
Hazardous Waste Division, HW-113
1200 Sixth Avenue
Seattle, WA, 98101
(206) 553-1200

ADDITIONAL READING

Barnett, Harold C. *Toxic Debts and the Superfund Dilemma.* Chapel Hill: University of North Carolina Press, 1994.

Church, Thomas W., and Robert T. Nakamura. *Cleaning Up the Mess: Implementation Strategies in Superfund.* Washington, D.C.: Brookings Institution, 1993.

Clean Sites, Inc. *A Remedy for Superfund: Designing a Better Way of Cleaning Up America.* Alexandria, Va.: Clean Sites, Inc., 1994.

Gay, Kathleen. *Silent Killers: Radon and Other Hazards.* New York: Franklin Watts, 1988.

Gore, Albert. *Earth in the Balance: Ecology and the Human Spirit.* Boston: Houghton Mifflin, 1992.

Huber, Peter. *Galileo's Revenge: Junk Science in the Courtroom.* New York: Basic Books, 1991.

Hadingham, Evan, and Janet Hadingham. *Garbage! Where It Comes From, Where It Goes.* New York: Simon and Schuster, 1990.

Kowalski, Kathleen M. *Hazardous Waste Sites.* Minneapolis: Lerner Publications, 1996.

Kronenwetter, Michael. *Managing Toxic Waste.* Englewood Cliffs, N.J.: Julian Messner, 1989.

Mazmanian, Daniel, and David Morell. *Beyond Superfailure: America's Toxics Policy for the 1990s.* Boulder, Colo.: Westview Press, 1992.

Ray, Dixie Lee, and Lou Guzzo. *Environmental Overkill: Whatever Happened to Common Sense?* Washington, D.C.: Regnery Gateway, 1993.

Setterberg, Fred, and Lonny Shavelson. *Toxic Nation: The Fight to Save Our Communities from Chemical Contamination.* New York: John Wiley and Sons, 1993.

U.S. Agency for Toxic Substances and Disease Registry. *Public Health Statements: What You Need to Know About Toxic Substances Commonly Found At Superfund Hazardous Waste Sites.* Atlanta: Public Health Service, U.S. Department of Health and Human Services, n.d.

U.S. Environmental Protection Agency. *Your World, My World: A Book for Young Environmentalists.* Washington, D.C.: U.S. Environmental Protection Agency, 1973.

Whelan, Elizabeth M. *Toxic Terror: The Truth behind the Cancer Scares.* Buffalo, N.Y.: Prometheus Books, 1993.

Woodburn, Judith. *The Toxic Waste Time Bomb.* Milwaukee: Gareth Stevens Publishing, 1992.

Zipko, Stephen J. *Toxic Threat: How Hazardous Substances Poison Our Lives.* Englewood Cliffs, N.J.: Julian Messner, 1986.

INDEX

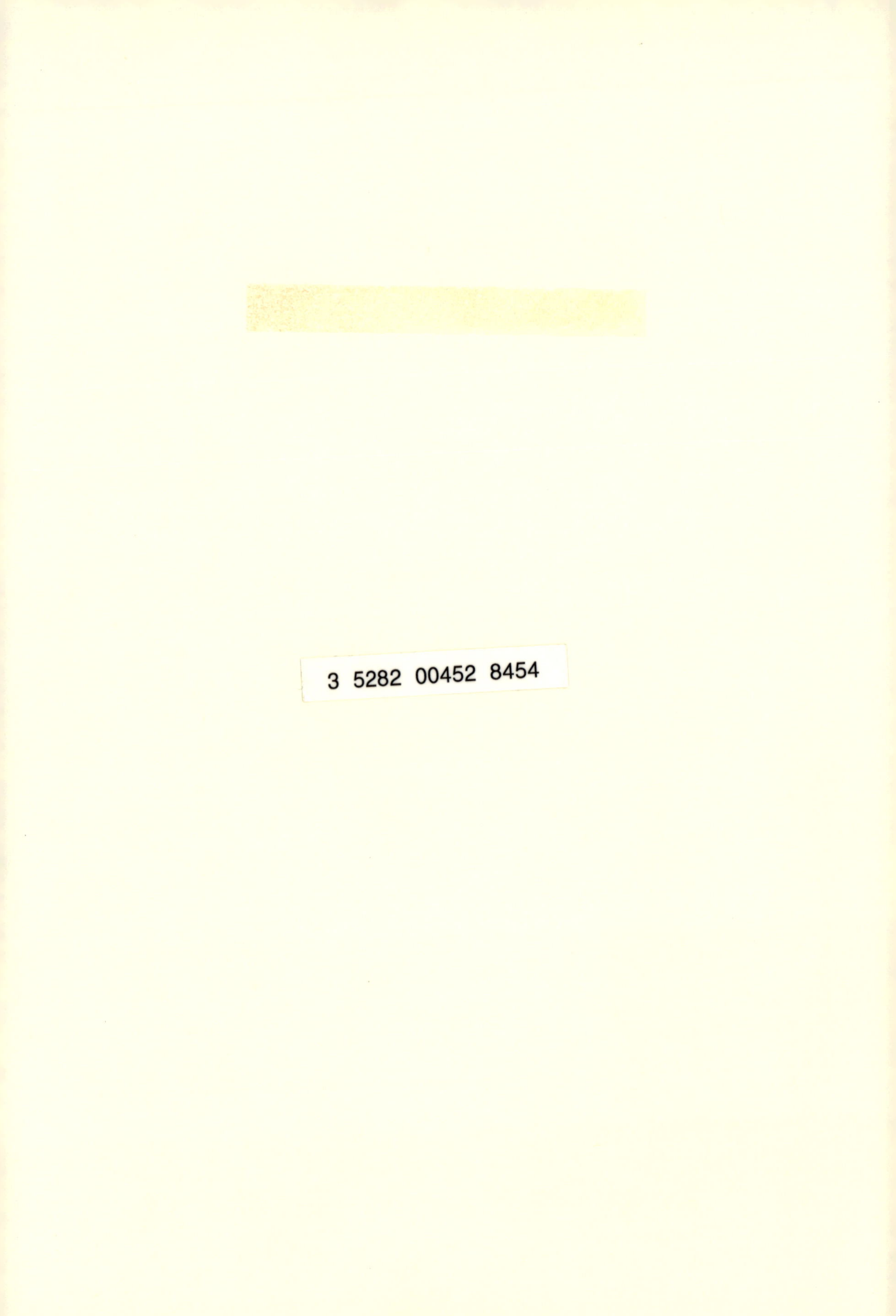